Hazardous Materials for First Responders

Fourth Edition

D0713586

Leslie Miller
Project Manager/Technical Writer

Clint Clausing
Senior Editor

ifsta

Validated by the International Fire Service Training Association

Published by
Fire Protection Publications • Oklahoma State University

RECYCLABLE

The International Fire Service Training Association

The International Fire Service Training Association (IFSTA) was established in 1934 as a *nonprofit educational association of fire fighting personnel who are dedicated to upgrading fire fighting techniques and safety through training.* To carry out the mission of IFSTA, Fire Protection Publications was established as an entity of Oklahoma State University. Fire Protection Publications' primary function is to publish and disseminate training texts as proposed and validated by IFSTA. As a secondary function, Fire Protection Publications researches, acquires, produces, and markets high-quality learning and teaching aids as consistent with IFSTA's mission.

The IFSTA Validation Conference is held the second full week in July. Committees of technical experts meet and work at the conference addressing the current standards of the National Fire Protection Association® and other standard-making groups as applicable. The Validation Conference brings together individuals from several related and allied fields, such as:

- Key fire department executives and training officers
- Educators from colleges and universities
- Representatives from governmental agencies
- Delegates of firefighter associations and industrial organizations

Committee members are not paid nor are they reimbursed for their expenses by IFSTA or Fire Protection Publications. They participate because of commitment to the fire service and its future through training. Being on a committee is prestigious in the fire service community, and committee members are acknowledged leaders in their fields. This unique feature provides a close relationship between the International Fire Service Training Association and fire protection agencies, which helps to correlate the efforts of all concerned.

IFSTA manuals are now the official teaching texts of most of the states and provinces of North America. Additionally, numerous U.S. and Canadian government agencies as well as other English-speaking countries have officially accepted the IFSTA manuals.

Copyright © 2010 by the Board of Regents, Oklahoma State University

All rights reserved. No part of this publication may be reproduced in any form without prior written permission from the publisher.

ISBN 978-0-87939-389-2   Library of Congress Control Number: 2010935225

Fourth Edition, First Printing, November 2010   *Printed in the United States of America*

10   9   8   7   6   5   4

If you need additional information concerning the International Fire Service Training Association (IFSTA) or Fire Protection Publications, contact:

Customer Service, Fire Protection Publications, Oklahoma State University
930 North Willis, Stillwater, OK 74078-8045
800-654-4055   Fax: 405-744-8204

For assistance with training materials, to recommend material for inclusion in an IFSTA manual, or to ask questions or comment on manual content, contact:

Editorial Department, Fire Protection Publications, Oklahoma State University
930 North Willis, Stillwater, OK 74078-8045
405-744-4111   Fax: 405-744-4112   E-mail: editors@osufpp.org

Oklahoma State University in compliance with Title VI of the Civil Rights Act of 1964 and Title IX of the Educational Amendments of 1972 (Higher Education Act) does not discriminate on the basis of race, color, national origin or sex in any of its policies, practices or procedures. This provision includes but is not limited to admissions, employment, financial aid and educational services.

Chapter Summary

Appendix

Table of Contents

List of Tables

Preface

This is the fourth edition of the IFSTA manual dealing with hazardous materials for first responders. It is intended as a primary text for all personnel seeking to qualify as Awareness- and/or Operations-Level responders to hazardous materials incidents and as a reference text for those who have already qualified. This manual addresses the competencies required by NFPA® 472, *Standard for Professional Competence of Responders to Hazardous Materials/Weapons of Mass Destruction Incidents* (2013 edition), for Awareness and Operations Levels. It also meets the requirements of the OSHA regulations in Title 29 *Code of Federal Regulations (CFR)* 1910.120, *Hazardous Waste Operations and Emergency Response (HAZWOPER)*, paragraph (q), for first responders at the Awareness and Operations Levels. This edition has been reorganized and updated, and it contains many new chapters, tables, photos, and illustrations.

Acknowledgement and special thanks are extended to the members of the material review committee who contributed their time, wisdom, and knowledge to the development of this manual.

IFSTA Hazardous Materials for First Responders, 4th Edition Validation Committee

Chair
Rich Mahaney

Mahaney Loss Prevention Services

MI TransCAER Coordinator

Canton, MI

Vice Chair
Phil Linder

Quantum Emergency Response

Vancouver, B.C. Canada

Secretary
Gary Allen

Tampa Fire Rescue

Tampa, Florida

Committee Members

Jimmie Leon Badgett

Hazardous Materials Specialist

Dallas County Fire & Rescue Service

Dallas, Texas

Steve George

Oklahoma Fire Service Training

Stillwater, OK

Chief Ed Hartin

Central Whidbey Island Fire & Rescue

Coupeville, WA

Michael Hildebrand

Hildebrand and Noll Associates, Inc.

Port Republic, Maryland

David W. Lewis

Maryland Fire and Rescue Institute

College Park, Maryland

Doug Weeks

City of Orange Fire Department (ret.)

Orange, CA

Much appreciation is given to the following individuals and organizations for contributing information, photographs, and technical assistance instrumental in the development of this manual:

Boca Raton Fire Rescue

CBRN Responder Training Facility, Fort Leonard Wood

Canadian Centre for Occupational Health and Safety

Fort Leonard Wood Fire Department

International Association of Fire Fighters

Iowa Fire Service Training Bureau

Mohave Museum of History and Arts

Moore Memorial Library, Texas City, TX

Moore (OK) Fire Department

Morning Pride Manufacturing, #1 Innovation Court, Dayton, OH 45414

MSA

New South Wales Fire Brigades

Oklahoma State Fire Service Training

Oklahoma Highway Patrol Bomb Squad

Owasso (OK) Fire Department

Stillwater (OK) Fire Department

Stillwater Animal Welfare

Chris Aguirre

J.K. Allread

Aaron Ansarov

Sherry Arasim

Danny Atchley

Lukas M. Atwell

Jocelyn Augustino

Michael E. Best

Robert E. Billen

Scott M. Biscuiti

Andrea Booher

Bill Branson

Joel G. Breman

Ben Brody

Gregory Bryan

Roy Callaway

Tom Clawson

Bo Cocannouer

Shawn Coffey

Gary Coppage

Rudy V. Cryer

Charles Csavossy

Angel Deilmer

John Demyan

Daniel Dery

Ken Drylie

Ray Elder

Jason Epley

Brent Erb

Jeremy Fairbanks

Mark D. Faram

Kevin Ferrara

Matthew Flynn

Jason Frost

Ronald Fury

Jim Gathany

Kevin Goodnight

Joe Gorman

Dusty Harkins

Clayton Hart

Win Henderson

Greg Henshall

Joan Hepler

Donny Howard

Chiaki Iramina

Steve Irby

James Isaacs

Ron Jeffers

Kevin Johnson

Scott Kim

Bradley A. Lail

Kurt Lamel

J.A. Lee II

Todd Lopez

Patsy Lynch

Ryan N. Marlar

Taylor Marr

Bob McMillan

Chris E. Mickal

David B. Moffitt

Adi Moncaz

Terrance Morrison Jr.

Phillip A. Nickerson Jr

Gerald L. Nino

Gregory G. Noll

Bob O'Donnell

Warren Peace

Stacy L. Pearsall

Todd Pendleton

William Powell

Christopher D. Reed

Mike Rieger

Liz Roll

Antonio Rosas

Paul Roszkowski

Mark Schultz

Ronald Shaw Jr

Jabob H. Smith

Kevin Stabinsky

William D. Stewart

Rhett Strain

Jeff Stroud

Joseph Terry

Josh Trent

Brad Tulley

Brian A. Tuthill

Jim Varhegyi

Fredrick P. Varney

August Vernon

Bryan West

Christopher J. Wiant

Kirk Worley

Sean Worrell

Wayne Yoder

Thanks also go to the many, many authors of the various government and noncopyrighted documents that were used throughout this manual from the following sources:

Health Canada

Los Alamos National Laboratory

Sandia National Laboratories

Transport Canada

Union Pacific Railroad

Oklahoma Highway Patrol Bomb Squad

Department of Fire Services, Commonwealth of MA

U.S. Agency for Toxic Substances and Disease Registry

U.S. Air Force

U.S. Army

U.S. Army Corps of Engineers

U.S. Bureau of Alcohol, Firearms, Tobacco and Explosives

U.S. Centers for Disease Control and Prevention

U.S. Center for Disease Control and Prevention Public Health Images Library (PHIL)

U.S. Coast Guard

U.S. Customs and Border Protection

U.S. Department of Agriculture

U.S. Department of Defense

U.S. Department of Energy

U.S. Department of Homeland Security

U.S. Department of Justice

U.S. Department of Transportation

U.S. Drug Enforcement Agency

U.S. Environmental Protection Agency

U.S. Federal Bureau of Investigation

U.S. Federal Emergency Management Agency

U.S. Fire Administration

U.S. Marines

U.S. National Institute for Occupational Safety and Health

U.S. National Nuclear Security Administration , Nevada Site Office

U.S. Navy

U.S. Nuclear Regulatory Commission

U.S. Occupational Safety and Health Administration

U.S. State Department

Special thanks go to Glen Rudner for his technical review and assistance. Also thanks to Rich Mahaney for providing most of the photos used in tables throughout the manual. Dennis Walus (www.detroitfiregroun-dimages.com) graciously provided the cover photo. Last, but certainly not least, gratitude is extended to the following members of the Fire Protection Publications staff whose contributions made the final publication of this manual possible.

Hazardous Materials for First Responders, 4th Edition, Project Team

Project Manager/Technical Writer
Leslie Miller, Senior Editor

Editor
Clint Clausing, Senior Editor

Photography
Jeff Fortney, Senior Editor

Production Manager
Ann Moffat, Coordinator, Publications Production

Technical Reviewer
Glen Rudner

Illustrators and Layout Designers
Errick Braggs, Senior Graphic Designer

Clint Parker, Senior Graphic Designer

Ruth Mudroch, Senior Graphic Designer

Editorial Staff
Ed Kirtley, IFSTA/Curriculum Projects Coordinator

Mike Sturzenbecker, Senior Editor

Libby Hieber, Senior Editor

Tara Gladden, Editorial Assistant

Curriculum Development
Melissa Noakes, Curriculum Developer

Andrea Haken, Curriculum Developer

Elkie Burnside, Curriculum Developer

Library Researcher/Copyrights
Susan F. Walker

The IFSTA Executive Board at the time of validation of the **Hazardous Materials for First Responders** manual was as follows:

IFSTA Executive Board

Chair
Jeffrey Morrissette
Commission on Fire Prevention and Control
Windsor Locks, Connecticut

Vice Chair
Paul Valentine
Mt. Prospect Fire Department
Mt. Prospect, Illinois

Executive Director
Mike Wieder
Fire Protection Publications
Stillwater, OK

Board Members

Stephen Ashbrock
Madeira & Indian Hill Fire Department
Cincinnati, OH

Steve Austin
Cumberland Valley Volunteer Fireman's
Association
Newark, DE

Roxanne Bercik
Los Angeles Fire Department
Long Beach, CA

Mary Cameli
City of Mesa Fire Department
Mesa, AZ

Bradd Clark
Owasso Fire Department
Owasso, OK

Dennis Compton
National Fallen Firefighter Foundation
Meza, AZ

Frank L. Cotton
Memphis Fire Department
Memphis, TN

George Dunkel
Special Districts Association of Oregon
Scappoose, OR

John Hoglund
Maryland Fire & Rescue Institute
College Park, MD

Wes Kitchel
Santa Rosa Fire Department
Santa Rosa, CA

Brett Lacey
Colorado Springs Fire Department
Colorado Springs, CO

Lori Moore-Merrell
International Association of Fire Fighters
Washington, DC

Ernest Mitchell
City of Pasadena Fire Department
Cerritos, CA

Introduction

Introduction Contents

Introduction

Hazardous materials are found in every jurisdiction, community, workplace and modern household. These substances possess a wide variety of harmful characteristics. Some can be quite deadly or destructive, and they may be used deliberately to cause harm by terrorists and other criminals. Because hazardous materials tend to complicate the emergency incidents in which they are involved, first responders (personnel who are likely to arrive first at an incident scene) must be alert to the presence of hazardous materials at incidents and take the proper precautions when they are. First responders must recognize and understand the hazards presented by various types of hazardous materials, and they must possess the skills necessary to address incidents involving them in a safe and effective manner.

Purpose and Scope

This book is written for emergency first responders who are mandated by law and/or called upon by necessity to prepare for and respond to hazardous materials and weapons of mass destruction (WMD) incidents. These first responders include the following individuals:

- Firefighters
- Law enforcement officers/personnel
- Emergency medical services personnel
- Military responders
- Industrial and transportation emergency response members
- Public works employees
- Utility workers
- Members of private industry
- Other emergency response professionals

The purpose of this book is to provide these first responders with the information they need to take appropriate initial actions at WMD incidents and hazardous materials spills or releases. Its scope is limited to giving detailed information about initial - and primarily defensive - operations. More advanced procedures require hazardous materials technicians who have specialized training.

Related regulations/standards are referenced in this book as applicable, but this book primarily addresses the training requirements of the following National Fire Protection Association® (NFPA®), Occupational Safety and Health Administration (OSHA), and Office for Domestic Preparedness (ODP) documents:

- NFPA® 472, *Standard for Professional Competence of Responders to Hazardous Materials/Weapons of Mass Destruction Incidents* (2013 edition), for the Awareness and Operations Levels (core competencies plus mission-specific competencies).

- OSHA regulations in Title 29 *Code of Federal Regulations (CFR)* 1910.120, *Hazardous Waste Operations and Emergency Response (HAZWOPER)*, paragraph (q), for first responders at the Awareness and Operational Levels

- Office for Domestic Preparedness, *Emergency Responder Guidelines, 2002*, Fire Service Awareness and Operational Levels for response to terrorist incidents involving weapons of mass destruction (WMD)

This book is designed to meet the requirements for NFPA®, OSHA, and ODP first responder Awareness and Operations Levels. It addresses the first responders' responsibilities to recognize the presence of hazardous materials and WMDs, secure the area, provide personnel protection, and request the assistance of trained technicians and law enforcement personnel when necessary. Additionally, it addresses the control of hazardous materials releases and WMDs using operations for which the responders have been trained.

Book Organization

This book is divided into three parts designed to meet the competencies of NFPA® 472. The first three chapters address Awareness-Level competencies. The next four chapters address Operations-Core competencies. The last seven chapters address Mission-Specific Competencies. The parts and chapters are as follows:

Part 1. Awareness-Level Competencies

Chapter 1 – Introduction to Hazardous Materials

Chapter 2 – Hazardous Materials Identification

Chapter 3 – Awareness-Level Actions at Hazardous Materials Incidents

Part 2. Operations-Core Competencies

Chapter 4 – Chemical Properties and Hazardous Materials Behavior

Chapter 5 – Incident Management

Chapter 6 – Strategic Goals and Tactical Objectives

Chapter 7 – Terrorist Attacks, Criminal Activities, and Disasters

Part 3. Mission-Specific Competencies

Chapter 8 – Personal Protective Equipment

Chapter 9 – Decontamination

Chapter 10 – Product Control

Chapter 11 – Air Monitoring and Sampling

Chapter 12 – Victim Rescue and Recovery

Chapter 13 – Evidence Preservation and Sampling

Chapter 14 – Illicit Laboratories

Learning objectives are provided at the beginning of each chapter to assist the reader in focusing on the appropriate topic and knowledge. NFPA® 472 competency numbers are referenced at the beginning of the chapters in which they are addressed. **Appendix A** contains a correlation guide that provides the page number(s) where each NFPA® 472 competency is addressed in the text.

Review questions are located at the end of each chapter to ensure that the reader has a good comprehension of the material in the chapter. The questions are based on the learning objectives. Please note that these questions should not be used for certification or course examinations.

Terminology

This manual is written with a global, international audience in mind. For this reason, it often uses general descriptive language in place of regional- or agency-specific terminology (often referred to as *jargon*). Additionally, the following three points should be remembered when reading this text:

1. The terms *emergency* and *incident* are often used interchangeably, with the understanding that the types of incidents addressed by this book are emergencies.

2. In order to keep sentences uncluttered and easy to read, the word *state* is used to represent both state and provincial level governments (or their equivalent). This usage is applied to this manual for the purposes of brevity and is not intended to address or show preference for only one nation's method of identifying regional governments within its borders.

3. NFPA® and OSHA have different terms for persons trained to the Awareness Level. NFPA® 472 refers to these individuals as *Awareness-Level personnel* whereas OSHA's 29 CFR 1910.120 uses the term *Awareness-Level responders*. When the term *first responder* is used in this manual, it generally refers to both Awareness- and Operations-Level responders as defined by OSHA. The authority having jurisdiction (AHJ) is responsible for defining the actions allowed by persons trained to the Awareness Level, depending on the standard to which they are trained.

Key Information

Various types of information in this book are given in shaded boxes marked by symbols or icons (What This Means To You boxes, sidebars, information, key information, and case histories). See the following definitions:

What This Means To You

These boxes take information presented in the text and synthesize it into an example of how the information is relevant to (or will be applied by) you, the intended audience.

Safety Alert

Safety alerts provide additional emphasis on matters of safety.

Case History

A case history analyzes an event. It can describe its development, action taken, investigation results, and lessons learned.

Information

Information boxes give facts that are complete in themselves but belong with the text discussion. It is information that may need more emphasis or separation. They can be summaries of points, examples, calculations, scenarios, or lists of advantages/disadvantages.

Information Plus

Information plus sidebars give additional relevant information that is more detailed, descriptive, or explanatory than that given in the text.

Key Information

Key information is a short piece of advice that accents the information in the accompanying text.

A key term is designed to emphasize key concepts, technical terms, or ideas that emergency responders need to know. They are listed at the beginning of each chapter and the definition is placed in the margin for easy reference. In the text, key terms will appear as bold, red words. An example of a key term is:

Toxic Inhalation Hazard (TIH) — Liquid or gas known to be a severe hazard to human health during transportation.

Three key signal words are found in the book: **WARNING, CAUTION,** and **NOTE.** Definitions and examples of each are as follows:

- **WARNING** indicates information that could result in death or serious injury to industrial fire brigade members. See the following example:

WARNING!
Deliberately using the human senses to detect the presence of hazardous materials is both unreliable and dangerous.

- **CAUTION** indicates important information or data that industrial fire brigade members need to be aware of in order to perform their duties safely. See the following example:

CAUTION
Intermodal freight containers can contain virtually anything, including extremely hazardous materials which may not be properly identified!

- **NOTE** indicates important operational information that helps explain why a particular recommendation is given or describes optional methods for certain procedures. See the following example:

NOTE: *Vapor* is a gaseous form of a substance that is normally in a solid or liquid state at room temperature and pressure. It is formed by evaporation from a liquid or sublimation from a solid.

Introduction to Hazardous Materials

Chapter Contents

Key Terms

Competencies

NFPA® 472:	5.2.2(8)(a)	5.2.3(1)(b)(ii)	5.2.3(8)(b)	5.2.3(8)(i)
4.2.1(1)	5.2.2(8)(b)	5.2.3(1)(b)(iii)	5.2.3(8)(c)	5.2.3(8)(j)
4.2.1(4)	5.2.2(8)(c)	5.2.3(1)(b)(iv)	5.2.3(8)(d)	5.4.3(1)
4.4.1(2)	5.2.2(8)(d)	5.2.3(1)(b)(v)	5.2.3(8)(e)	
4.4.1(3)(c)	5.2.3(1)(a)(iii)	5.2.3(1)(b)(vi)	5.2.3(8)(f)	
4.4.1(3)(d)	5.2.3(1)(a)(x)	5.2.3(7)	5.2.3(8)(g)	
5.2.2(8)	5.2.2(3)(a)	5.2.3(8)(a)	5.2.3(8)(h)	

Introduction to Hazardous Materials

Learning Objectives

1. Distinguish between hazardous materials incidents and other emergencies. [NFPA® 472, 4.2.1(4)]

2. Discuss the roles of Awareness-Level personnel and Operations-Level responders. [NFPA® 472, 4.4.1(2), 5.4.3(1)]

3. Describe the various types of hazardous materials hazards. [NFPA® 472, 4.4.1(3)(c), 5.2.2(3)(a), 5.2.2(8), 5.2.3(1)(a)(iii), 5.2.3(1)(a)(x), 5.2.3(1)(b)(ii-vi), 5.2.3(7), 5.2.3(8)(a-j)]

4. Explain each of the routes of entry. [NFPA® 472, 4.4.1(3)(d)]

5. Describe the U.S., Canadian, and Mexican hazardous materials regulations and definitions. [NFPA® 472, 4.2.1(1)]

6. Discuss hazardous materials incident statistics.

Chapter 1
Introduction to Hazardous Materials

Case History

In Bhopal, India, December 3, 1984, at an industrial plant owned by Union Carbide Corporation (UCC), large amounts of water entered a tank containing over 40 tons of methyl isocyanate (MIC). The resulting reaction increased the temperature inside the tank to over 400°F (200°C), raising the pressure to a level the tank was not designed to withstand. This forced the emergency venting of pressure from the MIC holding tank, releasing a large volume of toxic gases. The reaction sped up due to the presence of iron in corroding non-stainless steel pipelines. A mixture of poisonous gases flooded the city of Bhopal at night, causing massive panic as people woke up with a burning sensation in their lungs. Thousands died immediately due to the effects of the gas and many were trampled in the panic. Medical staff were unprepared for the number of victims, and they were not provided information about proper treatments to provide. Tens of thousands more suffered long-term effects from exposure. This incident is considered one of the worst industrial accidents in history.

Millions of tons of chemical substances, materials, and products are stored, manufactured, used, and transported throughout the world every year. However, in addition to their necessary and beneficial uses, many of these materials present considerable risks to the public and to the environment if they are uncontrolled or uncontained. Substances that possess harmful characteristics are called **hazardous materials** (or *haz mat*) in the United States and **dangerous goods** in Canada and other countries. When particularly dangerous hazardous materials such as certain chemical, biological, radiological, nuclear, or explosive (CBRNE) materials are used as weapons, they are sometimes referred to as **weapons of mass destruction (WMD)** because of their potential to cause mass casualties and damage.

A *hazardous materials (haz mat/WMD) incident* is an emergency involving a substance that poses an unreasonable risk to people, the environment, and/or property **(Figure 1.1, p. 10)**. It may involve a substance (product or chemical) that has been (or may be) released from a container or a substance that is on fire. The incident may be the result of an accident (such as a container of chemical falling off a forklift) or a deliberate attack (such as a terrorist attack using a deadly gas).

Haz mat and WMD incidents are often more complex than other types of emergency incidents **(Figure 1.2, p. 10)**. Hazardous materials can be dangerous in many different ways, sometimes even in very small quantities. Their

Hazardous Material — Any material that possesses an unreasonable risk to the health and safety of persons and/or the environment if it is not properly controlled during handling, storage, manufacture, processing, packaging, use, disposal, or transportation.

Dangerous Goods — Any product, substance, or organism included by its nature or by the regulation in any of the nine United Nations classifications of hazardous materials; used to describe hazardous materials in Canada and used in the U.S. and Canada for hazardous materials aboard aircraft.

Weapon of Mass Destruction (WMD) — Any weapon or device that is intended or has the capability to cause death or serious bodily injury to a significant number of people through the release, dissemination, or impact of one of the following means:

- Toxic or poisonous chemicals or their precursors
- A disease organism
- Radiation or radioactivity

Personal Protective Equipment (PPE) — General term for the equipment worn by fire and emergency services responders; includes helmets, coats, pants, boots, eye protection, gloves, protective hoods, self-contained breathing apparatus (SCBA), and personal alert safety system (PASS) devices. When working with hazardous materials, this may include Chemical Protective Clothing and Special Protective Clothing. Also called *bunker clothes, protective clothing, turnout clothing,* or *turnout gear,* and *full structural protective clothing.*

CAUTION

Hazardous materials incidents are not always clearly defined prior to the arrival of first responders. First responders must be constantly alert to the presence of hazardous materials and their possible effect on the incident. Whether involved or not, the mere presence of hazardous materials may change the dynamics of the incident.

Figure 1.1 Hazardous materials are so common that emergency incidents involving them may occur anywhere, anytime.

Figure 1.2 Haz mat/WMD incidents may be more complex and difficult to mitigate than other types of emergencies. *Courtesy of the U.S. Air Force, photo by Staff Sgt. Gary Coppage.*

hazards may be extremely difficult to contain and/or control, requiring specialized equipment, procedures, and **personal protective equipment (PPE)**. They may also be difficult to detect, requiring sophisticated monitoring and detection equipment to identify and predict their severity. It may not be possible to detect the presence of some chemicals due to lack of monitoring and detection equipment, and some are so toxic that there would not be time to identify them.

Human error, mechanical breakdowns/malfunctions, container failures, transportation accidents, or deliberate acts can be causes for incidents involving hazardous materials **(Figure 1.3)**. First responders must stay alert to

Figure 1.3 Many accidents involving hazardous materials occur during transportation. *Courtesy of Phil Linder.*

the potential that hazardous materials may be involved in fires, explosions, and criminal or terrorist activities. First responders must possess the skills necessary to address incidents involving hazardous materials in a safe and effective manner.

First responders must understand the role they play at hazardous materials incidents. They must know their limitations and realize when they cannot proceed any farther. In part, this role is established in government laws and national consensus standards that set forth the training requirements and response limitations imposed on personnel responding to these emergencies. These regulations affect how these materials are transported, used, stored, and disposed. Additionally, responders need a basic understanding of how hazardous materials can harm people and the environment.

This chapter explains the role of different North American government agencies in regulating hazardous materials as well as their different definitions for such materials. Many other countries have similar regulations. The chapter also defines the roles of first responders at haz mat incidents, and it provides hazardous materials incident statistics. It also discusses the hazards of hazardous materials and the routes of entry through which they enter the human body.

First Responder Roles

The United States (U.S.) **Occupational Safety and Health Administration (OSHA)** and the *U.S.* **Environmental Protection Agency (EPA)** require that responders to hazardous materials incidents meet specific training standards. The OSHA versions of these legislative mandates are outlined in paragraph (q) of Title 29 (Labor) **Code of Federal Regulations (*CFR*) 1910.120, Hazardous Waste Operations and Emergency Response (HAZWOPER)**. The training requirements found in 29 *CFR* 1910.120 (q) are included by reference in the EPA

Environmental Protection Agency (EPA) — U.S. government agency that creates and enforces laws designed to protect the air, water, and soil from contamination; responsible for researching and setting national standards for a variety of environmental programs.

Occupational Safety and Health Administration (OSHA) — U.S. federal agency that develops and enforces standards and regulations for occupational safety in the workplace.

Hazardous Waste Operations and Emergency Response (HAZWOPER) — U.S. regulations in Title 29 (Labor) *CFR* 1910.120 for cleanup operations involving hazardous substances and emergency response operations for releases of hazardous substances.

Authority Having Jurisdiction (AHJ) — Term used in codes and standards to identify the legal entity, such as a building or fire official, that has the statutory authority to enforce a code and to approve or require equipment. In the insurance industry it may refer to an insurance rating bureau or an insurance company inspection department.

National Fire Protection Association (NFPA®) — U.S. nonprofit educational and technical association located in Quincy, Massachusetts devoted to protecting life and property from fire by developing fire protection standards and educating the public.

Standard Operating Procedures (SOPs) — Standard methods or rules in which an organization or a fire department operates to carry out a routine function. Usually these procedures are written in a policies and procedures handbook and all firefighters should be well versed in their content. A SOP may specify the functional limitations of fire brigade members in performing emergency operations. Also called *Standard Operating Guidelines*.

regulations in Title 40 (Protection of Environment) *CFR* 311, *Worker Protection.* This EPA regulation provides protection to those responders not covered by an OSHA-approved State Occupational Health and Safety Plan. See **Appendix B,** OSHA Plan States, for a list of state-plan and non-state-plan states.

In addition to U.S. Government regulations, the **National Fire Protection Association (NFPA®)** has several consensus standards that apply to personnel who respond to hazardous materials emergencies. The requirements in these standards are recommendations, not laws or regulations, unless they are adopted as such by the **authority having jurisdiction (AHJ)**. However, because they are a national standard, they can be used as a basis for *accepted practice.* The NFPA®'s hazardous materials requirements are detailed in the following standards:

- NFPA® 472, S*tandard for Professional Competence of Responders to Hazardous Materials /Weapons of Mass Destruction Incidents* (2008)
- NFPA® 473, *Standard for Competencies for EMS Personnel Responding to Hazardous Materials/Weapons of Mass Destruction Incidents* (2008)

In Canada, the Ministry of Labour (in most provinces) or the Workers Compensation Board (WCB) in British Columbia are the regulatory bodies governing response to haz mat incidents and the training requirements for first responders. These provincial bodies also require employers to provide **standard operating procedures (SOPs)** or standard operating guidelines (SOGs) to protect their employees. Canadian firefighters and most emergency responders are trained to the same NFPA® standards as their U.S. counterparts. While Canada does not have the definitive equivalent of OSHA 29 *CFR* 1910.120, the minimum acceptable level of training for first responders is NFPA® 472.

Mexico has developed and implemented a variety of national laws dealing with the handling and regulation of hazardous materials. However, it does not currently have any national laws applying to the training of emergency haz mat first responders. Local jurisdictions may have their own training standards.

What This Means To You

If you are a first responder to haz mat incidents in the U.S., by law your employer must meet the requirements set forth in the HAZWOPER regulation (29 *CFR* 1910.120). If your AHJ has formally adopted the applicable NFPA® standards as law, your employer is required to meet them as well. If you belong to a volunteer fire and emergency services organization, you will also have to meet these regulations. Under 40 *CFR* 311, volunteers are considered employees.

If you are a first responder to haz mat incidents in Canada, your employer must provide you with standard operating procedures (or standard operating guidelines) and the training required by your province. If you are a firefighter, you must be trained in accordance with the requirements in NFPA® 472.

Mexico does not have any national laws applying to the training of haz mat first responders. Local jurisdictions may have their own standards.

NFPA® 472 and the OSHA regulations in 29 CFR 1910.120 identify two levels of training: **Awareness** and **Operations**. Per NFPA® 472, Operations Level responders may be trained to a set of core competencies (Operations Core) or beyond, incorporating mission-specific competencies (Operations Mission-Specific) for actions personnel may be trained to perform at haz mat/WMD incidents. The mission-specific competencies identified in NFPA® 472 are as follows:

- Personal Protective Equipment
- Mass Decontamination
- Technical Decontamination
- Evidence Preservation and Sampling
- Product Control
- Air Monitoring and Sampling
- Victim Rescue and Recovery
- Response to Illicit Laboratory Incidents

Awareness Level — Lowest level of training established by OSHA for personnel at hazardous materials incidents.

Operations Level — Level of training established by OSHA allowing first responders to take defensive actions at hazardous materials incidents.

Both documents also identify higher levels of response personnel who perform more complex operations. OSHA identifies three levels above the Operations Level: (1) *Hazardous Materials Technician,* (2) *Hazardous Materials Specialist,* and (3) *On Scene Incident Commander (OIC).* NFPA® 472 identifies five levels above the Operations Level: (1) *Hazardous Materials Technician* (plus three specialties), (2) *Hazardous Materials Branch Officer,* (3) *Hazardous Materials Branch Safety Officer,* (4) *Hazardous Materials Incident Commander,* and (5) *Private Sector Specialist Employee.* It is important to know and understand the responder's role at each of these levels; however, this book addresses the requirements of the *Awareness, Operations Core,* and *Operations Mission-Specific* competencies only.

It should be noted that OSHA considers personnel trained to each level (Awareness and Operations) to be *first responders,* whereas NFPA® 472 uses the terminology *personnel* as opposed to *responder* for the Awareness Level. When the term *first responder* is used throughout the manual, it is often used within the OSHA context, incorporating both Awareness and Operations levels.

Personnel trained to the Awareness and Operations Core Levels perform only defensive tasks at haz mat incidents. Operations Missions-Specific competencies allow responders to perform defensive and limited offensive actions. Hazardous materials technicians and specialists perform a full array of offensive and defensive actions.

Awareness-Level Personnel

Personnel who are trained and certified to the Awareness Level are individuals who, in the course of their normal duties, may be the first to arrive at or witness a haz mat/WMD incident **(Figure 1.4, p. 14)**. To summarize both OSHA and NFPA® requirements, individuals trained to the Awareness Level are expected to assume the following responsibilities when faced with an incident involving hazardous materials:

- Recognize the presence or potential presence of a hazardous material
- Recognize the type of container at a site and identify the material in it if possible

Figure 1.4 Awareness-Level personnel may be the first to witness and report haz mat incidents. Typically, they are at the scene when the accident occurs rather than being dispatched to the incident as emergency responders.

- Transmit information to an appropriate authority and call for appropriate assistance
- Identify actions to protect themselves and others from hazards
- Establish scene control by isolating the hazardous area and denying entry

Operations-Level Responders

Responders who are trained and certified to the Operations Level are individuals who respond to releases (or potential releases) of hazardous materials as part of their normal duties. This responder is expected to protect individuals, the environment, and property from the effects of the release in a defensive manner **(Figure 1.5).**

Figure 1.5 Operations-Level responders are dispatched to the scene to help mitigate the incident. *Courtesy of Danny Atchley.*

Responsibilities of the first responder at the Operations Level include the Awareness-Level responsibilities. The Operations Level adds confining a release in a defensive fashion from a safe distance. To summarize both OSHA and NFPA® requirements, first responders at the Operational Level must be able to perform the following actions:

- Identify the hazardous material(s) involved in an incident if possible
- Analyze an incident to determine the nature and extent of the problem
- Protect themselves, nearby persons, the environment, and property from the effects of a release
- Develop a defensive plan of action to address the problems presented by the incident (plan a response)
- Implement the planned response to **mitigate** or control a release from a safe distance (initiate defensive actions to lessen the harmful incident) and keep it from spreading
- Evaluate the progress of the actions taken to ensure that response objectives are safely met

Offensive Tasks Allowed by U.S. OSHA and Canada

U.S. OSHA and the Canadian government recognize that first responders at the Operations Level who have appropriate training (including demonstration of competencies and certification by employers), appropriate protective clothing, and adequate/appropriate resources can perform offensive operations involving flammable liquid and gas fire control of the following materials:

- Gasoline
- Diesel fuel
- Natural gas
- Liquefied petroleum gas (LPG)

Each fire and emergency services organization should have written procedures describing appropriate actions consistent with the level of training. A sample of such a guideline is found in **Appendix C,** Sample Written Guideline.

Hazardous Materials Hazards

To safely mitigate hazardous materials incidents, first responders must understand the variety of hazardous materials they may encounter, the potential health effects of the materials, and the physical **hazards** associated with them. Knowing some of these basic concepts will help prevent or reduce injury, loss of life, and environmental/property losses.

Many hazardous materials have major effects on human health. Exposures to hazardous materials may be **acute** (single exposure or several repeated exposures to a substance within a short time period) or **chronic** (long-term, reoccurring). Health effects can also be acute or chronic. **Acute health effects** are short-term effects that appear within hours or days, such as vomiting or diarrhea. **Chronic health effects** are long-term effects that may take years to appear, such as cancer.

Mitigate — (1) To cause to become less harsh or hostile; to make less severe, intense or painful; to alleviate. (2) To take actions to reduce or eliminate long-term risk to human life and property from natural, human-caused, and technological hazards and their effects. (3) One method of sizing up an emergency situation is to locate, isolate, and *mitigate.*

Acute — (1) Characterized by sharpness or severity; having rapid onset and a relatively short duration. (2) Single exposure (dose) or several repeated exposures to a substance within a short time period.

Acute Health Effects — Health effects that occur or develop rapidly after exposure to a substance.

Chronic — Of long duration, or recurring over a period of time; (opposite of acute).

Chronic Health Effects — Long-term health effects from either a one-time or repeated exposure to a substance.

Hazard — Condition, substance, or device that can directly cause injury or loss; the source of a risk.

Oxidation — Chemical process that occurs when a substance combines with oxygen; a common example is the formation of rust on metal.

Polymerization — Reactions in which two or more molecules chemically combine to form larger molecules. This reaction can often be violent.

Cryogen — Gas that is cooled to a very low temperature, usually below -130°F (-90°C), to change to a liquid. Also called refrigerated liquid and cryogenic liquid.

Corrosive Material — Gaseous, liquid, or solid material that can burn, irritate, or destroy human skin tissue and severely corrode steel.

Oxidizer — Any substance or material that yields oxygen readily and may stimulate the combustion of organic and inorganic matter.

Poison — Any material that when taken into the body is injurious to health.

The following have the potential to cause harm (with acute and/or chronic effects) at a hazardous materials incident:

- **T**hermal hazards
- **R**adiological hazards
- **A**sphyxiation hazards
- **C**hemical hazards
  - Poisons/toxins
  - Corrosives
  - Irritants
  - Convulsants
  - Carcinogens
  - Sensitizers/allergens
- **E**tiological/biological hazards
- **M**echanical hazards

Note: These are often referred to by the acronym, TRACEM.

Thermal Hazards

Thermal hazards are related to temperature extremes. Hazardous materials themselves can cause temperature extremes such as with elevated-temperature materials, exothermic reactions (sudden release of heat energy that might occur during **oxidation** or **polymerization**), fires, explosions, or cryogenic liquids. Thermal hazards can also be caused by conditions on the scene such as extreme external air temperature or the use of personal protective equipment. Cold hazards presented by cryogenic liquids and liquefied gases and the heat hazards presented by elevated-temperature materials, fires, or explosions are discussed in the sections that follow.

Low Temperatures

Cold exposure is a concern when dealing with cryogenic and *liquefied gases.* A liquefied gas (such as propane or carbon dioxide) is one that at the charging pressure is partially liquid at 70°F (21°C). A **cryogen** (sometimes called *refrigerated liquefied gas*) is a gas that turns into a liquid at or below -130°F (-90°C) at 14.7 psi (101 kPa) {1.01 bar}. Examples of cryogenic materials include liquid oxygen (LOX), nitrogen, helium, hydrogen, argon, and liquefied natural gas (LNG). These substances are commonly stored and transported in their liquid states. At these extremely cold temperatures, cryogens have the ability to instantly freeze materials (including human tissue) on contact. Some cryogens have other hazardous properties in addition to the cold hazard. An example of this type would be fluorine, which is also a **corrosive material**, an **oxidizer**, and a **poison**.

Cryogenic and liquefied gases may pool and transition from a liquid state to a vapor state slowly or rapidly depending on the product. A liquid spill or leak will boil into a much larger vapor cloud **(Figure 1.6).** These vapor clouds can be extremely dangerous if the vapors are flammable. Both types of liquids cause freeze burns, which are treated as cold injuries according to their severity. Any clothing saturated with a cryogenic material must be removed immediately. This action is particularly important if the vapors are flammable

Figure 1.6 A spill or leak of cryogenic liquids typically results in a vapor cloud. Not only are these materials very cold, but they can be flammable, as well, and responders should avoid coming into contact with them. *Courtesy of Steve Irby, Owasso Fire Department.*

Elevated Temperature Material — Material that when offered for transportation or transported in bulk packaging is (a) in a liquid phase and at a temperature at or above 212°F (100°C), (b) intentionally heated at or above its liquid phase flash point of 100°F (38°C), or (c) in a solid phase and at a temperature at or above 464°F (240°C).

WARNING!
Immediately remove any clothing saturated with a cryogenic material.

or oxidizers. A first responder cannot escape flames from clothing-trapped vapors if they ignite.

Working in extremely cold weather and cold injuries are discussed in Chapter 8, Personal Protective Equipment. The conditions and symptoms of cold exposure detailed in **Table 1.1** apply to any situation where cold exposure is a factor.

Another source of cold exposure is from anhydrous ammonia, which is sometimes used as a refrigerant in cold-storage facilities. This material, too, can cause cold injuries, and its vapors are toxic and may ignite.

Elevated Temperatures

Elevated-temperature materials such as molten sulfur and molten aluminum can present a thermal hazard in the form of heat **(Figure 1.7, p. 18).** Molten aluminum, for example, is generally shipped at temperatures above 1,300°F (704°C). First responders must be extremely cautious around these materials to avoid being burned. Molten aluminum and other high-temperature materials can ignite flammable and combustible materials (including wood). Working around or near elevated-temperature materials can increase the effect of wearing personal protective equipment due to high ambient air temperatures (see Chapter 8, Personal Protective Equipment). The U.S. Department of Transportation (DOT) defines an elevated-temperature material as one that when offered for transportation or transported in bulk packaging has one of the following properties:

**Table 1.1
Conditions and Symptoms of
Cold Exposure**

Condition	Symptoms
Frost Nip/Incipient Frostbite	Whitening or blanching of skin
Superficial Frostbite	• Waxy or white skin • Firm touch to outer layer of skin, underlying tissue is resilient (flexible)
Deep Frostbite	• Cold skin • Pale skin • Solid, hard skin • Black skin tissue
Systemic Hypothermia	• Shivering • Sleepiness, apathy listlessness • Core temperature of 95°F (35°C) or less • Slow pulse • Slow breathing • Glassy eyes • Unconsciousness • Freezing of extremities • Death

Figure 1.7 Materials marked as "Hot" are being transported at an elevated temperature and should be treated as burn hazards. *Courtesy of Rich Mahaney.*

- Liquid phase at a temperature at or above 212°F (100°C)
- Liquid phase with a flash point at or above 100°F (38°C) that is intentionally heated and offered for transportation or transported at or above its flash point
- Solid phase at a temperature at or above 464°F (240°C)

Emergency responders (particularly firefighters) also encounter temperature extremes caused by fires, steam, radiation, and the incendiary thermal effects of explosions (see Mechanical Trauma section). **Table 1.2** provides the three types (or degrees) of thermal burns with their symptoms.

Radiological Hazards

The potential for radiation exposure exists when first responders respond to incidents at medical centers, certain industrial operations, nuclear power plants, and research facilities. There is also the potential for exposure during terrorist attacks. Different types of radiation exist, and some are more energetic than others **(Figure 1.8).** The least energetic form of radiation is **nonionizing radiation** such as visible light and radio waves. The most energetic (and hazardous) form of radiation is **ionizing radiation**, and it is this type of radiation that is of greatest concern to first responders.

Types of Ionizing Radiation

The types of ionizing radiation are as follows:

- **Alpha** — Energetic, positively charged alpha particles (helium nuclei) emitted from the nucleus during radioactive decay that rapidly lose energy when passing through matter. They are commonly emitted in the radioactive decay of the heaviest radioactive elements such as uranium and radium as well as by some manmade elements. *Details:*

Nonionizing Radiation — Series of energy waves composed of oscillating electric and magnetic fields traveling at the speed of light. Examples: ultraviolet radiation, visible light, infrared radiation, microwaves, radio waves, and extremely low frequency radiation.

Ionizing Radiation — Radiation that has sufficient energy to remove electrons from atoms resulting in a chemical change in the atom.

Table 1.2
Thermal Burn Types and Symptoms

Thermal Burn Type	Symptoms
First Degree Burn — involves the first (top) layer of skin	• Redness • Tenderness (painful to touch) • Mild swelling
Second Degree Burn — involves the first two layers of skin	• Deep reddening of the skin • Pain • Blisters • Glossy appearance from leaking fluid • Possible loss of some skin
Third Degree Burn — penetrates the entire thickness of the skin and permanently destroys tissue	• Loss of skin layers • Often painless (Pain may be caused by patches of first-and second-degree burns, which often surround third-degree burns.) • Dry and leathery skin • Skin may appear charred or have patches which appear white, brown or black

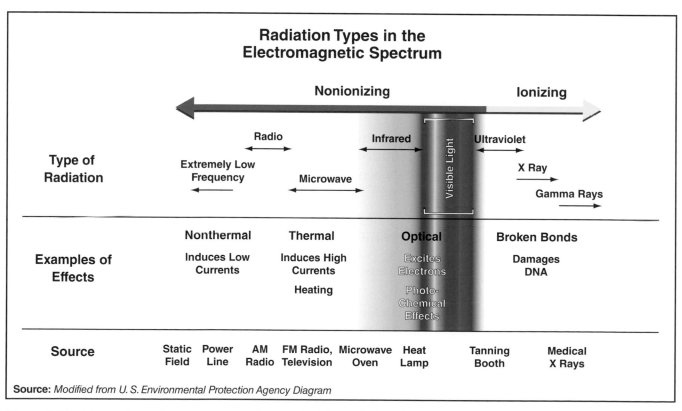

Source: *Modified from U. S. Environmental Protection Agency Diagram*

Figure 1.8 Ionizing radiation has more energy than nonionizing radiation. Generally speaking, ionizing radiation is of more concern to first responders because of the potential health effects associated with exposure.

— Alpha particles lose energy rapidly when travelling through matter and do not penetrate very deeply; however, they can cause damage over their short path through human tissue. They are usually completely blocked by the outer, dead layer of the human skin, so alpha-emitting radioisotopes are not a hazard outside the body. However, they can be very harmful if the material emitting the alpha particles is ingested or inhaled.

— Alpha particles can be stopped completely by a sheet of paper (**Figure 1.9**).

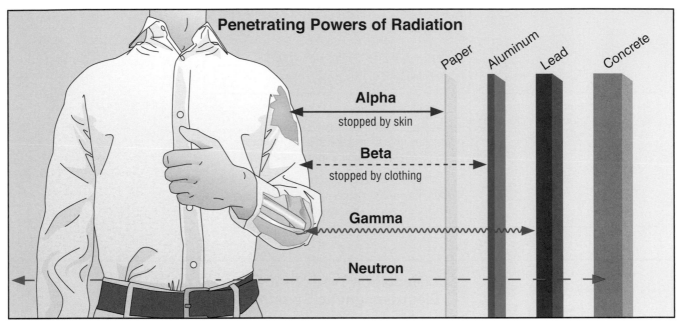

Source: *Modified from U.S. Environmental Protection Agency*

Figure 1.9 The penetrating powers of alpha and beta particles, gamma rays, and neutrons.

Electron — Minute component of an atom that possesses a negative charge.

Photon — Packet of electromagnetic energy.

- **Beta** — Fast-moving, positively or negatively charged **electrons** (beta particles) emitted from the nucleus during radioactive decay. Humans are exposed to beta particles from manufactured and natural sources such as tritium, carbon-14, and strontium-90. *Details:*

  — Beta particles are more penetrating than alpha particles but less damaging over equally traveled distances. Beta particles are capable of penetrating the skin and causing radiation damage; however, as with alpha emitters, beta emitters are generally more hazardous when they are inhaled or ingested.

  — Beta particles travel appreciable distances in air (up to 20 feet [6 m]) but can be reduced or stopped by a layer of clothing or by less than .08 an inch (two or three millimeters) of a substance such as aluminum (Figure 1.9).

- **Gamma** — High-energy **photons** (weightless packets of energy like visible light and X-rays). Gamma rays often accompany the emission of alpha or beta particles from a nucleus. They have neither a charge nor a mass but are very penetrating. One source of gamma radiation in the environment is naturally occurring potassium-40. Common industrial gamma emitting sources include cobalt-60, iridium-192 and cesium-137. *Details:*

— Gamma radiation can easily pass completely through the human body or be absorbed by tissue, thus constituting a radiation hazard for the entire body.

— Around two feet (about 0.6 m) of concrete, several feet (meters) of earth, or around two inches (about 50 mm) of lead may be required to stop the more energetic gamma radiation. Standard fire fighting protective clothing provides *no* protection against gamma radiation (Figure 1.9).

- **Neutron** — Ultrahigh energy particles that have a physical mass but have no electrical charge. Neutrons are highly penetrating (Figure 1.9). Fission reactions produce neutrons along with gamma radiation. Neutron radiation is difficult to measure in the field and is usually estimated based on gamma measurements. *Details:*

— Soil moisture density gauges, often used at construction sites, are a common source of neutron radiation. Neutrons may also be encountered in research laboratories or operating nuclear power plants.

— The health hazard that neutrons present arises from the fact that they cause the release of secondary radiation when they interact with the human body.

What's the Difference between Exposure to Radioactive Material and Radioactive Contamination?

Radioactive **contamination** occurs when a material that contains radioactive atoms is deposited on surfaces, skin, clothing, or any place where it is not desired. It is important to remember that radiation does *not* spread; rather, it is radioactive material/contamination that spreads. **Exposure** to radiation occurs when a person is near a radiation source, is exposed to the energy from that source, and biological damage occurs. Without biological damage, there is no exposure. For example, even if you touch or pick up a source of alpha radiation with your hand (not recommended), there is no exposure because there is no corresponding biological damage (since your skin protects you). Conversely, if someone were to pick up a source of beta or gamma radiation, an exposure will have taken place because there will be corresponding biological damage.

Exposure to radiation alone does *not* contaminate a person. Contamination only occurs when the radioactive material remains on a person or their clothing after coming into contact with the **contaminant**. A person can become contaminated externally, internally, or both. Radioactive material can enter the body via one or more routes of entry (see Routes of Entry section). An unprotected person contaminated with radioactive material receives radiation exposure until the source of radiation (radioactive material) is removed. Note the following situations:

- A person is *externally* contaminated (and receives external exposure) when radioactive material is on the skin or clothing.

- A person is *internally* contaminated (and receives internal exposure) when radioactive material is breathed, swallowed, or absorbed through wounds.

- The *environment* is contaminated when radioactive material is spread about or is unconfined. Environmental contamination is another potential source of external exposure.

Contamination — Condition of impurity resulting from a mixture or contact with foreign substance.

Exposure — Contact with a substance by swallowing, breathing, or touching the skin or eyes. Exposure may be short-term (acute exposure), of intermediate duration, or long-term (chronic exposure).

Contaminant — Any foreign substance that compromises the purity of a given substance.

Radiation Health Hazards

The effects of ionizing radiation occur at the cellular level. The human body is composed of many organs, and each organ of the body is composed of specialized cells. Ionizing radiation can affect the normal operation of these cells.

Radiation may cause damage to any material by ionizing the atoms in that material – changing the material's atomic structure. When atoms are ionized, the chemical properties of those atoms are altered. Radiation can damage a cell by ionizing the atoms and changing the resulting chemical behavior of the atoms and/or molecules in the cell. If a person receives a sufficiently high dose of radiation and many cells are damaged, in rare cases there may be observable health effects including genetic mutations and cancer.

The biological effects of ionizing radiation depend on how much and how fast a radiation dose is received. There are two categories of radiation doses: acute and chronic.

Acute doses. Exposure to radiation received in a short period of time is an acute dose. Some acute doses of radiation are permissible and have no long term health effects. Others can produce serious health effects that include reduced blood count, hair loss, nausea, vomiting, diarrhea, and fatigue. Extremely high levels of acute radiation exposure (such as those received by victims of a nuclear bomb) can result in death within a few hours, days, or weeks.

Chronic doses. Small amounts of radiation received over a long period of time are a chronic dose. The body is better equipped to handle a chronic dose of radiation than it is an acute radiation dose because the body has enough time to replace dead or nonfunctioning cells with healthy ones. Chronic doses do not result in the same detectable health effects seen with acute doses. However, there are studies that confirm that chronic exposure to radiation does cause cancer. Examples of chronic radiation doses include the everyday doses received from natural background radiation and those received by workers in nuclear and medical facilities.

The exposures likely to be encountered by first responders at most haz mat incidents are **very unlikely** to cause any health effects, especially if proper precautions are taken **(Figure 1.10).** Even at terrorist incidents, it is unlikely that first responders will encounter dangerous or lethal doses of radiation. However, it is important to monitor for radiation at any incident involving explosions or suspected terrorist attacks.

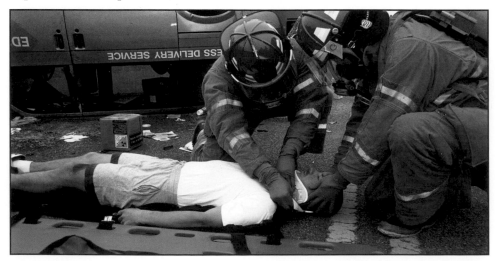

Figure 1.10 If first responders take the proper precautions when working incidents involving radiological materials, they are unlikely to experience adverse health effects. *Courtesy of Tom Clawson.*

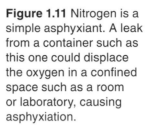

Figure 1.11 Nitrogen is a simple asphyxiant. A leak from a container such as this one could displace the oxygen in a confined space such as a room or laboratory, causing asphyxiation.

Asphyxiation Hazards

Asphyxiants are substances that affect the oxygenation of the body and generally lead to suffocation. Asphyxiants can be divided into two classes: simple and chemical. *Simple asphyxiants* are gases that displace the oxygen necessary for breathing **(Figure 1.11).** These gases dilute or displace the oxygen concentration below the level required by the human body. *Chemical asphyxiants* are substances that prohibit the body from using oxygen, and some of these chemicals may be used by terrorists for an attack. Even though oxygen is available, these substances starve the cells of the body for oxygen.

Asphyxiant — Any substance that prevents oxygen from combining in sufficient quantities with the blood or from being used by body tissues.

Chemical Hazards

Exposure to hazardous chemicals may produce a wide range of adverse health effects. The likelihood of an adverse health effect occurring and the severity of the effect depend on the following:

- Toxicity of the chemical
- Pathway or route of exposure
- Nature and extent of exposure
- Factors that affect the susceptibility of the exposed person such as age and the presence of certain chronic diseases

The sections that follow discuss types of hazardous chemicals or substances. These include toxins/poisons, irritants, convulsants, corrosives, carcinogens, and sensitizers/allergens.

Figure 1.12 Many chemicals have multiple hazards. Hydrogen fluoride anhydrous is a chemical that is both toxic and corrosive. It will do harm if it contacts the skin or gets into the body, especially if it is inhaled. *Courtesy of Rich Mahaney.*

DOT E11759
INHALATION HAZARD
HYDROGEN FLUORIDE ANHYDROUS

DOT 112S400W
SAFETY VALVE 300 PSI
TESTED 1999 DUE 2004
TANK 400 PSI
TESTED 1994 DUE 2004

1052
8

POISON

2 INCH HF COMP BRAKE SHOES

LIFT LUG HERE

JACK HERE

Systemic Effect — Something that affects an entire system rather than a single location or entity.

CAUTION

All personnel working at hazardous materials incidents must use appropriate personal protective equipment, including appropriate respiratory protection equipment.

Toxic Chemicals/Poisons

Toxic (poisonous) chemicals often produce injuries at the site where they come into contact with the body. A chemical injury at the site of contact (typically the skin and mucous membranes of the eyes, nose, mouth, or respiratory tract) is termed a *local toxic effect.* Irritant gases such as chlorine and ammonia can, for example, produce a localized toxic effect in the respiratory tract, while corrosive acids and bases can result in local damage to the skin **(Figure 1.12).** In addition, a toxic chemical may be absorbed into the bloodstream and distributed to other parts of the body, producing **systemic effects**. Many pesticides, for example, absorb through the skin, distribute to other sites in the body, and produce adverse effects such as seizures or cardiac, pulmonary, or other problems.

Many toxins (poisons) have fast-acting, acute toxic effects while others may have chronic effects that are not manifested for many years. Exposure to chemical compounds can result not only in the development of a single systemic effect but also in the development of multiple systemic effects or a combination of systemic and local effects. Some of these effects may be delayed.

Exposures to toxins can cause damage to organs or other parts of the body. Types of toxins, their target organs, and chemical examples are given in **Table 1.3.**

The methods by which poisons attack the body vary depending on the type of poison. Irritants and chemical asphyxiants interfere with oxygen flow to the lungs and the blood. Neurotoxins act on the body's nervous system by disrupting nerve impulses. Highly toxic materials may cause death or severe illness.

Eating and Drinking Can be Dangerous . . .

. . . on the scene of a hazardous material incident. If hazardous materials at an incident site contaminate food or water, the chemicals can be ingested into the body where they can cause harm. Therefore, never eat or drink in areas where hazardous materials may be present. Make sure that water comes from a clean source and is dispensed in disposable cups. Always place rehabilitation areas well away from any sources of contamination. Finally, wash your hands and be certain that you are completely decontaminated before eating or drinking.

**Table 1.3
Types of Toxins and Their Target Organs**

Toxin	Target Organ	Chemical Examples
Nephrotoxins	Kidney	Halogenated Hydrocarbons, Mercury, Carbon Tetrachloride
Hemotoxins	Blood	Carbon Monoxide, Cyanides, Benzene, Nitrates, Arsine, Naphthalene, Cocaine
Neurotoxins	Nervous System	Organophosphates, Mercury, Carbon Disulphide, Carbon Monoxide, Sarin
Hepatoxins	Liver	Alcohol, Carbon Tetrachloride, Trichloroethylene, Vinyl Chloride, Chlorinated HC
Immunotoxins	Immune System	Benzene, Polybrominated Biphenyls (PBBs), Polychlorinated Biphenyls (PCBs), Dioxins, Dieldrin
Endocrine Toxins	Endocrine System (including the pituitary, hypothalamus, thyroid adrenals, pancreas, thymus, ovaries, and testes)	Benzene, Cadmium, Chlordane, Chloroform, Ethanol, Kerosene, Iodine, Parathion
Musculoskeletal Toxins	Muscles/Bones	Fluorides, Sulfuric Acid, Phosphine
Respiratory Toxins	Lungs	Hydrogen Sulfide, Xylene, Ammonia, Boric Acid, Chlorine
Cutaneous Hazards	Skin	Gasoline, Xylene, Ketones, Chlorinated Compounds
Eye Hazards	Eyes	Organic Solvents, Corrosives, Acids
Mutagens	DNA	Aluminum Chloride, Beryllium, Dioxins
Teratogens	Embryo/Fetus	Lead, Lead Compounds, Benzene
Carcinogens	All	Tobacco Smoke, Benzene, Arsenic, Radon, Vinyl Chloride

Corrosives

Chemicals that destroy or burn living tissues and have destructive effects by virtue of their *corrosivity* (ability to cause corrosion, particularly to metals) are often called *corrosives*. With the exception of liquid and gas fuels, corrosives comprise the largest usage class (by volume) in industry. Corrosives are commonly divided into two broad categories: acids and bases (bases are sometimes called alkalis or caustics). However, it is important to note that some corrosives (such as hydrogen peroxide) are neither acids nor bases. The corrosivity of acids and bases is often measured or expressed in terms of **pH** (**Figure 1.13, p. 26**). Definitions are as follows:

- **Acid** — Any chemical that ionizes (breaks down) to yield hydrogen **ions** in water. Acids have pH values of 0 to 6.9. An acid may cause severe chemical burns to flesh and permanent eye damage. Contact with an acid typically causes immediate pain. Hydrochloric acid, nitric acid, and sulfuric acid are examples of common acids.

pH — Measure of acidity of an acid or the level of alkaline in a base.

Acid — Compound containing hydrogen that reacts with water to produce hydrogen ions; a proton donor; a liquid compound with a pH less than 7. Acidic chemicals are corrosive.

Ion — An atom which has lost or gained an electron and thus has a positive or negative charge.

Figure 1.13 The pH scale is used to express the corrosivity of acids and bases. A pH less than 7 is acidic, and a pH above 7 is basic. The lower or higher the number, the stronger the acid or base, respectively.

pH Scale

Concentration of Hydrogen Ions Compared to Distilled Water	pH Scale	Examples of Solutions at this pH
Acids	0	Strong Hydrofluoric Acid
Acids	1	Battery Acid
Acids	2	Vinegar
Acids	3	Orange Juice
Acids	4	Acid rain, Wine
Acids	5	Black Coffee
Acids	6	Milk
Neutral	7	Distilled Water
Bases	8	Seawater
Bases	9	Baking Soda
Bases	10	Milk of Magnesia
Bases	11	Ammonia
Bases	12	Lime
Bases	13	Lye
Bases	14	Sodium Hydroxide

Base — Corrosive water-soluble compound or substance containing group-forming hydroxide ions in water solution that reacts with an acid to form a salt; an alkaline (caustic) substance.

• **Base (alkalis)** — Water-soluble compound that breaks apart in water to form a negatively charged hydroxide ion. Bases react with an acid to form a salt by releasing an unshared pair of electrons to the acid or by receiving a proton (hydrogen ion) from the acid. Bases have pH values of 8 to 14. A base breaks down fatty skin tissues and can penetrate deeply into the body. They can also cause severe eye damage because they tend to adhere to the tissues in the eye, which makes bases difficult to remove. Bases often cause more eye damage than acids because of the longer duration of exposure. Contact with a base does not normally cause immediate pain. A common sign of exposure to a base is a greasy or slick feeling of the skin, which is caused by the breakdown of fatty tissues. Caustic soda, potassium hydroxide, and other alkaline materials commonly used in drain cleaners are examples of bases.

Is it Corrosive, Caustic, an Acid, a Base, or an Alkali?

Some experts make this differentiation: acids are *corrosive,* while bases are *caustic.* In the world of emergency response, however, both acids and bases are called *corrosives.* The U.S. Department of Transportation (DOT) and Transport Canada (TC), for example, do not differentiate between the two. Any materials that destroy metal or skin tissue are considered corrosives by these agencies.

The terms *base* and *alkali* are often used interchangeably, but some chemical dictionaries define alkalis as *strong bases.* Base solutions are usually referred to as *alkaline* rather than *basic*, but, again, the two terms are often used synonymously. Just be aware that if you hear the terms *caustic, alkali*, or *alkaline*, they are referring to bases or basic solutions.

In some cases, corrosives (particularly strong acids) can cause a fire or an explosion if they come in contact with combustibles because their corrosive actions can generate enough heat to start a fire. Some acids (for example, hydrochloric acid) can react with metal to form hydrogen gas (which is explosive). Additionally, acids and bases can react very violently when mixed together or water is added to them. This consideration is important during decontamination and spill cleanup.

Acids and bases can be toxic, flammable, reactive, and/or explosive and some are oxidizers (see Chapter 4, Hazardous Materials Properties). Because of the wide variety of hazards presented by corrosives, it is important that emergency responders do not focus solely on the corrosive properties when considering appropriate actions for managing incidents involving these materials.

Irritants

Irritants are toxins that cause temporary but sometimes severe inflammation to the eyes, skin, or respiratory system. Irritants often attack the mucous membranes of the body such as the surfaces of the eyes, nose, mouth, throat, and lungs.

Irritant/Irritating Material — Liquid or solid that upon contact with fire or exposure to air emits dangerous or intensely irritating fumes.

Convulsants

Convulsants are toxic materials that can cause convulsions (involuntary muscle contractions). Some chemicals considered to be convulsants are strychnine, organophosphates, carbamates, and infrequently used drugs such as picrotoxin. Death can result from asphyxiation or exhaustion.

Convulsant — Poison that causes an exposed individual to have convulsions.

Carcinogens

Carcinogens are cancer-causing agents. Examples of known or suspected carcinogenic hazardous materials are polyvinyl chloride, benzene, asbestos, some chlorinated hydrocarbons, arsenic, nickel, some pesticides, and many plastics **(Figure 1.14)**. Exact data is not available on the level and duration or dose of exposure needed for individual chemicals to cause cancer. However, exposures to only small amounts of some substances may have long-term consequences. Disease and complications can occur as long as 10 to 40 years after exposure.

Carcinogen — Cancer-producing substance.

Figure 1.14 Exposure to carcinogens like benzene can cause cancer after a long latency period.

Delay in Knowing Health Effects

Unlike corrosives that burn the skin on contact, some harmful substances do not hurt the body right away. In some cases, it may take many years for a chemical, agent, or substance to cause a disease like cancer. Because of this *delay* (sometimes called *latency period*), it can be difficult to establish a direct chain of cause and effect between an exposure to a particular substance and the resulting disease.

Delay in Knowing Health Effects (concluded)

The history of asbestos demonstrates how long it can take before enough evidence is gathered to produce action in the form of government intervention. Asbestos was first used in the U.S. in the early 1900s to insulate steam engines. It was not until the 1940s, however, that asbestos began to be used extensively. In particular, it was used in U.S. Navy shipyards to insulate the country's growing fleet of warships during World War II.

While some articles documenting the harmful effects of asbestos were published as early as the 1930s, it was not until the 1960s (15- to 40-year latency period) that studies began to show an unquestionably clear relationship between the inhalation of asbestos fibers and the development of lung cancer, asbestosis, and mesothelioma in groups such as U.S. Navy shipyard workers. As a result of these studies and a growing public awareness of the hazard, the U.S. government began regulating asbestos in the 1970s.

Many substances (acetaldehyde, chloroform, progesterone, and polychlorinated biphenyls [PCBs]) are listed by the U.S. Department of Health and Human Services as *reasonably anticipated to be carcinogens* or *suspected carcinogens* because the body of evidence concerning their chronic effects is still being gathered and evaluated. Saccharin, for example, was listed as a suspected carcinogen for nearly 20 years before it was removed from the list in 2000 due to a lack of evidence that it caused cancer. In the same year, diesel exhaust particulate was added to the list.

Our understanding of the health effects associated with chemical products and substances is often changing, and new products are continually being developed. First responders should keep in mind that chronic health effects of substances may not be known for many years, and what is considered safe today, may not be tomorrow.

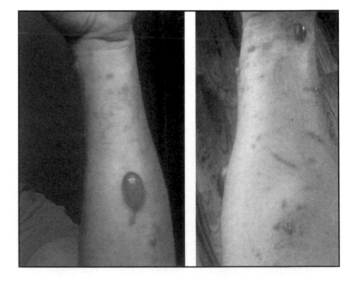

Figure 1.15 Urushiol, a chemical found in the sap of poison ivy, is an allergen that causes skin reactions in many people.

Sensitizers/Allergens

Allergen — Material that can cause an allergic reaction of the skin or respiratory system.

Allergens are substances that cause allergic reactions in people or animals. *Sensitizers* are chemicals that cause a substantial proportion of exposed people or animals to develop an allergic reaction after one or more exposures to the chemical. Common examples of sensitizers and allergens include latex, bleach, and urushiol (the chemical found in the sap of poison ivy, oak, and sumac)

(Figure 1.15). Some individuals exposed to a material may not be abnormally affected at first but may experience significant and dangerous effects when exposed to the material again.

Biological / Etiological Hazards

Biological (or *etiological*) *hazards* are microorganisms such as viruses or bacteria (or their toxins) that may cause severe, disabling disease or illness. Examples of these hazards are as follows:

- *Viruses* — Viruses are the simplest types of microorganisms that can only replicate themselves in the living cells of their hosts **(Figure 1.16).** Viruses do not respond to antibiotics.

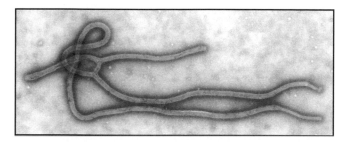

Figure 1.16 Ebola is a deadly virus. Viruses are unaffected by antibiotics. *Courtesy of the CDC Public Health Image Library.*

- *Bacteria* — Bacteria are microscopic, single-celled organisms **(Figure 1.17).** Most bacteria do not cause disease in people, but when they do, two different mechanisms are possible: invading the tissues or producing toxins (poisons).

- *Rickettsias* — Rickettsias are specialized bacteria that live and multiply in the gastrointestinal tract of arthropod carriers (such as ticks and fleas) **(Figure 1.18).** They are smaller than most bacteria, but larger than viruses.

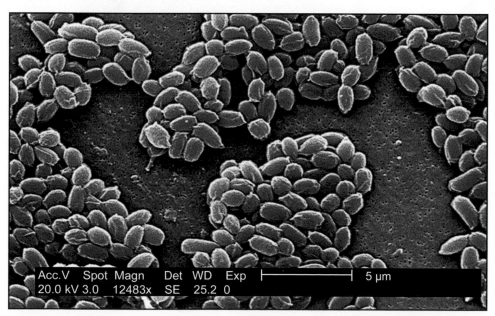

Acc.V	Spot	Magn	Det	WD	Exp	5 μm
20.0 kV	3.0	12483x	SE	25.2	0	

Figure 1.17 Anthrax is a bacteria. Bacterial infections can be treated by antibiotics. *Courtesy of the CDC Public Health Image Library.*

Figure 1.18 Ticks, fleas, and other arthropods may carry rickettsias. A bite from an infected carrier can transmit diseases such as Rocky Mountain Spotted Fever or Typhus. *Courtesy of the U.S. Department of Agriculture.*

Figure 1.19 Ricin, a biological toxin, is made from castor beans.

Infectious — Transmittable, able to infect people.

Contagious — Readily capable of being transmitted from one person to another by contact or close proximity.

They have properties that are similar to both. Like bacteria, they are single-celled organisms with their own metabolisms, and they are susceptible to broad-spectrum antibiotics. However, like viruses, they only grow in living cells. Most rickettsias are spread only through the bite of infected arthropods (such as ticks) and not through human contact.

• *Biological toxins* — Biological toxins are produced by living organisms; however, the biological organism itself is usually not harmful to people. Some biological toxins have been manufactured synthetically and/or genetically altered in laboratories for purposes of biological warfare **(Figure 1.19).** Examples include botulinum toxin and ricin; both are extremely lethal.

Infectious diseases are caused by the growth of microorganisms in the body and may or may not be **contagious.** (If a disease is contagious, it is readily communicable or easily transmitted from person to person.) Additionally, some biological hazards cause illness through their toxicity.

Examples of diseases associated with etiological hazards are malaria, acquired immunodeficiency syndrome (AIDS), tuberculosis, and typhoid. Exposure to biological hazards may occur in biological and medical laboratories or when dealing with people who are carriers of such diseases. Most of these diseases are carried in body fluids and are transmitted by contact with the fluids.

First responders may also be exposed to biological agents used as weapons in terrorist attacks and criminal activities. Biological weapons take the form of these disease-causing organisms and/or their toxins. Examples of potential biological weapons include:

• Smallpox (virus) **(Figure 1.20)**

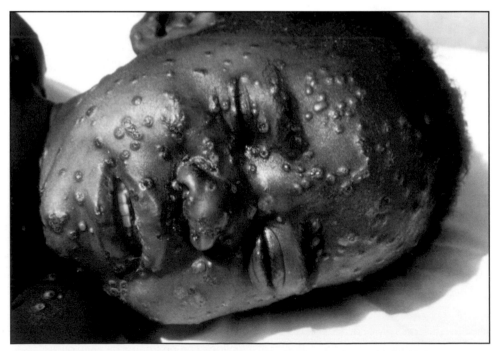

Figure 1.20 Experts fear that the smallpox virus could be used as a biological weapon. *Courtesy of the CDC Public Health Image Library.*

- Anthrax (bacteria)

- Botulism (toxin from the bacteria *Clostridium botulinum*)

Biological attacks could also be used to produce death and disease in animals and plants. The 2001 anthrax attacks in the United States were an example of a biological attack.

Mechanical Hazards

Mechanical hazards can cause trauma that occurs as a result of direct contact with an object. The two most common types are *striking* and *friction* exposures. Trauma can be mild, moderate, or severe and can occur in a single event.

In hazardous materials situations, a striking injury could be the result of an explosion caused by the failure of a pressurized container, a bomb (or **improvised explosive device [IED]**), or the **reactivity** of the hazardous material itself **(Figure 1.21)**. Friction injuries occur as a result of portions of the body rubbing against an abrasive surface, causing raw skin (abrasions), blisters, and burns. In hazardous material situations, contact between protective clothing and the skin sometimes causes friction injuries.

Improvised Explosive Device (IED) — Device that is categorized by its container and the way it is initiated; usually homemade, constructed for a specific target, and contained in almost anything.

Reactivity / Instability — Ability of two or more chemicals to react and release energy and the ease with which this reaction takes place.

Figure 1.21 Explosions such as the one caused by this car bomb can cause serious mechanical injuries to first responders. *Courtesy of the U.S. Department of Defense, photo by Ken Drylie, civilian.*

An explosion can cause the following four hazards (three mechanical and one thermal):

- *Blast-pressure wave (shock wave)* — Gases being released rapidly create a shock wave that travels outward from the center. As the wave increases in distance, the strength decreases. This blast-pressure wave is the primary reason for injuries and damage.

- *Shrapnel fragmentation* — Small pieces of debris thrown from a container or structure that ruptures from containment or restricted blast pressure. Shrapnel may be thrown over a wide area and great distances (fragmentation), causing personal injury and other types of damage to surrounding structures or objects. Shrapnel can result in bruises, punctures, or even avulsions (part of the body being torn away) when the container or pieces of the container strike a person.

- *Seismic effect* — Earth vibration similar to an earthquake. When a blast occurs at or near ground level, the air blast creates a ground shock or crater. As shock waves move across or underground, a seismic disturbance is formed. The distance the shock wave travels depends on the type and size of the explosion and type of soil.

- *Incendiary thermal effect* — During an explosion, thermal heat energy in the form of a fireball is the result of burning combustible gases or flammable vapors and ambient air at very high temperatures. The thermal heat fireball is present for a limited time after the explosive event.

Routes of Entry

The Centers for Disease Control (CDC) and the National Institute for Occupational Safety and Health (NIOSH) list the following three main **routes of entry** through which hazardous materials can enter the body and cause harm (see information box for others) **(Figure 1.22):**

- *Inhalation* — Process of taking in materials by breathing through the nose or mouth. Hazardous vapors, smoke, gases, liquid aerosols, fumes, and suspended dusts may be inhaled into the body. When a hazardous material presents an inhalation threat, respiratory protection is required.

- *Ingestion* — Process of taking in materials through the mouth by means other than simple inhalation. Taking a pill is a simple example of how a chemical might be deliberately ingested. However, poor hygiene after handling a hazardous material can lead to its ingestion accidentally. Other examples:

  — Chemical residue on the hands can be transferred to food and then ingested while eating. Hand washing is very important to prevent accidental ingestion of hazardous materials.

  — Ingestion can also occur when particles of insoluble materials become trapped in the mucous membranes and are swallowed after being cleared from the respiratory tract.

- *Skin contact (also includes contact with mucous membranes)* — Process of taking in materials when a chemical or hazardous material (in any state — solid, liquid, or gas) contacts the skin or exposed surface of the body (such as the mucous membranes of the eyes, nose, or mouth). If the skin is damaged or abraded, it is easier for substances to absorb into the body through the breaks in the skin's protective barrier. Toxic substances are also more readily absorbed into the body when the skin is wet (particularly from sweat). When this contact occurs, one of the following four things is likely to occur:

  — The skin and its associated film of lipid (fat) act as an effective barrier against penetration, injury, or other forms of disturbance (that is, the skin prevents any harm from being done).

  — The substance reacts with the skin surface and causes primary irritation or dermatitis (such as might happen when the skin is burned by an acid).

  — The substance penetrates the skin and conjugates with tissue protein, resulting in skin sensitization (meaning the substance can cause an allergic reaction such as what many people experience when contacting poison ivy).

Routes of Entry Pathways by which hazardous materials get into (or affect) the human body; commonly listed routes are inhalation, ingestion, skin contact, injection, and absorption.

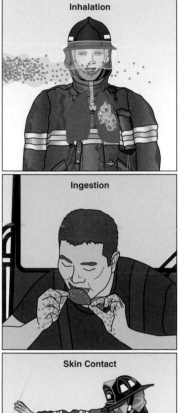

Figure 1.22 Ingestion, inhalation, and skin contact are the primary routes by which hazardous substances enter the body. In this context, skin contact also includes absorption of chemicals via the mucous membranes such as the eyes.

— The substance penetrates the skin, enters the blood stream, and acts as a potential systemic poison. This penetration is most likely to occur through two main absorption pathways: skin cells *(transepidermal)*; hair follicles and sweat glands *(pilosebaceous)*.

Other Routes of Entry

While this manual discusses the routes of entry as described by CDC and NIOSH, many sources designate four main routes of entry: *inhalation* and *ingestion* (the same as the CDC definitions given earlier) plus *injection* and *absorption* (terms used in place of *skin contact*). These entry routes are defined as follows:

- *Injection* — Process of taking in materials through a puncture or break in the skin. Protection from injection must be a consideration when dealing with any sort of contaminated (or potentially contaminated) objects easily capable of cutting or puncturing the skin (such as broken glass, nails, sharp metal edges, tools like utility knives, or other sharps).

- *Absorption* — Process of taking in materials through the skin or eyes. Some materials pass easily through the mucous membranes or areas of the body where the skin is the thinnest, allowing the least resistance to penetration. The eyes, nose, mouth, wrists, neck, ears, hands, groin, and underarms are areas of concern. Many poisons are easily absorbed into the body system in this manner. Others can enter the system easily through the unknowing act of touching a contaminated finger to one's eye.

Be Aware!

Some chemicals may have multiple routes of entry. For example, toluene (a solvent) can cause moderate irritation to the skin through skin contact, but it can also cause dizziness, lack of coordination, coma, and even respiratory failure when inhaled in sufficient concentrations. Other chemicals with multiple routes of entry include methyl ethyl ketone (MEK), benzene, and other solvents **(Figure 1.23)**.

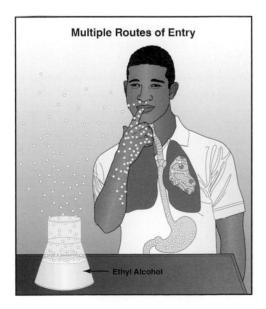

Figure 1.23 Many chemicals have multiple routes of entry. First responders must ensure that all routes are protected against exposure.

Comprehensive Environmental Response, Compensation, and Liability Act (CERCLA) — U.S. law that created a tax on the chemical and petroleum industries and provided broad federal authority to respond directly to releases or threatened releases of hazardous substances that may endanger public health or the environment.

Hazardous Materials Regulations and Definitions

This section explains the roles of different North American government agencies in regulating hazardous materials, as well as their definitions for such materials. Laws and regulations are often viewed negatively as too restrictive or limiting. For this reason, it is important for today's first responder to realize that the regulations affecting hazardous materials emergency response have been developed in response to a long string of hazardous materials emergencies, disasters, and environmental damage incidents. For example, the Love Canal environmental emergency and disaster directly led to the enactment of the **Comprehensive Environmental Response, Compensation, and Liability Act (CERCLA)** of 1980 (often referred to as the *Superfund*) (see case history box).

In 1986, fire service professional organizations representing both labor and management testified before the U.S. Congress and requested inclusion of emergency responders in the provisions of the Superfund Amendment and Reauthorization Act (SARA). Emergency responders requested this inclusion based on a history of harmful and deadly incidents that have affected the emergency response community. In response to the emergency response community's request, Congress directed both OSHA and EPA to include emergency responders in the 29 *CFR* 1910.120 and 40 *CFR* 311 regulations (see First Responder Roles section). In fact, the team that authored 29 *CFR* 1910.120 included individuals with strong fire-fighting and emergency-response backgrounds.

Why Do We Have All These Regulations?
Love Canal, Niagara Falls (1978)

The Love Canal saga began nearly 100 years before the environmental nightmare came to the world's attention. In the 1890s, industrialist William T. Love devised a plan to dig a canal around Niagara Falls. The canal would allow marine traffic around the falls, provide a water source for inexpensive hydroelectric power, and create a distinct boundary for a model planned industrial community. The U.S. economy entered a sharp decline shortly after the project began, and the development ceased. The open canal was publicly auctioned and by 1920 was in use as a local land fill and swimming area.

In 1942, Hooker Chemical Company (later purchased by Occidental Chemical) purchased the site and used it as a waste disposal site for its Niagara Falls Plant. Between 1942 and 1954, Hooker dumped an estimated 22,000+ tons (22 352+ tonnes) of chemical waste into the 3,000-foot long, 60-foot wide, 40-foot deep (194 m long, 18 m wide, 12 m deep) canal and several 25-foot (8 m) deep trenches built near the canal. In 1952, the local school board asked Hooker to sell a small portion of the filled canal to the board for the site of a new school. Hooker initially declined, and the school board subsequently threatened to forcibly acquire the section of the canal.

In 1953, Hooker agreed to provide the district with the property for $1 provided the school district took the entire canal site, and Hooker was relieved of any liability for the site. Hooker was allowed to continue dumping on the site until construction began. The deed-transfer paperwork carried specific warnings to the school board about not disturbing parts of the site. The school board built a school on the site, and the surrounding area was developed.

In 1978, a series of news reports identified local health issues and the presence of high levels of toxic materials including dioxins and polychlorinated biphenyls (PCBs)**(Figure 1.24).** The reports led to the evacuation of the entire area, the declaration of a federal emergency, and the development of the Comprehensive Environmental Response, Compensation, and Liability Act (CERCLA).

Figure 1.24 Love Canal was used as a chemical waste disposal site before being sold to a local school board for development in 1953. The resulting health problems eventually led to evacuation of the area and passage of the Comprehensive Environmental Response, Compensation, and Liability Act (CERCLA). *Courtesy of the U.S. Environmental Protection Agency.*

Torrey Canyon tank ship, England and France (March 18, 1967)

On March 18, 1967, one of the world's first oil supertankers *(Torrey Canyon)* ran aground off the southern coast of England. The mishap, caused by a series of navigational miscalculations, resulted in the release of approximately 31,000,000 gallons (125 000 000 L) of Kuwaiti oil from the *Torrey Canyon*. The spill caused extensive damage to marine life and shorelines of France and England. Additionally, since no one had ever planned for an event like this one, mitigation techniques only worsened the damage.

Attempts to mitigate the spill used over 10,000 tons (10 160 tonnes) of dispersing agents that became more toxic than the oil itself. Additionally, the wreckage and oil remaining in the ship were bombed and napalmed in attempts to burn the oil. Aviation gas was even dumped into the spill in an attempt to get it to burn more completely.

As a result of the incident, many countries instituted national plans for dealing with spills of large quantities of oil off their shores. The U.S. passed legislation providing for such planning, and designated the U.S Coast Guard (USCG) as the agency responsible for protecting U.S. coastlines from oil spills and other hazardous materials emergencies. The Coast Guard also placed in service what are called *USCG Strike Teams.*

Kansas City, MO (November 29, 1988)

Kansas City (MO) Fire Department Engine Companies 41 and 30 responded to a report of a pickup truck fire at a highway construction site in the early morning hours of November 29, 1988. Security guards who phoned in

Why Do We Have All These Regulations? (continued)

the call reported explosives were stored on site, and that information was communicated to the responding units. Arriving units found multiple fires on the site including two smoldering truck trailers that were loaded with explosives. The trailers' contents (which were not required to be marked or labeled) contained a total of nearly 50,000 pounds (22 680 kg) of ammonium nitrate, fuel-oil-mixture-based explosives. Other explosives stored on site were labeled. Shortly after arrival, the smoldering trailers detonated **(Figure 1.25).** The first blast killed the six firefighters on scene instantly and destroyed the two pumping apparatus. The explosion left an 80-foot (24 m) diameter and 8-foot (2.4 m) deep crater.

Nearly 10 years later, U.S. Department of Transportation regulations regarding when placards can be removed from vehicles were changed as a direct result of the incident.

Figure 1.25 U.S. Department of Transportation regulations regarding placards were changed after an unmarked trailer containing 50,000 pounds (22 680 kg) of ammonium nitrate exploded in Kansas City in 1988 killing six firefighters. *Courtesy of Ray Elder.*

Kingman, AZ (July 5, 1973)

Two workers had been preparing a railcar for unloading at the Kingman Doxol Gas plant. During the process, a leak was detected and as workers attempted to stop it, a fire ignited, seriously burning one of the workers and killing the other **(Figure 1.26).** The Kingman Fire Department was dispatched at 13:57, and first arriving units were on scene at 14:00. Fire department personnel initially deployed handlines and worked to secure water supply for deck-gun operations from the closest hydrant (approximately 1,200 feet [366 m] away).

Figure 1.26 After this tank car exploded in Kingman, Arizona, in 1973, the U.S. Department of Transportation changed its thermal protection standards for railcars. *Courtesy of Mohave Museum of History and Arts.*

Less than 20 minutes from the time the fire started, the railcar tank shell failed, releasing the propane contents in what has come to be known as a *BLEVE (boiling liquid expanding vapor explosion)*. The BLEVE killed four fire fighters instantly and burned seven others so badly that they succumbed to their injuries in the following days. One member of the fire department who had climbed into a truck to talk on the radio suffered severe burns but survived.

As a result of the incident, the Department of Transportation required all railcars in flammable gas service at the time to be retrofitted with thermal protection that would provide significant protection of the tank during similar fire conditions. Today's flammable gas railcars are thermally protected to protect the tank from 100 minutes of exposure to a 1,600°F fire (871°C) and at least 30 minutes of 2,200°F (1 204°C) impingement as a direct result of this incident.

Texas City, TX (April 16, 1947)

The greatest industrial disaster in U.S. history occurred in April of 1947 when a French ship, the *Grandcamp,* ignited during loading in Texas City, Texas. The ship was already loaded with tobacco, twine, cotton, other commodities, and 2,300 tons (2 337 tonnes) ammonium nitrate fertilizer when workers were completing the loading process in Hold 4 of the ship. The hold contained more than 800 tons (813 tonnes) of ammonium nitrate fertilizer and was being loaded with more when a fire ignited (thought to be caused by workers smoking and other lax safety practices). The crew attempted to fight the fire but was driven from the cargo hold rapidly by thick smoke. The crew sounded a dock alarm at about 08:30.

By 08:45, the Texas City Fire Department was on scene and had deployed hoselines. By 09:00, extensive flames were coming from the ship, and at 09:12 the ship disintegrated, killing all 27 firefighters of the Texas City Department and 34 vessel crew members and causing additional explosions and fires at nearby refineries and on adjacent ships. The initial blast sent the *Grandcamp's* 3,200-pound (1 452 kg) anchor flying over 1.6 miles (2.6 km) inland. The initial explosion and subsequent events killed over 550 people and injured well over 3,000 **(Figure 1.27).**

Figure 1.27 Over 550 people were killed and well over 3,000 injured in Texas City, Texas, in 1947 when a ship carrying ammonium nitrate fertilizer exploded at dock. *Courtesy of Moore Memorial Library, 1701 9th Avenue North, Texas City, TX 77590.*

Why Do We Have All These Regulations? (concluded)

As a result of this incident, the USCG Board of Investigation recommended establishment of a federal office to do the following:

- Collect, evaluate, and disseminate information on fire prevention and extinguishment on board merchant vessels.

- Prepare and publish a fire prevention and fire extinguishment manual for use on board merchant vessels.

- Establish and operate a fire-fighting school for training of key operating personnel of merchant ship operators and stevedores.

- Conduct other related marine safety activities.

It also served to increase awareness that fire departments in general are at great risk when responding to hazardous materials incidents and fires.

Shreveport, LA (September 17, 1984)

On September 17, 1984, the Shreveport (LA) Fire Department responded to an anhydrous ammonia leak at Dixie Cold Storage. While working to control the leak in the warehouse, a spark ignited the ammonia. The explosion and flash fire severely burned two firefighters who were working in the area in chemical protective clothing. One firefighter died from his injuries a few days later. As a direct result of the incident, the NFPA® standards on chemical protective clothing now address the hazards of flash fires and require garments to be constructed of materials that will not contribute to injuries in similar situations.

U.S. Regulations/Definitions

In the U.S., the four main agencies involved in the regulation of hazardous materials and/or wastes at the federal level are as follows:

- Department of Transportation (DOT)
- Environmental Protection Agency (EPA)
- Department of Labor (DOL)
- Nuclear Regulatory Commission (NRC)

In addition, agencies with a more minor emphasis on haz mat regulations are also introduced in the sections that follow.

Department of Transportation (DOT)

The DOT issues transportation regulations in Title 49 (Transportation) *CFR*. There are seven volumes of transportation regulations. These legally binding regulations are enforced at the federal, state, and local levels. The regulations specifically governing the transportation of hazardous materials found in Title 49 *CFR* are sometimes referred to as the *Hazardous Materials Regulations* or *HMR.* They address the transportation of hazardous materials in all modes: air, highway, pipeline, rail, and water.

The purpose of the HMR is to provide adequate protection against the risks to life, property, and the environment inherent in transporting hazardous materials in commerce by improving the regulatory and enforcement authority of the Secretary of Transportation. The DOT Pipeline and Hazardous Materials

Safety Administration (PHMSA) agency carries out these duties through a program of regulation, enforcement, emergency response education and training, and data collection and analysis.

Environmental Protection Agency (EPA)

The EPA is responsible for researching and setting national standards for a variety of environmental programs. It delegates the responsibility for issuing permits, monitoring, and enforcing compliance to states and tribes. It also works closely with other federal agencies, state and local governments, and Indian tribes to develop and enforce regulations under existing environmental laws. The EPA also works with industries and all levels of government in a wide variety of voluntary pollution prevention programs and energy conservation efforts.

Several pieces of environmental legislation are of particular interest to first responders. The first is the Comprehensive Environmental Response, Compensation, and Liability Act (CERCLA), commonly known as the *Superfund Act,* which was enacted by Congress on December 11, 1980. This law created a tax on the chemical and petroleum industries and provided broad federal authority to respond directly to releases or threatened releases of hazardous substances that may endanger public health or the environment **(Figure 1.28).** Over 5 years, $1.6 billion was collected, and the tax went to a trust fund for cleaning up abandoned or uncontrolled hazardous waste sites. CERCLA was responsible for the following actions:

Figure 1.28 Passage of the Comprehensive Environmental Response, Compensation, and Liability Act (CERCLA) allowed the EPA to begin cleaning up abandoned and uncontrolled hazardous waste sites across the country. *Courtesy of the U.S. Environmental Protection Agency.*

- Established prohibitions and requirements concerning closed and abandoned hazardous waste sites
- Provided for liability of persons responsible for releases of hazardous waste at these sites
- Established a trust fund to provide for cleanup when no responsible party could be identified

The law authorizes the following two kinds of response actions:

1. Short-term removals where actions may be taken to address releases or threatened releases requiring prompt response
2. Long-term remedial response actions that permanently and significantly reduce the dangers associated with releases or threats of releases of hazardous substances that are serious, but not immediately life-threatening (see EPA's National Priorities List for more information)

CERCLA also enabled the revision of the National Contingency Plan (NCP). The NCP provided the guidelines and procedures needed to respond to releases and threatened releases of hazardous substances, pollutants, or contaminants. CERCLA was amended by the Superfund Amendments and Reauthorization Act (SARA) on October 17, 1986.

SARA reflected the EPA's experience in administering the complex Superfund program during its first 6 years and made several important changes and additions to the program. SARA was responsible for the following actions:

- Stressed the importance of permanent remedies and innovative treatment technologies in cleaning up hazardous waste sites
- Required Superfund actions to consider the standards and requirements found in other state and federal environmental laws and regulations
- Provided new enforcement authorities and settlement tools
- Increased state involvement in every phase of the Superfund program
- Increased the focus on human health problems posed by hazardous waste sites
- Encouraged greater citizen participation in making decisions on how sites should be cleaned up
- Increased the size of the trust fund to $8.5 billion

The Emergency Planning and Community Right-to-Know Act (EPCRA) (also known as Title III of SARA) was enacted by Congress as the national legislation on community safety. This law was designated to help local communities protect public health, safety, and the environment from chemical hazards. To implement EPCRA, Congress required each state to appoint a State Emergency Response Commission (SERC). The SERCs were required to divide their states into Emergency Planning Districts and to name a **Local Emergency Planning Committee (LEPC)** for each district. Broad representation by emergency responders, health officials, government and media representatives, community groups, industrial facilities, and emergency managers ensures that all necessary elements of the planning process are represented.

Local Emergency Planning Committee (LEPC) — Required by SARA Title III, LEPCs are composed of local officials, citizens, and industry representatives with the task of designing, reviewing, and updating a comprehensive emergency plan for an emergency planning district; plans may address hazardous materials inventories, hazardous material response training, and assessment of local response capabilities.

LEPCs are structured in many different ways and their Planning District determines their role, responsibilities and capabilities. LEPCs can be at the local level, combining several jurisdictions into a planning district, or just have one at the state level. They often assume the planning and/or correlation of emergency activities in some communities. One of the primary benefits of an active LEPC is opening the line of communications between the private and public sector. Some LEPCs are very effective and work closely with emergency responders, others function in name only. The LEPC can be very beneficial to emergency responders by sharing chemical information, contact numbers, etc. with them.

The Resource Conservation and Recovery Act (RCRA) gave the EPA the authority to control hazardous waste and hold the responsible party accountable for the hazardous materials they produce (from their creation to their final disposal, often referred to as *cradle-to-grave*). This control includes the generation, transportation, treatment, storage, and disposal of hazardous waste. RCRA also set forth a framework for the management of nonhazardous wastes.

The 1986 amendments to RCRA enabled the EPA to address environmental problems that could result from underground tanks storing petroleum and other hazardous substances. RCRA focuses only on active and future facilities and does not address abandoned or historical sites.

The Toxic Substances Control Act (TSCA) of 1976 was enacted by Congress to give the EPA the ability to track the 75,000 industrial chemicals currently produced or imported into the U.S. The EPA repeatedly screens these chemicals and can require reporting or testing of those that may pose an environmental or human-health hazard. The EPA can ban the manufacture and import of those chemicals that pose an unreasonable risk.

Also, the EPA has mechanisms in place to track the thousands of new chemicals that industry develops each year with either unknown or dangerous characteristics. The EPA then can control these chemicals as necessary to protect human health and the environment. TSCA supplements other federal statutes, including the Clean Air Act and the Toxic Release Inventory under EPCRA. The EPA issues legislation to protect the environment in Title 40 *CFR*.

Department of Labor (DOL)

The DOL is responsible for overseeing U.S. labor laws. The U.S. Congress passed the Occupational Safety and Health (OSH) Act in 1970. One year later a new department, the Occupational Safety and Health Administration (OSHA), was created to oversee compliance with the Act under the jurisdiction of the DOL.

OSHA issues legislation relating to worker safety under Title 29 *CFR*. OSHA legislation of interest to first responders includes the *HAZWOPER* regulation (29 *CFR* 1910.120) discussed earlier as well as the *Hazard Communication* regulation (29 *CFR* 1910.1200) and the *Process Safety Management of Highly Hazardous Chemicals* regulation (29 *CFR* 1910.119).

The Hazard Communication Standard (HCS) is designed to ensure that information about chemical hazards and associated protective measures is disseminated to workers and employers. This dissemination is accomplished by requiring chemical manufacturers and importers to evaluate the hazards of the chemicals they produce or import and provide information about them through labels on shipped containers and safety data sheets (SDSs, formerly called material safety data sheets or MSDSs).

The Process Safety Management (PSM) of Highly Hazardous Chemicals (HHCs) standard is intended to prevent or minimize the consequences of a catastrophic release of toxic, reactive, flammable, or explosive HHCs from a process. A process is any activity or combination of activities including any use, storage, manufacturing, handling, or on-site movement of HHCs. A *process* includes any group of vessels that are interconnected and separate vessels that are located such that a HHC could be involved in a potential release.

Nuclear Regulatory Commission

The NRC regulates U.S. commercial nuclear power plants and the civilian use of nuclear materials as well as the possession, use, storage, and transfer of radioactive materials through Title 10 (Energy) *CFR* 20, *Standards for Protection Against Radiation*. The NRC's primary mission is to protect the public's health and safety and the environment from the effects of radiation from nuclear reactors, materials, and waste facilities. Title 10 *CFR* 20 includes information on the following items:

- Radiation dose limits for workers and members of the public
- Requirements for monitoring and labeling of radioactive materials
- Requirements for posting of radiation areas
- Requirements for reporting the theft or loss of radioactive materials
- Tables of individual radionuclide exposure limits

Other Agencies

Several other U.S. agencies are involved with hazardous materials:

- ***Department of Energy (DOE)*** — Manages the national nuclear research and defense programs, including the storage of high-level nuclear waste **(Figure 1.29).** The DOE oversees the National Nuclear Security Administration (NNSA), which provides the main capability for responding to nuclear or radiological incidents within the U.S. and abroad. The NNSA also provides operational planning and training to counter both domestic and international nuclear terrorism. The Office of Civilian Radioactive Waste Management, which manages and disposes of high-level radioactive waste and spent nuclear fuel in the U.S., also falls under the DOE.

Figure 1.29 The U.S. Department of Energy oversees the Office of Civilian Radioactive Waste Management and the National Nuclear Security Administration, both of which play a role in the management of nuclear materials in the U.S., including nuclear waste. *Courtesy of the National Nuclear Security Administration, Nevada Site Office.*

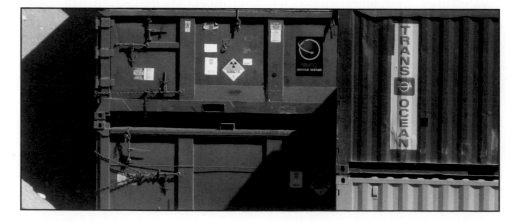

- *Department of Homeland Security (DHS)* — Has three primary missions: (1) prevent terrorist attacks within the U.S., (2) reduce America's vulnerability to terrorism, and (3) minimize the damage from potential attacks and natural disasters. DHS was created in the aftermath of the September 11, 2001, terrorist attacks and assumes primary responsibility for ensuring that emergency response professionals are prepared for any situation in the event of a terrorist attack, natural disaster, or other large-scale emergency. *Details:*

  — Responsibilities include providing a coordinated, comprehensive federal response to any large-scale crisis and mounting a swift and effective recovery effort.

  — The Federal Emergency Management Agency (FEMA) and U.S. Coast Guard (USCG) are just a few of the agencies that were located within DHS following the 9/11 attacks **(Figure 1.30).**

Figure 1.30 The U.S. Department of Homeland Security was created in response to the 9/11 attacks. *Courtesy of FEMA News Photos, photo by Mike Rieger.*

- *Consumer Product Safety Commission (CPSC)* — Oversees and enforces compliance with the Federal Hazardous Substances Act (FHSA), which requires that certain hazardous household products (hazardous substances) carry cautionary labeling to alert consumers to the potential hazards that those products present and inform them of the measures needed to protect themselves from those hazards. Under this Act, the CPSC also has the authority to ban a product if the hazardous substance it contains is so hazardous that the cautionary labeling required by the Act is inadequate to protect the public.

- *Department of Defense Explosives Safety Board (DDESB), Department of Defense (DOD)* — Provides oversight of the development, manufacture, testing, maintenance, demilitarization, handling, transportation, and storage of explosives, including chemical agents on DOD facilities worldwide.

- *Bureau of Alcohol, Tobacco, Firearms and Explosives (ATF), Department of Treasury* — Enforces the federal laws and regulations relating to alcohol, tobacco products, firearms, explosives, and arson.

- *Department of Justice (DOJ)* — Assigns primary responsibility for operational response to threats or acts of terrorism within U.S. territory to the Federal Bureau of Investigation (FBI). The FBI then operates as the on-scene manager for the federal government. It is ultimately the lead agency on terrorist incident scenes. *FBI duties:*

  — Investigates the theft of hazardous materials

  — Collects evidence for crimes

  — Prosecutes criminal violations of federal hazardous materials laws and regulations

U.S. Agencies

Table 1.4 lists the main U.S. agencies involved in the regulation of hazardous materials, their spheres of responsibility, and associated pieces of legislation. The table also includes the hazardous material terms and definitions coming from the respective regulations that are of primary concern to first responders. First responders need to be aware of the contexts in which these different terms are used.

Canadian Regulations/Definitions

In Canada the four main agencies that are involved in the regulation of hazardous materials and/or wastes at the national level are as follows:

- Transport Canada (TC)
- Environment Canada
- Health Canada
- Canadian Nuclear Safety Commission (CNSC)

Transport Canada

TC is the focal point for the national program to promote public safety during the transportation of dangerous goods (the Canadian term for hazardous materials in transport). The department's Transport Dangerous Goods (TDG) Directorate serves as the major source of regulatory development, information, and guidance on dangerous goods transport for the public, industry employees, and government employees, particularly in regards to the Transportation of Dangerous Goods Act, the Canadian equivalent of the U.S. HMR. The Directorate also operates the Canadian Transport Emergency Centre (CANUTEC) (see Emergency Response Centers section).

Environment Canada

Environment Canada shares with Health Canada the task of assessing and managing the risks associated with toxic substances. Under the Canadian Environmental Protection Act (CEPA), 1999, the potential risks of environmental pollutants and toxic substances are evaluated. The Act also addresses pollution prevention and the protection of the environment and human health. In order to distinguish new substances from existing ones and prescribe reporting requirements for new substances, Environment Canada has established the following substance inventories or lists:

Agency	Sphere of Responsibility	Important Legislation	Hazardous Material Terms/Definitions
Department of Transportation (DOT) Research and Special Programs Administration (RSPA)	Transportation Safety	Title 49 (Transportation) *CFR* 100-185 Hazardous Materials Regulations (HMR)	***Hazardous Material:*** a substance or material (including hazardous wastes, marine pollutants, and elevated temperature materials) that has been determined by the U.S. Secretary of Transportation to be capable of posing an unreasonable risk to health, safety, and property when transported in commerce and which has been so designated*
Environmental Protection Agency (EPA)	Public Health and the Environment	Title 40 (Protection of Environment) *CFR* 302.4 Designation of Hazardous Substances	***Hazardous Substance:*** a chemical that if released into the environment above a certain amount must be reported and, depending on the threat to the environment, federal involvement in handling the incident can be authorized
		40 *CFR* 355 Superfund Amendments and Reauthorization Act (SARA)	***Extremely Hazardous Substance:*** any chemical that must be reported to the appropriate authorities if released above the threshold reporting quantity** ***Toxic Chemical:*** one whose total emission or release must be reported annually by owners and operators of certain facilities that manufacture, process, or otherwise use a listed toxic chemical***
		40 *CFR* 261 Resource Conservation and Recovery Act (RCRA)	***Hazardous Wastes:*** chemicals that are regulated under the Resource Conservation and Recovery Act (40 *CFR* 261.33 provides a list of hazardous wastes.)
Department of Labor (DOL) Occupational Safety and Health Administration (OSHA)	Worker Safety	29 (Labor) *CFR* 1910.1200 Hazard Communications	***Hazardous Chemical:*** any chemical that would be a risk to employees if exposed in the workplace (Hazardous chemicals cover a broader group of chemicals than the other chemical lists.)
		29 *CFR* 1910.120 Hazardous Waste Operations and Emergency Response (HAZWOPER)	***Hazardous Substance:*** every chemical regulated by the U.S. DOT and EPA.
		29 *CFR* 1910.119 Process Safety Management of Highly Hazardous Chemicals	***Highly Hazardous Chemicals:*** those chemicals that possess toxic, reactive, flammable, or explosive properties (A list of these chemicals is published in Appendix A of 29 *CFR* 1910.119.)

Continued

Table 1.4 (concluded)

Agency	Sphere of Responsibility	Important Legislation	Hazardous Material Terms/Definitions
Consumer Product Safety Commission (CPSC)	Hazardous Household Products (chemical products intended for consumers)	Title 16 (Commercial Practices) *CFR* 1500 Hazardous Substances and Articles Federal Hazardous Substances Act (FHSA)	***Hazardous Substance:*** any substance or mixture of substances that is toxic; corrosive; an irritant; a strong sensitizer; flammable or combustible; or generates pressure through decomposition, heat, or other means and if such substance or mixture of substances may cause substantial personal injury or substantial illness during or as a proximate result of any customary or reasonably foreseeable handling or use, including reasonably foreseeable ingestion by children. ***Also:*** any radioactive substance if, with respect to such substance as used in a particular class of article or as packaged, the Commission determines by regulation that the substance is sufficiently hazardous to require labeling in accordance with the Act in order to protect the public health****
Nuclear Regulatory Commission (NRC)	Radioactive Materials (use, storage, and transfer)	Title 10 (Energy) *CFR* 20 Standards for Protection Against Radiation	

* DOT uses the term hazardous materials to cover 9 hazard classes, some of which have subcategories called divisions. DOT includes in its regulations hazardous substances and hazardous wastes, both of which are regulated by the U.S. EPA if their inherent properties would not otherwise be covered. The different DOT hazard classes are discussed in Chapter 3, Hazardous Materials Identification.

** Each substance has a threshold reporting quantity. The list of extremely hazardous substances is identified in Title III of SARA of 1986 (see 40 CFR 355).

*** The list of toxic chemicals is provided in Title III of SARA (see 40 CFR 355). The EPA regulates these materials because of public health and safety concerns. While regulatory authority is granted under the Resource Conservation and Recovery Act, the DOT regulates the transport of these materials.

**** The complete definition of hazardous substance as found in the FHSA contains five parts (A–E) and includes such items as toys and other articles intended for use by children. Only parts A and C are cited here in their entirety.

- Domestic Substances List
- Export Control List
- National Pollutant Release Inventory
- Non-Domestic Substances List
- Priority Substances List
- Toxic Substances List
- Waste or other matter that may be disposed of at sea

Health Canada

Health Canada provides national leadership to develop health policy, enforce health regulations, promote disease prevention, and enhance healthy living for all Canadians. The Minister of Health has total or partial responsibility for administration of the following Acts:

- **Hazardous Products Act** — Controls the sale, advertising, and importation of hazardous products used by consumers in the workplace that are not covered by other acts and listed as prohibited or restricted products. The Act covers consumer products that are poisonous, toxic, flammable, explosive, corrosive, infectious, oxidizing, and reactive. It also covers workplace hazardous materials; products intended for domestic or personal use, gardening, sports, or other recreational activities; and products for lifesaving or children (such as toys, games, and equipment), which pose or are likely to pose a hazard to public health and safety because of their design, construction, or contents.

- **Pest Control Products Act (and regulations)** — Intended to protect people and the environment from risks posed by pesticides, which include a variety of products such as insecticides, herbicides, and fungicides. Any pesticide imported, sold, or used in Canada must first be registered under this Act, which is administered by the Pest Management Regulatory Agency of Health Canada.

Canadian Nuclear Safety Commission

The Canadian Nuclear Safety Commission can be best described as the watchdog over the use of nuclear energy and materials in Canada. In addition to nuclear power plants and nuclear research facilities, the CNSC regulates numerous other uses of nuclear material. Some examples include radioisotopes used in the treatment of cancer, the operation of uranium mines and refineries, and the use of radioactive sources for oil exploration and in instruments such as precipitation measurement devices.

Canadian Agencies

Table 1.5, p. 48 lists the main Canadian agencies, their spheres of responsibility, important legislation, and the hazardous materials terms and definitions associated with the legislation. **Table 1.6, p. 49** provides a brief summary of the regulatory programs administered by other agencies of the Canadian government relating to chemical substances.

Mexican Regulations/Definitions

The three main agencies that are involved in the regulation of hazardous materials and/or wastes at the national level in Mexico are as follows:

- **Secretaría de Comunicaciones y Transportes (SCT)** — Ministry of Communications and Transport

- **Secretaría de Medio Ambiente y Recursos Naturales (SEMARNAT)** — Ministry of Environment and Natural Resources

- **Secretaría del Trabajo y Previsión Social (STPS)** — Ministry of Labor and Social Welfare

Secretaría de Comunicaciones y Transportes

The Ministry of Communications and Transport is responsible for publishing and maintaining the official Mexican standards (NOMs) covering the *Mexican Hazardous Materials Land Transportation Regulation*. The Mexican NOMs are

Table 1.5
Main Canadian Agencies Involved in the
Regulation of Hazardous Materials

Agency	Sphere of Responsibility	Important Legislation	Hazardous Material Terms/Definitions
Transport Canada (TC) Transport Dangerous Goods (TDG) Directorate	Transportation Safety	Transportation of Dangerous Goods Act	***Dangerous Goods:*** any product, substance, or organism included by its nature, or by the regulation, in any of the classes listed in the schedule of the nine United Nations (UN) Classes of Hazardous Materials*
Environment Canada	Public Health and the Environment	Canadian Environmental Protection Act 1999	***Toxic Substance:*** a substance that if it is entering or may enter the environment in a quantity or concentration or under conditions that it (a) Has or may have an immediate or long-term harmful effect on the environment or its biological diversity (b) Constitutes or may constitute a danger to the environment on which life depends (c) Constitutes or may constitute a danger in Canada to human life or health
Transboundary Movement Division	Transportation of Hazardous Waste	Canadian Environmental Protection Act, 1999 Export and Import of Hazardous Wastes Regulations (EIHWR)	***Hazardous Waste:*** any substance specified in Parts I, II, III, or IV of the List of Hazardous Wastes Requiring Export or Import Notification in Schedule III, *(déchets dangereux);* or any product, substance, or organism that is dangerous goods, as defined in Section 2 of the Transportation of Dangerous Goods Act, 1992, that is no longer used for its original purpose and that is recyclable material or intended for treatment or disposal, including storage prior to treatment or disposal, but does not include a product, substance or organism that is (i) Household in origin (ii) Returned directly to a manufacturer or supplier of the product, substance, or organism for reprocessing, repackaging, or resale, including a product, substance or organism that is (A) Defective or otherwise not usable for its original purpose (B) In surplus quantities but still usable for its original purpose (iii) Included in Class 1 or 7 of the *Transportation of Dangerous Goods Regulations*

Continued

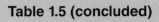

Table 1.5 (concluded)

Agency	Sphere of Responsibility	Important Legislation	Hazardous Material Terms/Definitions
Health Canada	Worker Safety	Hazardous Product Act	***Hazardous Product:*** any product, material, or substance that is, or contains, a poisonous, toxic, flammable, explosive, corrosive, infectious, oxidizing, or reactive product, material, or substance (or other product, material or substance of a similar nature) that the Governor in Council is satisfied is (or is likely to be) a danger to the health or safety of the public
Workplace Hazardous Materials Information System Division (WHMIS Division)	Worker Safety/ Chemicals Intended for the Workplace	Hazardous Product Act Workplace Hazardous Materials Information System (WHMIS)	***Controlled Product:*** any product, material, or substance specified by the regulations to be included in any of the classes listed in Schedule II of the Hazardous Product Act

* Internationally, hazardous materials in transport are generally referred to as dangerous goods.

Table 1.6
Other Canadian Agencies Involved with Hazardous Materials

Canadian Authority	Program
Health Canada, Consumer Products Division	Chemicals on the retail market
Natural Resources Canada, Explosives Regulatory Division	Explosives
Environment Canada, Waste Management and Remediation	Management of hazardous wastes, assessment and remediation of contaminated sites, and the control of waste disposal at sea
Canadian Nuclear Safety Commission	Nuclear substances
Health Canada, Pest Management Regulatory Agency	Pesticides
Health Canada, Radiation Protection Bureau	Radioactive substances
National Energy Board	Transportation of chemical products (oil and natural gas) via pipeline

fairly consistent with the *United Nations Recommendations on the Transport of Dangerous Goods (UN Recommendations).* Since the U.S. Hazardous Materials Regulations are also based on the *UN Recommendations,* the HMR and the Mexican regulations are fairly consistent. The significant differences between the two are discussed in Chapter 2, Hazardous Materials Identification.

Secretaría de Medio Ambiente y Recursos Naturales

The Ministry of Environment and Natural Resources is roughly equivalent to the U.S. EPA. Its main purpose is to create a state environmental protection policy and oversee the Federal General Law of Ecological Equilibrium and the Protection of the Environment (LGEEPA). Included in LGEEPA are regulations for manufacturing, handling, storage, and disposal of hazardous wastes.

Secretaría del Trabajo y Previsión Social

Labor is the responsibility of the Ministry of Labor and Social Welfare, and there is one labor law in Mexico: Ley Federal del Trabajo or LFT (Federal Labor Law). In the Federal Labor Law, there are two regulations: (1) Regulation for Safety, Health, and Environment in the Workplace (RFSHMAT) and (2) Regulation for the Inspection and Application of Sanctions for Violations of Labor Legislation. Of particular interest to the emergency responder are the following NOMs outlining the requirements and responsibilities set forth in RFSHMAT:

- *System for the Identification and Communication of Hazards and Risks for Dangerous Chemical Substances in the Workplace (NOM-018-STPS-2000)* — Equivalent to the U.S. Hazard Communication Standard in that it sets forth requirements dealing with chemical labels, employee training and communication, and material safety data sheets; also provides a list of substances divided by classification and grade of risk

- *Signs and Colors for Safety and Health, and Identification of Risk of Accidents by Fluids Conducted in Pipes (NOM-026-STPS-1998)* — Spells out the color and signage requirements for pipelines carrying various hazardous materials

- *Health and Safety Conditions in the Workplace for the Handling, Transport, and Storage of Hazardous Chemical Substances (NOM-005-STPS-1998)* — Details the requirements for handling, transport, and warehousing of hazardous materials, including programs relating to confined spaces, personal protective equipment, training, and medical monitoring as well as maintenance of machinery, containers, installations, and other equipment

Mexican Agencies

Table 1.7, p. 52 lists the Mexican agencies, their spheres of responsibility, and important legislation. It also covers the hazardous material terms and definitions with which first responders need to be familiar.

What This Means To You

Don't get caught up in the specific definitions of different haz mat terms. To you, as an emergency responder, they are all *hazardous*, and that means potentially dangerous. However, since you may hear other terms being used, be aware that the location where you find a hazardous material and how it is being used may determine what it is called for government purposes.

For example, when xylene is being transported in the U.S., the DOT regulates it, and it would be called a *hazardous material.* (If it were being transported in Canada, it would be called a *dangerous good.*) In the industry (or place of employment) where it is being used or manufactured, it becomes subject to the OSHA requirements protecting employees who work with it, and it would be considered a *hazardous chemical.* If it were marketed to consumers for purchase and use, it would fall subject to the Consumer Product Safety Commission, and it would be called a *hazardous substance.*

If at any point xylene was accidentally discharged from its packaging into the environment, it would become a hazardous substance as regulated by the EPA. When xylene completes its useful life in a plant or workplace and must be disposed of (in any manner), it becomes a **hazardous waste** and would be subject to both the EPA and DOT regulations (during transport). Additionally, if xylene was used to kill or injure a large number of people in a terrorist attack, it might be called a *weapon of mass destruction* by federal law enforcement authorities such as the FBI (see Chapter 7, Terrorist Attacks, Criminal Activities, and Disasters).

Hazardous Wastes — Discarded materials regulated by the Environmental Protection Agency because of public health and safety concerns. Regulatory authority is granted under the Resource Conservation and Recovery Act.

Hazardous Materials Incident Statistics

Haz mat incidents occur frequently. It is likely that all emergency first responders will have to deal with hazardous materials at some point in their careers. In fact, haz mat spills, releases, and incidents are so common that there are several different U.S. government agencies maintaining databases to track them, including the NRC, EPA, DOT, OSHA, and ATSDR.

Because certain hazardous materials are more common than others, they are statistically more likely to be involved in incidents and accidents. Additionally, clandestine, illegal methamphetamine labs with their variety of hazardous products are problems in many jurisdictions. Records have shown that the majority of haz mat incidents involve the following products (not necessarily in this order):

- Flammable/combustible liquids (petroleum products, paint products, resins, adhesives, etc.)
- Corrosives (sulfuric acid, hydrochloric acid, sodium hydroxide, etc.)
- Anhydrous ammonia
- Chlorine

Many incidents occur while hazardous materials are being transported, and statistics indicate that the majority of haz mat transportation incidents occur while the materials are being transported via highway rather than by air, rail, or water. **Table 1.8, p. 53** shows the numbers of incidents divided by mode (or means of transport) reported to the U.S. DOT between the years 1999 and 2008. **Table 1.9, p. 54** shows transportation incident statistics by commodity for 2008. Note that the majority of incidents involved flammable and combustible liquids, with corrosives being the second highest.

Table 1.7
Mexican Agencies Involved in the Regulation of Hazardous Materials

Agency	Sphere of Responsibility	Important Legislation	Hazardous Material Terms/Definitions
Secretaría de Communicaciones y Tranportes Ministry of Communications and Transportation	Transportation Safety	Mexican Hazardous Materials Land Transportation Regulation NOM-004-SCT-2000: System of Identification of Units Designated for the Transport of Hazardous Substances, Materials, and Wastes NOM-005-SCT 2000: Emergency Information for the Transport of Hazardous Substances, Materials, and Wastes	
Secretaría de Medio Ambiente y Recursos Naturales Ministry of the Environment and Natural Resources	Public Health and the Environment	La Ley General de Equilibrio Ecológico y Protección al Ambiente: Federal General Law of Ecological Equilibrium and the Protection of the Environment (LGEEPA) Regulation of LGEEPA in the area of hazardous wastes	
Secretaría del Trabajo y Previsión Social Ministry of Labor and Social Welfare	Worker Safety/ Labor	NOM-018-STPS-2000: System for the Identification and Communication of Hazards and Risks for Dangerous Chemical Substances in the Workplace Communications Wastes	***Sustancias qui'micas peligrosas (dangerous chemical substances):*** those chemicals that through their physical and chemical properties upon being handled, transported, stored, or processed present the possibility of fire, explosion, toxicity, reactivity, radioactivity, corrosive action, or harmful biological action and can effect the health of the persons exposed or cause damage to installations and equipment
		NOM-005-STPS-1998: Health and Safety Conditions in the Workplace for the Handling, Transport, and Storage of Hazardous Chemical Substances	***Sustancias to'xicas (toxic substances):*** those chemicals in solid, liquid, or gaseous state that can cause death or damage to health if they are absorbed by the worker even in relatively small amounts
		NOM-026-STPS-1998: Signs and Colors for Safety and Health, and Identification of Risk of Accidents by Fluids Conducted in Pipes	***Fluidos de bajo riesgo (dangerous fluids):*** those liquids and gases that can cause injury or illness on the job because of their intrinsic hazards such as flammables, unstable combustibles that can cause explosion, irritants, corrosives, toxics, reactives, radioactives, biological agents, or those that are subjected to extreme pressure or temperature as part of a process

Table 1.8
U.S. Department of Transportation
Hazardous Materials Information System
Incidents By Mode and Incident Year

	1999	2000	2001	2002	2003	2004	2005	2006	2007	2008	Total
Air	1,582	1,419	1,083	732	750	993	1,655	2,411	1,555	1,276	13,456
Highway	14,953	15,063	15,804	13,502	13,594	13,071	13,461	17,157	16,904	14,780	148,289
Rail	1,073	1,058	899	870	802	765	745	704	750	750	8,416
Water	8	17	6	10	10	17	69	68	61	98	364
Total	**17,616**	**17,557**	**17,792**	**15,114**	**15,156**	**14,846**	**15,930**	**20,340**	**19,270**	**16,904**	**170,525**

Some statistics are also kept for releases at fixed facilities and pipelines. **Table 1.10, p. 55** shows a pipeline accident summary by commodity provided by the U.S. Office of Pipeline and Hazardous Materials Safety Administration (PHMSA) for the year 2002. While there were no fatalities related to these incidents, there were 10 injuries. Almost all pipcline incidents involved flammable or combustible liquids with crude oil topping the list. First responders should also be aware that many of the pipelines operate at very high pressures (800-1200 pounds per square inch [psi]). If emergency responders are called to the scene of a pipeline incident, they need to be concerned about flammability hazards.

The U.S. **Agency for Toxic Substances and Disease Registry (ATSDR)** maintains the Hazardous Substances Emergency Events Surveillance (HSEES) database. In 2009, 14 state health departments had cooperative agreements with ATSDR to participate in HSEES: Colorado, Florida, Iowa, Louisiana, Michigan, Minnesota, New Jersey, New York, North Carolina, Oregon, Texas, Utah, Washington, and Wisconsin. **Table 1.11, p. 56** provides insight into the hazardous materials involved in reported incidents for both fixed facilities and transportation in these states. Carbon monoxide, nitrogen oxide, sulfur dioxide, and ammonia were the most common materials reported in 2005.

Agency for Toxic Substances and Disease Registry (ATSDR) — Lead U.S. public health agency responsible for implementing the health-related provisions of the Comprehensive Environmental Response, Compensation, and Liability Act (CERCLA); charged with assessing health hazards at specific hazardous waste sites, helping to prevent or reduce exposure and the illnesses that result, and increasing knowledge and understanding of the health effects that may result from exposure to hazardous substances.

Table 1.9
U.S. Department of Transportation
Hazardous Materials Information System
Summary of Incidents by Hazard Class, 2008

Hazard Class	Incidents	Injuries Hospitalized	Injuries Non-Hospitalized	Fatalities	Damages
Flammable-Combustible Liquid	8,111	2	28	4	27,956,644
Corrosive Material	4,313	8	78	2	7,290,739
Combustible Liquid	1,212	1	0	0	7,724,092
Nonflammable Compressed Gas	907	0	21	0	1,445,913
Miscellaneous Hazardous Material	552	0	3	0	1,469,563
Oxidizer	403	1	17	0	1,899,910
Flammable Compressed Gas	380	1	8	3	2,924,409
Poisonous Materials	272	1	6	0	307,498
Other Regulated Material, Class D	224	0	0	0	110,865
Organic Peroxide	154	0	0	0	49,622
Flammable Solid	59	0	4	0	365,252
Infectious Substance (Etiologic)	51	0	0	0	2,566
Poisonous Gas	41	3	36	0	267,393
Explosive No Blast Hazard	34	0	1	0	90,000
Dangerous When Wet Material	13	0	0	0	11,450
Radioactive Material	10	0	0	0	28,700
Flammable Solid (Pre 1991)	9	0	0	0	4,550
Explosive Mass Explosion Hazard	8	0	0	0	306,481
Spontaneously Combustible	8	0	0	0	4,400
Very Insensitive Explosive	3	0	0	0	141,481
Explosive Fire Hazard	2	0	0	0	2,900
Explosives, Class C	1	0	0	0	0
Total-2008	**16,767**	**17**	**202**	**9**	**$52,404,428**

Note: Due to multiple Hazard Classes being involved in a single incident, the totals above may not correspond to the totals in the other reports.

Source: Hazardous Materials Information System, U.S. Department of Transportation. Data as of 09/03/2009

Table 1.10
Hazardous Liquid Pipeline Accident Summary by Commodity, 2002

Commodity	Number of Accidents	Percent of Total Accidents	Barrels Lost	Property Damages	Injuries/ Fatalities
Crude Oil	39	27.27	12,010	$10,793,431	0
Gasoline	13	9.09	6,445	$2,548,656	0
Diesel Fuel	8	6.2	3,445	$2,548,656	0
Propane	9	6.29	12,001	$324,388	0
Anhydrous Ammonia	1	0.69	28	$6,839	0
Ethylene	2	1.39	660	$100,296	0
Butane	4	2.79	28	$26,543	1
Fuel Oil	6	4.19	890	$1,066,876	0
Jet Fuel (JP8)	2	1.39	286	$115,900	0
Natural Gasoline	2	1.39	228	$50,000	0
Unleaded Gasoline	8	5.59	4,444	$2,303,643	0
Jet A Turbine Fuel	1	0.69	19	$330,000	0
Carbon Dioxide	2	1.39	3,912	$10,430	0
Gasoline and Diesel	1	0.69	950	$76,000	0
No Data	5	3.49	231	$599,946	0
TOTAL	**100**		**41,712**	**$19,129,883**	**1**

Table 1.11
ATSDR
Most Frequently Released Substances
Hazardous Substances Emergency Events Surveillance, 2005

Substance Name	No. of Releases	% of Releases	Rank
Carbon Monoxide	702	6.10	1
Nitrogen Oxide (nox) (includes Oxides of Nitrogen)	665	5.78	2
Sulfur Dioxide	570	4.95	3
Ammonia	528	4.59	4
Paint not otherwise specified	441	3.83	5
Volatile Organic Compounds not otherwise specified	343	2.98	6
Sulfuric Acid	262	2.28	7
Sodium Hydroxide	254	2.21	8
Chlorine	239	2.08	9
Hydrochloric Acid	238	2.07	10
Benzene	145	1.26	11
Methamphetamine Chemicals not otherwise specified	139	1.21	12
Mercury	130	1.13	13
Ethylene Glycol	128	1.11	14
Hydrogen Sulfide	117	1.02	15
Nitric Oxide	115	1.00	16
Nitrogen Dioxide	92	0.80	17
Acetone	86	0.75	18
Polychlorinated Biphenyls	77	0.67	19
Butadiene	74	0.64	20
Vinyl Chloride	74	0.64	21
Sodium Hypochlorite	65	0.56	22
Acid not otherwise specified	63	0.55	23
Toluene	61	0.53	24
Freon	58	0.50	25

Continued

Table 1.11 (continued)

Substance Name	No. of Releases	% of Releases	Rank
Resin not otherwise specified	57	0.50	26
Solvent not otherwise specified	56	0.49	27
Hydrogen Peroxide	52	0.45	28
Asbestos	52	0.45	29
Iodine	52	0.45	30
Ethylene	46	0.40	31
Hydrochloric (Muriatic) Acid	45	0.39	32
Propylene	45	0.39	33
Methanol not otherwise specified	44	0.38	34
Nitric Acid	44	0.38	35
Antifreeze	43	0.37	36
Pesticide not otherwise specified	42	0.37	37
Propane	42	0.37	38
Pseudoephedrine	42	0.37	39
Diesel Fuel	42	0.37	40
Phosphoric Acid	41	0.36	41
Potassium Hydroxide	40	0.35	42
Xylene	40	0.35	43
Alumina, Activated	40	0.35	44
Alcohol not otherwise specified	39	0.34	45
Chloroform	36	0.31	46
Methamphetamine	36	0.31	47
Ethylene Oxide	35	0.30	48
Ethyl Ether	35	0.30	49
Isopropyl Alcohol	33	0.29	50

Continued

Table 1.11 (concluded)

Substance Name	No. of Releases	% of Releases	Rank
Hydrofluoric Acid	32	0.28	51
Nitrous Oxide	32	0.28	52
Ethyl Alcohol	32	0.28	53
Phosphorus	31	0.27	54
Flammable Liquid not otherwise specified	30	0.26	55

Summary

First responders must understand their roles and limitations at haz mat emergencies. These are, in part, spelled out by government laws such as U.S. OSHA 29 *CFR* 1910.120, *Hazardous Waste Operations and Emergency Response (HAZWOPER)* and consensus standards such as NFPA® 472, *Standard for Professional Competence of Responders to Hazardous Materials/Weapons of Mass Destruction Incidents* (2008), and NFPA® 473, *Standard for Competencies for EMS Personnel Responding to Hazardous Materials/Weapons of Mass Destruction Incidents* (2008). Additionally, numerous government agencies and government regulations play important roles in how hazardous materials are manufactured, used, transported, and disposed of. It is important for first responders to be familiar with this legal framework in order to understand how and why hazardous materials are handled, packaged, used, and transported the way they are.

Most commonly, haz mat incidents involve flammable and combustible liquids, corrosives, anhydrous ammonia, and chlorine. However, first responders must be prepared to deal with emergencies involving any of the thousands of industrial chemical products in use today as well as biological, nuclear, and explosive materials that might be used in a terror attack. Responders must also understand the ways in which these materials can cause harm.

Review Questions

1. How are hazardous materials incidents different from other types of emergencies?

2. What are Awareness-Level personnel's responsibilities at a hazardous materials incident?

3. What are Operations-Level responders' responsibilities at a hazardous materials incident?

4. What are the four types of ionizing radiation? Describe each briefly.

5. The likelihood and severity of an adverse health effect resulting from a chemical exposure are dependent upon which factors?

6. What is the difference between an acid and a base?

7. Describe the different types of etiological hazards.

8. What four hazards can be caused by an explosion?

9. What are the three main routes of entry as defined by the U.S. CDC?

10. What are the main agencies involved in the regulation of hazardous materials in the U.S., Canada, and Mexico?

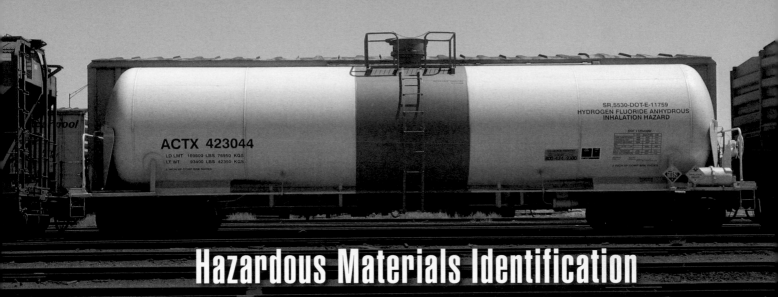

Hazardous Materials Identification

Chapter Contents

Divider page photo courtesy of Rich Mahaney.

Key Terms

Competencies

NFPA® 472:	4.2.1(7)(d)	4.2.1(10)(e)	4.2.1(17)	5.2.1	5.2.1.1.3(1)
4.2.1	4.2.1(7)(e)	4.2.1(10)(f)	4.2.1(18)	5.2.1.1	5.2.1.1.3(2)
4.2.1(2)	4.2.1(7)(f)	4.2.1(10)(g)	4.2.1(19)	5.2.1.1.1(1)	5.2.1.1.3(3)
4.2.1(3)	4.2.1(8)	4.2.1(11)	4.2.1(20)	5.2.1.1.1(2)	5.2.1.1.3(4)
4.2.1(5)	4.2.1(9)	4.2.1(12)	4.2.2(1)	5.2.1.1.1(3)	5.2.1.1.3(5)
4.2.1(6)	4.2.1(10)(a)	4.2.1(13)	4.2.2(2)	5.2.1.1.2(1)	5.2.1.1.3(6)
4.2.1(7)(a)	4.2.1(10)(b)	4.2.1(14)	4.2.2(3)	5.2.1.1.2(2)	5.2.1.1.3(7)
4.2.1(7)(b)	4.2.1(10)(c)	4.2.1(15)	5.1.2.2(1)(a)	5.2.1.1.2(3)(a)	5.2.1.1.4(1)
4.2.1(7)(c)	4.2.1(10)(d)	4.2.1(16)	5.1.2.2(1)(b)	5.2.1.1.2(3)(b)	5.2.1.1.4(2)

Competencies (continued)

5.2.1.1.4(3)	5.2.1.1.6(3)	5.2.1.2.3	5.2.1.3.2(5)	5.2.2(3)(b)	5.2.2(3)(j)
5.2.1.1.5(1)	5.2.1.1.6(4)	5.2.1.3.1(1)	5.2.1.3.2(6)	5.2.2(3)(c)	5.2.2(5)
5.2.1.1.5(2)	5.2.1.1.6(5)	5.2.1.3.1(2)	5.2.1.3.3	5.2.2(3)(d)	5.2.4(3)
5.2.1.1.5(3)	5.2.1.2	5.2.1.3.1(3)	5.2.1.5	5.2.2(3)(e)	5.2.4(4)
5.2.1.1.5(4)	5.2.1.2.1(1)	5.2.1.3.2(1)	5.2.1.6	5.2.2(3)(f)	5.3.1(4)
5.2.1.1.5(5)	5.2.1.2.1(2)	5.2.1.3.2(2)	5.2.2(1)	5.2.2(3)(g)	
5.2.1.1.6(1)	5.2.1.2.1(3)	5.2.1.3.2(3)	5.2.2(2)	5.2.2(3)(h)	
5.2.1.1.6(2)	5.2.1.2.2	5.2.1.3.2(4)	5.2.2(3)(a)	5.2.2(3)(i)	

Learning Objectives

1. Identify the seven clues to the presence of hazardous materials. [NFPA® 472, 4.2.1(7)(a-f), 4.2.2, 4.2.2(1), 5.2.1.1]

2. Discuss the occupancy types, locations, and pre-incident surveys that may indicate hazardous materials. [NFPA® 472, 4.2.1(5), 4.2.1(13)]

3. Describe the container shapes that may contain hazardous materials. [NFPA® 472, 4.2.1(6), 5.2.1.1.1(1-3), 5.2.1.1.2(1-2), 5.2.1.1.2(3)(a-b), 5.2.1.1.3(1-7), 5.2.1.1.4(1-3), 5.2.1.1.5(1-5), 5.2.1.1.6(1-5)]

4. Identify placards, labels, and markings that designate the presence of hazardous materials. [NFPA® 472, 4.2.1(2-3), 4.2.1(7)(a), 4.2.1(9), 4.2.2(2), 5.2.1.2, 5.2.1.2.1(1-3), 5.2.1.3.3, 5.2.2(1)]

5. Describe the other markings and colors that may indicate the presence of hazardous materials. [NFPA® 472, 4.2.1(7)(f), 4.2.1(8), 5.2.1.2, 5.2.1.2.2, 5.2.1.3.1(1-3), 5.2.1.3.2(1-6)]

6. Explain the written resources available to indicate the presence of hazardous materials. [NFPA® 472, 4.2.1(10)(a-g), 4.2.2(3), 5.2.1.5, 5.2.2(2), 5.2.2(3)(a-j), 5.2.2(5)]

7. Discuss the limitations of using the senses to determine the presence or absence of hazardous materials. [NFPA® 472, 4.2.1(11), 4.2.1(12), 5.2.2(3)(d)]

8. Discuss monitoring and detection devices. [NFPA® 472, 5.2.4(3), 5.2.4(4)]

9. Analyze scenarios to detect the presence of hazardous materials. [NFPA® 472, 4.2.1, 5.2.1; Learning Activity 2-1]

10. Interpret representative shipping papers. [NFPA® 472, 4.2.1(10); Learning Activity 2-2]

11. Interpret a safety data sheet (SDS). [472, 4.2.1(10); Learning Activity 2-3]

12. Explain how to identify terrorist attacks and illicit laboratories. [NFPA® 472, 4.2.1(14-20), 5.2.1.6, 5.3.1(4)]

Chapter 2
Hazardous Materials Identification

Case History

On January 19, 2003, a 32-year-old volunteer firefighter in Texas died while fighting a structure fire at a specialized vehicle restoration shop. Soon after beginning interior attack operations, the fire intensified and rolled over the heads of the 4-member crew. Within minutes, the nozzleman had to exit the building due to burning hands and, another firefighter took the nozzle. As he was exiting, an air horn was sounded warning the crew to exit the building. Two of the three remaining crew members made it to safety. Less than a minute after they exited, a nitrous oxide cylinder that was attached to a race car in the building exploded **(Figure 2.1).** A Rapid Intervention Team (RIT) was assembled to rescue the missing firefighter (the victim). The RIT made two attempts to rescue the victim but had to exit because of the intensity of the fire. After approximately 40 minutes of master stream application, three teams entered the structure and found the victim lying near the office door. The alarm for his Personal Alert Safety System (PASS) device was functioning but was not audible due to his prone position.

The preliminary autopsy findings indicated that the victim had received significant blast injuries. Both eardrums were ruptured and there was concussive damage to his lungs. A subsequent NIOSH investigation determined that the nitrous oxide cylinder ruptured and exploded with a force equivalent to as much as 4 pounds (2 kg) of TNT. At a distance of 10 feet (3 m) from the exploding cylinder, the firefighter could have been exposed to a shock wave of up to 30 psi, well above the threshold level for eardrum rupture and internal lung damage. The effects of a blast overpressure shock wave are increased when explosions occur in closed or confined spaces such as inside a building or a vehicle. In addition, blast waves are reflected by solid surfaces. Thus a person standing next to a wall or vehicle may suffer increased primary blast injury.

Source: NIOSH

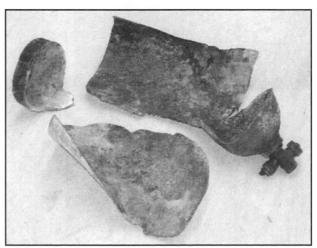

Figure 2.1 Shards of the nitrous oxide cylinder that exploded, killing a firefighter. *Courtesy of NIOSH.*

Awareness-level personnel and first responders must be able to detect and identify the presence of hazardous materials. Historically, the failure of responders to recognize the presence of hazardous materials at accidents, fires, spills, and other emergencies has caused unnecessary casualties. Haz mat incidents can be controlled only when the personnel involved have sufficient information to make informed decisions. Part of this information is learned from size-up of all the materials that may be hazardous. The time and effort devoted to a positive identification of the contents of buildings, vehicles, and containers result in greater safety for first responders and the community.

Once the presence of a hazardous material is detected, first responders can use a number of resources to identify the material and its hazards and specify recommended protective measures. Once first responders have determined the properties of substances, then they can perform tasks confidently and evaluate changing conditions accurately. First responders must be diligent and observant of the hazardous materials present at every emergency.

Some clues to the presence of hazardous materials may be easily identified from a distance while others require responders to be much closer. The closer responders need to be in order to identify the material, the greater their chances of being in an area where they could be exposed to its harmful effects.

The seven clues to the presence of hazardous materials are as follows:

1. Occupancy Types, Locations, and Pre-Incident Surveys

2. Container shapes

3. Transportation placards, labels, and markings

4. Other markings and colors (non-transportation)

5. Written resources

6. Senses

7. Monitoring and detection devices

This chapter will discuss these seven clues in detail. The order of the clues also represents, in general, an increasing level of risk to responders **(Figure 2.2).** It will also describe how to recognize a terrorist attack and identify the weapon of mass destruction (WMD) or device involved in the attack.

Prepare for the Unexpected

While these clues suggest many ways to identify hazardous materials involved in incidents, they do have limitations. First responders may encounter situations where they are unable to clearly see placards, markings, labels, and signs from a safe distance. Identifying markings can be destroyed in the incident. Inventories may change from those identified during pre-incident surveys, or containers may be improperly labeled. Mixed loads in transportation incidents may not be marked at all **(Figure 2.3).** Shipping papers may be inaccessible. First responders must always be prepared for the unexpected, including recognizing terrorist attacks.

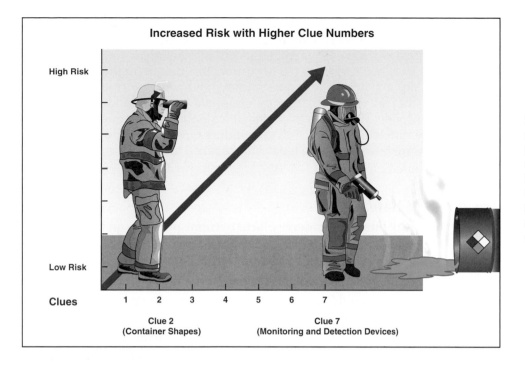

Increased Risk with Higher Clue Numbers

High Risk

Low Risk

Clues   1   2   3   4   5   6   7

Clue 2
(Container Shapes)

Clue 7
(Monitoring and Detection Devices)

Figure 2.2 The risk to responders increases as they move closer to the hazardous material. It is much safer to identify a material from a distance based on a container shape than it is to physically sample the substance with a detection device.

Figure 2.3 Small quantities and mixed loads of hazardous materials may be transported without any placards or other markings on the vehicle or outer container.

Clue 1: Occupancy Types, Locations, and Pre-incident Surveys

Simply stated, hazardous materials may be found anywhere. Not all locations or occupancies are as obvious as a local manufacturing plant, and responders may have little or no warning of materials being transported through their jurisdictions by road, rail, or waterway. However, pre-incident surveys and the occupancy type for a particular structure may provide the first clue to the emergency responder that hazardous materials may be involved in an incident. The location and occupancy of an incident may also be an indicator that terrorism is involved.

Verbal Reports

First evidence that hazardous materials are present at an emergency often comes from a knowledgeable or responsible person at the site (such as the facility manager). This person may have vital information about the events that led to the emergency, materials involved, and humans or property exposed. Whether a telecommunicator/dispatcher questions this person over the telephone or first responders question the person at the scene, emergency personnel must be prepared to make maximum use of this resource.

Pre-incident Surveys

The very nature of haz mat incidents places extreme pressure on first responders to make decisions quickly and accurately. First responders can reduce the number of on-site decisions by conducting **pre-incident surveys** (also called *preplans*) and being familiar with local emergency response plans (see Chapter 3, Predetermined Procedures and Emergency Response Plans). With the groundwork laid, first responders can concentrate on the situation and operate more safely and efficiently. Planning reduces oversights, confusion, and duplication of efforts, and it results in a desirable outcome. Furthermore, pre-incident surveys identify the following items:

Pre-Incident Survey — Survey of a facility or location made before an emergency occurs in order to prepare for an appropriate emergency response. Sometimes called Preplan.

- Exposures (people, property, and environment)
- Types, quantities, and locations of hazardous materials in the area
- Dangers of the hazardous materials
- Building features (location of fixed fire-suppression systems, etc.)
- Site characteristics
- Possible access/egress difficulties
- Inherent limitations of the responding organizations to control certain types of haz mat emergencies
- Twenty-four hour telephone numbers of responsible parties and site experts

Planning is an ongoing process that includes reviewing surveys and updating them regularly. Pre-incident surveys are not always accurate, however, because inventories, businesses, and other factors may change without notice. Compliance with existing reporting rules and regulations cannot be guaranteed. First responders must always expect to find the unexpected.

Occupancies

Occupancy — (1) General fire service term for a building, structure, or residency. (2) Building code classification based on the use to which owners or tenants put buildings or portions of buildings. Regulated by the various building and fire codes. Also called Occupancy Classification.

Certain **occupancies** are always highly probable locations for finding significant quantities of hazardous materials, including the following:

- Fuel storage facilities
- Gas/service stations (and convenience stores) **(Figure 2.4)**
- Paint supply stores
- Plant nurseries, garden centers, and agricultural facilities
- Pest control and lawn care companies
- Medical facilities
- Photo processing laboratories

Figure 2.4 Service stations are an occupancy where hazardous materials are likely to be involved in non-medical emergency incidents.

Figure 2.5 Large quantities of hazardous materials may pass through busy port facilities, making them a common location for haz mat incidents. *Courtesy of U.S. Customs and Border Protection, photo by Charles Csavossy.*

- Dry cleaners
- Plastics and high-technology factories
- Metal-plating businesses
- Mercantile concerns (hardware stores, groceries stores, certain department stores)
- Chemistry (and other) laboratories in educational facilities (including high schools)
- Lumberyards
- Feed/farm stores
- Veterinary clinics
- Print shops
- Warehouses
- Industrial and utility plants
- Port shipping facilities (with changing cargo hazards) **(Figure 2.5)**
- Treatment storage disposal (TSD) facilities
- Abandoned facilities that may have contained or used hazardous materials

Residential occupancies are not exempt from danger because hazardous chemicals in the form of drain cleaners, pesticides, fertilizers, paint products, and flammable liquids (such as gasoline) are common household products. In rural areas, propane tanks often provide fuel for heating, and farms may have large quantities of dangerous products such as pesticides and anhydrous ammonia. Any building with a fume hood exhaust stack (or stacks) on the roof (such as a high school or medical office building) probably has a functioning laboratory inside **(Figure 2.6).**

Figure 2.6 The presence of fume hood exhaust stacks on the roof or exterior of a building is a good indicator that hazardous materials are used inside.

Locations

Local experience with transportation accidents also indicates where to expect haz mat incidents. Ports, docks or piers, railroad sidings, airplane hangars, truck terminals, and other places of material transfer (such as trucking warehouses) are also likely locations for haz mat accidents. Each of the following modes of transportation has particular locations where accidents may occur more frequently:

- *Roadways*
  — Designated truck routes
  — Blind intersections
  — Poorly marked or poorly engineered interchanges
  — Areas frequently congested by traffic
  — Heavily traveled roads
  — Sharp turns
  — Steep grades
  — Highway interchanges and ramps
  — Bridges and tunnels

- *Railways*
  — Depots, terminals, and switch or classification yards
  — Sections of poorly laid or poorly maintained tracks
  — Steep grades and severe curves
  — Shunts and sidings
  — Uncontrolled crossings
  — Loading and unloading facilities
  — Bridges, trestles, and tunnels **(Figure 2.7)**

Figure 2.7 Locations that experience more transportation accidents in general, such as railway bridges and trestles, are also more likely to be involved in haz mat incidents. *Courtesy of Phil Linder.*

- *Waterways*
  — Difficult passages at bends or other threats to navigation
  — Bridges and other crossings

- — Piers and docks
- — Shallow areas
- — Locks
- — Loading/unloading stations
- *Airways*
  - — Fueling ramps
  - — Repair and maintenance hangars
  - — Freight terminals
  - — Crop duster planes and supplies
- *Pipelines*
  - — Exposed crossings over waterways or roads
  - — Pumping stations
  - — Construction and demolition sites
  - — Intermediate or final storage facilities

Other locations may be identified by consulting with local law enforcement officials to determine problem spots based on traffic studies.

First responders should also pay attention to the water level in rivers and tidal areas and be aware of the following facts:

- Many accidents occur because flow volume and tidal conditions were not considered. These flow and tidal variances affect clearance under bridges, many of which also have pipelines, water mains, gas lines, and the like attached to them.

- Occupancies in low-lying areas that may be affected by flood conditions must have a contingency plan to isolate and protect hazardous materials.

- Tidal and flow conditions are constantly changing. Areas that were once considered safe may become compromised by change of tide direction, flow rate, back eddies, etc.

- Once a material reaches an outside water source, it becomes a moving incident and is extremely difficult to contain, confine, and mitigate.

First responders should be familiar with the types of haz mat shipments that come through their jurisdictions. For example, farming communities may be more likely to see tanks of anhydrous ammonia passing through, whereas a port serving an industrial complex with many refineries might see more petroleum products.

Terrorist Targets

Certain occupancies are more likely to be targeted for terrorist attacks than others. First responders must be able to identify those locations where an attack has the potential to do the greatest harm and predict the consequences of such an attack **(Figure 2.8, p. 70).** *Harm* should be defined in terms of the following concerns (and others that are similar):

- Killing or injuring persons
- Causing panic and/or disruption

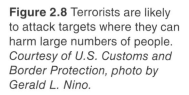

Figure 2.8 Terrorists are likely to attack targets where they can harm large numbers of people. *Courtesy of U.S. Customs and Border Protection, photo by Gerald L. Nino.*

- Damaging the economy
- Destroying property
- Demoralizing the community

When the goal is to kill as many people as possible, any location or occupancy that has large public gatherings (such as football stadiums, sports arenas, theaters, and shopping malls) might become a potential target. Terrorists might also target places with historical, economic, or symbolic significance such as local monuments, high-profile buildings, or high-traffic bridges **(Figure 2.9).** The following are examples of potential terrorist targets:

Figure 2.9 Terrorists may target local monuments, high-profile buildings, and high-traffic bridges. *Courtesy of U.S. Customs and Border Protection, photo by Charles Csavossy.*

- *Mass Transportation* — Airports, ferry terminals and buildings, maritime port facilities, planes, subways, buses, commuter trains, mass transit stations
- *Critical Infrastructure* — Dams, water treatment facilities, power plants, electrical substations, nuclear power plants, trans-oceanic cable landings, telecommunication switch centers (telecom hotels), financial institutions, rail and road bridges, tunnels, levees, liquefied natural gas (LNG) terminals, natural gas (NG) compressor stations, petroleum pumping stations, petroleum storage tank farms

- *Areas of Public Assembly and Recreation* — Convention centers, hotels, casinos, shopping malls, stadiums, theme parks
- *High Profile Buildings and Locations* — Monuments, buildings/structures of historic or national significance, high-rise buildings
- *Industrial Sites* — Chemical manufacturing facilities, shipping facilities, warehouses
- *Educational Sites* — Colleges, universities, community colleges, vocational/training facilities, primary and secondary schools
- *Medical and Science Facilities* — Hospitals, clinics, nuclear research labs, other research facilities, non-power nuclear reactors, national health stockpile sites

Incidents reported at these occupancies should be scrutinized closely for potential terrorist involvement. If **terrorism** is suspected, notify law enforcement authorities immediately.

Terrorism — Unlawful use of force or violence against persons or property for the purpose of intimidating or coercing a government, the civilian population or any segment thereof, in furtherance of political or social objectives; defined by the U.S. Federal Bureau of Investigation (FBI).

Clue 2: Container Shapes

While first responders may recognize the location of an incident or type of occupancy as one that handles hazardous materials, the presence of certain storage vessels (as in pressure storage vessels or hollow utensils used for holding liquids), tanks, containers, packages, or vehicles can alert responders to their presence with certainty. These containers can provide useful information about the materials inside, so it is important for first responders to recognize the shapes of the different types of **packaging** and containers in which hazardous materials are stored and transported **(Figure 2.10).**

Packaging — Term used by the U.S. Department of Transportation to describe shipping containers and their markings, labels, and/or placards.

Figure 2.10 Containers' shapes can tell first responders a great deal about the hazardous materials that might be inside. *Courtesy of Rich Mahaney.*

Types of containers can be categorized in the following ways:

- *Bulk and nonbulk* — refers to capacity as defined by the U.S. Department of Transportation (DOT) and Transport Canada (TC)
- *Pressure and nonpressure* — refers to the design of the container based on the internal pressure
- *Bulk-capacity fixed-facility containment systems and transportation packaging* — refers to the facility or mode

What This Means To You

As an Awareness-Level responder, you need to be able to identify the typical container shapes (fixed-facility tanks, transportation tanks, and bulk and nonbulk containers) that may contain hazardous materials. Here is a hint: If it looks like it might contain *something* — it probably does. The next step is to determine if that *something* is likely to be hazardous and take appropriate actions from there. For example, if you witness a highway accident involving a cargo tank truck that is now spilling a liquid, you should recognize the liquid as a potentially hazardous material and not go forward to assist the driver in the cab until it is determined it is safe to do so.

As an Operations-Level responder, you need to be able to identify not only each of the specific types of containers discussed in this section but also what materials they are likely to contain. If you are the first to arrive on the scene of a highway accident involving a cargo tank truck, you should be able to recognize a corrosive liquid tank and immediately have a better understanding of the potential hazards associated with the liquid and how to act accordingly. This information also helps you make an exact identification of the material through other methods.

Bulk Packaging — Packaging, other than a vessel or barge, including transport vehicle or freight container, in which hazardous materials are loaded with no intermediate form of containment and which has (a) a maximum capacity greater than 119 gallons (450 L) as a receptacle for a liquid, (b) maximum net mass greater than 882 pounds (400 kg) and a maximum capacity greater than 119 gallons (450 L) as a receptacle for a solid, or (c) water capacity greater than 1,000 pounds (454 kg) as a receptacle for a gas.

The sections that follow discuss the following types of containers in both pressure and nonpressure varieties:

- Bulk-capacity containment systems at fixed facilities
- Bulk transportation packaging
- Nonbulk containers in general

What Is Bulk/Nonbulk Packaging?

Bulk packaging refers to a packaging, other than that on a vessel (ship) or barge, in which materials are loaded with no intermediate form of containment. This packaging type includes a transport vehicle or freight container such as a cargo tank, railcar, or portable tank. Intermediate bulk containers (IBCs) and intermodal (IM) containers are also examples. To be considered bulk packaging, one of the following criteria must be met:

- Maximum capacity is greater than 119 gallons (450 L) as a receptacle for a liquid
- Maximum net mass is greater than 882 pounds (400 kg) or maximum capacity is greater than 119 gallons (450 L) as a receptacle for a solid
- Water capacity is 1,001 pounds (454 kg) or greater as a receptacle for a gas

Nonbulk packaging is packaging that is smaller than the minimum criteria established for bulk packaging. Drums, boxes, carboys, and bags are examples. Composite packages (packages with an outer packaging and an inner receptacle) and combination packages (multiple packages grouped together in a single outer container such as bottles of acid packed inside a cardboard box) may also be classified as nonbulk packaging.

Nonbulk Packaging — Package that has the following characteristics: (a) maximum capacity of 119 gallons (450 L) or less as a receptacle for a liquid, (b) maximum net mass of 882 pounds (400 kg) or less and a maximum capacity of 119 gallons (450 L) or less as a receptacle for a solid, and (c) water capacity of 1,000 pounds (454 kg) or less as a receptacle for a gas.

Bulk-Capacity Fixed-Facility Containers

Containers at fixed facilities include the following:

- Buildings
- Machinery
- Aboveground storage tanks
- Underground storage tanks

- Pipelines
- Open piles or bins
- Storage cabinets
- Reactors
- Vats
- Other fixed, on-site containers

This section focuses on storage tanks holding bulk quantities of hazardous materials. Nonbulk packages that may be found at fixed facilities are discussed in the Nonbulk Packaging section. Pipeline identification and other labeling and marking systems used to identify the materials in storage cabinets, bins, vats, and the like are discussed in the section, Other Markings and Colors.

In general, aboveground storage tanks are divided into two major categories:

1. *Nonpressure tanks (also called atmospheric tanks)* — If these tanks are storing any product, they will normally have a small amount of pressure (up to 0.5 psi [3.45 kPa] {0.03 bar}) inside.

2. *Pressure tanks* — These tanks are divided into the following two categories:

   — Low-pressure storage tanks that have pressures between 0.5 psi to 15 psi (3.45 kPa to 103 kPa) {0.03 bar to 1.03 bar}

   — Pressure vessels that have pressures above 15 psi (103 kPa) {1.03 bar}

Underground storage tanks may be atmospheric or pressurized. **Cryogenic liquid storage tanks** have varying pressures, but some can be very high. They are usually heavily insulated with a vacuum in the space between the outer and inner shells.

Cryogenic Liquid Storage Tank — Heavily insulated, vacuum-jacketed tanks used to store cryogenic liquids, equipped with safety-relief valves and rupture disks.

Pressure Measurements and Terms

The following are terms with which Operations-Level first responders should be familiar:

- *Pounds per square inch (psi)* — Unit for measuring pressure in the English or Customary System. The International System of Units (SI) equivalent is kilopascal (kPa). Another equivalent unit of pressure is bar *(1 bar = 14.5038 psi) {1 bar = 100 kPa}*. This book uses the units of *psi* as the primary measurement of pressure. Metric equivalent pressures expressed in *bar* and *kPa* are also given.

- *Atmospheric pressure* — Force exerted by the weight of the atmosphere at the surface of the earth. Atmospheric pressure is greatest at low altitudes; consequently, its pressure at sea level is used as a standard. At sea level, the atmosphere exerts a pressure of 14.7 psi (101 kPa) {1.01 bar}. A common method of measuring atmospheric pressure is to compare the weight of the atmosphere with the weight of a column of mercury: the greater the atmospheric pressure, the taller the column of mercury. A pressure of 1 psi (7 kPa) {0.069 bar} makes the column of mercury about 2.04 inches (52 mm) tall. At sea level, then, the column of mercury is 2.04 × 14.7, or 29.9 inches tall (760 mm). See the diagram in **Figure 2.11.**

- *Pound-force per square inch gauge (psig)* — Unit of pressure relative to the surrounding atmosphere. For example, at sea level, a reading of 30 psig on a tire gauge represents an absolute pressure of 44.7 psi because the gauge was calibrated to zero in atmospheric pressure of approximately 14.7.

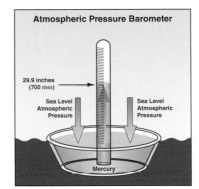

Figure 2.11 Barometers measure atmospheric pressure. Because pressure varies depending on altitude, the pressure at sea level is used as a standard.

Nonpressure/Atmospheric Storage Tanks

Nonpressure/atmospheric storage tanks are designed to hold contents under little pressure. The maximum pressure under which an atmospheric tank is capable of holding its contents is 0.5 psi (3.45 kPa) {1.03 bar}. Common types of atmospheric tanks are horizontal tanks, ordinary cone roof tanks, floating roof tanks, lifter roof tanks, and vapordome roof tanks **(Figure 2.12)**. **Table 2.1** provides pictures and examples of various atmospheric storage tanks.

Figure 2.12 The vents around the rim of this tank identify it as a covered top floating roof tank, a type of atmospheric storage tank. *Courtesy of Rich Mahaney.*

Atmospheric Storage Tank Failures

Catastrophic failures of aboveground atmospheric storage tanks can occur when flammable vapors in a tank explode and break either the shell-to-bottom or side seam. These failures have caused tanks to rip open and (in rare cases) hurtle through the air. Shell-to-bottom seam failures are more common among old storage tanks. Steel storage tanks built before 1950 generally do not conform to current industry standards for explosion and fire venting situations. A properly designed and maintained storage tank will break along the shell-to-top seam, which is more likely to limit the fire to the damaged tank and prevent the contents from spilling.

Atmospheric tanks used for storage of flammable and combustible liquids should be designed to fail along the shell-to-roof seam when an explosion occurs in the tank. This feature prevents the tank from propelling upward or splitting along the side.

Many safety issues arise once atmospheric tanks become involved in or are exposed to fire. Emergency response planning is essential to prevent injuries or deaths caused by the special problems presented by tank fires and emergencies.

Pressure Storage Tanks

Pressure tanks are designed to hold contents under pressure. The NFPA® uses the term *pressure tank* to cover both low-pressure storage tanks and pressure vessels (with higher pressures) **(Figure 2.13, p. 77)**. Low-pressure storage tanks have operating pressures from 0.5 to 15 psi (3.45 kPa to 103 kPa) {0.03 bar to 1.03 bar}. Pressure vessels (including many large cryogenic liquid storage tanks) have pressures of 15+ psi (103 kPa) {1.03 bar} or greater. **Table 2.2, p. 78** provides pictures and examples of various pressure tanks.

Table 2.1
Atmospheric/Nonpressure Storage Tanks

Tank Type	Descriptions
	**Horizontal Tank** Cylindrical tanks sitting on legs, blocks, cement pads, or something similar; typically constructed of steel with flat ends. Horizontal tanks are commonly used for bulk storage in conjunction with fuel-dispensing operations. Old tanks (pre-1950s) have bolted seams, whereas new tanks are generally welded. A horizontal tank supported by unprotected steel supports or stilts (prohibited by most current fire codes) may fail quickly during fire conditions. **Contents:** Flammable and combustible liquids, corrosives, poisons, etc.
	**Cone Roof Tank** Have cone-shaped, pointed roofs with weak roof-to-shell seams that break when or if the container becomes overpressurized. When it is partially full, the remaining portion of the tank contains a potentially dangerous vapor space. **Contents:** Flammable, combustible, and corrosive liquids
	**Open Top Floating Roof Tank** Large-capacity, aboveground holding tanks. They are usually much wider than they are tall. As with all floating roof tanks, the roof actually floats on the surface of the liquid and moves up and down depending on the liquid's level. This roof eliminates the potentially dangerous vapor space found in cone roof tanks. A fabric or rubber seal around the circumference of the roof provides a weather-tight seal. **Contents:** Flammable and combustible liquids
\n\nVents around rim provide differentiation from Cone Roof Tanks	**Covered Top Floating Roof Tank** Have fixed cone roofs with either a pan or deck-type float inside that rides directly on the product surface. This tank is a combination of the open top floating roof tank and the ordinary cone roof tank. **Contents:** Flammable and combustible liquids

Continued

Table 2.1 (concluded)

Tank Type	Descriptions

Covered Top Floating Roof Tank with Geodesic Dome

Floating roof tanks covered by geodesic domes are used to store flammable liquids.

Lifter Roof Tank

Have roofs that float within a series of vertical guides that allow only a few feet (meters) of travel. The roof is designed so that when the vapor pressure exceeds a designated limit, the roof lifts slightly and relieves the excess pressure.

Contents: Flammable and combustible liquids

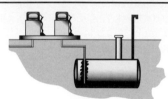

Vapordome Roof Tank

Vertical storage tanks that have lightweight aluminum geodesic domes on their tops. Attached to the underside of the dome is a flexible diaphragm that moves in conjunction with changes in vapor pressure.

Contents: Combustible liquids of medium volatility and other nonhazardous materials

Fill Connections Cover

Atmospheric Underground Storage Tank

Constructed of steel, fiberglass, or steel with a fiberglass coating. Underground tanks will have more than 10 percent of their surface areas underground. They can be buried under a building or driveway or adjacent to the occupancy.

This tank has fill and vent connections located near the tank. Vents, fill points, and occupancy type (gas/service stations, private garages, and fleet maintenance stations) provide visual clues.

Many commercial and private tanks have been abandoned, some with product still in them. These tanks are presenting major problems to many communities.

Contents: Petroleum products

Rare and technically are not "tanks." First responders should be aware that some natural and manmade caverns are used to store natural gas. The locations of such caverns should be noted in local emergency response plans.

Figure 2.13 This liquid oxygen tank is both a cryogenic storage tank (product below - 130 degrees) and a pressure tank because the storage tank pressure is over 15psi. *Courtesy of Rich Mahaney.*

Bulk Transportation Containers

It is important for first responders to recognize the most common types of bulk transportation containers. These containers can be divided into three main categories determined by the mode of transportation as follows:

- Tank cars (railroad)

- Cargo tank trucks (highway)

- **Intermodal containers** (highway, railroad, or marine vessel)

This section also discusses *cargo vessels* (ships) that transport hazardous materials via waterways and *unit loading devices* used in air transportation, which technically are not considered bulk packaging by the U.S. DOT's definition. Intermediate bulk containers, ton containers/cylinders, and storage bladders are also discussed.

NOTE: The standards discussed in the following sections are minimum requirements. These may be modified or exceeded. For example, the vapor recovery system of the MC/306 and 406 cargo tanks are typically located on the right side of the tank, however, there may be exceptions to this rule.

Railroad Cars

Railroads are used to transport a variety of hazardous materials in various types of cars. Tank cars carry the bulk of the hazardous materials transported by rail. Some railroad tank cars have capacities in excess of 30,000 gallons (113 562 L) **(Figure 2.14).** Because of the large quantities these cars hold, a sudden, pressurized/liquefied material release could overwhelm the capabilities of most responding organizations. By recognizing distinctive railroad cars, first responders can begin the identification process from the greatest possible distance. The type of car gives clues as to what material may be within as well as the material's weight and volume.

Intermodal Container — Freight container designed and constructed to be used interchangeably in two or more modes of transport.

Figure 2.14 Railroad tank cars have extremely large capacities (some in excess of 30,000 gallons [113 562 L]). Haz mat releases from tank cars often require assistance from experts and resources outside of local response agencies. *Courtesy of Rich Mahaney.*

Table 2.2
Low-Pressure Storage Tanks and Pressure Vessels

Tank/Vessel Type	Descriptions
	**Dome Roof Tank** Generally classified as low-pressure tanks with operating pressures as high as 15 psi (103 kPa). They have domes on their tops. **Contents:** Flammable liquids, combustible liquids, fertilizers, solvents, etc.
	Spheroid Tank Low-pressure storage tanks. They can store 3,000,000 gallons (11 356 200 L) or more of liquid. **Contents:** Liquefied petroleum gas (LPG), methane, and some flammable liquids such as gasoline and crude oil
	Noded Spheroid Tank Low-pressure storage tanks. They are similar in use to spheroid tanks, but they can be substantially larger and flatter in shape. These tanks are held together by a series of internal ties and supports that reduce stresses on the external shells. **Contents:** LPG, methane, and some flammable liquids such as gasoline and crude oil
	Horizontal Pressure Vessel* Have high pressures and capacities from 500 to over 40,000 gallons (1 893 L to over 151 416 L). They have rounded ends and are not usually insulated. They usually are painted white or some other highly reflective color. **Contents:** LPG, anhydrous ammonia, vinyl chloride, butane, ethane, compressed natural gas (CNG), chlorine, hydrogen chloride, and other similar products
	Spherical Pressure Vessel Have high pressures and capacities up to 600,000 gallons (2 271 240 L). They are often supported off the ground by a series of concrete or steel legs. They usually are painted white or some other highly reflective color. **Contents:** Liquefied petroleum gases and vinyl chloride

Continued

Table 2.2 (concluded)

Tank/Vessel Type	Descriptions
	**Cryogenic-Liquid Storage Tank** Insulated, vacuum-jacketed tanks with safety-relief valves and rupture disks. Capacities can range from 300 to 400,000 gallons (1 136 L to 1 514 160 L). Pressures vary according to the materials stored and their uses. **Contents:** Cryogenic carbon dioxide, liquid oxygen, liquid nitrogen, etc.

* It is becoming more common for horizontal propane tanks to be buried underground. Underground residential tanks usually have capacities of 500 or 1,000 gallons (1 893 L or 3 785 L). Once buried, the tank may be noticeable only because of a small access dome protruding a few inches (millimeters) above the ground.

Tank cars are divided into the following three main categories:

- Low-pressure or general service tank cars (sometimes called nonpressure tank cars)
- Pressure tank cars
- Cryogenic liquid tank cars

Several other types of railroad cars may also carry hazardous materials:

- Hopper cars (including pneumatically unloaded hopper cars)
- Boxcars
- Flat cars transporting other containers of hazardous materials (including intermodal containers, see Intermodal Container section)
- Special service (or specialized) cars

NOTE: The source for most of the following information on railroad tank cars is courtesy of *A General Guide to Tank Cars*, prepared by the Union Pacific Railroad, April 2003.

Low-pressure tank car. Low-pressure tank cars (known as *general service* or *nonpressure* tank cars) transport hazardous and nonhazardous materials with vapor pressures below 25 psi (172 kPa) {1.7 bar} at 105 to 115°F (41°C to 46°C) **(Figure 2.15, p. 80).** Tank test pressures for low-pressure tank cars are 60 and 100 psi (414 kPa and 689 kPa) {4.1 bar and 6.9 bar}. Capacities range from 4,000 to 45,000 gallons (15 142 L to 170 343 L).

Low-pressure tank cars transport a variety of hazardous materials such as flammable liquids, flammable solids, reactive liquids, reactive solids, oxidizers, organic peroxides, poisons, irritants, and corrosive materials. They also transport nonhazardous materials such as fruit and vegetable juices, wine and other alcoholic beverages, tomato paste, and other agricultural products.

Figure 2.15 Low-pressure tank cars have vapor pressures below 25 psi (172 kPa) {1.7 bar} at 105 to 115°F (41°C to 46°C), and their capacities range from 4,000 to 45,000 gallons (15 142 L to 170 343 L). *Courtesy of Rich Mahaney.*

Manway — (1) Hole through which a person may go to gain access to an underground or enclosed structure. (2) Openings usually equipped with removable, lockable covers large enough to admit a person into a tank trailer, tank car, or dry bulk trailer. Also called Manhole.

Figure 2.16 Tanks cars with multiple fittings visible at the top and/or bottom of the car are low-pressure tank cars.

Low-pressure tank cars are cylindrical with rounded ends (heads). They have at least one **manway** for access to the tank's interior. Fittings for loading/unloading, pressure and/or vacuum relief, gauging, and other purposes are visible at the top and/or bottom of the car **(Figure 2.16).**

For many years, one method for identifying low-pressure tank cars was to look for multiple fittings and equipment on top of the tank car. However, some new DOT 111 tank cars now enclose some or all of those fittings inside a protective housing similar to a pressure car (see next section). First responders must now look at the top of the car, and if a single protective housing is present, they must verify if it is a high-pressure tank car or a DOT 111 tank car by identifying the DOT specifications stenciled on the right-hand side of the car **(Figures 2.17 a and b).**

Low-pressure tank cars may be compartmentalized with up to six compartments **(Figure 2.18).** Each compartment is constructed as a separate and distinct tank with its own set of fittings. Each compartment may have a different capacity and transport a different commodity.

Pressure tank car. Pressure tank cars typically transport flammable, non-flammable, and poisonous gases at pressures greater than 25 psi (172 kPa) {1.7 bar} at 68°F (20°C) **(Figure 2.19).** They also transport flammable liquids and liquified compressed gases. Tank test pressures from these tank cars range from 100 to 600 psi (689 kPa, to 137 kPa) {6.9 bar to 41.2 bar}. Pressure tank car capacities range from 4,000 to 45,000 gallons (15 142 L to 170 343 L).

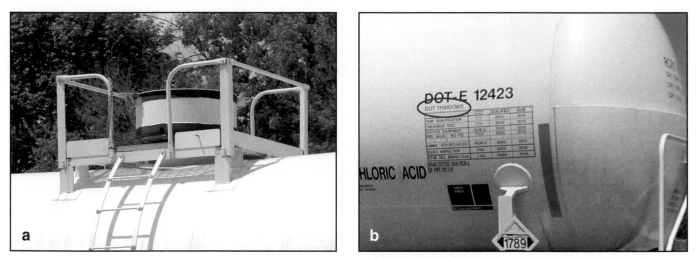

Figure 2.17a and b (a) Some low-pressure tank cars have fittings enclosed in a protective housing. (b) To differentiate these cars from pressure tank cars, responders must check for the DOT specifications stenciled on the right-hand side of the car. *Courtesy of Rich Mahaney.*

Figure 2.18 Some tank cars have multiple compartments. Each may contain a separate product. *Courtesy of Rich Mahaney.*

Figure 2.19 Pressure tank cars typically transport hazardous materials at pressures greater than 25 psi (172 kPa) {1.7 bar} at 68F (20°C). Contents are usually gases or flammable liquids. *Courtesy of Rich Mahaney.*

Pressure tank cars are cylindrical, noncompartmentalized metal (steel or aluminum) tanks with rounded ends (heads). They typically are top-loading cars with their fittings (loading/unloading, pressure-relief, and gauging) located inside the protective housings mounted on the manway cover plates in the top center of the tanks. Pressure tank cars typically have all fittings out of sight under the single protective housings on top of the tanks. New

Figure 2.20 Because pressure tank cars often carry extremely hazardous materials, newer cars are often equipped with GPS tracking devices and anti-tampering mechanisms. *Courtesy of Rich Mahaney.*

pressure tank cars are designed with greater accident protection features and will withstand greater damage without leaking. They feature thicker walls, a lower profile of protective housing, and higher tank test pressures. New pressure tank cars are significantly heavier than old cars, and they are also equipped with GPS tracking devices and anti-tampering mechanisms **(Figure 2.20)**.

Pressure tank cars may be insulated and/or thermally protected. Those without insulation and without jacketed thermal protection may have at least the top two-thirds of the tanks painted white.

Cryogenic liquid tank car. Cryogenic liquid tank cars carry low-pressure (usually below 25 psi [172 kPa] {1.7 bar}) refrigerated liquids (-130°F and below [-90°C and below]). Materials found in these tanks include argon, hydrogen, nitrogen, and oxygen. Liquefied natural gas (LNG) and ethylene may be found at somewhat higher pressures. Fittings for loading/unloading, pressure relief, and venting are in ground-level cabinets on the sides of the car or the end of the car **(Figure 2.21)**.

Figure 2.21 Cryogenic liquid tank cars can be identified by ground-level cabinets on the sides or end of the car. *Courtesy of Rich Mahaney.*

A cryogenic liquid tank car is in the *tank-within-a-tank* category with a stainless steel inner tank supported within a strong outer tank. The space between the inner tank and outer tank is filled with insulation. This space is also kept under a vacuum. The combination of insulation and vacuum protects the contents from ambient temperatures for only 30 days. The shipper tracks these time-sensitive shipments.

Other railroad cars. Other railroad cars include hopper cars and miscellaneous cars such as boxcars and gondolas (see **Table 2.3** for examples). Descriptions of these cars are as follows:

- ***Covered hopper cars*** — often used to transport dry bulk materials such as grain, calcium carbide, ammonium nitrate, and cement

- ***Uncovered (or open top) hopper cars*** — may carry coal, sand, gravel, or rocks

- ***Pneumatically unloaded hopper cars*** — unloaded by air pressure and used to transport dry bulk loads such as ammonium nitrate fertilizer, dry caustic soda, plastic pellets, and cement. Pressure ratings during unloading range from 20 to 80 psi (138 kPa to 552 kPa) {1.4 bar to 5.5 bar}.

Table 2.3
Other Railroad Cars

Covered Hopper Car	Open Top Hopper

Carries: Calcium carbide, cement, and grain	**Carries:** Coal, rock, gravel, and sand
Miscellaneous: Boxcar	Miscellaneous: Gondola

Carries: All types of materials and finished goods	**Carries:** Sand, rolled steel, and other materials that do not require protection from the weather
Miscellaneous: Flat Bed Car with Intermodal Containers	Pneumatically Unloaded Hopper Car

Carries: 1 ton containers, intermodal containers (shown), large vehicles, and other commodities that do not require protection from the weather	**Carries:** Dry caustic soda, ammonium nitrate fertilizer, other fine-powdered materials, plastic pellets, and flour

- ● *Miscellaneous cars* — Boxcars and gondolas are often used to carry other containers of hazardous materials. These cars can include mixed cargos of a variety of products in different types of packaging.

NOTE: Cars may be fumigated, presenting additional hazards.

Cargo Tanks

Highway vehicles that transport hazardous materials include cargo tanks (also called *tank motor vehicles,* and *tank trucks*), dry bulk containers, compressed gas tube trailers, and mixed load containers (also called *box trucks* or *dry van trucks*) **(Figure 2.22).** These vehicles transport all types of hazardous materials in a wide range of quantities.

Figure 2.22 Cargo tanks carry a wide range of hazardous materials. Different construction features enable first responders to identify the different types of trucks and types of hazardous materials being transported. *Courtesy of Rich Mahaney.*

Cargo tank trucks are recognizable because they have construction features, fittings, attachments, or shapes characteristic of their uses. Even if first responders recognize one of the cargo tank trucks described in this section, the process of positive identification must proceed from placards to shipping papers or other formal sources of information. **Table 2.4** provides brief descriptions and illustrations of various cargo tank trucks and trailers.

Cargo tank trucks are commonly used to transport bulk amounts of hazardous materials via roadway. Most cargo tank trucks that haul hazardous materials are designed to meet government tank-safety specifications. These specifications set minimum tank construction material thicknesses, required safety features, and maximum allowable working pressures. The two specifications currently in use are the motor carrier (MC) standards and DOT standards. Cargo tank trucks built to a given specification are designated using the *MC* or *DOT/TC* initials followed by a three-digit number identifying the specification (such as *MC 306* and *DOT/TC 406*). Emergency responders can recognize these cargo tank trucks by their required construction features, fittings, attachments, and shapes, such as **ring stiffeners** on corrosive liquid tanks and bolted manways on high pressure tanks.

Tanks not constructed to meet one of the common MC or DOT/TC specifications are commonly referred to as *nonspec* tanks. Nonspec tanks may haul hazardous materials if the tank was designed for a specific purpose and exempted from the DOT/TC requirements. Nonhazardous materials may be hauled in either nonspec cargo tank trucks or cargo tank trucks that meet a designated specification **(Figure 2.23).**

Ring Stiffener — Circumferential tank shell stiffener that helps to maintain the tank's cross section.

Figure 2.23 Not all cargo tanks carry hazardous materials, so it is important to look for other clues such as placards and markings.

Table 2.4
Cargo Tank Trucks

Nonpressure Liquid Tank	Descriptions
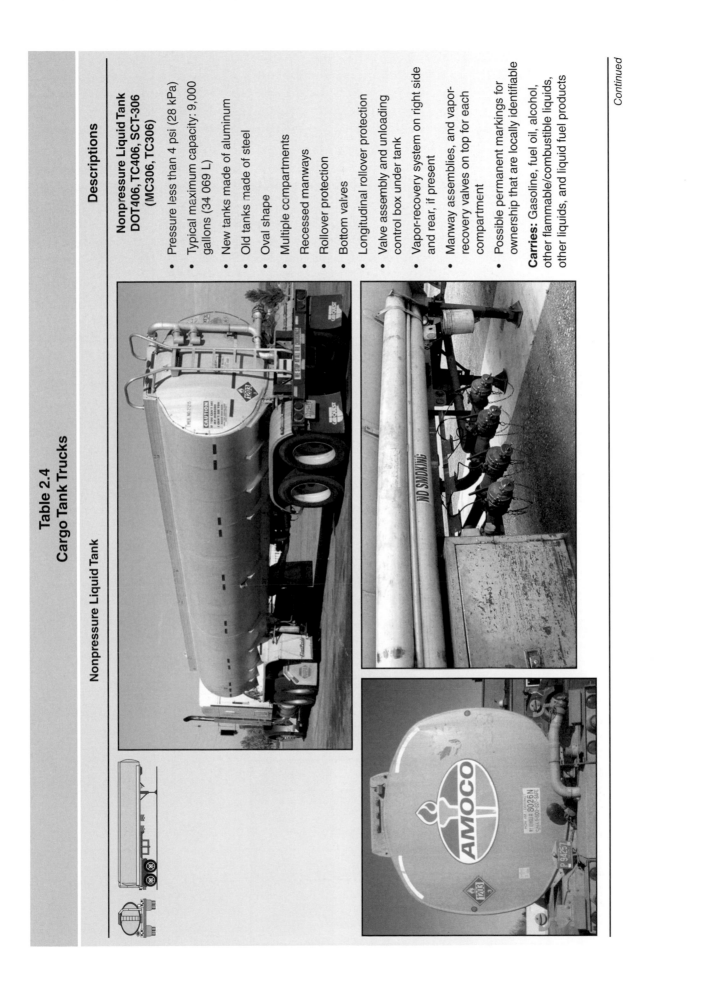	**Nonpressure Liquid Tank DOT406, TC406, SCT-306 (MC306, TC306)** • Pressure less than 4 psi (28 kPa) • Typical maximum capacity: 9,000 gallons (34 069 L) • New tanks made of aluminum • Old tanks made of steel • Oval shape • Multiple compartments • Recessed manways • Rollover protection • Bottom valves • Longitudinal rollover protection • Valve assembly and unloading control box under tank • Vapor-recovery system on right side and rear, if present • Manway assemblies, and vapor-recovery valves on top for each compartment • Possible permanent markings for ownership that are locally identifiable **Carries:** Gasoline, fuel oil, alcohol, other flammable/combustible liquids, other liquids, and liquid fuel products

Continued

Table 2.4 (continued)

Low-Pressure Chemical Tank

Descriptions

Low-Pressure Chemical Tank
DOT407, TC407, SCT-307
(MC307, TC307)

- Pressure under 40 psi (172 kPa to 276 kPa)
- Typical maximum capacity: 7,000 gallons (26 498 L) [per NFPA]
- Rubber lined or steel
- Typically double shell
- Stiffening rings may be visible or covered
- Circumferential rollover protection at each end
- Single or multiple compartments
- Single- or double-top manway assembly protected by a flash box that also provides rollover protection
- Single-outlet discharge piping at midship or rear
- Fusible plugs, frangible disks, or vents outside the flash box on top of the tank
- Drain hose from the flash box down the side of the tank
- Rounded or horse-shoe shaped ends

Carries: Flammable liquids, combustible liquids, acids, caustics, and poisons

Continued

Table 2.4 (continued)

Corrosive Liquid Tank

Descriptions

Corrosive Liquid Tank
DOT412, TC412, SCT-412
(MC312, TC312)

- Pressure less than 75 psi (517 kPa)
- Typical maximum capacity: 7,000 gallons (26 498 L) [per NFPA]
- Rubber lined or steel
- Typically single compartment
- Small-diameter round shape
- Exterior stiffening rings may be visible on uninsulated tanks
- Rear or middle top-loading/unloading station with exterior piping extending to the bottom of the tank
- Splashguard serving as rollover protection around valve assembly
- Additional circumferential rollover protection at front of tank
- Flange-type rupture disk vent either inside or outside the splashguard
- May have discoloration around loading/unloading area or area painted or coated with corrosive-resistant material
- Permanent ownership markings that are locally identifiable

Carries: Corrosive liquids (usually acids)

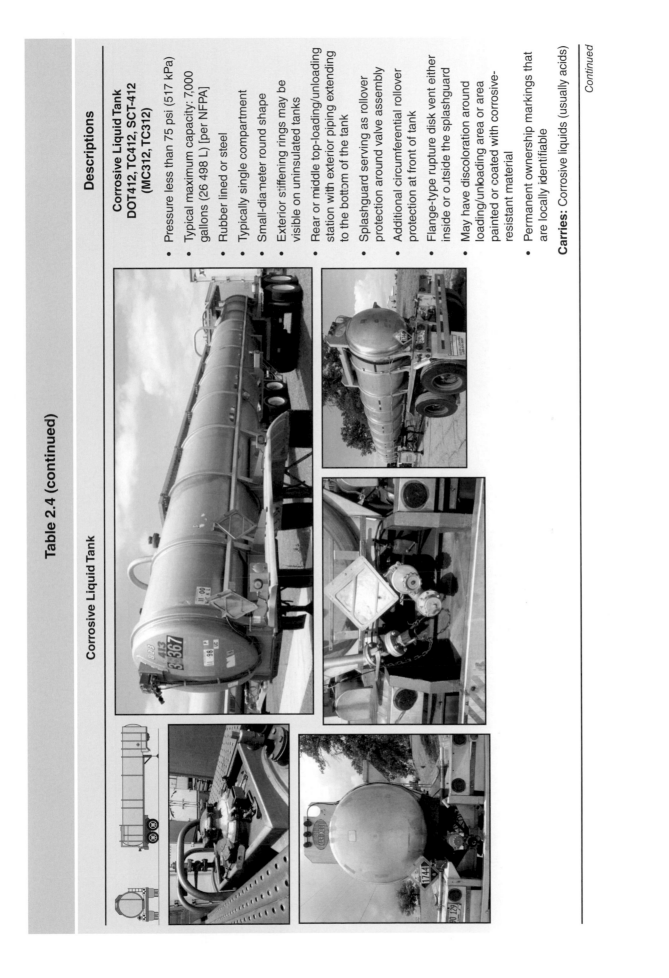

Continued

Table 2.4 (continued)

High-Pressure Tank

Descriptions

High-Pressure Tank
MC-331, TC331, SCT-331

- Pressure above 100 psi (689 kPa)
- Typical maximum capacity: 11,500 gallons (43 532 L)
- Single steel compartment
- Noninsulated
- Bolted manway at front or rear
- Internal and rear outlet valves
- Typically painted white or other reflective color
- Large hemispherical heads on both ends
- Guard cage around the bottom loading/unloading piping
- Uninsulated tanks, single-shell vessels
- Permanent markings such as *FLAMMABLE GAS, COMPRESSED GAS,* or identifiable manufacturer or distributor names

Carries: Pressurized gases and liquids, anhydrous ammonia, propane, butane, and other gases that have been liquefied under pressure

High-Pressure Bobtail Tank

Used for local delivery of liquefied petroleum gas and anhydrous ammonia

Continued

Table 2.4 (continued)

Cryogenic Liquid Tank

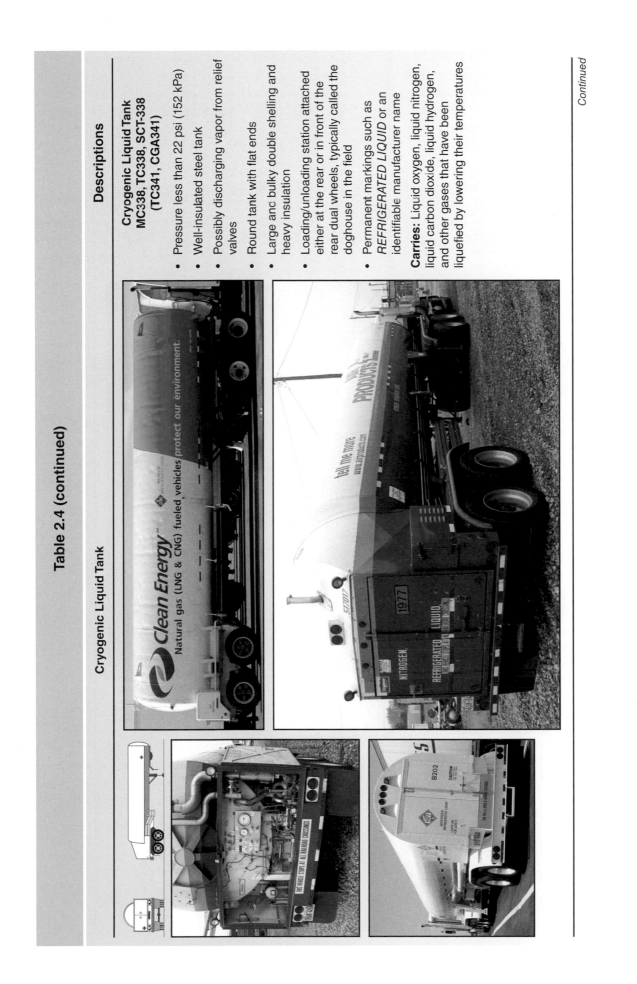

Descriptions
Cryogenic Liquid Tank MC338, TC338, SCT-338 (TC341, CGA341)

- Pressure less than 22 psi (152 kPa)
- Well-insulated steel tank
- Possibly discharging vapor from relief valves
- Round tank with flat ends
- Large and bulky double shelling and heavy insulation
- Loading/unloading station attached either at the rear or in front of the rear dual wheels, typically called the doghouse in the field
- Permanent markings such as *REFRIGERATED LIQUID* or an identifiable manufacturer name

Carries: Liquid oxygen, liquid nitrogen, liquid carbon dioxide, liquid hydrogen, and other gases that have been liquefied by lowering their temperatures

Continued

Table 2.4 (continued)

Compressed-Gas/Tube Trailer

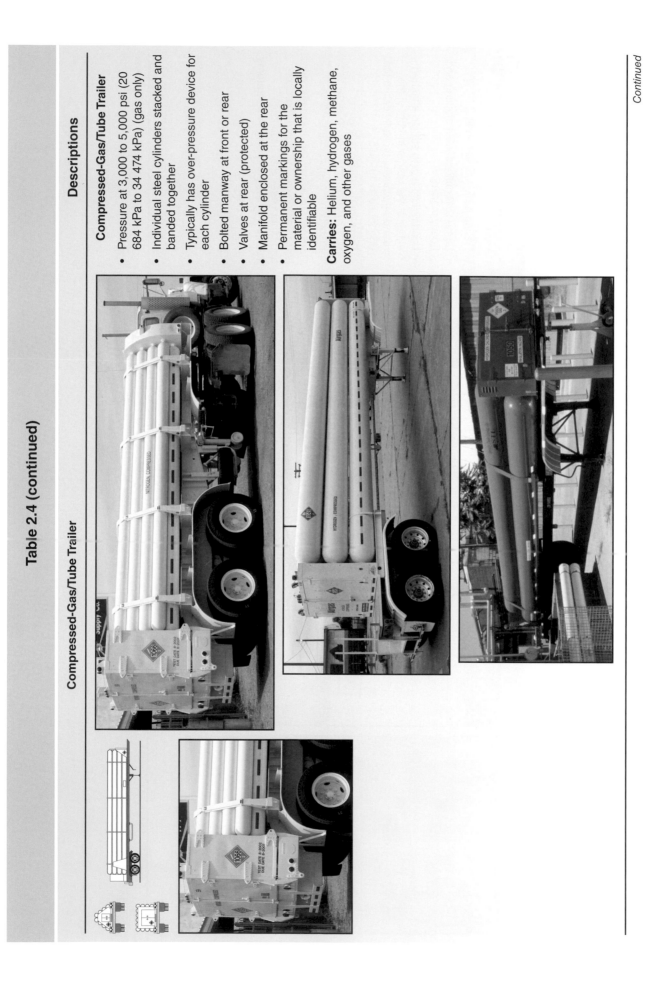

Descriptions

Compressed-Gas/Tube Trailer

- Pressure at 3,000 to 5,000 psi (20 684 kPa to 34 474 kPa) (gas only)
- Individual steel cylinders stacked and banded together
- Typically has over-pressure device for each cylinder
- Bolted manway at front or rear
- Valves at rear (protected)
- Manifold enclosed at the rear
- Permanent markings for the material or ownership that is locally identifiable

Carries: Helium, hydrogen, methane, oxygen, and other gases

Continued

Table 2.4 (concluded)

Dry Bulk Cargo Trailer

Descriptions

Dry Bulk Cargo Trailer

- Pressure less than 22 psi (152 kPa)
- Typically not under pressure
- Bottom valves
- Shapes vary, but has V-shaped bottom-unloading compartments
- Rear-mounted, auxiliary-engine-powered compressor or tractor-mounted power-take-off air compressor
- Air-assisted, exterior loading and bottom unloading pipes
- Top manway assemblies

Carries: Calcium carbide, oxidizers, corrosive solids, cement, plastic pellets, and fertilizers

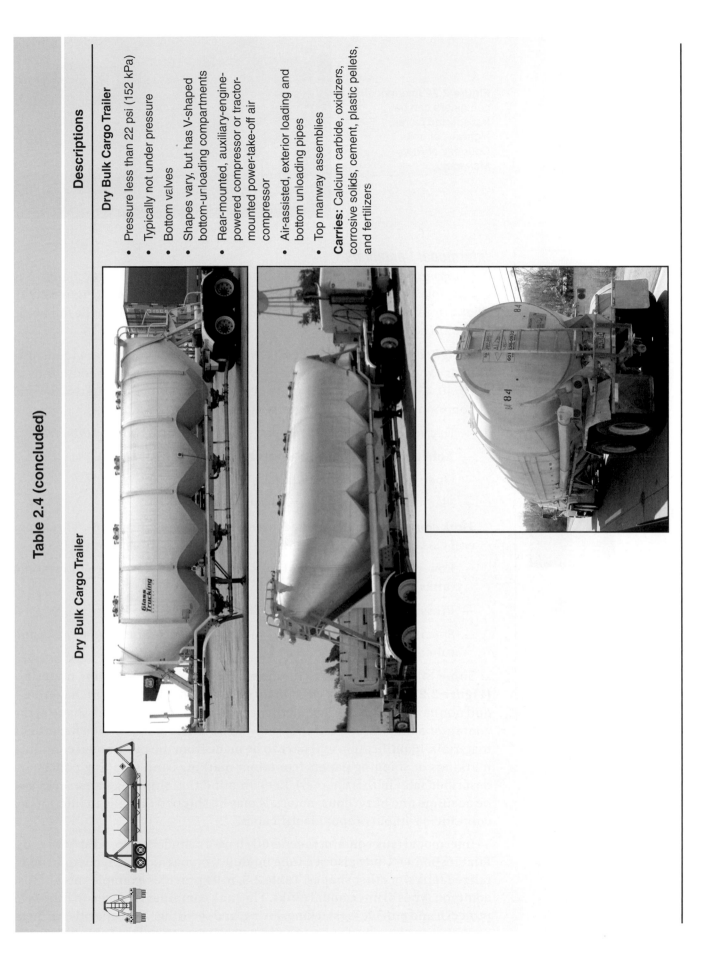

Figure 2.24 Intermodal containers can be transported by rail, highway, or ship. *Courtesy of Rich Mahaney.*

Figure 2.25 Freight containers come in a variety of different types including refrigerated containers (also called *reefers*), flat containers, dry van containers (sometimes called *box containers*), and open top containers.

Refrigerated Intermodal Container — Cargo container having its own refrigeration unit; also called a *reefer.*

Intermodal Containers

An *intermodal container* is a freight container that is used interchangeably in multiple modes of transport such as rail, highway, and ship **(Figure 2.24).** The various types of intermodal containers can be divided into the following two main categories:

1. *Freight containers* — Transport a wide range of products, from foodstuffs to dry goods. They come in a variety of types and sizes, most commonly in 20, 40, 45, 48, and 53-foot (6 m, 12 m, 14 m, 15 m, and 16 m) lengths. Several common types of freight containers are **(Figure 2.25):**

   — Dry van intermodal containers (sometimes called *box containers*)

   — **Refrigerated intermodal containers** (also called *reefers*)

   — Open top intermodal containers

   — Flat intermodal containers of various sorts

2. *Tank container*s — Also called *intermodal tanks* **(Figure 2.26).** Three general classifications of intermodal tank containers are as follows:

   — Low-pressure intermodal tanks (also called nonpressure intermodal tanks)

   — Pressure intermodal tanks

   — Specialized intermodal tanks such as cryogenic intermodal tanks and tube modules

Some intermodal freight containers may contain hazardous materials **(Figure 2.27).** Others may contain mixed loads that include both hazardous and nonhazardous materials. With many freight containers, the shape of the container alone will not tell first responders whether it contains hazardous materials. Identification will have to be made from the intermodal container markings or shipping papers (container markings and shipping papers are described later in this chapter). Keep in mind that shipping papers may not be accurate and hazardous materials may be shipped illegally in intermodal containers without proper identification.

Intermodal tank containers generally have a cylinder enclosed at both ends. First responders may also see tube modules, cryogenic tanks, compartmentalized tanks, or other shapes. **Table 2.5, p. 94** provides examples of the most common types of intermodal tanks. The tank container is placed in frames to protect it and provide for stacking, lifting, and securing. The capacities of these containers ordinarily do not exceed 6,340 gallons (24 000 L) **(Table 2.6, p. 95).**

Figure 2.26 Intermodal tank containers carry a variety of hazardous materials. Sometimes construction features enable identification of the type of tank, but it is often necessary to look at the tank's specification information. *Courtesy of Rich Mahaney.*

Figure 2.27 This intermodal freight container has placards indicating it contains hazardous materials. Many freight containers may carry mixed loads of non-hazardous and hazardous materials that will not have placards or markings. *Courtesy of Rich Mahaney.*

Low-pressure intermodal tank. This tank is the most common intermodal tank used in transportation. Even though they are often called *nonpressure intermodal tanks,* these tanks may have pressures as high as 100 psi (689 kPa) {6.9 bar}. They are also called *intermodal portable tanks* or *IM portable tanks.* The two common groups of low-pressure/nonpressure intermodal tank containers are as follows:

1. *IM 101 portable tanks* — Built to withstand a working pressure of 25.4 to 100 psi (175 kPa to 689 kPa) {1.75 bar to 6.9 bar}. They transport both hazardous and nonhazardous materials. Internationally, they are called *International Maritime Organization (IMO) Type 1 tank containers* **(Figure 2.28).**

> **CAUTION**
> Intermodal freight containers can contain virtually anything, including extremely hazardous materials which may not be properly identified!

Figure 2.28 Though called nonpressure intermodal tanks, IM 101 portable tanks may have pressures from 25.4 to 100 psi (175 kPa to 689 kPa) {1.75 bar to 6.9 bar}. *Courtesy of Rich Mahaney.*

Table 2.5
Intermodal Tanks

Tank Type	Descriptions
	Nonpressure Intermodal Tank • IM-101: 25.4 to 100 psi (175 kPa to 689 kPa) • IM-102: 14.5 to 25.4 psi (100 kPa to 175 kPa) **Contents:** Liquids or solids (both hazardous and nonhazardous)
	Pressure Intermodal Tank 100 to 500 psi (689 kPa to 3 447 kPa) **Contents:** Liquefied gases, liquefied petroleum gas, anhydrous ammonia, and other liquids
	Cryogenic Intermodal Tank **Contents:** Refrigerated liquid gases, argon, oxygen, helium
	Tube Module Intermodal Container **Contents:** Gases in high-pressure cylinders (3,000 or 5,000 psi [20 684 kPa or 34 474 kPa]) mounted in the frame

2. *IM 102 portable tanks* — Designed to handle maximum allowable working pressures of 14.5 to 25.4 psi (169 kPa to 175 kPa) {1.69 to 1.75 bar}. These containers are gradually being removed from service. They transport materials such as alcohols, pesticides, resins, industrial solvents, and flammables with flash points between 32 and 140°F (0°C to 60°C). Most commonly, they transport nonregulated materials (those not specifically covered by regulations) such as food commodities. Internationally, they are called *IMO Type 2 tank containers* (**Figure 2.29**).

Table 2.6
Intermodal Tank Container Descriptions

Specification	Materials Transported	Capacity	Design Pressure
IM 101 Portable Tank	Hazardous and nonhazardous materials, including toxics, corrosives, and flammables with flash points below 32°F (0°C)	Normally range from 5,000 to 6,300 gallons (18 927 to 23 848 L)	25.4 to 100 psi (175 to 689 kPa) {1.75 to 6.89 bar}
IM 102 Portable Tank	Whiskey, alcohols, some corrosives, pesticides, insecticides, resins, industrial solvents, and flammables with flash points ranging from 32 to 140°F (0 to 60° C)	Normally range from 5,000 to 6,300 gallons (18 927 to 23 848 L)	14.5 to 25.4 psi (100 to 175 kPa) {1 to 1.75 bar}
Spec. 51 Portable Tank	Liquefied gases such as LPG, anhydrous ammonia, high vapor pressure flammable liquids, pyrophoric liquids (such as aluminum alkyls), and other highly regulated materials	Normally range from 4,500 to 5,500 gallons (17 034 to 0 820 L)	100 to 500 psi (689 to 3 447kPa) {6.89 to 34.5 bar}

Figure 2.29 IM 102 portable tanks have pressures from 14.5 to 25.4 psi (100 kPa to 175 kPa) {1.69 bar to 1.75 bar} (less than IM 101s). It is difficult to tell the difference between IM 101 and IM 102 tanks without looking at specification markings. *Courtesy of Rich Mahaney.*

Pressure intermodal tank. A pressure intermodal tank container is less common in transport. It is designed for working pressures of 100 to 500 psi (689 kPa to 3 447 kPa) {6.9 to 34.5 bar} and usually transports liquefied gases under pressure. DOT classifies this tank as *Spec. 51,* while internationally it is known as an *IMO Type 5 tank container* (**Figure 2.30, p. 96**).

Specialized intermodal tank or container. There are several types of specialized intermodal tank containers. Cryogenic liquid tank containers carry refrigerated liquid gases, argon, oxygen, and helium. Cryogenic-type containers are built to IMO Type 7 specifications (**Figure 2.31, p. 96**). The tube module transports gases in high-pressure cylinders (3,000 to 5,000 psi [20 684 kPa to 34 474 kPa] {207 bar to 345 bar}). Dry bulk intermodal containers carry materials such as fertilizer and cement (**Figure 2.32, p. 96**).

Figure 2.30 Spec. 51 tanks, internationally known as IMO Type 5 tank containers, usually transport liquefied gases under pressure. *Courtesy of Rich Mahaney.*

Figure 2.31 Cryogenic liquid tank containers carry refrigerated liquid gases and are built to IMO Type 7 specifications. *Courtesy of Rich Mahaney.*

Figure 2.32 Dry bulk intermodal containers carry many different products including materials such as fertilizer. *Courtesy of Rich Mahaney.*

Vessel Cargo Carriers

It is estimated that over 90 percent of the world's cargo is transported by marine vessels, and the amount of cargo transported by vessels is expected to continue to increase. Hazardous materials incidents involving vessels can be minor (such as a small spill that occurs at a port during loading or unloading) or major, such as a spill contaminating miles (kilometers) of river or coastline waters or a large spill inside a ship. Statistics on oil spills show that most spills result from routine operations such as loading and unloading, which normally occur in ports or at oil/chemical terminals. The majority of these operational spills are small, with some 91 percent involving quantities less than 7 tons (7.1 tonnes). First responders need to be aware of vessel types and cargos that are likely to contain hazardous materials.

Tanker. A vessel that exclusively carries liquid products in bulk is generally known as *tanker* or *tank vessel* **(Figure 2.33).** Modern tankers are capable of transporting very large quantities of liquid products. Tankers often carry a variety of products in segregated tanks. Tankers can be divided into the following three general categories:

Figure 2.33 Tankers transport very large quantities of liquid products such as crude oil or finished petroleum products.

1. *Petroleum carriers* — Transport crude or finished petroleum products. They range in size from 200-foot (61 m), 15,000-barrel coastal tankers of 2,000 deadweight tons to 1,200-foot (366 m), 3,680,000-barrel ultra-large crude carriers of 480,000 deadweight tons. Details:

   — When entering the U.S. and Canada, the operator of any tank vessel carrying petroleum products is required by law to maintain vessel emergency response plans that identify and ensure the availability of both a salvage company with expertise and equipment and a company with pollution incident response capabilities in the area(s) in which the vessel operates.

   — The availability of preplanned resources should not be overlooked during a marine fire-fighting or emergency response.

2. *Chemical carriers* — Transport multiple commodities; these carriers are sometimes nicknamed floating drugstores. They may carry oils, solvents, gasoline, sulfur, and other commodities (many classified as hazardous materials) in 30 to 58 separate tanks. Each tank usually has its own pump (and piping), so the deck of a chemical carrier typically has a maze of piping **(Figure 2.34).** About 3,000 chemical carriers are in operation worldwide, and they have the lowest average annual loss rate of all cargo vessel types. Chemical carriers are not required to carry placards. The only way to positively identify a chemical cargo is to ask the master or mate (captain or first officer) or obtain the cargo plan that identifies where each commodity is stowed on the vessel.

CAUTION
Chemical carriers are not required to carry placards.

Figure 2.34 Chemical carriers typically have mazes of piping and may carry oils, solvents, gasoline, sulfur, and other hazardous materials in 30 to 58 separate tanks.

3. ***Liquefied flammable gas carriers*** — Transport liquefied natural gas (LNG) and liquefied petroleum gas (LPG) (propane and butane for example), and generally use large insulated spherical tanks for product storage **(Figure 2.35).** However, other configurations of gas carriers look very similar to ordinary tankers. The tanks are isolated within the vessel's hull by cofferdams (empty spaces between compartments) designed to contain low-volume leakage from the tanks. LPG carriers are usually identified by a large number of pressure vessels. Cargo piping is located above the main deck so that any piping leaks vent to the atmosphere rather than inside the vessel. Details:

Figure 2.35 Liquefied flammable gas carriers transport liquefied natural gas (LNG) and liquefied petroleum gas (LPG). Some use large insulated spherical tanks for product storage; others look very similar to ordinary tankers.

— In U.S. ports that handle LNG and LPG carriers, the Captain of the Port is required to maintain LNG/LPG vessel management and emergency contingency plans.

— In Canada, each port handling hazardous shipments is required to conduct an evaluation that defines all threats to the port and environment and prepare contingency plans to manage emergencies. These plans are consulted for area-specific guidance in handling emergencies involving these vessels.

Cargo Vessel. The size of commercial cargo vessels can be overwhelming. Cargo vessels are typically 500 to 900 feet (152 m to 274 m) in length, 50 to 130 feet (15 m to 40 m) in beam, and have hold depths from 40 to 60 feet (12 m to 18 m). Cargo is shipped in the following four vessel types:

1. ***Bulk carrier*** — Can be either liquid bulk (tanker) or dry bulk carrier.

— Dry bulk vessels carry products such as coal, wood chips, grain, iron ore, sand, gravel, salt, grain, and fertilizers. The cargo is loaded directly into a hold without packaging, much like liquid in a tanker. Some of these cargoes generate dust (for example, grain), creating the possibility of an explosion.

— The two primary liquid bulk cargoes are chemicals (that may or may not be flammable) and liquid hydrocarbons. Liquid hydrocarbons carried in bulk include crude oils and refined oil products such as diesel fuel, gasoline, lubricating oils, and kerosene. These products vary widely in their characteristics, and some can be very volatile. The hazards are similar to those found at any petrochemical refinery or bulk storage facility.

2. **Break bulk carrier** — Has large holds to accommodate a wide range of products such as vehicles, pallets of metal bars, liquids in drums, or items in bags, boxes, and crates.

3. **Container vessel** — Carries cargo in standard containers that measure 8 feet (2.4 m) wide with varying heights and lengths. Container vessels may transport intermodal tanks (each enclosed in an open framework with standard container-size dimensions) **(Figure 2.36)**.

Figure 2.36 Container vessels transport intermodal containers, including intermodal tanks.

Figure 2.37 Roll-on/roll-off vessels have large stern and side ramp structures that are lowered to allow vehicles to be driven on and off the vessel.

4. **Roll-on/roll-off vessel** — Has large stern and side ramp structures that are lowered to allow vehicles to be driven on and off the vessel **(Figure 2.37)**. This vessel can be visualized as a floating, moving, multilevel parking garage.

Barge. *Barges* are typically box-shaped, flat-decked vessels used for transporting cargo. Towing or pushing vessels are usually used to move barges because they are not self-propelled. Virtually anything can be transported on a barge. Some barges are configured as floating barracks for military or construction crews; some are designed as bulk oil and chemical tankers **(Figure 2.38, p. 100)**. Other barges carry LNG in cylinders that may not be visible until a person is aboard. Barges may serve as floating warehouses with hazardous goods, vehicles, or rail cars inside.

Unit Loading Devices

Unit loading devices (ULDs) are containers and aircraft pallets used to consolidate air cargo into a single, transportable unit. ULDs are designed and shaped to fit into the various decks and compartments of airplanes (commercial cargo planes), and in some cases they may be stacked. Hazardous materials may be shipped in ULDs provided they are in accordance with U.S. Federal Aviation Administration (FAA) and DOT regulations **(Figure 2.39, p. 100)**. ULDs containing hazardous materials are required to be labeled or placarded in accordance with Title 49 (Transportation) CFR 172.512 (a) and (b).

Figure 2.38a and b (a) Barges can travel waterways that large vessels cannot. They are very versatile in their cargos, and some are designed to carry specific hazardous materials. (b) It may not be possible to tell from a distance if hazardous materials are being transported on a barge. *Courtesy of Rich Mahaney.*

Figure 2.39 Unit loading devices (ULDs) are used to consolidate air cargo into a single, transportable unit. ULDs containing hazardous materials must be appropriately placarded and labeled. *Courtesy of John Demyan.*

Figure 2.40 Intermediate bulk containers come in a variety of styles and may carry both solid materials and fluids. This is a rigid intermediate bulk container. *Courtesy of Rich Mahaney.*

Intermediate Bulk Containers

Per the U.S. DOT, an *intermediate bulk container (IBC)* is either rigid or flexible portable packaging (other than a cylinder or portable tank) designed for mechanical handling **(Figure 2.40).** Design standards for IBCs in the U.S., Canada, and Mexico are based on *United Nations Recommendations on the Transportation of Dangerous Goods.* The maximum capacity of an IBC is not more than three 3 cubic meters (3,000 L, 793 gal, or 106 ft³). The minimum capacity is not less than 0.45 cubic meters (450 L, 119 gal, or 15.9 ft³) or a maximum net mass of not less than 400 kilograms (882 lbs).

NOTE: There is no weight limit on solid products.

IBCs are authorized to transport a wide variety of materials and hazard classes:

- Aviation fuel (turbine engine)
- Gasoline
- Hydrochloric acid
- Methanol
- Toluene
- Corrosive liquids
- Solid materials in powder, flake, or granular forms

IBCs are divided into two types: flexible intermediate bulk containers (FIBCs) and rigid intermediate bulk containers (RIBCs). Both types are often called *totes*, although only FIBCs are truly totes.

Flexible intermediate bulk container (FIBC). FIBCs are sometimes called *bulk bags, bulk sacks, supersacks, big bags, tote bags,* or *totes.* They are flexible, collapsible bags or sacks that are used to carry both solid materials and fluids **(Figure 2.41).** The designs of FIBCs are as varied as the products they carry. Often the bags used to transport wet or hazardous materials are lined with polypropylene or some other high-strength fabric. Others may be constructed of multiwall paper or other textiles. A common-sized supersack can carry the equivalent of four to five 55-gallon (208 L) drums and (depending on design and the material inside) be stacked one on top of another. Sometimes FIBCs are transported inside a rigid exterior container made of corrugated board or wood.

Figure 2.41 Sometimes called totes or supersacks, some flexible intermediate bulk containers carry the equivalent of four to five 55-gallon (208 L) drums of solids or liquids. *Courtesy of Rich Mahaney.*

Rigid intermediate bulk container (RIBC). RIBCs are typically made of steel, aluminum, wood, fiberboard, or plastic; and they are often designed to be stacked. RIBCs can contain both solid materials and liquids. Some liquid containers may look like smaller versions of intermodal nonpressure tanks with metal or plastic tanks inside rectangular box frames **(Figures 2.42).** Other RIBCs may be large, square or rectangular boxes or bins **(Figure 2.43).** Rigid portable tanks may be used to carry liquids, fertilizers, solvents, and other chemicals; and they may have capacities up to 400 gallons (1 514 L) and pressures up to 100 psi (689 kPa) {6.9 bar}.

Figure 2.42 This small tank is also a rigid intermediate bulk container (RIBC). *Courtesy of Rich Mahaney.*

Figure 2.43 RIBCs may be large, square or rectangular boxes or bins carrying liquids, fertilizers, solvents, or other chemicals with capacities up to 400 gallons (1 514 L). *Courtesy of Rich Mahaney.*

Figure 2.44 Ton containers often contain chlorine, but they may also contain other products such as sulfur dioxide, anhydrous ammonia, and other gases. *Courtesy of Rich Mahaney.*

Figure 2.45 Ton containers may have convex or concave ends and have two valves in the center of one end, one above the other. *Courtesy of Rich Mahaney.*

Ton Containers

Ton containers are tanks that have capacities of 1 short ton or approximately 2,000 pounds (907 kg or 0.9 tonne). Typically stored on their sides, the ends (heads) of the containers are convex or concave and have two valves in the center of one end, one above the other **(Figure 2.44).** One valve connects to a tube going into the liquid space; the other valve connects to a tube going into the vapor space above **(Figure 2.45).** They also have pressure-relief devices (fusible plugs) in case of fire or exposure to elevated temperatures. Ton containers commonly contain chlorine and are often found at water treatment plants, commercial swimming pools, etc. Ton containers may also contain other products such as sulfur dioxide, anhydrous ammonia, or Freon® refrigerant.

Leaks from ton containers require special equipment (and Technician-Level training) to patch, which is the case for most of the containers in this chapter. Always ensure responders and civilians are evacuated to a safe distance to avoid the vapor cloud that escapes from these containers.

Nonbulk Packaging

Containers that are used to transport smaller quantities of hazardous materials than bulk or IBCs are called *nonbulk packaging*. The majority of these incidents occur during highway transport or routine use. **Table 2.7** shows common types of nonbulk packaging including the following types of containers:

- Bags
- Carboys and jerry cans
- Cylinders
- Drums
- Dewar flasks (cryogenic liquids)

Containers for Radioactive Materials

All shipments of radioactive materials (sometimes called *RAM*) must be packaged and transported according to strict regulations. These regulations protect the public, transportation workers, and the environment from potential exposure to radiation. The type of packaging used to transport radioactive materials is determined by the activity, type, and form of the material to

CAUTION

Structural fire-fighting gear does not provide adequate protection against the hazardous materials commonly stored in ton containers.

Table 2.7
Nonbulk Packaging

Package Type	Descriptions
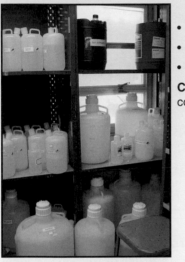	**Bags** • Made of paper, plastic, film, textiles, woven material, or others • Sizes vary **Contents:** Explosives, flammable solids, oxidizers, organic peroxides, fertilizers, pesticides, and other regulated materials.
	**Carboys and Jerrycans** • Made of glass or plastic • Often encased in a basket or box • Sizes vary **Contents:** Flammable and combustible liquids, corrosives.
	Cylinders • Presssures higher than 40 psi (276 kPa) {2.76 bar} but vary • Sizes range from lecture bottle size to very large **Contents:** Compressed gases.

Continued

Table 2.7 (concluded)

Package Type	Descriptions
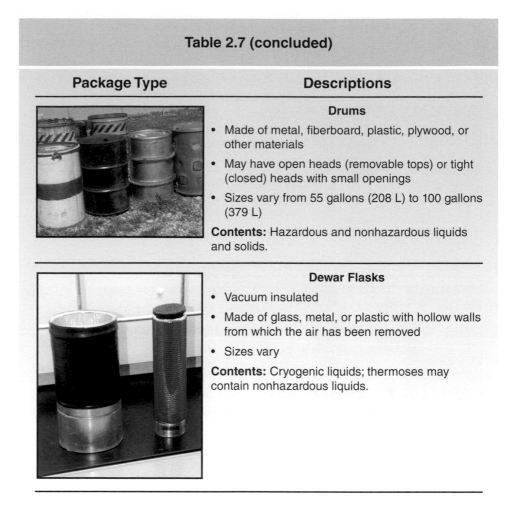	**Drums** • Made of metal, fiberboard, plastic, plywood, or other materials • May have open heads (removable tops) or tight (closed) heads with small openings • Sizes vary from 55 gallons (208 L) to 100 gallons (379 L) **Contents:** Hazardous and nonhazardous liquids and solids.
	Dewar Flasks • Vacuum insulated • Made of glass, metal, or plastic with hollow walls from which the air has been removed • Sizes vary **Contents:** Cryogenic liquids; thermoses may contain nonhazardous liquids.

be shipped. Depending upon these factors, radioactive material is shipped in one of the following five basic types listed in order of increasing level of radioactive hazard:

1. ***Excepted*** — Packaging used for transportation of materials that have very limited radioactivity such as articles manufactured from natural or depleted uranium or natural thorium. Excepted packagings are only used to transport materials with extremely low levels of radioactivity that present no risk to the public or environment. Excepted packaging is not marked or labeled as such. *Other information:*

   — Empty packaging is also excepted.

   — Because of its low risk, excepted packaging is exempt from several labeling and documentation requirements.

2. ***Industrial*** — Container that retains and protects the contents during normal transportation activities. Materials that present limited hazard to the public and the environment are shipped in these packages. Industrial packages are not identified as such on the packages or shipping papers. *Material examples:*

   — Slightly contaminated clothing

   — Laboratory samples

   — Smoke detectors

3. ***Type A*** — Packages that must demonstrate their ability to withstand a series of tests without releasing their contents. The package and shipping papers will have the words *Type A* on them. Regulations require that the package protect its contents and maintain sufficient shielding under conditions normally encountered during transportation. Radioactive materials with relatively high specific activity levels are shipped in Type A packages **(Figure 2.46a)**. *Material examples:*

   — Radiopharmaceuticals (radioactive materials for medical use)

   — Certain regulatory qualified industrial products

Figure 2.46a Type A packages transport materials with relatively high specific activity levels such as radioactive materials for medical use or certain industrial products like density gauges. *Courtesy of Tom Clawson.*

4. ***Type B*** — Packages must not only demonstrate their ability to withstand tests simulating normal shipping conditions, but they must also withstand severe accident conditions without releasing their contents. Type B packages are identified as such on the package itself as well as on shipping papers. The size of these packages can range from small containers to those weighing over 100 tons (102 tonnes). These large, heavy packages provide shielding against radiation. Radioactive materials that exceed the limits of Type A package requirements must be shipped in Type B packages **(Figure 2.46b)**. *Material examples:*

   — Materials that would present a radiation hazard to the public or the environment if there were a major release

   — Materials with high levels of radioactivity such as spent fuel from nuclear power plants

5. ***Type C*** — Very rare packages used for high-activity materials (including plutonium) transported by aircraft **(Figure 2.46c)**. They are designed to withstand severe accident conditions associated with air transport without loss of containment or significant increase in external radiation levels. The Type C package performance requirements are significantly more stringent than those for Type B packages.

Figure 2.46b Type B packages are designed to withstand severe accident conditions. Type B packages contain materials with high levels of radioactivity that would present a radiation hazard to the public or the environment if there were a major release. *Courtesy of the National Nuclear Security Administration, Nevada Site Office.*

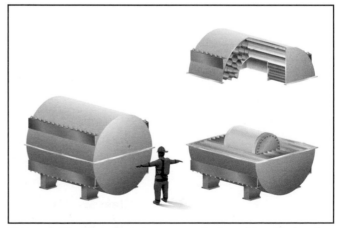

Figure 2.46c Type C packages are designed to withstand severe accident conditions associated with air transport. Type C packages are very rare, and first responders are unlikely to encounter them. *Courtesy of the U.S. Air Force.*

Clue 3: Transportation Placards, Labels, and Markings

The U.S., Canada, and Mexico have all adopted the *Transport of Dangerous Goods – Model Regulations*, published by the United Nations (also known as the *U.N. Recommendations*). Therefore, with a few country-specific variations, the majority of the placards, labels, and markings used to identify hazardous materials during transport are very similar in all three countries.

Placards, labels, and markings based on the UN system for classifying and identifying transported hazardous materials are discussed in this section, along with common intermodal markings and railcar specification markings. Placards, labels, markings, and colors associated with other systems (such as NFPA® 704, *Standard System for the Identification of the Hazards of Materials for Emergency Response* [2001], and military markings) are discussed in the next section, Clue 4: Other Markings and Colors.

UN Recommendations on the Transport of Dangerous Goods

The *UN Recommendations* provides a uniform basis for development of *harmonized regulations* for all modes of transport in order to facilitate trade and the safe, efficient transport of hazardous materials. The *UN Recommendations* establishes minimum requirements applicable to the transport of hazardous materials by all modes of transport, covering all aspects of transportation necessary to provide international uniformity. The publication includes a comprehensive criteria-based classification system for substances that pose a significant hazard in transportation as well as standards for packaging and multimodal tanks used to transport hazardous materials. The publication also includes a system of communicating the hazards of substances in transport through hazard communication requirements. This system addresses labeling and marking of packages, placarding of tanks and freight units, and documentation and emergency response information required to accompany each shipment.

UN Hazard Classes

Under the UN system, nine hazard classes are used to categorize hazardous materials:

Class 1: Explosives

Class 2: Gases

Class 3: Flammable liquids

Class 4: Flammable solids, substances liable to spontaneous combustion, substances that emit flammable gases on contact with water

Class 5: Oxidizing substances and organic peroxides

Class 6: Toxic and infectious substances

Class 7: Radioactive materials

Class 8: Corrosive substances

Class 9: Miscellaneous dangerous substances and articles

To avoid redundancy, a detailed explanation of these nine major hazard classes is given in the U.S. Transportation Placards, Labels, and Markings section under DOT Placards. While there may be minor variances between the UN and the U.S. class definitions, because most North American first responders primarily deal with DOT or Transport Canada (TC) placards, labels, and markings, the unique UN placards are not detailed here. Examples of the UN class placards with brief explanations are found in **Appendix D,** UN Class Placards and Labels.

Four-Digit Identification Numbers

The UN has also developed a system of four-digit identification numbers that is used in conjunction with illustrated placards in North America. Each individual hazardous material is assigned a unique four-digit number. This number will often be displayed on placards, orange panels, and certain markings in association with materials being transported in cargo tanks, portable tanks, tank cars, or other containers and packages.

The 4-digit ID number must be displayed on bulk containers in one of the three ways illustrated by **Figure 2.47.** In North America, the numbers must be displayed on the following containers/packages:

- Rail tank cars
- Cargo tank trucks
- Portable tanks
- Bulk packages
- Vehicle containers containing large quantities (at least 8,820 lbs or 4 000 kg) of hazardous materials
- Certain nonbulk packages (for example, poisonous gases in specified amounts)

Figure 2.47 Examples of how UN numbers will be displayed on bulk containers (such as cargo tank trucks and rail tank cars) and certain nonbulk packages.

Previously, orange panels may have been preceded by the letters NA for North America or UN for United Nations **(Figure 2.48).** This practice is now obsolete.

The *Emergency Response Guidebook (ERG)* provides a key to the four-digit identification numbers in the yellow-bordered section (see Chapter 3, Awareness-Level Actions at Hazardous Materials Incidents). Therefore, if the four-digit identification number of a hazardous material is identified, first responders can use the *ERG* to determine appropriate initial response information based on the material involved. The four-digit identification number will also appear on shipping papers, and it should match the numbers displayed on the exteriors of tanks or shipping containers.

It should be noted that common reference materials such as the *ERG* do not list all 4-digit UN Identification numbers. For example, the *ERG* does not list any numbers below 1000, such as 0331 **(Figure 2.49).** In the U.S., the entire list is included in 49 *CFR* 172.101.

Figure 2.48 The four-digit UN identification number can be used to identify hazardous materials in the *Emergency Response Guidebook. Courtesy of Rich Mahaney.*

Figure 2.49 The *ERG* does not list numbers below 1000, however, numbers lower than 1000 are included in 49 *CFR* 172.101.

0331

Careful!

Don't be confused by an orange placard with two sets of numbers on intermodal tanks and containers. The four-digit ID number is on the bottom. The top number is a hazard identification number (or code) required under European and some South American regulations **(Figure 2.50)**. These numbers indicate the following hazards:

2 — Emission of a gas due to pressure or chemical reaction

3 — Flammability of liquids (vapors) and gases or self-heating liquid

4 — Flammability of solids or self-heating solid

5 — Oxidizing (fire intensifying) effect

6 — Toxicity or risk of infection

7 — Radioactivity

8 — Corrosivity

9 — Miscellaneous Dangerous Substance

Doubling a number (such as 33, 44, or 88) indicates an intensification of that particular hazard. When the hazard associated with a material is adequately indicated by a single number, it is followed by a *zero* (such as 30, 40, or 60). A hazard identification code prefixed by the letter X (such as X88) indicates that the material will react dangerously with water. When 9 appears as a second or third digit, this may present a risk of spontaneous violent reaction.

Figure 2.50 The UN ID number is the orange panel on the bottom. The top number is a hazard identification number (code) that some European and South American regulations require. *Courtesy of Rich Mahaney.*

U.S. Transportation Placards, Labels, and Markings

The UN system forms the basis for the DOT regulations. DOT classifies hazardous materials according to their primary danger and assigns standardized symbols to identify the classes. DOT regulations cover some additional categories of substances, including other regulated materials (ORM-Ds), materials of trade (MOTs), and fumigated loads.

Shreveport, Louisiana Anhydrous Ammonia Incident

Placarded materials may have many hazards not reflected by the placard classification. A specific example is anhydrous ammonia; it is placarded in the U.S. as a nonflammable gas. However, under certain conditions (particularly indoors where vapors can become concentrated), it will burn.

In 1984, in Shreveport, Louisiana, two hazardous materials response team members entered a cold-storage facility to stop a leak of anhydrous ammonia. With a lower explosive limit (LEL) of 16 percent and a flammable range of 16 to 25 percent (see Chapter 4, Hazardous Materials Properties and Behavior, Flammability section), the vapors inside the facility reached a flammable concentration. When a spark ignited the vapors, one team member was killed and the other was seriously burned. Anhydrous ammonia can ignite even though it is not classified as a flammable gas by the DOT.

In other countries, anhydrous ammonia is classified as a corrosive (caustic) liquid and a poison gas because of its chemical effects. Inhaling concentrated vapors can kill a person even though it is not classified as an inhalation hazard by the DOT!

DOT Placards

A *placard* is a diamond-shaped, color-coded sign provided by shippers to identify the materials in transportation containers. Each of the nine hazard classes has a specific placard that identifies the class of the material and assists the responder in identifying the hazards associated with the product. A material's hazard class is indicated either by its class (or division) number or name. The hazard class or division number must be displayed in the lower corner of placards corresponding to the primary hazard class of a material. **Figure 2.51** provides the required dimensions of DOT placards and summarizes the information conveyed by them. Placards may be found on the following types of containers:

- Bulk packages

- Rail tank cars

- Cargo tank vehicles

- Portable tanks

- Unit load devices containing hazardous materials over 640 cubic feet (18 m^3) in capacity

- Certain nonbulk containers

Specifically, placards are required on any container transporting any quantity of the materials listed in **Table 2.8, p. 112** (Placarding Table 1 of DOT Chart 14). When the aggregate gross weight of all hazardous materials in non-bulk packages covered in **Table 2.9, p. 112** is less than 454 kg (1,001 lbs), no placard is required on a transport vehicle or freight container when transported by highway or rail. Unfortunately, improperly marked, unmarked, and otherwise illegal shipments are common. These shipments may include incompatible products, products that contravene local, state/provincial, and federal laws, and waste products shipped and disposed of without permit.

Table 2.10, p. 113 provides an explanation of the DOT hazard classes and subdivisions, the Hazardous Material Regulation (HMR) references for each, and their associated placards. The following is a list of important facts that relate to placards:

- A placard is not required for shipments of infectious substances, ORM-Ds, MOTs, limited quantities, small-quantity packages, radioactive materials (white label I or yellow label II; see Table 2.11), or combustible liquids in nonbulk packaging (see information box).

- Some private agriculture and military vehicles may not have placards, even though they are carrying significant quantities of hazardous materials. For example, farmers may carry fertilizer, pesticides, and fuel between fields of their farms or to and from their farms without any placarding.

- The hazard class or division number corresponding to the primary or subsidiary hazard class of a material must be displayed in the lower corner of a placard **(Figure 2.52, p. 112).**

- Other than Class 7 or the *DANGEROUS* placard, text indicating a hazard (for example, the word *FLAMMABLE*) is not required. Text may be omitted from the Oxygen placard only if the specific ID number is displayed.

- The shipper is required to provide placards. Drivers may not know what they are carrying or may have varying degrees of information about the hazardous materials in their vehicles.

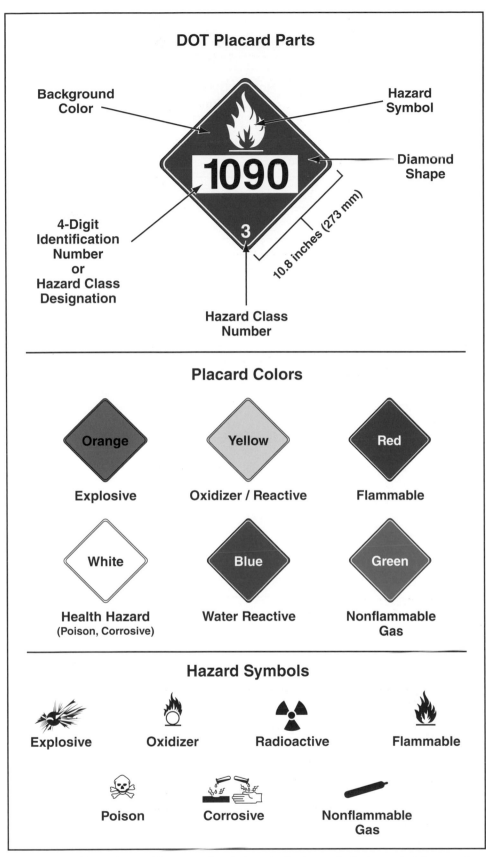

Figure 2.51 Placards provide many visual clues to the hazards that a material presents.

Table 2.8
Materials Requiring Placarding Regardless of Quantity

Category of Material (Hazard Class or division number and additional description, as appropriate)	Placard name
1.1	EXPLOSIVES 1.1
1.2	EXPLOSIVES 1.2
1.3	EXPLOSIVES 1.3
2.3	POISON GAS
4.3	DANGEROUS WHEN WET
5.2 (Organic peroxide, Type B, liquid or solid, temperature controlled)	ORGANIC PEROXIDE
6.1 (Materials poisonous by inhalation)	POISON INHALATION HAZARD
7 (Radioactive Yellow III label only)	RADIOACTIVE[1]

[1] RADIOACTIVE placard also required for exclusive use shipments of low specific activity material and surface contaminated objects transported in accordance with §173.427(a)(6).

Table 2.9
Materials in Excess of 1,001 Pounds (454 kg) Requiring Placarding

Hazard Class or Division	Placard name
1.4	Explosives 1.4
1.5	Explosives 1.5
1.6	Explosives 1.6
2.1	Flammable Gas
2.2	Nonflammable Gas
3	Flammable
Combustible Liquid	Combustible
4.1	Flammable Solid
4.2	Spontaneously Combustible
5.1	Oxidizer
5.2 (Other than organic peroxide, Type B, liquid or solid, temperature controlled)	Organic Peroxide
6.1 (Other than materials poisonous by inhalation)	Poison
6.2	None
8	Corrosive
9	Class 9 [See 49 *CFR*, Part 172, subpart E, paragraph 172.504 (f)(9)]
ORM-D	None

Source: U.S. Department of Transportation, Chart 14s, Hazardous Materials Marking, Labeling & Placarding Guide, Table 2.

Figure 2.52 All primary and subsidiary placards must have the hazard class or division number displayed. The placard with the four-digit UN ID number is always the primary placard. *Courtesy of the U.S. Air Force.*

Table 2.10
U.S. DOT Placard Hazard Classes and Divisions

Class 1: Explosives (49 *CFR* 173.50)

An **explosive** is any substance or article (including a device) that is designed to function by explosion (that is, an extremely rapid release of gas and heat) or (by chemical reaction within itself) is able to function in a similar manner even if not designed to function by explosion.

Explosive placards will have a *compatibility group* letter on them, which is a designated alphabetical letter used to categorize different types of explosive substances and articles for purposes of stowage and segregation. However, it is the division number that is of primary concern to first responders.

The primary hazards of explosives are thermal (heat) and mechanical, but may include the following:

- Blast pressure wave
- Shrapnel fragmentation
- Incendiary thermal effect
- Seismic effect
- Chemical hazards from the production of toxic gases and vapors
- Ability to self-contaminate with age, which increases their sensitivity and instability
- Sensitivity to shock and friction

Compatibility Group Letter will vary

EXPLOSIVES 1.1A 1	**Division 1.1** — Explosives that have a mass explosion hazard. A mass explosion is one that affects almost the entire load instantaneously. *Examples:* dynamite, mines, wetted mercury fulminate
EXPLOSIVES 1.2B 1	**Division 1.2** — Explosives that have a projection hazard but not a mass explosion hazard. *Examples:* detonation cord, rockets (with bursting charge), flares, fireworks
EXPLOSIVES 1.3L 1	**Division 1.3** — Explosives that have a fire hazard and either a minor blast hazard or a minor projection hazard or both, but not a mass explosion hazard. *Examples:* liquid-fueled rocket motors, smokeless powder, practice grenades, aerial flares
1.4 EXPLOSIVES B 1	**Division 1.4** — Explosives that present a minor explosion hazard. The explosive effects are largely confined to the package and no projection of fragments of appreciable size or range is expected. An external fire must not cause virtually instantaneous explosion of almost the entire contents of the package. *Examples:* signal cartridges, cap type primers, igniter fuses, fireworks
1.5 BLASTING AGENT D 1	**Division 1.5** — Substances that have a mass explosion hazard but are so insensitive that there is very little probability of initiation or of transition from burning to detonation under normal conditions of transport. **Examples:** prilled ammonium nitrate fertilizer/fuel oil (ANFO) mixtures and blasting agents
1.6 EXPLOSIVES N 1	**Division 1.6** — Extremely insensitive articles that do not have a mass explosive hazard. This division is comprised of articles that contain only extremely insensitive detonating substances and that demonstrate a negligible probability of accidental initiation or propagation.

Continued

Table 2.10 (continued)

Class 2: Gases (49 *CFR* 173.115)

DOT defines *gas* as a material that has a vapor pressure greater than 43.5 psi (300 kPa) at 122°F (50°C) or is completely gaseous at 68°F (20°C) at a standard pressure of 14.7 psi (101.3 kPa).

NOTE: The DOT definition for gas is much more specific than the definition provided in Chapter 2, Hazardous Materials Properties and Hazards.

The potential hazards of gases may include thermal, asphyxiation, chemical, and mechanical hazards:

- Thermal hazards (heat) from fires, particularly associated with Division 2.1 and oxygen
- Thermal hazards (cold) associated with exposure to cryogens in Division 2.2
- Asphyxiation caused by leaking/released gases displacing oxygen in a confined space
- Chemical hazards from toxic and/or corrosive gases and vapors, particularly associated with Division 2.3
- Mechanical hazards from a boiling liquid expanding vapor explosion (BLEVE) for containers exposed to heat or flame
- Mechanical hazards from a ruptured cylinder rocketing after exposure to heat or flame

FLAMMABLE GAS 2	**Division 2.1: Flammable Gas** — Consists of any material that is a gas at 68°F (20°C) or less at normal atmospheric pressure or a material that has a boiling point of 68°F (20°C) or less at normal atmospheric pressure and that (1) Is ignitable at normal atmospheric pressure when in a mixture of 13 percent or less by volume with air, or (2) Has a flammable range at normal atmospheric pressure with air of at least 12 percent, regardless of the lower limit. *Examples:* compressed hydrogen, isobutene, methane, and propane
NON-FLAMMABLE GAS 2	**Division 2.2: Nonflammable, Nonpoisonous Gas** — Nonflammable, nonpoisonous compressed gas, including compressed gas, liquefied gas, pressurized cryogenic gas, and compressed gas in solution, asphyxiant gas and oxidizing gas; means any material (or mixture) which exerts in the packaging an absolute pressure of 40.6 psi (280 kPa) or greater at 68ºF (20ºC) and does not meet the definition of Divisions 2.1 or 2.3. *Examples:* carbon dioxide, helium, compressed neon, refrigerated liquid nitrogen, cryogenic argon, anhydrous ammonia, pepper spray
INHALATION HAZARD 2	**Division 2.3: Gas Poisonous by Inhalation** — Material that is a gas at 68°F (20°C) or less and a pressure of 14.7 psi (101.3 kPa) (a material that has a boiling point of 68°F [20°C] or less at 14.7 psi [101.3 kPa]), and that is known to be so toxic to humans as to pose a hazard to health during transportation; or (in the absence of adequate data on human toxicity) is presumed to be toxic to humans because of specific test criteria on laboratory animals. Division 2.3 has *ERG*-designated hazard zones associated with it, determined by the concentration of gas in the air: • Hazard Zone A — LC_{50} less than or equal to 200 ppm • Hazard Zone B — LC_{50} greater than 200 ppm and less than or equal to 1,000 ppm • Hazard Zone C — LC_{50} greater than 1,000 ppm and less than or equal to 3,000 ppm • Hazard Zone D — LC_{50} greater than 3,000 ppm and less than or equal to 5,000 ppm *Examples:* cyanide, diphosgene, germane, phosphine, selenium hexafluoride, and hydrocyanic acid, arsine, cyanogen chloride, phosgene, chlorine
OXYGEN 2	**Oxygen Placard** — Oxygen is not a separate division under Class 2, but first responders may see this oxygen placard on containers with 1,001 lbs (454 kg) or more gross weight of either compressed gas or refrigerated liquid.

Continued

Table 2.10 (continued)

Class 3: Flammable and Combustible Liquids (49 *CFR* 173.120)

A *flammable liquid* is generally a liquid having a flash point of not more than 140ºF (60ºC), or any material in a liquid state with a flash point at or above 100ºF (37.8ºC) that is intentionally heated and offered for transportation or transported at or above its flash point in a bulk packaging.

A *combustible liquid* is any liquid that does not meet the definition of any other hazard class and has a flash point above 141°F (60.5°C) and below 200 °F (93 °C). A flammable liquid with a flash point at or above 100°F (38°C) that does not meet the definition of any other hazard class may be reclassified as a combustible liquid. This provision does not apply to transportation by vessel or aircraft, except where other means of transportation is impracticable. An elevated temperature material that meets the definition of a Class 3 material because it is intentionally heated and offered for transportation or transported at or above its flash point may not be reclassified as a combustible liquid.

The primary hazards of flammable and combustible liquids are thermal, asphyxiation, chemical, and mechanical, and may include the following:

- Thermal hazards (heat) from fires and vapor explosions
- Asphyxiation from heavier than air vapors displacing oxygen in low-lying, and/or confined spaces
- Chemical hazards from toxic and/or corrosive gases and vapors
- Chemical hazards from the production of toxic and/or corrosive gases and vapors during fires
- Mechanical hazards from a BLEVE, for containers exposed to heat or flame
- Mechanical hazards caused by a vapor explosion
- Vapors that can mix with air and travel great distances to an ignition source
- Environmental hazards (pollution) caused by runoff from fire control

When responding to a transportation incident, first responders must keep in mind that a flammable liquid placard can indicate a product with a flash point as high as 140°F (60°C).

	Flammable Placard *Examples:* gasoline, methyl ethyl ketone
	Gasoline Placard — May be used in the place of a flammable placard on a cargo tank or a portable tank being used to transport gasoline by highway
	Combustible Placard *Examples:* diesel, fuel oils, pine oil
	Fuel Oil Placard — May be used in place of a combustible placard on a cargo tank or portable tank being used to transport fuel oil by highway.

Class 4: Flammable Solids, Spontaneously Combustible Materials, and Dangerous-When-Wet Materials (49 *CFR* 173.124)

This class is divided into three divisions: 4.1 Flammable Solids, 4.2 Spontaneously Combustible Materials, and 4.3 Dangerous When Wet (see definitions below).

First responders must be aware that fires involving Class 4 materials may be extremely difficult to extinguish. The primary hazards of Class 4 materials are thermal, chemical, and mechanical and may also include the following hazards:

Continued

Table 2.10 (continued)

Class 4 (continued)

- Thermal hazards (heat) from fires that may start spontaneously or upon contact with air or water
- Thermal hazards (heat) from fires and vapor explosions
- Thermal hazards (heat) from molten substances
- Chemical hazards from irritating, corrosive, and/or highly toxic gases and vapors produced by fire or decomposition
- Severe chemical burns
- Mechanical effects from unexpected, violent chemical reactions and explosions
- Mechanical hazards from a BLEVE, for containers exposed to heat or flame (or if contaminated with water, particularly for Division 4.3)
- Production of hydrogen gas from contact with metal
- Production of corrosive solutions on contact with water, for Division 4.3
- May spontaneously reignite after fire is extinguished
- Environmental hazards (pollution) caused by runoff from fire control

FLAMMABLE SOLID 4	**Division 4.1: Flammable Solid Material** — Includes (1) wetted explosives, (2) self-reactive materials that can undergo a strongly exothermal decomposition, and (3) readily combustible solids that may cause a fire through friction, certain metal powders that can be ignited and react over the whole length of a sample in 10 minutes or less, or readily combustible solids that burn faster than 2.2 mm/second: • **Wetted explosives:** Explosives with their explosive properties suppressed by wetting with sufficient alcohol, plasticizers, or water • **Self-reactive materials:** Materials liable to undergo a strong exothermic decomposition at normal or elevated temperatures due to excessively high transport temperatures or to contamination • **Readily combustible solids:** Solids that may ignite through friction or any metal powders that can be ignited *Examples:* phosphorus heptasulfide, paraformaldehyde, magnesium
SPONTANEOUSLY COMBUSTIBLE 4	**Division 4.2: Spontaneous Combustible Material** — Includes (1) a pyrophoric material (liquid or solid) that, without an external ignition source, can ignite within 5 minutes after coming in contact with air and (2) a self-heating material that, when in contact with air and without an energy supply, is liable to self-heat *Examples:* sodium sulfide, potassium sulfide, phosphorus (white or yellow, dry), aluminum and magnesium alkyls, charcoal briquettes
DANGEROUS WHEN WET 4	**Division 4.3: Dangerous-When-Wet Material** — Material that, by contact with water, is liable to become spontaneously flammable or to release flammable or toxic gas at a rate greater than 1 liter per kilogram of the material per hour *Examples:* magnesium powder, lithium, ethyldichlorosilane, calcium carbide, potassium

Class 5: Oxidizers and Organic Peroxides (49 *CFR* 173.127 and 128)

This class is divided into two divisions: 5.1 Oxidizers and 5.2 Organic Peroxides (see definitions below). Oxygen supports combustion, so the primary hazards of Class 5 materials are fires and explosions with their associated thermal and mechanical hazards:

- Thermal hazards (heat) from fires that may explode or burn extremely hot and fast
- Explosive reactions to contact with hydrocarbons (fuels)
- Chemical hazards from toxic gases, vapors, and dust
- Chemicals hazards from toxic products of combustion
- Chemical burns
- Ignition of combustibles (including paper, cloth, wood, etc.)

Continued

Table 2.10 (continued)

Class 5: (continued)

- Mechanical hazards from violent reactions and explosions
- Accumulation of toxic fumes and dusts in confined spaces
- Sensitivity to heat, friction, shock, and/or contamination with other materials

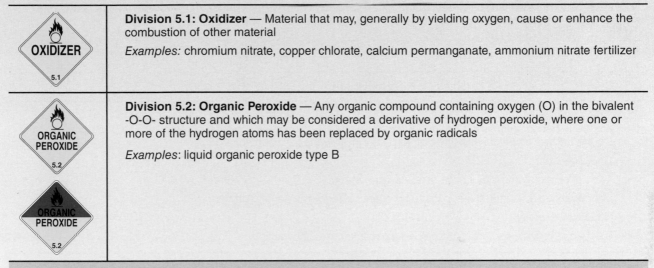

OXIDIZER 5.1	**Division 5.1: Oxidizer** — Material that may, generally by yielding oxygen, cause or enhance the combustion of other material *Examples:* chromium nitrate, copper chlorate, calcium permanganate, ammonium nitrate fertilizer
ORGANIC PEROXIDE 5.2 / ORGANIC PEROXIDE 5.2	**Division 5.2: Organic Peroxide** — Any organic compound containing oxygen (O) in the bivalent -O-O- structure and which may be considered a derivative of hydrogen peroxide, where one or more of the hydrogen atoms has been replaced by organic radicals *Examples*: liquid organic peroxide type B

Class 6: Poison (Toxic) and Poison Inhalation Hazard (49 *CFR* 173.132 and 134)

A poisonous material is a material, other than a gas, that is known to be toxic to humans. The primary hazards of Class 6 materials are chemical and thermal and may include the following:

- Toxic effects due to exposure via all routes of entry
- Chemicals hazards from toxic and/or corrosive products of combustion
- Thermal effects (heat) from substances transported in molten form
- Flammability and its associated thermal hazards (heat) from fires

POISON 6	**Division 6.1: Poisonous Material** — Material, other than a gas, that is known to be so toxic to humans as to afford a hazard to health during transportation or that is presumed to be toxic to humans based on toxicity tests on laboratory animals *Examples:* aniline, arsenic, mustard agents, nerve agents, hydrogen cyanide, most riot control agents
No Placard for Division 6.2, see labels	**Division 6.2: Infectious Substance** — Material known to contain or suspected of containing a pathogen. A pathogen is a virus or microorganism (including its viruses, plasmids, or other genetic elements, if any) or a proteinaceous infectious particle (prion) that has the potential to cause disease in humans or animals.
INHALATION HAZARD 6	**Inhalation Hazard Placard** — Used for any quantity of Division 6.1, Zones A or B inhalation hazard only (see Division 2.3 for hazard zones)
PG III 6	**PG III** — For Division 6.1, packing group III* (PG III) materials, a POISON placard may be modified to display the text "PG III" below the mid line of the placard rather than the word "POISON" **A packing group is a DOT packaging category based on the degree of danger presented by the hazardous material. Packing Group I indicates great danger; Packing Group II, medium danger; and Packing Group III, minor danger. The PG III placard, then, might be used for materials that are not as dangerous as those that would be placarded with the "POISON" placard*

Continued

Table 2.10 (concluded)

Class 7: Radioactive Materials (49 *CFR* 173.403)

A radioactive material means any material having a specific activity greater than 70 becquerels per gram (0.002 microcurie per gram). The primary hazard of Class 7 materials is radiological, including burns and biological effects.

Radioactive Placard — Is required on certain shipments of radioactive materials; vehicles with this placard are carrying "highway route controlled quantities" of radioactive materials and must follow prescribed, predetermined transportation routes

Examples: solid thorium nitrate, uranium hexafluoride

Class 8: Corrosive Materials (49 *CFR* 173.136)

A corrosive material means a liquid or solid that causes full thickness destruction of human skin at the site of contact within a specific period of time or a liquid that has a severe corrosion rate on steel or aluminum. The primary hazards of Class 8 materials are chemical and thermal, and may include the following hazards:

- Chemical burns
- Toxic effects due to exposure via all routes of entry
- Thermal effects, including fire, caused by chemical reactions generating heat
- Reactivity to water
- Mechanical effects caused by BLEVEs and violent chemical reactions

Corrosive Placard

Examples: battery fluid, chromic acid solution, soda lime, sulfuric acid, hydrochloric acid (muriatic acid), sodium hydroxide, potassium hydroxide

Class 9: Miscellaneous Dangerous Goods (49 *CFR* 173.140)

A miscellaneous dangerous good is a material that (1) has an anesthetic, noxious, or other similar property that could cause extreme annoyance or discomfort to flight crew members and would prevent their correct performance of assigned duties; (2) is a hazardous substance or a hazardous waste; or (3) is an elevated temperature material; or (4) is a marine pollutant.

Miscellaneous dangerous goods will primarily have thermal and chemical hazards. For example, polychlorinated biphenyls (PCBs) are carcinogenic, while elevated temperature materials may present some thermal hazards. However, hazardous wastes may present any of the hazards associated with the materials in normal use.

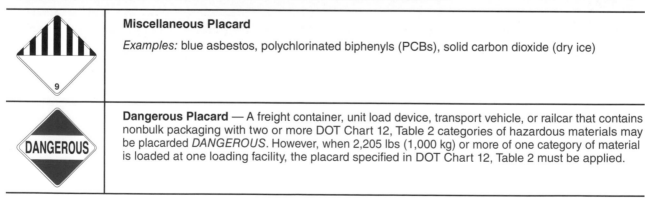

Miscellaneous Placard

Examples: blue asbestos, polychlorinated biphenyls (PCBs), solid carbon dioxide (dry ice)

Dangerous Placard — A freight container, unit load device, transport vehicle, or railcar that contains nonbulk packaging with two or more DOT Chart 12, Table 2 categories of hazardous materials may be placarded *DANGEROUS*. However, when 2,205 lbs (1,000 kg) or more of one category of material is loaded at one loading facility, the placard specified in DOT Chart 12, Table 2 must be applied.

ⓘ

Other Regulated Materials (ORM-Ds) and Materials of Trade (MOTs)

ORM-Ds are consumer commodities that present a limited hazard during transportation due to their form, quantity, and packaging. No placards are required for ORM-Ds, but they are otherwise subject to the requirements of the Hazardous Materials Regulations (HMR). *Examples:* consumer commodities, small arms cartridges.

A *MOT* is a hazardous material, other than a hazardous waste, that is carried on a motor vehicle for the purposes listed. Placards, shipping papers, emergency response information, and formal record keeping and training are not required for them. MOT purposes:

● To protect the health and safety of motor vehicle operators or passengers. Examples: insect repellant, fire extinguishers, and self-contained breathing apparatus (SCBA)

● To support the operation or maintenance of motor vehicles, including its auxiliary equipment. Examples: spare batteries, gasoline, and engine starting fluid

● To directly support principal businesses (by private motor carriers) that are not transportation. Examples: lawn care, pest control, plumbing, welding, painting, and door-to-door sales

Many ORM-Ds (such as hairspray) may qualify as MOTs. However, self-reactive materials, poison inhalation hazard materials, and hazardous wastes are *never* eligible to qualify as MOTs.

DOT Labels

Basically, DOT-required labels provide the same information as vehicle placards **(Figure 2.53).** Labels on packages communicate the hazards posed by the material in the event the package spills from the transport vehicle. Labels are 3.9-inch (100 mm), square-on-point diamonds, which may or may not have written text that identifies the hazardous material within the packaging. Class 7 Radioactive labels must always contain text.

Figure 2.53 DOT-required labels provide the same basic information as vehicle placards, but they are used on nonbulk packaging such as drums, boxes, bags, and other small containers. *Courtesy of Rich Mahaney.*

First responders must be familiar with the pictogram and hazard class or division number for the material. Packaging will contain a primary label and a subsidiary label for materials that meet the definition of more than one hazard class. In **Figure 2.54,** the toxic label is the primary label, while the flammable liquid label is the subsidiary. DOT regulations require that subsidiary labels have the class number displayed. The regulations governing the use of labels are contained in Title 49 *CFR.*

Figure 2.54 The toxic label is the primary label (higher and to the left), while the flammable liquid label is the subsidiary. *Courtesy of Rich Mahaney.*

Table 2.11 provides examples of the unique DOT labels. Other labels for the nine hazard classes and subdivisions are essentially the same as the placards shown earlier in Table 2.10.

DOT Markings

By the DOT definition, a *marking* is a descriptive name, an identification number, a weight, or a specification and includes instructions, cautions, or UN marks (or combinations thereof) required on outer packagings of hazardous materials. This section, however, shows only those markings found on DOT Chart 13. Markings on intermodal containers, tank cars, and other packaging are discussed in later sections. **Table 2.12, p. 123** shows the DOT Chart 13 markings.

Table 2.11
Unique U.S. DOT Labels

Subsidiary Risk Labels

Subsidiary risk labels may be used for the following classes: Explosives, Flammable Gases, Flammable Liquids, Flammable Solids, Corrosives, Oxidizers, Poisons, Spontaneously Combustible Materials, and Dangerous-When-Wet Materials.

Class 1: Explosives

Explosive Subsidiary Risk Label

Class 3: Flammable Liquid

Flammable Liquid Label — Marks packages containing flammable liquids.

Examples: gasoline, methyl ethyl ketone

Class 6: Poison (Toxic), Poison Inhalation Hazard, Infectious Substance

Infectious Substances Label — Marks packages with infectious substances (viable micro-organism, or its toxin, which causes or may cause disease in humans or animals).

This label may be used to mark packages of Class 6.2 materials as defined in 49 *CFR* 172.432.

Examples: anthrax, hepatitis B virus, *escherichia* coli (e coli)

Biohazard Label — Marks bulk packaging containing a regulated medical waste as defined in 49 *CFR* 173.134(a)(5).

Examples: used needles/syringes, human blood or blood products, human tissue or anatomical waste, carcasses of animals intentionally infected with human pathogens for medical research

Etiological Agents Label — Marks packages containing etiologic agents transported in interstate traffic per 42 *CFR* 72.3 and 72.6

Examples: rabies virus, rickettsia, Ebola virus, salmonella bacteria

Continued

Table 2.11 (concluded)

Class 7: Radioactive Materials

Packages of radioactive materials must be labeled on two opposite sides, with a distinctive warning label. Each of the three label categories — RADIOACTIVE WHITE-I, RADIOACTIVE YELLOW-II, or RADIOACTIVE YELLOW-III — bears the unique trefoil symbol for radiation.

Class 7 Radioactive I, II, and III labels must always contain the following additional information:

- Isotope name
- Radioactive activity

Radioactive II and III labels will also provide the transport Index (TI) indicating the degree of control to be excercised by the carrier during transportation. The number in the transport index box indicates the maximum radiation level measured (in mrem/hr) at one meter from the surface of the package. Packages with the Radioactive I label have a Transport Index of 0.

	Radioactive I Label — Label with an all-white background color that indicates that the external radiation level is low and no special stowage controls or handling are required.
	Radioactive II Label — Upper half of the label is yellow, which indicates that the package has an external radiation level or fissile (nuclear safety criticality) characteristic that requires consideration during stowage in transportation.
	Radioactive III Label — Yellow label with three red stripes indicates the transport vehicle must be placarded RADIOACTIVE.
	Fissile Label — Used on containers of fissile materials (materials capable of undergoing fission such as uranium-233, uranium-235, and plutonium-239). The Criticality Safety Index (CSI) must be listed on this label. The CSI is used to provide control over the accumulation of packages, overpacks, or freight containers containing fissile material.
	Empty Label — Used on containers that have been emptied of their radioactive materials, but still contain residual radioactivity.

Aircraft Labels

	Danger - Cargo Aircraft Only — Used to indicate materials that cannot be transported on passenger aircraft.

Table 2.12
Unique U.S. DOT Markings

Marking	Description
HOT	**Hot Marking** — Has the same dimensions as a placard and is used on elevated temperature materials. *Note:* Bulk containers of molten aluminum or molten sulfur must be marked MOLTEN ALUMINUM or MOLTEN SULFUR, respectively.
(Marine pollutant symbol)	**Marine Pollutant Marking** — Must be displayed on packages of substances designated as marine pollutants. *Examples:* cadmium compounds, copper cyanide, mercury based pesticides
INHALATION HAZARD	**Inhalation Hazard Marking** — Used to mark materials that are poisonous by inhalation. *Examples:* anhydrous ammonia, methyl bromide, hydrogen cyanide, hydrogen sulfide
DANGER ☠ THIS UNIT IS UNDER FUMIGATION WITH _____ APPLIED ON Date Time Ventilated on Ventilated by DO NOT ENTER	**Fumigant Marking** — Warning affixed on or near each door of a transport vehicle, freight container, or railcar in which the lading has been fumigated or is undergoing fumigation with any material. The vehicle, container, or railcar is considered a package containing a hazardous material unless it has been sufficiently aerated so that it does not pose a risk to health and safety.
(Orientation arrows)	**Orientation Markings** — Markings used to designate the orientation of the package. Sometimes these markings will be accompanied by words such as "this side up."
CONSUMER COMMODITY ORM-D	**ORM-D** — Used on packages of ORM-D materials. *Examples:* consumer commodities, small arms cartridges
CONSUMER COMMODITY ORM-D-AIR	**ORM-D-AIR** — Used on packages of ORM-D materials shipped via air.
OVERPACK	**Inner Packaging** — Used on authorized packages containing hazardous materials being transported in an overpack as defined in 49 *CFR* 171.8 and 49 *CFR* 173.25 (a) (4).
(Keep away from heat symbol) keep away from heat	**Keep Away From Heat Marking** — Used for aircraft transportation of packages containing self-reactive substances of Division 4.1 or organic peroxides of Division 5.2.
UN3373	**Category B Biological Substances** — Diagnostic and clinical specimens that do not cause permanent disability or life threatening or fatal disease to humans or animals when exposure occurs.
(Excepted quantity symbol)	**Excepted Quantity** — Excepted quantities of hazardous materials. The "*" must be replaced by the primary hazard class, or when assigned, the division of each of the hazardous materials contained in the package. The "**" must be replaced by the name of the shipper or consignee if not shown elsewhere on the package.
(IBC stacking symbols) ...kg max	**Marking of IBCs** — For IBCs not designed for stacking, the figure "0" and the symbol for IBCs not capable of being stacked must be displayed. For IBCs designed for stacking, the maximum permitted stacking load applicable (in kilograms) when the IBC is in use must be included with the symbol for IBCs capable of being stacked.

Canadian Transportation Placards, Labels, and Markings

Transport Canada (TC) and the Dangerous Goods Act govern transportation placards, labels, and markings in Canada. Like the U.S. HMR, the Dangerous Goods Act is based on the *UN Recommendations* and, therefore, is very similar. The nine hazard classes are identical. **Table 2.13** provides Canadian placards, labels, and markings divided by class. There are some differences, however, between Canadian and U.S. placards, labels, and markings such as the following:

• Most Canadian transport placards do not have any signal words written on them.

• Labels and markings may be in both English and French.

• Canada requires a unique placard for anhydrous ammonia.

Mexican Transportation Placards, Labels, and Markings

Like Canada and the U.S., Mexican transportation placards, labels, and markings are based on the *UN Recommendations* and have the same hazard classes and subdivisions. In fact, Canadian and Mexican placards and labels are virtually the same. However, because international regulations authorize the insertion of text (other than the class or division number) in the space below the symbol as long as the text relates to the nature of the hazard or precautions to be taken in handling, placards and labels in Mexico may have text that is in Spanish **(Figure 2.55).** Likewise, information provided on markings is likely to be written in Spanish. English-speaking first responders in Mexico or along the U.S./Mexican border should familiarize themselves with the more common Spanish hazard warning terms such as *peligro* (danger).

Figure 2.55 Placards and labels in Mexico may have text that is written in Spanish. English-speaking responders should still be able to recognize symbols, shapes, and colors that provide them with information about the hazards associated with the contents of the package or container.

Table 2.13
Canadian Transportation Placards, Labels, and Markings

Class 1: Explosives

Placard and Label	**Class 1.1** — Mass explosion hazard
Placard and Label	**Class 1.2** — Projection hazard but not a mass explosion hazard
	Class 1.3 — Fire hazard and either a minor blast hazard or a minor projection hazard or both but not a mass explosion hazard
1.4 **Placard and Label**	**Class 1.4** — No significant hazard beyond the package in the event of ignition or initiation during transport * = Compatibility group letter
1.5 **Placard and Label**	**Class 1.5** — Very insensitive substances with a mass explosion hazard
1.6 **Placard and Label**	**Class 1.6** — Extremely insensitive articles with no mass explosion hazard

Class 2: Gases

Placard and Label	**Class 2.1 — Flammable Gases**
Placard and Label	**Class 2.2 — Nonflammable and nontoxic Gases**

Continued

Table 2.13 (continued)

Class 2: Gases (continued)

	Class 2.3 — Toxic Gases
1005 2 **Placard and Label**	**Anhydrous Ammonia**
2 **Placard and Label**	**Oxidizing Gases**

Class 3: Flammable Liquids

3 **Placard and Label**	**Class 3 — Flammable Liquids**

Class 4: Flammable Solids, Substances Liable to Spontaneous Combustion, and Substances that on Contact with Water Emit Flammable Gases (Water-Reative Substances)

Placard and Label	**Class 4.1 — Flammable Solids**
4 **Placard and Label**	**Class 4.2 — Substances Liable to Spontaneous Combustion**
4 **Placard and Label**	**Class 4.3 — Water-Reactive Substances**

Class 5: Oxidizing Substances and Organic Peroxides

5.1 **Placard and Label**	**Class 5.1 — Oxidizing Substances**

Continued

Table 2.13 (continued)

Class 5: Oxidizing Substances and Organic Peroxides (continued)

Placard and Label	**Class 5.2 — Organic Peroxides**

Class 6: Toxic and Infectious Substances

Placard and Label	**Class 6.1 — Toxic Substances**
Label Only	**Class 6.2 — Infectious Substances** Text: INFECTIOUS In case of damage or leakage, Immediately notify local authorities AND INFECTIEUX En cas de Dommage ou de fuite communiquer Immédiatement avec les autorités locales ET CANUTEC 613-996-6666
Placard Only	**Class 6.2 — Infectious Substances**

Class 7: Radioactive Materials

Label and Optional Placard	**Class 7 — Radioactive Materials** **Category I** — White RADIOACTIVE CONTENTS......................CONTENU ACTIVITY.......................... ACTIVITÉ
Label and Optional Placard	**Class 7 — Radioactive Materials** **Category II** — Yellow RADIOACTIVE CONTENTS......................CONTENU ACTIVITY..........................ACTIVITÉ INDICE DE TRANSPORT INDEX

Continued

Table 2.13 (concluded)

Class 7: Radioactive Materials (continued)

Label and Optional Placard	**Class 7 — Radioactive Materials** **Category III** — Yellow RADIOACTIVE CONTENTS......................CONTENU ACTIVITY.........................ACTIVITÉ INDICE DE TRANSPORT INDEX
Placard	**Class 7 — Radioactive Materials** The word RADIOACTIVE is optional.

Class 8: Corrosives

Placard and Label	**Class 8 — Corrosives**

Class 9: Miscellaneous Products, Substances, or Organisms

Placard and Label	**Class 9 — Miscellaneous Products, Substances, or Organisms**

Other Placards, Labels, and Markings

	Danger Placard
	Elevated Temperature Sign
	Fumigation Sign Text is in both English and French
	Marine Pollutant Mark The text is MARINE POLLUTANT or POLLUANT MARIN.

Some differences between the Mexican transportation regulations and the U.S. HMR are as follows:

- The official Mexican standards do not authorize the use of the *DANGEROUS* placard since NOM-004 does not include provisions for its use. However, *PELIGROSO (DANGEROUS)* placards may still be seen in Mexico.

- Package markings are consistent except that the proper shipping name is provided in Spanish in addition to English. NOM-002-SCT2/1994 provides the official Mexican proper shipping names.

- The *HOT* mark used for elevated temperature materials in the U.S. is not authorized in Mexico. In Mexico, the elevated temperature mark provided in the *UN Recommendations* must be used, which is the same as the Canadian elevated temperature mark.

- The Mexican regulations do not require the marine pollutants mark for surface transportation.

- The Mexican standards incorporate provisions for consumer commodities but do not authorize the use of the ORM-D description as a package marking.

- The Mexican standard regarding the classification of flammable liquids does not incorporate provisions for combustible liquids. Combustible liquid requirements only apply in the U.S.

- Like Canada, subsidiary placards and labels in Mexico will not have the class number in the bottom corner.

Other North American Highway Vehicle Identification Markings

In addition to 4-digit identification numbers, highway transportation vehicles may have other identification markings. These markings may include company names, logos, specific tank colors (for certain tanks), stenciled commodity names (such as *Liquefied Petroleum Gas – LPG*), and manufacturers' specification plates. Specification plates provide information about the standards to which the container/tank was built **(Figure 2.56).** These plates are usually found on the roadside/driver's side of the vehicle.

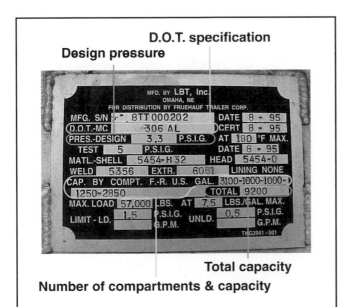

Figure 2.56
Manufacturers' specification plates provide a wealth of information about the standards to which the container/tank was built including pressure, capacity, number of compartments, and DOT specification. *Courtesy of Rich Mahaney.*

North American Railroad Tank Car Markings

There may be a variety of markings on railroad tank cars that responders can use to gain valuable information about the tank and its contents such as the following:

- Initials (reporting marks) and number
- Capacity stencil
- Specification marking

The *ERG* provides a key to these markings in the railcar identification chart, and more information is provided in sections that follow. Additionally, manufacturers' names on cars may provide some contact information. Dedicated railcars transporting a single material should have the name of that material painted on the car. Likewise, more than 250 hazardous materials transported by rail are required to have their names stenciled on the sides of the car in 4-inch (102 mm) letters, primarily shipped in high pressure tank cars.

Reporting Marks

Reporting marks (also called *initials and numbers*) may be used to get information about the car's contents from the railroad's computer or the shipper. Tank cars (like all other freight cars) are marked with their own unique sets of reporting marks. These reporting marks should match the initials and numbers provided on the shipping papers for the car. They are stenciled on both sides (to the left when facing the side of the car) and both ends (upper center) of the tank car tank **(Figure 2.57a and b).** Some shippers and car owners also stencil the top of the car with the car's initials and numbers to help identify the car in case an accident turns it on its side.

Reporting Marks —
Combination of letters and numbers stenciled on rail tank cars that may be used to get information about the car's contents from the railroad's computer or the shipper. Also called Initials and Numbers.

Figure 2.57a and b
The reporting marks are highlighted on these tank cars. Reporting marks can be used to identify the specific car in order to get information about the car's contents from the shipper or shipping papers. *Courtesy of Rich Mahaney.*

Capacity Stencil

The **capacity stencil** shows the volume of the tank car tank. The volume in gallons (and sometimes liters) is stenciled on both ends of the car under the car's initials and number. The volume in pounds (and sometimes kilograms) is stenciled on the sides of the car under the car's initials and number. The term *load limit* may be used to mean the same thing as *capacity*. For certain tank cars, the water capacity (water weight) of the tank, in pounds (and typically kilograms) is stenciled on the sides of the tank near the center of the car **(Figure 2.58).**

Capacity Stencil — Number stenciled on the exterior of tank cars to indicated the volume of the tank.

Figure 2.58 The load limit (volume in pounds and kilograms) is located beneath the reporting marks. The number beneath it represents the weight of the car when empty. The capacity stencil (volume in gallons) can be seen on the end of the tank beneath the reporting marks. *Courtesy of Rich Mahaney.*

Specification Marking

The **specification marking** indicates the standards to which a tank car was built. The marking is stenciled on both sides of the tank. When facing the side of the car, the marking will be to the right (opposite from the initials and number) **(Figure 2.59).** The specification marking is also stamped into the tank heads where it is not readily visible. First responders can also get specification information from the railroad, shipper, car owner, or the Association of American Railroads by using the car's initials and number. **Figure 2.60** provides a brief explanation of tank car specification markings.

Specification Marking — Stencil on the exterior of tank cars indicating the standards to which the tank car was built; specification markings may also be found on intermodal containers and cargo tank trucks.

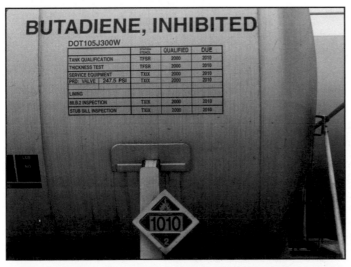

Figure 2.59 Specification markings are located on the right side of the tank car (opposite the reporting marks) and indicate the standards to which a tank car was built. *Courtesy of Rich Mahaney.*

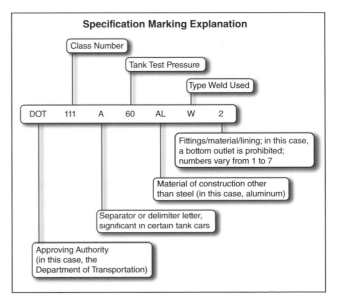

Figure 2.60 Key to tank car specification markings. First responders are most likely to need the DOT class number.

Figure 2.61 Like tank car reporting marks, intermodal reporting marks (initials and numbers) identify the specific container. *Courtesy of Rich Mahaney.*

Figure 2.62 Specification information indicates this is an IMO 1/ IM 101 container.

International Intermodal Container/Tank Markings

In addition to DOT-required placards, intermodal tanks and containers are marked with initials (reporting marks) and tank numbers **(Figure 2.61).** These markings are generally found on the right-hand side of the tank or container as the emergency responder faces it from either the sides or the ends. The markings are either on the tank/container or the frame. As with tank car reporting marks, emergency responders can use this information in conjunction with the shipping papers or computer data to identify and verify the contents of the tank or container. Marks on intermodal containers can also provide specification information **(Figure 2.62).** Intermodal containers may also have proper shipping names stenciled on them.

Clue 4: Other Markings and Colors

In addition to DOT placards, labels, and markings, a number of other markings, marking systems, labels, labeling systems, colors, color-codes, and signs may indicate the presence of hazardous materials at fixed facilities, on pipelines, on piping systems, and on other containers. These other markings may be as simple as the word *chlorine* stenciled on the outside of a fixed-facility tank or as complicated as a site-specific hazard communication system using a unique combination of labels, placards, emergency contact information, and color codes **(Figure 2.63).** Some containers may be marked with special information, for example, *non-odorized,* meaning that the product will not have a smell. Some fixed-facility containers may have identification numbers that correspond to site or emergency plans that provide details on the product, quantity, and other pertinent information.

First responders need to be familiar with some of the more widely used specialized marking systems for hazardous materials. The sections that follow highlight the most common specialized systems in North America, including the following:

- NFPA® 704

CAUTION

Read the container and understand all of the information provided!

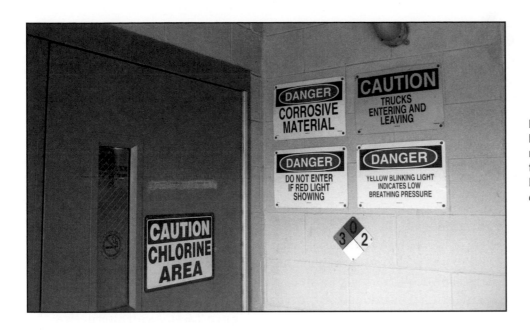

Figure 2.63 Fixed facilities may have a variety of other signs, markings, and color-codes that indicate the presence of hazardous materials. *Courtesy of Rich Mahaney.*

- Common hazardous communication labels
- International Organization for Standardization (ISO) safety symbols
- Globally harmonized system symbols
- Military markings
- Pipeline identifications
- Signal words
- CAS® numbers
- American Petroleum Institute (API) markings
- Pesticide labels
- Color codes

NFPA® 704 System

The information in NFPA® 704, *Standard System for the Identification of the Hazards of Materials for Emergency Response,* gives a widely recognized method for indicating the presence of hazardous materials at commercial, manufacturing, institutional, and other fixed-storage facilities. Use of this system is commonly required by local ordinances for all occupancies that contain hazardous materials. It is designed to alert emergency responders to health, flammability, instability, and related hazards (specifically, oxidizers and water-reactive materials) that may present as short-term, acute exposures resulting from a fire, spill, or similar emergency. The NFPA® 704 system is *not* designed for the following situations or hazards:

- Transportation
- General public use
- Nonemergency occupational exposures
- Explosives and blasting agents, including commercial explosive materials
- Chronic health hazards
- Biological agents and other similar hazards

NFPA® 704 Limitations

NFPA® 704 markings provide very useful information but the system does have its limitations. For example, an NFPA® diamond does not tell exactly what chemical or chemicals may be present in specific quantities. Nor does it tell exactly where hazardous materials may be located when the sign is used for a building, structure, or area (such as a storage yard) rather than an individual container. Positive identification of the materials needs to be made through other means such as container markings, employee information, company records, and pre-incident surveys.

Specifically, the NFPA® 704 system uses a rating system of numbers from *0* to *4*. The number *0* indicates a minimal hazard, whereas the number *4* indicates a severe hazard. The rating is assigned to three categories: health, flammability, and instability. The rating numbers are arranged on a diamond-shaped marker or sign. The health rating is located on the blue background, the flammability hazard rating is positioned on the red background, and the instability hazard rating appears on a yellow background **(Figure 2.64).** As an alternative, the backgrounds for each of these rating positions may be any contrasting color, and the numbers (*0* to *4*) may be represented by the appropriate color (blue, red, and yellow).

Figure 2.64 Key and layout of the NFPA® 704 hazard identification system.

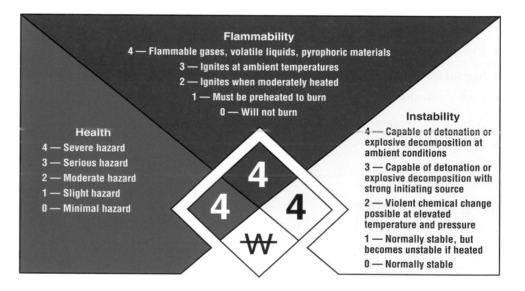

Special hazards are located in the six o'clock position and have no specified background color; however, white is most commonly used. Only two special hazard symbols are presently authorized for use in this position by the NFPA®: *W*, indicating unusual reactivity with water, and *OX*, indicating that the material is an oxidizer. However, first responders may see other symbols in the white quadrant on old placards, including the trefoil radiation symbol. If more than one special hazard is present, multiple symbols may be seen.

U.S. Hazard Communications Labels and Markings

OSHA's Hazard Communication Standard (HCS) requires employers to identify hazards in the workplace and train employees how to recognize those hazards. It also requires the employer to ensure that all hazardous material containers are labeled, tagged, or marked with the identity of the substances contained in them along with appropriate hazard warnings. The standard does not specify what system (or systems) of identification must be used, leaving that to be determined by individual employers. First responders, then, may encounter a variety of different (and sometimes unique) labeling and marking systems in their jurisdictions **(Figure 2.65).** Conducting pre-incident surveys should assist responders in identifying and understanding these systems.

Figure 2.65 (left)
The OSHA Hazard Communications Standard requires employers to identify hazards in the workplace. First responders may encounter a variety of different identification systems used by employers in their area.

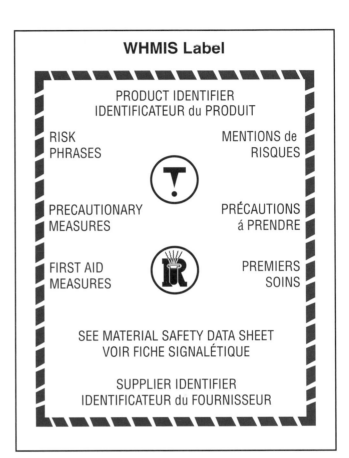

WHMIS Label

PRODUCT IDENTIFIER
IDENTIFICATEUR du PRODUIT

RISK PHRASES

MENTIONS de RISQUES

PRECAUTIONARY MEASURES

PRÉCAUTIONS á PRENDRE

FIRST AID MEASURES

PREMIERS SOINS

SEE MATERIAL SAFETY DATA SHEET
VOIR FICHE SIGNALÉTIQUE

SUPPLIER IDENTIFIER
IDENTIFICATEUR du FOURNISSEUR

Figure 2.66 (right)
An example of a blank Canadian supplier label. Completed supplier labels must be provided on all controlled products received at Canadian workplaces.

Canadian Workplace Hazardous Materials Information System

Like the U.S. HCS, the Canadian Workplace Hazardous Materials Information System (WHMIS) requires that hazardous products be appropriately labeled and marked. It also spells out requirements for **safety data sheets** (*SDSs*, also known as *material safety data sheets*, or *MSDSs*). As with the HCS, there are different ways for Canadian employers to meet the requirements of WHMIS; however, two types of labels will most commonly be used: the supplier label **(Figure 2.66)** and the workplace label. These labels will include information such as the product name, a statement that an SDS is available, and other information that will vary depending on the type of label (supplier labels will included information about the supplier). **Table 2.14, p. 136** shows the WHMIS Symbols and Hazard Classes.

Mexican Hazard Communication System

Mexico's equivalent to HCS is NOM-018-STPS-2000. It, too, requires employers to ensure that hazardous chemical substances in the workplace are appropriately and adequately labeled. Essentially, it adopts NFPA® 704 and a related hazard communication label system as the official label and marking systems.

Safety Data Sheet (SDS) —
A sixteen-section information sheet provided by a chemical product's manufacturer or importer that contains information such as the chemical composition, physical and chemical properties, health and safety hazards, emergency response procedures, and transportation and regulatory information of the specified material. The SDS is the globally harmonized version of the *material safety data sheet*, a similar form required by 29 *CFR* 1910.1200, OSHA's Hazard Communications Standard.

Table 2.14
WHMIS Symbols and Hazard Classes

Symbol	Hazard Class	Description
	Class A: **Compressed Gas**	Contents under high pressure; cylinder may explode or burst when heated, dropped, or damaged
	Class B: **Flammable and Combustible Material**	May catch fire when exposed to heat, spark, or flame; may burst into flames
	Class C: **Oxidizing Material**	May cause fire or explosion when in contact with wood, fuels, or other combustible material
	Class D, Division 1: **Poisonous and Infectious Material:** **Immediate and serious toxic effects**	Poisonous substance; a single exposure may be fatal or cause serious or permanent damage to health
	Class D, Division 2: **Poisonous and Infectious Material:** **Other toxic effects**	Poisonous substance; may cause irritation; repeated exposure may cause cancer, birth defects, or other permanent damage
	Class D, Division 3: **Poisonous and Infectious Material:** **Biohazardous infectious materials**	May cause disease or serious illness; drastic exposures may result in death
	Class E: **Corrosive Material**	Can cause burns to eyes, skin, or respiratory system
	Class F: **Dangerously Reactive Material**	May react violently, causing explosion, fire, or release of toxic gases when exposed to light, heat, vibration, or extreme temperatures

Source: WHMIS = Canadian Workplace Hazardous Materials Information System. Table adapted from Canadian Centre for Occupational Health and Safety (CCOHS) with pictograms from Health Canada.

However, employers can opt to use alternative systems so long as they comply with the objectives and purpose of the standard and are authorized by the Secretary of Labor and Social Welfare.

NOM-026-STPS-1998 ("Signs and Colors for Safety and Health") authorizes the use of some ISO safety symbols (ISO-3864, "Safety Colors and Safety Signs") on signs to communicate hazard information. General caution symbols in Mexico are triangular rather than round like those in Canada (WHMIS) or rectangular as typically found in the U.S.

Manufacturers' Labels and Signal Words

Under the HCS, chemical manufacturers and importers are required to provide appropriate labels on their product containers. Manufacturers' labels provide a variety of information to first responders, including the name of the product, manufacturer's contact information, and precautionary hazard warnings. These labels may also provide directions for use and handling, names of active ingredients, first aid instructions, and other pertinent information.

Under the Federal Hazardous Substances Act (FHSA), labels on products destined for consumer households must incorporate one of the following four **signal words** to indicate the degree of hazard associated with the product **(Figure 2.67):**

Signal Word — Government-mandated warnings provided on product labels that indicate the level of toxicity, for example CAUTION, WARNING, or DANGER.

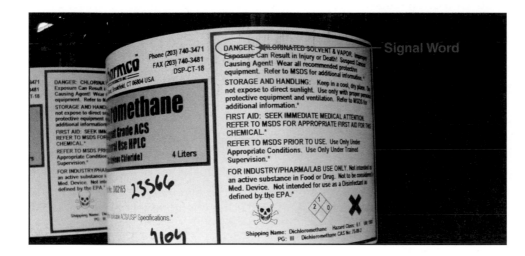

Figure 2.67 Signal words such as *CAUTION*, *WARNING*, and *DANGER* indicate the degree of hazard associated with the product. *DANGER* indicates the highest degree of hazard. An additional word, *POISON*, is required on highly toxic materials such as pesticides.

- *CAUTION* — Indicates the product may have minor health effects (such as eye or skin irritation)

- *WARNING* — Indicates the product has moderate hazards such as significant health effects or flammability

- *DANGER* — Indicates the highest degree of hazard (used on products that have potentially severe or deadly effects); also used on products that explode when exposed to heat

- *POISON* — Required in addition to DANGER on the labels of highly toxic materials such as pesticides

The FHSA also requires labels to provide other information such as the following:

- Name and business address of the manufacturer, packer, distributor, or seller
- Common or chemical name of each hazardous ingredient
- Affirmative statement of the principal hazard or hazards that the product presents, for example, *Flammable, Harmful if Swallowed, Causes Burns, Vapor Harmful*, and the like
- Precautionary statements telling users what they must do or what actions they must avoid to protect themselves
- Instructions for first-aid treatment where it is appropriate in the event the product injures someone
- Instructions for consumers to follow to protect themselves when a product requires special care in handling or storage
- Statement, *Keep out of the reach of children*

The EPA mandates that appropriate signal words (CAUTION, WARNING, DANGER, or DANGER/POISON) be provided on the labels of pesticides. Most chemical manufacturers and importers incorporate these signal words into their labels based on the American National Standard Institute's (ANSI) "Hazardous Industrial Chemicals - Precautionary Labeling" document (ANSI Z129.1-2000) regardless of whether the product is intended for consumers or not.

CAS® Numbers

Chemical Abstract Service® (CAS®, a division of the American Chemical Society) registry numbers (often called **CAS® numbers**, CAS® #s, or CAS® RNs) are unique numerical identifiers assigned to individual chemicals and chemical compounds, polymers, mixtures, and alloys. They may also be assigned to biological sequences. Over 100 million chemical substances and biological sequences have been registered. Most chemical databases are searchable by CAS® number, and they are typically included on safety data sheets (see Safety Data Sheets section) and other chemical reference materials such as the *NIOSH Pocket Guide*.

Other Symbols and Signs

Other hazard communication symbols that first responders should be able to recognize are shown in **Table 2.15.** There are many other hazard communication symbols as well. Every facility may have its own system and its own symbols, signs, and markings.

The EPA requires a warning label on any containers, transformers, or capacitors that contain polychlorinated biphenyl (PCB), which is considered hazardous because it may cause cancer. **Figure 2.68a** shows a typical U.S. PCB warning label, whereas **Figure 2.68b** shows a Canadian PCB warning label.

ISO Safety Symbols

ISO defines the design criteria for international safety signs in their standard, ISO-3864. These symbols are being used more frequently in the U.S. in conjunction with OSHA-required hazard signs (designed per ANSI Standard Z535.4,

CAS® Number — Unique number assigned to a chemical substance or biological sequence by the American Chemical Society's Chemical Abstract Service® registry; used widely to provide quick identification of the material's physical properties and safety information.

Figure 2.68a and b (a) U.S. PCB warning labels are found on containers, transformers, or capacitors that contain polychlorinated biphenyls. (b) This example is one of several different styles of Canadian PCB warning labels.

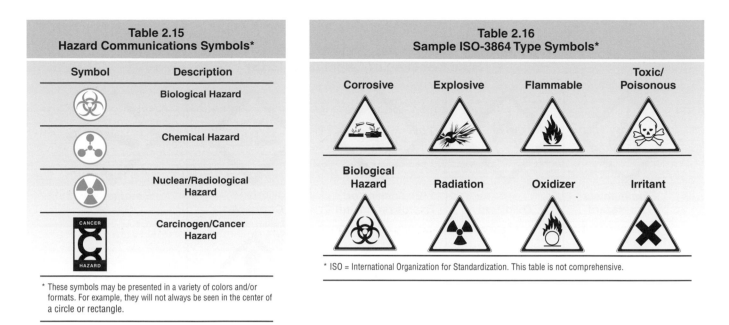

Table 2.15
Hazard Communications Symbols*

Symbol	Description
	Biological Hazard
	Chemical Hazard
	Nuclear/Radiological Hazard
	Carcinogen/Cancer Hazard

* These symbols may be presented in a variety of colors and/or formats. For example, they will not always be seen in the center of a circle or rectangle.

Table 2.16
Sample ISO-3864 Type Symbols*

Corrosive   Explosive   Flammable   Toxic/Poisonous

Biological Hazard   Radiation   Oxidizer   Irritant

* ISO = International Organization for Standardization. This table is not comprehensive.

"Product Safety Signs and Labels") as well as in Mexico, so first responders should be able to recognize the more common symbols that are used to indicate hazardous materials **(Table 2.16).**

Globally Harmonized System Symbols

The U.S. and many other countries throughout the world have developed a Globally Harmonized System of Classification and Labeling of Chemicals (GHS). The purpose of GHS is to promote common, consistent criteria for classifying chemicals according to their health, physical, and environmental hazards and encourage the use of compatible hazard labels, SDSs for employees, and other hazard communication information based on the resulting classifications.

Several key harmonized information elements of GHS are as follows:

- Uniform classification of hazardous substances and mixtures
- Hazard communications: labeling standards
  — Allocation of label elements
  — Symbols and pictograms **(Table 2.17, p. 140)**
  — Signal words: *DANGER* (most severe hazard categories) and *WARNING* (less severe hazard categories)
  — Hazard statements
  — Precautionary statements and pictograms
  — Product and supplier identification
  — Multiple hazards and precedence of information
  — Arrangements for presenting GHS label elements
  — Special labeling arrangements
- Hazard communications: safety data sheet (SDS) content and format

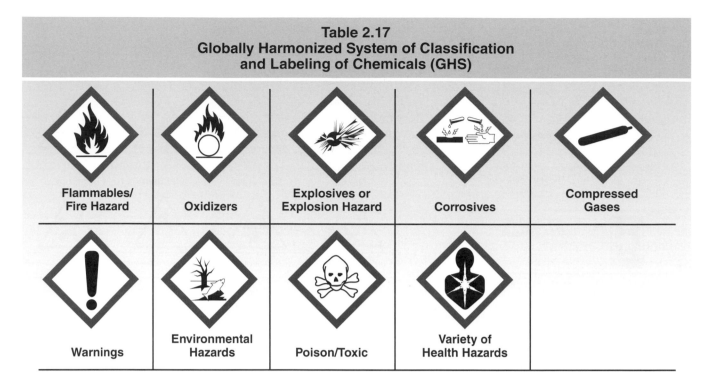

Table 2.17
Globally Harmonized System of Classification
and Labeling of Chemicals (GHS)

Flammables/ Fire Hazard	Oxidizers	Explosives or Explosion Hazard	Corrosives	Compressed Gases
Warnings	Environmental Hazards	Poison/Toxic	Variety of Health Hazards	

Figure 2.69 Unique military markings can be found on fixed facilities and vehicles, but the military placard system is not uniform, and secure locations that contain hazardous materials may not be marked. *Courtesy of Rich Mahaney.*

CAUTION

The military ships some hazardous materials and chemicals by common carrier. When this is done they are not required to be marked with DOT and TC transportation markings.

Military Markings

The U.S. and Canadian military services have their own marking systems for hazardous materials and chemicals in addition to DOT and TC transportation markings **(Figure 2.69).** These markings are used on fixed facilities, and they may be seen on military vehicles, although they are not required. Caution must be exercised, however, because the military placard system is not necessarily uniform. For security reasons, some buildings and areas that store hazardous materials may not be marked. **Table 2.18, p. 142** provides the U.S. and Canadian military markings for explosive ordnance and fire hazards, chemical hazards, and PPE requirements.

Pipeline Identification

Many types of materials, particularly petroleum varieties, are transported across both the U.S. and Canada in an extensive network of pipelines, most of which are buried in the ground. The U.S. DOT Pipeline and Hazardous Materials

Safety Administration (PHMSA) regulates pipelines that carry hazardous materials across state borders, navigable waterways, and federal lands in the U.S. In Canada, the Canadian National Energy Board regulates oil and natural gas pipelines.

Where pipelines cross under (or over) roads, railroads, and waterways, pipeline companies must provide markers. They must be in sufficient numbers along the rest of the pipeline to identify the pipe's location. However, first responders should be aware that pipeline markers do not always mark the exact location of the pipeline, and they should not assume that the pipeline runs in a perfectly straight line between markers. Pipeline markers in the U.S. and Canada include the signal words *CAUTION, WARNING,* or *DANGER* (representing an increasing level of hazard) and contain information describing the transported commodity and the name and emergency telephone number of the carrier **(Figure 2.70).**

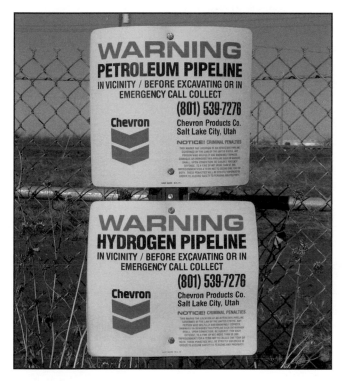

Figure 2.70 Pipeline markers in the U.S. and Canada include signal words, information describing the transported commodity, and the name and emergency telephone number of the carrier. *Courtesy of Rich Mahaney.*

Pesticide Labels

The EPA regulates the manufacture and labeling of pesticides. Each EPA label must contain the manufacturer's name for the pesticide and one of the following signal words: *DANGER/POISON, WARNING,* or *CAUTION.* The words *DANGER/POISON* are used for highly toxic materials, *WARNING* means moderate toxicity, and *CAUTION* is used for chemicals with relatively low toxicity **(Figure 2.71, p. 143).** The words *EXTREMELY FLAMMABLE* are also displayed on the label if the contents have a flash point below 80°F (27°C).

The label also lists an EPA registration number. This number normally is used to obtain information about the product from the manufacturer's 24-hour emergency contact. Another requirement is an establishment number that identifies the manufacturing facility. Other information that may be found on these labels includes routes of entry into the body, precautionary statements (such as *Keep out of the reach of children*), active ingredients, requirements for

Table 2.18
U.S. and Canadian Military Symbols

Symbol	Fire (Ordnance) Divisions
1	**Division 1: Mass Explosion** Fire Division 1 indicates the greatest hazard. This division is equivalent to DOT/UN Class 1.1 Explosives Division **Also, this exact symbol may be used for:** **Division 5: Mass Explosion — very insensitive explosives (blasting agents)** This division is equivalent to DOT/UN Class 1.5 Explosives Division
2	**Division 2: Explosion with Fragment Hazard** This division is equivalent to DOT/UN Class 1.2 Explosives Division **Also, this exact symbol may be used for:** **Division 6: Nonmass Explosion — extremely insensitive ammunition** This division is equivalent to DOT/UN Class 1.6 Explosives Division
3	**Division 3: Mass Fire** This division is equivalent to DOT/UN Class 1.3 Explosives Division
4	**Division 4: Moderate Fire — no blast** This division is equivalent to DOT/UN Class 1.4 Explosives Division

Symbol	Chemical Hazards
"Red You're Dead"	**Wear Full Protective Clothing (Set One)** Indicates the presence of highly toxic chemical agents that may cause death or serious damage to body functions.
"Yellow You're Mellow"	**Wear Full Protective Clothing (Set Two)** Indicates the presence of harassing agents (riot control agents and smokes).
"White is Bright"	**Wear Full Protective Clothing (Set Three)** Indicates the presence of white phosphorus and other spontaneously combustible material.

Continued

Table 2.18 (concluded)

Symbol	Chemical Hazards
	Wear Breathing Apparatus Indicates the presence of incendiary and readily flammable chemical agents that present an intense heat hazard. This hazard and sign may be present with any of the other fire or chemical hazards/symbols.
	Apply No Water Indicates a dangerous reaction will occur if water is used in an attempt to extinguish the fire. This symbol may be posted together with any of the other hazard symbols.

Symbol	Supplemental Chemical Hazards
G	**G-Type Nerve Agents** — persistent and nonpersistent nerve agents *Examples: sarin (GB), tabun (GA), soman (GD)*
VX	**VX Nerve Agents** — persistent and nonpersistent V-nerve agents *Example: V-agents (VE, VG, VS)*
BZ	**Incapacitating Nerve Agent** *Examples: lacrymatory agent (BBC), vomiting agent (DM)*
H	**H-Type Mustard Agent/Blister Agent** *Example: persistent mustard/lewisite mixture (HL)*
L	**Lewisite Blister Agent** *Examples: nonpersistent choking agent (PFIB), nonpersistent blood agent (SA)*

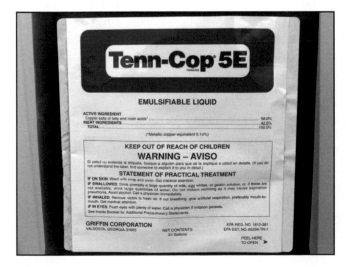

Figure 2.71 According to EPA requirements, pesticide labels must provide the name of the pesticide, the appropriate signal word, a precautionary statement, a hazard statement, and a list of active ingredients. Canadian pesticide labels will also have a PCP (Pest Control Products) Act number.

storage and disposal, first aid information, antidotes for poisoning (if known), and hazard statements indicating that the product poses an environmental hazard.

Materials originating in Canada carry a Pest Control Products (PCP) Act number. The Canadian Transport Emergency Centre (CANUTEC) operated by TC provides information about these materials when given the PCP registration number. Canadian products also have the same signal words and required information as the U.S.

Color Codes

Colors can sometimes provide clues to the nature of hazardous materials in North America. For example, even if a DOT placard is too far away to clearly read the number, a first responder can deduce that the material inside is some kind of oxidizer if the placard background color is yellow. If the placard color is red, the material is flammable. Pre-incident surveys can assist in identifying color systems used by local industries, for example, to identify materials used in piping systems.

ANSI Z535.1 sets forth the following safety color code that is recommended for use in the U.S. and Canada:

- Red — Means *Danger* or *Stop*; is used on containers of flammable liquids, emergency stop bars, stop buttons, and fire-protection equipment

- Orange — Means *Warning*; is used on hazardous machinery with parts that can crush or cut or energized equipment

- Yellow — Means *Caution*; solid yellow, yellow and black stripes, or yellow and black checkers may be used to indicate physical hazards such as tripping hazards; also used on containers of corrosive or unstable materials

- Green — Marks safety equipment such as first-aid stations, safety showers, and exit routes

- Blue — Marks safety information signage such as labels or markings indicating the type of required personal protective equipment (PPE)

Clue 5: Written Resources

A variety of written resources are available to assist responders in identifying hazardous materials at both fixed facilities and transportation incidents. Fixed facilities should have safety data sheets, inventory records, and other facility documents in addition to signs, markings, container shapes, and other labels. At transportation incidents, first responders should be able to use the current *ERG* as well as shipping papers.

Shipping Papers

Bill of Lading — Shipping paper used by the trucking industry (and others) indicating origin, destination, route, and product; placed in the cab of every truck tractor. This document establishes the terms of a contract between shippers and transportation companies; serves as a document of title, contract of carriage, and receipt for goods.

Shipments of hazardous materials must be accompanied by shipping papers that describe them. The information can be provided on a **bill of lading**, waybill, or similar document. The general location and type of paperwork change according to the mode of transport **(Table 2.19).** However, the exact location of the documents varies. The exceptions are hazardous waste shipments, which must be accompanied by a Uniform Hazardous Waste Manifest document. Instructions for describing hazardous materials are provided in the DOT/TC regulations as follows:

Table 2.19
Shipping Paper Identification

Transportation Mode	Shipping Paper Name	Location of Papers	Party Responsible
Air	Air Bill	Cockpit	Pilot
Highway	Bill of Lading	Vehicle Cab	Driver
Rail	Trainlist/Consist	Engine (or Caboose)	Conductor
Water	Dangerous Cargo Manifest	Bridge or Pilot House	Captain or Master

- UN Number
- Proper shipping name of the material
- Hazard class represented by the material
- Packing group assigned to the material
- Quantity of material

When Operations-Level responders know that a close approach to an incident is safe, they can then examine the cargo shipping papers. Responders may need to check with the responsible party in order to locate these documents. If the responsible party is not carrying them, responders will need to check the appropriate locations. In trucks and airplanes, these papers are placed near the driver or pilot. On ships and barges, the papers are placed on the bridge or in the pilothouse of a controlling tugboat.

The train crew should have train consists (entire train's cargo lists), train list, and/or wheel reports. Look for the train crew first, as they should have the paperwork (train list), however, if they cannot be located, contact the railroad through their emergency phone number for a copy of the train list. It is possible there may also be a copy of the current train list in the engine. On the train list, most railroad companies will count and list their train cars from the front of the train to the back. During pre-incident surveys, the location of the papers (and how to read them) for a specific rail line can be determined.

Figure 2.72, p. 146 provides a summary of shipping paper requirements. The Basic Description provided in shipping papers will follow a sequence best remembered by the acronym, **ISHP: I** = **I**dentification Number, **S** = Proper **S**hipping Name, **H** = **H**azard Class or Division, and **P** = **P**acking Group."

Standard Transportation Commodity Code Numbers

Railroads and railroad paper work also use Standard Transportation Commodity Code numbers (STCC numbers) to identify chemicals. These are a seven digit number. If the seven digit number starts with 48, it is a hazardous waste. If it starts with 49, it's a hazardous material. These numbers may also be found in some hazardous materials reference sources.

Figure 2.72 Shipping paper requirements. Note the X placed in the column captioned HM for hazardous material.

Transborder shipments between the U.S. and Mexico are accompanied by shipping documents in both English and Spanish. To satisfy the emergency response information requirements in the U.S. or Mexico, a shipper may attach a copy of the appropriate guide page from the current *ERG* to the shipping papers. The information must be provided in Spanish when the material is shipped to Mexico and in English when shipped to the U.S. so that emergency responders in each country will be able to understand the appropriate initial response procedures in the event of a hazardous material release. **Table 2.20** shows the format of Mexico's emergency sheet for the transport of hazardous materials and wastes required by NOM-005-SCT2/1994.

Table 2.20
Emergency Sheet for the Transport of Hazardous Materials and Wastes

1. Company Name and Address	3. Product or Residue Commercial Name:	6. Carrier Company
• Manufacturer • Importer • Consumer • Distributor • Generator		
2. Shipper's Emergency Phone and Fax Numbers	Chemical Name:	7. Emergency Phone and Fax
	4. Class	
	5. UN No. of Material	

8. Physical State	9. Physical and Chemical Properties	10. Report to National Emergency System and Hazardous Materials Specific Authorities: Federal Highway Police, Fire Department, Red Cross, etc.

11. Personal Protection Equipment

In Case of Accident
- Stop engine
- Set signals in danger zone
- Keep unnecessary people away; out of the danger zone

12. Risks
    If this happens
	13. Actions
	    Do this

14. Intoxication/Exposure	15.
16. Pollution	17.
18. Medical Information	19.
20. Spills/Leakages	21.
22. Fire/Explosion	23.
24. Name	Signature Position Phone

25. This sheet must be on an easily accessible place to be used in case of emergency and must be filled in its entirety.

Safety Data Sheets

A safety data sheet (SDS), similar to a material safety data sheet (MSDS), is a detailed information bulletin prepared by the manufacturer or importer of a chemical that provides specific information about the product. Both OSHA and Canadian regulations had mandated requirements for MSDS contents; however, both countries are switching to the Globally Harmonized System (GHS) SDS format. Until the switch is complete, responders may see MSDSs developed to American National Standards Institute (ANSI) standards, OSHA MSDS standards, Canadian MSDS standards, or the new GHS standard.

SDSs are often the best sources of detailed information about a particular material to which firefighters have access. First responders can acquire them from the manufacturer of the material, the supplier, the shipper, an emergency response center such as CHEMTREC®, or the facility hazard communication plan **(Figure 2.73).** They are sometimes attached to shipping papers and containers as well.

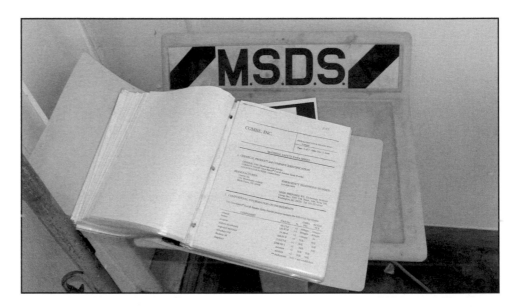

Figure 2.73 In the U.S., employers are required to provide employees access to safety data sheets for all the hazardous materials in their workplace.

The GHS for Hazard Classification and Communication specifies minimum information to be provided on SDSs. These sheets are being used worldwide. SDSs must include the following sections:

1: Identification

2: Hazard(s) identification

3: Composition / information on ingredients

4: First aid measures

5: Firefighting measures

6: Accidental release measures

7: Handling and storage

8: Exposure controls / personal protection

9: Physical and chemical properties

10: Stability and reactivity

11: Toxicological information

12: Ecological information

13: Disposal considerations

14: Transport information

15: Regulatory information

16: Other information

Emergency Response Guidebook (ERG)

The current *ERG* was developed jointly by TC, DOT, and the Secretaría de Co-municaciones y Transportes (SCT) of Mexico (with the collaboration of Centro de Información Química para Emergencias [CIQUIME] of Argentina) for use by firefighters, law enforcement, and other emergency services personnel who may be the first to arrive at the scene of a transportation incident involving dangerous goods/hazardous materials **(Figure 2.74).** The *ERG* is primarily a guide to aid emergency responders in quickly identifying the specific or ge-neric hazards of materials involved in an emergency incident and protecting themselves and the general public during the initial response phase of the incident. For information on how to use the *ERG*, see Chapter 3.

Figure 2.74 When provided with the four-digit UN identification number, first responders can quickly identify a material using the *ERG*. *Courtesy of Rich Mahaney.*

The *ERG* does not address all possible circumstances that may be associ-ated with a dangerous goods/hazardous materials incident. It is primarily designed for use at a dangerous goods/hazardous materials incident occur-ring on a highway or railroad. There may be limited value in its application at fixed-facility locations.

Operations-Level responders at the scene of a haz mat incident should seek additional, specific information about any material in question as soon as possible. The information received by contacting the appropriate emergency response agency, calling the emergency response number on the shipping document, or consulting the information on or accompanying the shipping document may be more specific and accurate than the guidebook in providing guidance for the materials involved.

Inventory Records and Facility Documents

The HCS requires U.S. employers to maintain Chemical Inventory Lists (CILs) of all their hazardous substances. Because CILs usually contain information about the locations of materials within a facility, they can be useful tools in identifying containers that may have damaged or missing labels or markings (such as a label or marking made illegible because of fire damage). Several other documents and records may provide information about hazardous materials at a facility such as the following:

- Shipping and receiving documents
- Inventory records
- Risk management and hazardous communication plans
- Chemical inventory reports (known as *Tier II reports*)

Another potential source of information is the Local Emergency Planning Committee (LEPC). Emergency response plans developed by LEPCs are a good source of information for emergency response organizations. The LEPC was designed to provide a forum for emergency management agencies, responders, industry, and the public to work together to evaluate, understand, and communicate chemical hazards in the community and develop appropriate emergency plans in case of accidental release of these chemicals.

Clue 6: Senses

WARNING!
Deliberately using the human senses to detect the presence of hazardous materials is both unreliable and dangerous.

Olfactory Fatigue — Gradual inability of a person to detect odors after initial exposure; may be extremely rapid in the case of some toxins such as hydrogen sulfide.

Vision is definitely the safest of the five senses to use in the detection of a hazardous material. While it may be perfectly safe to observe an overturned cargo tank from a distance through binoculars, emergency responders have to come into close or actual physical contact with a hazardous material (or its mists, vapors, dusts, or fumes, and the like) in order to hear, smell, taste, or feel a release. While many products release odors well below dangerous levels, there is a good chance that if personnel are this close to a hazardous material, they are too close for safety's sake. It should also be noted that many hazardous materials are invisible, have no odor, and cannot readily be detected by the senses. Hydrogen sulfide may cause **olfactory fatigue** (in other words, you may cease to smell it even though it is still present). However, any smells, tastes, or symptoms reported by victims and witnesses may prove to be helpful. Warning properties of chemicals include visible gas clouds, pungent odors, and irritating fumes.

First responders should be aware of visual/physical chemical indicators that provide tangible evidence of the presence of hazardous materials. Unusual noises (such as the hiss of a gas escaping a valve at high pressure) may also alert responders to the presence of hazards. Some hazardous materials have odorants added to them to aid in detection; for example, the distinct odor normally associated with natural gas (an odorless gas) is actually caused by mercaptan, an additive.

Direct visible evidence that physical and/or chemical actions and reactions are taking place include the following:

- Spreading vapor cloud or smoke **(Figure 2.75)**
- Unusual colored smoke
- Flames
- Gloves melting
- Changes in vegetation
- Container deterioration
- Containers bulging
- Sick humans
- Dead or dying birds, animals, insects, or fish
- Discoloration of valves or piping

Physical actions are processes that do not change the elemental composition of the materials involved. Several indications of a physical action are as follows:

- Rainbow sheen on water surfaces **(Figure 2.76)**
- Wavy vapors over a volatile liquid
- Frost or ice buildup near a leak
- Containers deformed by the force of an accident
- Activated pressure-relief devices
- Pinging or popping of heat-exposed vessels

Figure 2.75 Spreading smoke or a vapor cloud is a visual indicator that a chemical reaction is taking place.

Figure 2.76 A rainbow sheen on water surfaces is a good indication that a hazardous material is present. *Courtesy of FEMA News Photos, photo by Liz Roll.*

Chemical reactions convert one substance to another. Visual and sensory evidence of chemical reactions include the following:

Exothermic — Chemical reaction between two or more materials that changes the materials and produces heat, flames, and toxic smoke.

Endothermic — Chemical reaction involving the absorption of heat energy.

- **Exothermic** heat
- Unusual or unexpected temperature drop (cold) (from an **endothermic** reaction)
- Extraordinary fire conditions
- Peeling or discoloration of a container's finish
- Spattering or boiling of unheated materials
- Distinctively colored vapor clouds
- Smoking or self-igniting materials
- Unexpected deterioration of equipment
- Peculiar smells
- Unexplained changes in ordinary materials
- Symptoms of chemical exposure

Physical signs and symptoms of chemical exposure may also indicate the presence of hazardous materials. Symptoms can occur separately or in clusters, depending on the chemical. Symptoms of chemical exposure include the following:

- *Changes in respiration* — Difficult breathing, increase or decrease in respiration rate, tightness of the chest, irritation of the nose and throat, and/or respiratory arrest
- *Changes in level of consciousness* — Dizziness, lightheadedness, drowsiness, confusion, fainting, and/or unconsciousness
- *Abdominal distress* — Nausea, vomiting, and/or cramping
- *Change in activity level* — Fatigue, weakness, stupor, hyperactivity, restlessness, anxiety, giddiness, and/or faulty judgment
- *Visual disturbances* — Double vision, blurred vision, cloudy vision, burning of the eyes, and/or dilated or constricted pupils
- *Skin changes* — Burning sensations, reddening, paleness, fever, and/or chills
- *Changes in excretion or thirst* — Uncontrolled tears, profuse sweating, mucus flowing from the nose, diarrhea, frequent urination, bloody stool, and/or intense thirst
- *Pain* — Headache, muscle ache, stomachache, chest pain, and/or localized pain at sites of substance contact

Clue 7: Monitoring and Detection Devices

Monitoring and detection devices can be useful in determining the presence of hazardous materials as well as the concentration(s) present. They can also be used to determine the scope of the incident. As with the senses, effectively using the monitoring and detection devices requires actual contact with the hazardous material (or its mists, dusts, vapors, or fumes) in order to measure it, and it is therefore outside the scope of action for Awareness-Level personnel. Chapter 11, Air Monitoring and Sampling, addresses monitoring and detection devices for Operations-Level responders.

Identification of Terrorist Attacks and Illicit Laboratories

Response to a terrorist incident is essentially the same as that for response to other haz mat incidents; however, there are critical differences that must be understood by first responders. For example, terrorist incident sites are crime scenes; therefore, law enforcement organizations must be notified and included in the initial response. Other differences between terrorist and haz mat incidents include the following:

- Size and complexity
- Number of casualties
- Presence of extremely hazardous materials
- Potential for armed resistance
- Booby traps
- Necessity of crime scene preservation
- Secondary devices
- Higher level of risk from contaminated victims, structural collapse hazards, and other dangers **(Figure 2.77)**

Figure 2.77 Scenes of terrorist attacks may present a higher degree of risk to responders than other types of more routine emergencies. *Courtesy of FEMA News Photos, photo by Mike Rieger.*

Also, most hazardous materials incidents do not specifically target people (although it is possible to encounter a deliberate release of hazardous materials, for example, in illegal dumping). Terrorists specifically target the public, first responders, or both.

Because of these important differences between terrorist incidents and other haz mat emergencies, first responders must try to identify incidents involving terrorism as quickly as possible. The following are a few examples of situations that can cue the responder to consider the possibility of a terrorist attack:

- Report of two or more medical emergencies in public locations such as a shopping mall, transportation hub, mass transit system, office building, assembly occupancy, or other public buildings
- Unusually large number of people with similar signs and symptoms arriving at physicians' offices or medical emergency rooms

- Reported explosion at a movie theater, department store, office building, government building, or a location with historical or symbolic significance

Additional information can provide clues as to the type of attack. Chemical, biological, radiological, nuclear, and explosive attacks (CBRNE attacks) each have their own unique indicators with which responders should be familiar. **Table 2.21** provides a very brief summary of these differences. Monitoring and detection devices also play an important role in determining which of these materials may be present at the incident scene. If criminal or terrorist activity is suspected at an incident, first responders must quickly forward that information to law enforcement representatives. CBRNE attacks are described in the sections that follow. In addition, information about illicit labs and secondary devices is provided.

NOTE: The information in this section is designed to assist a responder in identifying a terrorist attack. More in depth information about terrorist attacks is provided in Chapter 7, Terrorist Attacks, Criminal Activities, and Disasters.

Table 2.21
Terrorist Attacks at a Glance

Chemical Attack	Biological Attack
• Victims in a concentrated area • Symptoms immediate (seconds to hours after exposure) • Symptoms very similar (SLUDGEM) • May have observable features such as chemical residue, dead foliage, dead animals/insects, and pungent odor	• Victims dispersed over a wide area • Symptons delayed (days — weeks after exposure) • Symptoms most likely vague and flu-like • No observable features
Explosive Attack	**Radiological Attack**
• Explosion self evident (debris field, fire, etc.) • Victims in a concentrated area • Mechanical and thermal injuries • Potential radiation and chemical agent risk — monitoring for both is necessary	• Explosion self evident (debris field, fire, etc.) • Victims in a concentrated area • Mechanical and thermal injuries initially, radiological symptoms (if any) will likely be delayed • Radiation detected through monitoring

Chemical Warfare Agent — Chemical substance that is intended for use in warfare or terrorist activities to kill, seriously injure, or seriously incapacitate people through its physiological effects.

Toxic Industrial Material (TIM)/Toxic Industrial Chemical (TIC) — Industrial chemical that is toxic at certain concentration, is readily available, and could be used by terrorists to deliberately kill, injury, or incapacitate people.

Chemical Attack Indicators

Chemical attacks may utilize **chemical warfare agents** (nerve agents, blister agents, blood agents, choking agents, and riot control agents) as well as **toxic industrial materials/toxic industrial chemicals (TIMs/TICs)**. Chemical attacks usually result in readily observable features including signs and symptoms that develop very rapidly. Chemical attack indicators include:

- Warning or threat of an attack or received intelligence
- Presence of hazardous materials or laboratory equipment that is not relevant to the occupancy
- Intentional release of hazardous materials

- Unexplained patterns of sudden onset of similar, nontraumatic illnesses or deaths (the pattern could be geographic, by employer, or associated with agent dissemination methods)

- Unexplained odors or tastes that are out of character with the surroundings

- Multiple individuals exhibiting unexplained signs of skin, eye, or airway irritation

- Unexplained bomb or munition-like material, especially if it contains a liquid

- Unexplained vapor clouds, mists, and plumes, particularly if they are not consistent with their surroundings

- Multiple individuals exhibiting unexplained health problems such as nausea, vomiting, twitching, tightness in chest, sweating, pinpoint pupils (miosis), runny nose (rhinorrhea), disorientation, difficulty breathing, or convulsions **(Figure 2.78)**

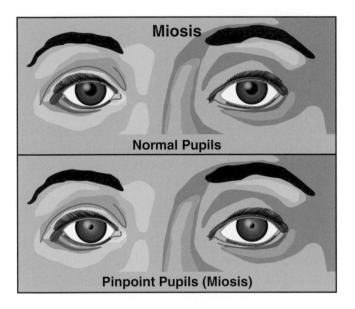

Figure 2.78 Exposure to chemical agents may cause miosis (pinpoint pupils).

- Unexplained deaths and/or mass casualties

- Casualties distributed downwind (outdoors) or near ventilation systems (indoors)

- Multiple individuals experiencing blisters and/or rashes

- Trees, shrubs, bushes, food crops, and/or lawns that are dead (not just a patch of dead weeds), discolored, abnormal in appearance, or withered (not under drought conditions)

- Surfaces exhibiting oily droplets or films and unexplained oily film on water surfaces

- Abnormal number of sick or dead birds, animals, and/or fish **(Figure 2.79)**

- Unusual security, locks, bars on windows, covered windows, and barbed wire enclosures

Figure 2.79 Chemical attacks kill people and animals indiscriminately. This photo was taken after a chemical attack in Halabja, Iraq. *Courtesy of Sayeed Janbozorgi, published under the Creative Commons Atrribution ShareAlike 3.0 License, http:// creativecommons.org/licenses/ by-sa/3.0/.*

SLUDGEM or DUMBELS

Some people teach the symptoms of exposure to chemical warfare agents with the acronyms *SLUDGEM* or *DUMBELS*:

- **S**alivation (drooling)
- **L**acrimation (tearing)
- **U**rination
- **D**efecation
- **G**astrointestinal upset/aggravation (cramping)
- **E**mesis (vomiting)
- **M**iosis (pinpointed pupils) or **M**uscular twitching/spasms

or

- **D**efecation
- **U**rination
- **M**iosis or **M**uscular twitching
- **B**ronchospasm (wheezing)
- **E**mesis
- **L**achrimation
- **S**alivation

Virus — Simplest type of microorganism that can only replicate itself in the living cells of their hosts.

TIMs/TICs used as chemical weapons may be identified through traditional methods such as identification of occupancy types and locations; container shapes; hazardous materials placards, labels, and markings; written resources; sensory indicators; and use of monitoring and detection devices.

Biological Attack Indicators

Biological attacks utilize **viruses**, **bacteria**, **rickettsia**, and/or **biological toxins**. The effects of biological attacks may not be readily noticeable. Signs and symptoms may take many days to develop **(Figure 2.80).** Biological attack indicators include:

Bacteria — Microscopic, single-celled organisms.

- Warning or threat of an attack or received intelligence
- Presentation of specific unusual diseases (such as smallpox)
- Unusual number of sick or dying people or animals (often of different species)

Rickettsia — Specialized bacteria that live and multiply in the gastrointestinal tract of arthropod carriers (such as ticks and fleas).

- Multiple casualties with similar signs or symptoms
- Dissemination of unscheduled or unusual spray **(Figure 2.81)**
- Abandoned spray devices (devices may have no distinct odors)
- Non-endemic illness for the geographic area (for example, Venezuelan equine encephalitis in Europe)

Biological Toxin — Poison produced by living organisms.

- Casualty distribution aligned with wind direction
- Electronic tracking of signs and symptoms (**syndromic surveillance**) reported to hospitals, pharmacies, and other health care organizations

Syndromic Surveillance — Surveillance using health-related data that precede diagnosis and signal a sufficient probability of a case or an outbreak to warrant further public health response.

- Illnesses associated with a common source of food, water, or location
- Large numbers of people exhibiting flu-like symptoms during non-flu months

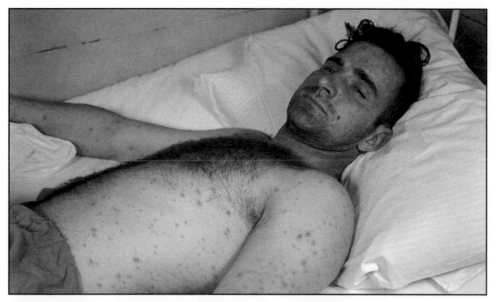

Figure 2.80 Symptoms of a biological attack (such as smallpox) may take days to develop. *Courtesy of the CDC Public Health Image Library.*

Figure 2.81 Unscheduled or unusual spray activity (for example, over a crowded metropolitan area) could be a sign of a biological attack. *Courtesy of the U.S. Department of Agriculture.*

Depending on the agent used and the scope of an incident, emergency medical services (EMS) responders and health-care personnel may be first to realize that there has been a biological attack. In some cases there may be reliable evidence to implicate terrorist activity such as a witness to an attack or the discovery of an appropriate delivery system (such as finding a contaminated dissemination device from which an infectious agent is subsequently isolated and identified). If a biological attack is suspected, first responders should immediately notify their local health care agency.

Radiological Attack Indicators

Radiological attacks utilize weapons that release radiological materials, most likely in the form of dust or powder. Dispersal may be accomplished by including the material in a bomb or explosive device, i.e., a **radiological dispersal device** (RDD), sometimes called a *dirty bomb*. Radiological attack indicators include:

- Warning or threat of an attack or received intelligence
- Individuals exhibiting signs and symptoms of radiation exposure
- Radiological materials packaging left unattended or abandoned in public locations **(Figure 2.82)**
- Suspicious packages that appear to weigh more than they should (such packages may contain lead to shield a radiation source)
- Activation of radiation detection devices, with or without an explosion
- Material that is hot or seems to emit heat without any sign of an external heat source
- Glowing material (strongly radioactive material may emit or cause radioluminescence)

Radiological Dispersal Device (RDD) — Device that spreads radioactive contamination over the widest possible area by detonating conventional high explosives wrapped with radioactive material.

Figure 2.82 Unattended radiological packaging left in public places may be evidence of radiological attack. *Courtesy of Tom Clawson.*

Nuclear Attack Indicators

Nuclear attacks are the intentional detonation of a nuclear weapon. Nuclear attack indicators include:

- Warning or threat of an attack or received intelligence
- Mushroom cloud **(Figure 2.83)**
- Exceptionally large/powerful explosion
- Electromagnetic pulse (EMP)

Figure 2.83 A mushroom cloud is an indicator of nuclear attack. *Courtesy of the U.S. Department of Energy.*

Figure 2.84 The majority of terrorist attacks utilize conventional weapons such explosives and incendiary devices like the IEDs pictured here. *Courtesy of the U.S. Department of Defense.*

Explosive/Incendiary Attack Indicators

The majority of terrorist attacks involve the use of explosive materials and incendiary devices **(Figure 2.84).** Typically, these are considered conventional attacks, however, when used to inflict high casualties and large-scale damage (such as with a car or truck bomb destroying an occupied building), explosives may be classified as weapons of mass destruction. Explosives may also be used to disseminate chemical, biological, and radiological materials.

Explosive/incendiary attack indicators include the following:

- Warning or threat of an attack or received intelligence
- Reports of an explosion
- Explosion
- Accelerant odors (gasoline smells and other similar odors)
- Multiple fires or explosions
- Incendiary device or bomb components (such as broken glass from a Molotov cocktail or wreckage of a car bomb)
- Unexpectedly heavy burning or high temperatures
- Unusually fast burning fires
- Unusually colored smoke or flames
- Presence of propane or other flammable gas cylinders in unusual locations
- Unattended packages/backpacks/objects left in high traffic/public areas
- Fragmentation damage/injury

- Damage that exceeds the level usually seen during gas explosions, including shattered reinforced concrete or bent structural steel **(Figure 2.85)**
- Crater(s) **(Figure 2.86)**
- Scattering of small metal objects such as nuts, bolts, and/or nails used as shrapnel

Figure 2.85 Car and truck bombs can do greater damage than accidental gas explosions. Indicators include including shattered reinforced concrete and bent structural steel. *Courtesy of U.S. Air Force, photo by Senior Airman Sean Worrell.*

Figure 2.86 Craters also indicate the use of explosives. *Courtesy of the U.S. Department of Defense.*

Illicit Laboratory Indicators

Illicit laboratories are established to produce or manufacture illegal or controlled substances such as drugs, chemical warfare agents, explosives, or biological agents. Illegal clandestine labs can be found virtually anywhere. They may be located in abandoned buildings, hotel rooms, rural farms, urban apartments, rental storage units, or upscale residential neighborhoods. Illegal **methamphetamine (meth) labs** can be so portable that they have even been found in campgrounds, highway rest stops, and vehicles **(Figure 2.87).**

Illicit — Illegal, unlawful.

Meth Lab — Illegal clandestine laboratory established to produce illegal methamphetamine (meth).

Figure 2.87 Meth labs can be very portable. This is an example of a small box lab. *Courtesy of MSA.*

These labs can present numerous threats to first responders. In addition to the chemical, explosive, or biological hazards present, responders may face armed resistance, booby traps, and/or other weapons when responding to emergencies at illegal labs. Many of the products used in clandestine labs are toxic, explosive, and/or highly flammable.

While it may be difficult to tell the difference between explosive, chemical agent, and drug labs, the chemicals and materials present will typically give clues as to what the lab is producing. For example, if unusual quantities of over-the-counter allergy medications containing pseudoephedrine are present, it's probably a meth lab. If **agar** plates are present, it's probably a biological lab. However, extreme caution must be exercised at incidents involving **any** illicit laboratory. If an illicit lab is discovered or suspected, Awareness-Level personnel should withdraw and contact law enforcement authorities.

On rare occasions, it may be easy to identify an illicit lab. More typically, operators of illicit labs usually try to hide them since they are engaged in illegal activities. In these cases the clues are often subtle, especially from the exterior.

Exterior clues to the presence of an illicit lab include the following:

- Blacked-out windows
- Discarded chemical containers
- Booby traps
- Hidden or disguised entrances
- Inappropriate levels of protection and security
- Excessive amounts of trash **(Figure 2.88)**

Agar — Gelatinous or jelly-like substance used to grow bacterial cultures.

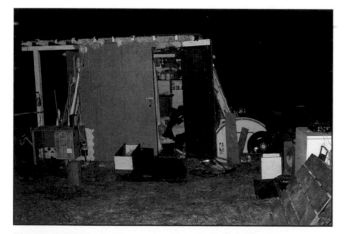

Figure 2.88 Excessive amounts of trash around a property can also indicate illegal activity. *Courtesy of MSA.*

Figure 2.89 Unusual flasks, glassware, and lab equipment should immediately raise suspicions of an illicit lab. *Courtesy of MSA.*

Interior indicators that a lab is in operation include:

- Covered windows in an occupied building
- Chemical odors
- Flasks and other glassware **(Figure 2.89)**
- Unusual heat sources
- Containers of unknown substances

- Pill packages or blister packs (for pseudoephedrine and/or antihistamines such as Sudafed®)

- Large quantities of lithium batteries (lithium is used in the manufacture of meth)

- Propane bottles (used to store anhydrous ammonia)

NOTE: Identification of specific types of illicit laboratories is covered in more detail in Chapter 14, Illicit Laboratories.

Secondary Attacks and Booby Traps

The use of secondary devices at terrorist attacks or illicit laboratories is always a possibility. Secondary devices are often designed to affect an ongoing emergency response in order to create more chaos and injure responders and bystanders. Booby traps may be set to protect illicit laboratories **(Figure 2.90).** Usually, secondary devices are explosives of some kind, most likely an improvised explosive device (IED). Booby traps utilizing other weapons are also possible, and some may use chemical, biological, or radiological materials. Some may use animals such as snakes or guard dogs. **Secondary devices** may also be deployed as a diversionary tactic to route emergency responders away from the primary attack area.

Secondary devices will be hidden or camouflaged. The devices may be detonated by a time delay, but radio-controlled devices, cell phone-activated devices, and other activation devices are also used. In some cases, an obvious IED may be used to lure personnel to a specific area where a less obvious IED is hidden. Contact law enforcement and/or explosive ordnance disposal (EOD)/bomb squad personnel any time booby traps or secondary devices are found or suspected.

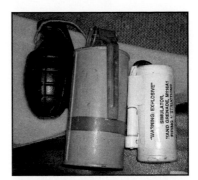

Figure 2.90 Operators of illicit labs may set booby traps. These grenades were confiscated from one such lab. *Courtesy of U.S. Drug Enforcement Administration.*

Secondary Device — Bomb placed at the scene of an ongoing emergency response that is intended to cause casualties among responders; secondary explosive devices are designed to explode after a primary explosion or other major emergency response event has attracted large numbers of responders to the scene.

Use of Secondary Devices

Bombings on January 27, 1997 outside the Atlanta Northside Family Planning Service and on February 21, 1997 at the Otherside Lounge in Atlanta both had secondary explosive devices placed in proximity to the primary explosion. In another incident, a Puerto Rican peace officer was reported to have been killed by a flashlight bomb rigged with an improvised motion switch at a crime scene.

In other parts of the world, usage of secondary devices varies. In Israel, secondary devices have only been used rarely; however, they are commonly used in Iraq. Regardless of historical patterns, secondary devices are part of the terrorist's toolbox. **A good rule of thumb is: if there has already been one explosion (or one device has been found), always expect another.**

Guidelines for protecting against possible secondary devices include the following:

- Anticipate the presence of a secondary device at any suspicious incident.

- Visually search for a secondary device (or anything suspicious) before moving into the incident area.

- Limit the numbers of emergency response personnel to those performing critical tasks (rescue) until the area has been checked and confirmed that no additional devices are present.

- Avoid touching or moving anything that may conceal an explosive device (including items such as backpacks and purses).

- Manage the scene with cordons, boundaries, and scene control zones.

- Evacuate victims and nonessential personnel as quickly as possible.

- Preserve the scene as much as possible for evidence collection and crime investigation.

While secondary devices and booby traps can be disguised as almost anything, responders should look for things that may seem out of place **(Figure 2.91)** If anything suspicious is found, responders should note the item, treat the item with appropriate caution, notify appropriate personnel (law enforcement/Explosive Ordnance Disposal/bomb squad personnel), and evacuate the area immediately. Responders should be very cautious of any item(s) that arouse curiosity, including the following:

- Containers with unknown liquids or materials

- Unusual devices or containers with electronic components such as wires, circuit boards, cellular phones, antennas and other items attached or exposed

- Devices containing quantities of fuses, fireworks, match heads, black powder, smokeless powder, incendiary materials or other unusual materials

- Materials attached to or surrounding an item such as nails, bolts, drill bits, marbles, etc. that could be used for shrapnel **(Figure 2.92)**

- Ordnance such as blasting caps, detcord, military explosives, commercial explosives, grenades, etc.

- Any combination of the above described items

WARNING!

First responders should NEVER approach or move suspicious objects. Notify appropriate personnel (law enforcement/ Explosive Ordnance Disposal/bomb squad personnel) and evacuate the area immediately.

Figure 2.91 While secondary devices like IEDs will be concealed, responders should look for anything that arouses their attention or seems out of place. *Courtesy of the U.S. Department of Defense, photo by Staff Sgt. Stacy L. Pearsall.*

Figure 2.92 This replica of an actual IED that detonated in Israel shows the materials that were added to the device to increase shrapnel injuries.

Summary

First responders must be able to identify the presence of hazardous materials at emergency incidents. By paying attention to and using the seven clues to the presence of hazardous materials — occupancy; container shape; transportation placards, labels, and markings; other markings and colors; written resources; senses; and monitoring and detection devices — first responders can take the first steps towards successful mitigation of a hazardous material incident.

However, responders need to be aware that making a correct identification may be difficult despite knowledge and recognition of these clues. Fires or explosions can destroy shipping papers, labels, and other markings. Shipments may contain mixed loads or quantities of materials so small that placards are not required. Facilities may not be in compliance with regulations requiring SDSs, and mistakes can be made in labeling and placarding. For that matter, responders may be unable to get close enough to the material or container to make an accurate identification. For these and other reasons, responders must always be prepared to face the unexpected and deal with the unknown.

First responders must also be able to recognize when an incident may be the result of a terrorist attack. Attacks involving chemicals, biological materials, radiological materials, and explosives will have features that differ according to the type of attack. Similarly, illicit labs used for manufacturing drugs, chemical agents, biological agents, and explosives will have different materials and processes that should enable responders to identify the type of lab present.

Review Questions

1. What are the seven clues to the presence of hazardous materials?

2. What are some of the highly probable locations for finding significant quantities of hazardous materials?

3. What are the criteria for bulk packaging?

4. What are the nine UN hazard classes?

5. On DOT placards, what does the color orange indicate?

6. What are the required sections of safety data sheet (SDS)?

7. What are some symptoms of chemical exposure?

8. What are several indicators of a chemical attack?

9. What are several indicators of a radiological attack?

10. What are some exterior clues to the presence of an illicit lab?

Awareness-Level Actions at Hazardous Materials Incident

Chapter Contents

Divider page photo courtesy of Rich Mahaney.

Key Terms

Competencies

NFPA® 472:	4.4.1(3)(b)	4.4.1(5)(b)	4.4.1(6)(c)	4.4.1(10)	5.2.2(4)(b)
4.2.3(1)	4.4.1(4)(a)	4.4.1(5)(c)	4.4.1(7)	4.4.1(11)	5.2.2(4)(c)
4.2.3(2)	4.4.1(4)(b)	4.4.1(5)(d)	4.4.1(8)	4.4.1(12)	5.2.2(5)
4.4.1(1)	4.4.1(4)(c)	4.4.1(6)(a)	4.4.1(9)(a)	4.4.2	5.2.3(1)
4.4.1(3)(a)	4.4.1(5)(a)	4.4.1(6)(b)	4.4.1(9)(b)	5.2.2(4)(a)	5.2.3(1)(b)(i)

Learning Objectives

1. Discuss predetermined procedures and emergency response plans. [NFPA® 472, 4.4.1(1)]

2. Describe notification requirements. [NFPA® 472, 4.4.2]

3. Discuss the use of the *Emergency Response Guidebook (ERG)*. [NFPA® 472, 4.2.3(1-2), 4.4.1(3) (a-b), 4.4.1(4)(a-c), 4.4.1(5)(a-d), 4.4.1(6)(b-c), 4.4.1(7), 4.4.1(8), 4.4.1(9)(a-b), 4.4.1(10), 5.2.2(4)(a-c), 5.2.2(5), 5.2.3(1), 5.2.3(1)(b)(i)]

4. Obtain information about a hazardous material using the *ERG*. [NFPA® 472, 4.2.3, 4.4.2; Skill Sheet 3-1]

5. Describe isolation and discuss denial of entry. [NFPA® 472, 4.4.1(6)(a), 4.4.1(11)]

6. Discuss terrorist incidents. [NFPA® 472, 4.4.1(12)]

Chapter 3

Awareness-Level Actions at Hazardous Materials Incidents

Case History

In June, 2007, two homemade car bombs were placed strategically on the streets outside Tiger Tiger, a large night club in London, England. The homemade bombs were constructed of petrol cans, gas canisters, nails, and mobile phone-based triggers. However, before either bomb was detonated, both were detected and disarmed. An ambulance crew spotted the first vehicle when the crew responded to a minor incident at the nightclub. They reported suspicious fumes to local police. Police responded to the scene, and the bomb was disarmed.

The second vehicle was towed to a car impound after being ticketed for illegal parking. After learning of the first car bomb, the staff where the vehicle was impounded noticed similar strange fumes. They took the appropriate precautions to isolate the vehicle and then notified police, who were able to disarm the second bomb. The decisive actions of both the ambulance crew and the impound yard staff helped prevent potential severe mass casualties.

Awareness-Level personnel have an important role at haz mat incidents, and their initial actions can affect the course of the incident for better or worse **(Figure 3.1, p. 168)**. Awareness-Level personnel are expected to assume the following responsibilities when faced with an incident involving hazardous materials:

- Recognize the presence or potential presence of a hazardous material

- Recognize the type of container at a site and identify the material in it if possible

- Transmit information to an appropriate authority and call for appropriate assistance

- Identify actions to protect themselves and others from hazards

- Establish scene control by isolating the hazardous area and denying entry

Persons trained to this level may not be emergency responders. Rather, they may be industrial personnel, public or private utility workers, or other personnel who work with hazardous materials, or work in areas where hazardous materials are used or transported **(Figure 3.2, p. 168)**. They may lack the training necessary to take many actions at emergency incidents beyond the initial stages of getting themselves out of harm's way, getting others out of harm's way, calling for help, and preventing others from entering the hazardous area.

This chapter discusses roles and responsibilities of Awareness-Level personnel at emergency incidents (as opposed to incidental releases for which they may be trained to mitigate without calling for additional assistance). For more information about response to **incidental releases**, see 29 *CFR* 1910.120.

Incidental Release — Spill or release of a hazardous material where the substance can be absorbed, neutralized, or otherwise controlled at the time of release by employees in the immediate release area, or by maintenance personnel who are not considered to be emergency responders.

Figure 3.1 The actions of Awareness-Level personnel at haz mat incidents can be very beneficial. Calling for help quickly and isolating the area can save lives. *Courtesy of Rich Mahaney.*

Figure 3.2 Utility workers or others who work out in the community may be trained to the Awareness Level. These individuals are often the first to witness an accident or incident involving hazardous materials.

This chapter will discuss:

- Predetermined procedures, also known as SOPs
- Emergency response plans
- Notification requirements for Awareness-Level personnel
- How to use the *Emergency Response Guidebook*
- Basic procedures to isolate the incident
- Steps that should be taken in case an incident involves terrorism or criminal activity

NOTE: All of the actions discussed in this chapter are applicable to Operations-Level responders, as well.

Predetermined Procedures and Emergency Response Plans

Most emergency services organizations and other agencies have predetermined procedures for nearly every conceivable type of emergency that can occur. These plans are known as the organization's *standard operating procedures (SOPs), standard operating guidelines (SOGs), or operating instructions (OIs)* (collectively referred to as SOPs throughout the rest of this manual). SOPs should spell out the role of Awareness-Level personnel at emergency incidents, including those involving hazardous materials. Personnel should be aware of what actions they are expected to take at such incidents.

Additionally, OSHA 29 *CFR* 1910.120(q)(2) requires emergency response organizations in the U.S. to develop emergency response plans that must cover the following elements:

- Pre-emergency planning and coordination with outside parties
- Personnel roles, lines of authority, training, and communication
- Emergency recognition and prevention
- Safe distances and places of refuge
- Site security and control
- Evacuation routes and procedures
- Decontamination
- Emergency medical treatment and first aid
- Emergency alerting and response procedures
- Critiques of response and follow up
- PPE and emergency response equipment

For emergency responders trained beyond Awareness Level, SOPs provide a standard set of actions that are the core of every hazardous material incident plan. Responders need to comply with any procedures set forth in the local emergency response plan. While the procedures may vary considerably in different localities, the principles are usually the same. However, responders must still incorporate specific regulatory requirements such as those required by OSHA. SOPs must be written and their use required in order for them to be effective. Individuals must be trained in their use and disciplined when they do not follow the procedures. First responders must know the location of their agency's emergency response plan and written SOPs **(Figure 3.3)**.

Figure 3.3 First responders must be familiar with their agency's emergency response plan and SOPs. *Courtesy of the U.S. Air Force.*

Even though there are obvious variations in haz mat incidents, they all have some similarities. These similarities are the basis for predetermined procedures. These procedures have a built-in flexibility that allows, with reasonable justification, adjustments when unforeseen circumstances occur. The first units that reach the scene usually initiate the predetermined actions. They do not replace size-up, decisions based on professional judgment, evaluation, or command. In addition, there may be several predetermined procedures from which to choose, depending on incident severity, location, and the ability of first-in units to achieve control.

Following predetermined procedures and the provisions of the emergency response plan reduces chaos on the haz mat scene. All resources can be used in a coordinated effort to rescue victims, stabilize the incident, and protect the environment and property. Operational procedures that are standardized, clearly written, and mandated to each department/organization member establish accountability and increase command and control effectiveness.

These predetermined procedures also help prevent duplication of effort and uncoordinated operations because all positions are assigned and covered. In addition, predetermined procedures describe assumption and transfer of command, communications procedures, and tactical procedures.

Notification Requirements

Predetermined procedures, such as SOPs and the emergency response plan, should define roles in the notification process. For example, for Awareness-Level personnel, notification may be as simple as dialing 9-1-1 to report an incident and get emergency help dispatched **(Figure 3.4)**. Fixed-facility responders may have their own internal procedures to follow such as calling for an internal fire brigade or haz mat response team. If criminal or terrorist activity is suspected, personnel should notify law enforcement immediately. Information about notification requirements for Ops-Level responders is covered in Chapter 6, Strategic Goals and Tactical Objectives, Notification section.

Figure 3.4 Awareness-Level personnel must know who to call for help in case of emergency. *Courtesy of U.S. Army Corps of Engineers, photo by Rob Haynes.*

Using the Emergency Response Guidebook

The *ERG* is primarily a guide to aid emergency responders in quickly identifying the specific or generic hazards of materials involved in an emergency incident and protecting themselves and the general public during the initial response phase of the incident. The *ERG* does not address all possible circumstances that may be associated with a dangerous goods/hazardous materials incident. It is primarily designed for use at hazardous materials incidents occurring on a highway or railroad. Isolation and protective distances in the *ERG* are based on conditions commonly associated with transportation incidents in open areas and may be of limited value when applied to fixed-facility locations or in urban settings.

Awareness-Level personnel and first responders can locate the appropriate initial action guide page in several different ways:

- Identify the four-digit U.N. identification number on a placard or shipping papers and then look up the appropriate guide in the yellow-bordered pages of the guidebook.

- Use the name of the material involved (if known) in the blue-bordered section of the guidebook (see *ERG* Material Name Index section). Many chemical names differ only by a few letters, so **exact spelling is important** when using this method **(Figure 3.5)**.

- Identify the transportation placard of the material and then use the three-digit guide code associated with the placard in Table of Placards and Initial Response Guide to Use On-Scene located in the front of the *ERG*.

- Use the container profiles provided in the white pages in the front of the book. First responders can identify container shapes, and then reference the guide number to the orange-bordered page provided in the nearest circle **(Figure 3.6)**.

Figure 3.5 The exact spelling of a chemical name is very important when using the *Emergency Response Guidebook*. Misspellings can cause inappropriate (and potentially dangerous) actions to be taken. *Courtesy of Rich Mahaney.*

Figure 3.6 As a last resort, if placards or 4-digit ID numbers are not visible, first responders can use container profiles to identify the proper *ERG* guidebook page. *Courtesy of Rich Mahaney.*

Using the four-digit ID number or the chemical name allows responders to locate the most specific initial action guide. **Skill Sheet 3-1** provides examples of how to use the *ERG*. The sections that follow describe the design and layout of the *ERG*.

Multiple Information Sources

First responders at the scene of a haz mat incident should always seek additional specific information about any material in question as soon as possible. Do not rely on the *ERG* alone. The information received by contacting the appropriate emergency response agency, calling the emergency response number on the shipping document, or consulting the accompanying shipping document may be more specific and accurate than the guidebook in providing guidance for the materials involved.

When consulting chemical reference sources for information about a particular substance, more than one reference source should be consulted to ensure information is complete and accurate. Reference books may be written for a specific purpose (such as compiling information about the most dangerous workplace chemicals), and many chemicals may be left out. Absence from one reference book does *not* mean that the substance is safe. Check multiple sources.

ERG ID Number Index (Yellow Pages)

The yellow-bordered pages of the *ERG* provide an index list of hazardous materials in numerical order of ID number. This index displays the four-digit UN/NA ID number of the material followed by its assigned emergency response Guide and the material's name.

Toxic Inhalation Hazard (TIH) — Volatile liquid or gas known to be a severe hazard to human health during transportation.

The purpose of the yellow section in the *ERG* is to enable first responders to quickly identify the Guide (orange-bordered pages) to consult for the ID number of the substance involved. If a material in the yellow or blue index is highlighted, it means that it releases gases that are **toxic inhalation hazard (TIH)** materials. These materials require the application of additional emergency response distances.

ERG Material Name Index (Blue Pages)

The blue-bordered pages of the *ERG* provide an index of dangerous goods in alphabetical order by material name so that the first responder can quickly identify the Guide to consult for the name of the material involved. This list displays the name of the material followed by its assigned emergency response Guide and four-digit ID number **(Figure 3.7)**.

ERG Initial Action Guides (Orange Pages)

The orange-bordered section of the book is the most important because it provides safety recommendations and general hazards information. It comprises a total of 62 individual guides presented in a two-page format **(Figure 3.8)**. The left-hand page provides safety related information, whereas the right-hand page provides emergency response guidance and activities for fire situations, spill or leak incidents, and first aid. Each Guide is designed to cover a group of materials that possess similar chemical and toxicological characteristics. The Guide title identifies the general hazards of the dangerous goods covered.

Each Guide is divided into three main sections:

- *Potential Hazards*
- *Public Safety*
- *Emergency Response*

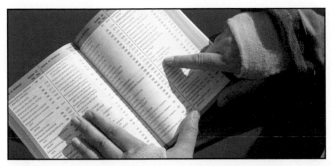

Figure 3.7 The blue pages of the *ERG* provide an index of hazardous materials in alphabetical order.

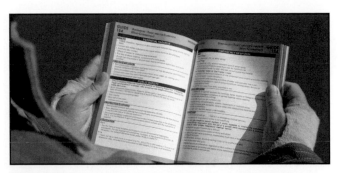

Figure 3.8 The orange-bordered guide pages provide safety recommendations, general hazard information, and basic emergency response actions.

GUIDE 117 GASES - TOXIC - FLAMMABLE (EXTREME HAZARD) ERG2008

POTENTIAL HAZARDS

HEALTH
- TOXIC; Extremely Hazardous.
- May be fatal if inhaled or absorbed through skin.
- Initial odor may be irritating or foul and may deaden your sense of smell.
- Contact with gas or liquefied gas may cause burns, severe injury and/or frostbite.
- Fire will produce irritating, corrosive and/or toxic gases.
- Runoff from fire control may cause pollution.

FIRE OR EXPLOSION
- These materials are extremely flammable.
- May form explosive mixtures with air.
- May be ignited by heat, sparks or flames.
- Vapors from liquefied gas are initially heavier than air and spread along ground.
- Vapors may travel to source of ignition and flash back.
- Runoff may create fire or explosion hazard.
- Cylinders exposed to fire may vent and release toxic and flammable gas through pressure relief devices.
- Containers may explode when heated.
- Ruptured cylinders may rocket.

PUBLIC SAFETY
- CALL Emergency Response Telephone Number on Shipping Paper first. If Shipping Paper not available or no answer, refer to appropriate telephone number listed on the inside back cover.
- As an immediate precautionary measure, isolate spill or leak area for at least 100 meters (330 feet) in all directions.
- Keep unauthorized personnel away.
- Stay upwind.
- Many gases are heavier than air and will spread along ground and collect in low or confined areas (sewers, basements, tanks).
- Keep out of low areas.
- Ventilate closed spaces before entering.

PROTECTIVE CLOTHING
- Wear positive pressure self-contained breathing apparatus (SCBA).
- Wear chemical protective clothing that is specifically recommended by the manufacturer. It may provide little or no thermal protection.
- Structural firefighters' protective clothing provides limited protection in fire situations ONLY; it is not effective in spill situations where direct contact with the substance is possible.

EVACUATION
Spill
- See Table 1 - Initial Isolation and Protective Action Distances.

Fire
- If tank, rail car or tank truck is involved in a fire, ISOLATE for 1600 meters (1 mile) in all directions; also, consider initial evacuation for 1600 meters (1 mile) in all directions.

Page 180

GUIDE 117 GASES - TOXIC - FLAMMABLE (EXTREME HAZARD) ERG2008

POTENTIAL HAZARDS

HEALTH
- TOXIC; Extremely Hazardous.
- May be fatal if inhaled or absorbed through skin.
- Initial odor may be irritating or foul and may deaden your sense of smell.
- Contact with gas or liquefied gas may cause burns, severe injury and/or frostbite.
- Fire will produce irritating, corrosive and/or toxic gases.
- Runoff from fire control may cause pollution.

FIRE OR EXPLOSION
- These materials are extremely flammable.
- May form explosive mixtures with air.
- May be ignited by heat, sparks or flames.
- Vapors from liquefied gas are initially heavier than air and spread along ground.
- Vapors may travel to source of ignition and flash back.
- Runoff may create fire or explosion hazard.
- Cylinders exposed to fire may vent and release toxic and flammable gas through pressure relief devices.
- Containers may explode when heated.
- Ruptured cylinders may rocket.

PUBLIC SAFETY
- CALL Emergency Response Telephone Number on Shipping Paper first. If Shipping Paper not available or no answer, refer to appropriate telephone number listed on the inside back cover.
- As an immediate precautionary measure, isolate spill or leak area for at least 100 meters (330 feet) in all directions.
- Keep unauthorized personnel away.
- Stay upwind.
- Many gases are heavier than air and will spread along ground and collect in low or confined areas (sewers, basements, tanks).
- Keep out of low areas.
- Ventilate closed spaces before entering.

PROTECTIVE CLOTHING
- Wear positive pressure self-contained breathing apparatus (SCBA).
- Wear chemical protective clothing that is specifically recommended by the manufacturer. It may provide little or no thermal protection.
- Structural firefighters' protective clothing provides limited protection in fire situations ONLY; it is not effective in spill situations where direct contact with the substance is possible.

EVACUATION
Spill
- See Table 1 - Initial Isolation and Protective Action Distances.

Fire
- If tank, rail car or tank truck is involved in a fire, ISOLATE for 1600 meters (1 mile) in all directions; also, consider initial evacuation for 1600 meters (1 mile) in all directions.

Page 180

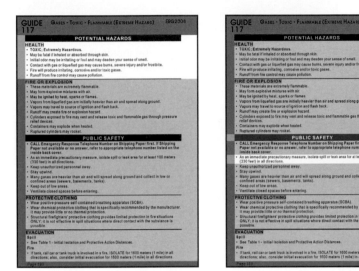

Figure 3.9 (left) The *Potential Hazards* section describes health hazards, with the highest potential hazard listed first.

Figure 3.10 (right) Initial isolation distances are provided as bulleted points in the *Public Safety* section.

Potential Hazards Section

The *Potential Hazards* section describes potential hazards that the material may display **(Figure 3.9)**. Two types of hazards are addressed in separate sections: health hazards and fire or explosion hazards. The highest potential hazard is listed first. This section should be consulted first because it will assist in making decisions regarding the protection of individuals at the incident. Examples of information provided include: ***TOXIC; may be fatal if inhaled or absorbed through skin***, or **EXTREMELY FLAMMABLE.**

Public Safety Section

The *Public Safety* section provides general information regarding immediate isolation of the incident site and recommended type of protective clothing and respiratory protection. This section also lists suggested evacuation distances for small and large spills and for fire situations (which include distances for fragmentation hazards for tanks that might explode).

Isolation distances are provided in the bullet points immediately below the Public Safety section heading **(Figure 3.10)**. The **initial isolation distance** is a distance within which all persons should be considered for evacuation in all directions from the haz mat spill or leak source **(Figure 3.11, p. 174)**. If safe to do so, Awareness-Level personnel and Ops-Level responders should evacuate people from the hazard area to this safe distance (at a minimum) **(Figure 3.12, p. 174)**. They should then prevent others from entering this area by denying entry/access and securing the scene.

Initial Isolation Distance — Distance within which all persons are considered for evacuation in all directions from a hazardous materials incident.

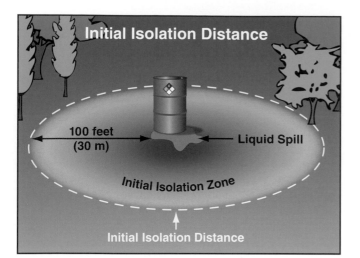

Figure 3.11 The *initial isolation distance* is the distance within which all persons should be considered for evacuation in all directions.

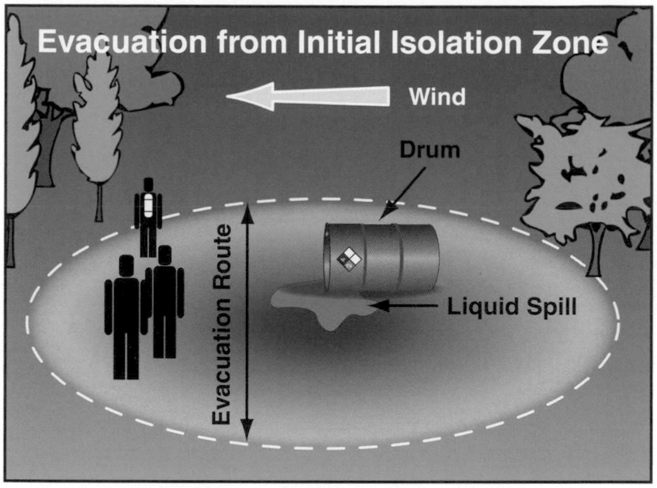

Figure 3.12 When downwind from the incident, evacuation from the initial isolation zone should be conducted at right angles to the prevailing wind direction, if possible.

Figure 3.13 (left) The *Protective Clothing* section recommends the type of personal protective clothing and equipment that should be worn.

Figure 3.14 (right) Evacuation distances for known spills or fires are provided in the *Evacuation* section. These distances may differ from the ones provided in the *Public Safety* section. In this case, when a spill is involved, the user is referred to the *Table of Initial Isolation and Protective Action Distances* (green-bordered pages) for more information.

GUIDE 117 GASES - TOXIC - FLAMMABLE (EXTREME HAZARD) ERG2008

POTENTIAL HAZARDS

HEALTH
- TOXIC; Extremely Hazardous.
- May be fatal if inhaled or absorbed through skin.
- Initial odor may be irritating or foul and may deaden your sense of smell.
- Contact with gas or liquefied gas may cause burns, severe injury and/or frostbite.
- Fire will produce irritating, corrosive and/or toxic gases.
- Runoff from fire control may cause pollution.

FIRE OR EXPLOSION
- These materials are extremely flammable.
- May form explosive mixtures with air.
- May be ignited by heat, sparks or flames.
- Vapors from liquefied gas are initially heavier than air and spread along ground.
- Vapors may travel to source of ignition and flash back.
- Runoff may create fire or explosion hazard.
- Cylinders exposed to fire may vent and release toxic and flammable gas through pressure relief devices.
- Containers may explode when heated.
- Ruptured cylinders may rocket.

PUBLIC SAFETY
- CALL Emergency Response Telephone Number on Shipping Paper first. If Shipping Paper not available or no answer, refer to appropriate telephone number listed on the inside back cover.
- As an immediate precautionary measure, isolate spill or leak area for at least 100 meters (330 feet) in all directions.
- Keep unauthorized personnel away.
- Stay upwind.
- Many gases are heavier than air and will spread along ground and collect in low or confined areas (sewers, basements, tanks).
- Keep out of low areas.
- Ventilate closed spaces before entering.

PROTECTIVE CLOTHING
- Wear positive pressure self-contained breathing apparatus (SCBA).
- Wear chemical protective clothing that is specifically recommended by the manufacturer. It may provide little or no thermal protection.
- Structural firefighters' protective clothing provides limited protection in fire situations ONLY; it is not effective in spill situations where direct contact with the substance is possible.

EVACUATION
Spill
- See Table 1 - Initial Isolation and Protective Action Distances.

Fire
- If tank, rail car or tank truck is involved in a fire, ISOLATE for 1600 meters (1 mile) in all directions; also, consider initial evacuation for 1600 meters (1 mile) in all directions.

Page 180

GUIDE 117 GASES - TOXIC - FLAMMABLE (EXTREME HAZARD) ERG2008

POTENTIAL HAZARDS

HEALTH
- TOXIC; Extremely Hazardous.
- May be fatal if inhaled or absorbed through skin.
- Initial odor may be irritating or foul and may deaden your sense of smell.
- Contact with gas or liquefied gas may cause burns, severe injury and/or frostbite.
- Fire will produce irritating, corrosive and/or toxic gases.
- Runoff from fire control may cause pollution.

FIRE OR EXPLOSION
- These materials are extremely flammable.
- May form explosive mixtures with air.
- May be ignited by heat, sparks or flames.
- Vapors from liquefied gas are initially heavier than air and spread along ground.
- Vapors may travel to source of ignition and flash back.
- Runoff may create fire or explosion hazard.
- Cylinders exposed to fire may vent and release toxic and flammable gas through pressure relief devices.
- Containers may explode when heated.
- Ruptured cylinders may rocket.

PUBLIC SAFETY
- CALL Emergency Response Telephone Number on Shipping Paper first. If Shipping Paper not available or no answer, refer to appropriate telephone number listed on the inside back cover.
- As an immediate precautionary measure, isolate spill or leak area for at least 100 meters (330 feet) in all directions.
- Keep unauthorized personnel away.
- Stay upwind.
- Many gases are heavier than air and will spread along ground and collect in low or confined areas (sewers, basements, tanks).
- Keep out of low areas.
- Ventilate closed spaces before entering.

PROTECTIVE CLOTHING
- Wear positive pressure self-contained breathing apparatus (SCBA).
- Wear chemical protective clothing that is specifically recommended by the manufacturer. It may provide little or no thermal protection.
- Structural firefighters' protective clothing provides limited protection in fire situations ONLY; it is not effective in spill situations where direct contact with the substance is possible.

EVACUATION
Spill
- See Table 1 - Initial Isolation and Protective Action Distances.

Fire
- If tank, rail car or tank truck is involved in a fire, ISOLATE for 1600 meters (1 mile) in all directions; also, consider initial evacuation for 1600 meters (1 mile) in all directions.

Page 180

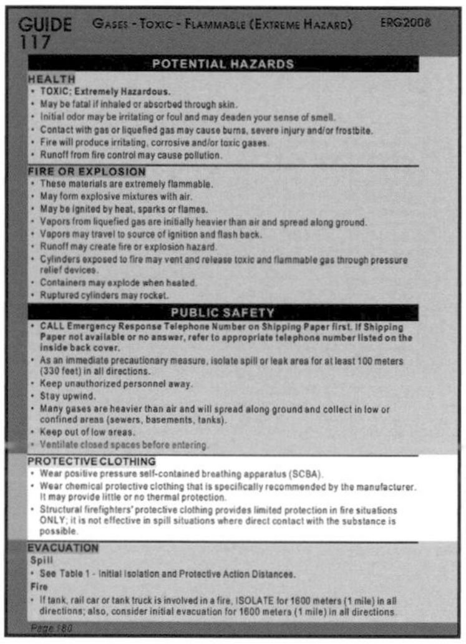

Street Clothes — Clothing that is anything other than chemical protective clothing or structural firefighters' protective clothing, including work uniforms and ordinary civilian clothing.

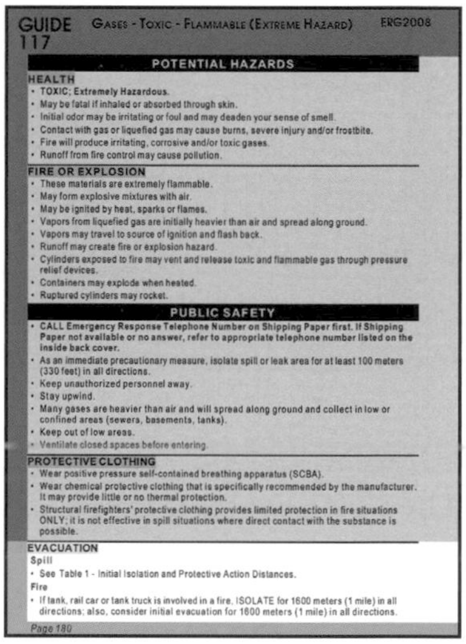

Structural Firefighters' Protective Clothing — General term for the equipment worn by fire and emergency services responders; includes helmets, coats, pants, boots, eye protection, gloves, protective hoods, self-contained breathing apparatus (SCBA), and personal alert safety system (PASS) devices.

Protective Clothing Section. This section recommends the type of personal protective clothing and equipment that should be worn at incidents involving these products **(Figure 3.13)**. Examples include the following:

- **Street clothing** and work uniforms
- **Structural firefighters' protective clothing** (also called bunker gear or turnouts)
- Positive pressure **self-contained breathing apparatus (SCBA)**
- **Chemical protective clothing (CPC)**

Awareness-Level personnel will not have protective clothing beyond street clothes and/or work uniforms and cannot work in hazardous areas. Personnel expected to use protective clothing and equipment such as firefighter protective equipment, SCBA, and CPC should be trained to Ops-Level. More information about protective clothing is provided in Chapter 8, Personal Protective Equipment.

Evacuation Section. This section provides **evacuation** recommendations for spills/large spills and fires **(Figure 3.14)**. When the material is a green-highlighted chemical in the yellow-bordered and blue-bordered pages, this section also directs the reader to consult the tables on the green-bordered pages listing TIH materials and water-reactive materials (see *ERG Table of Initial Isolation and Protective Action Distances [Green Pages]* section). Awareness-Level personnel will probably not be involved in evacuations beyond the initial isolation phase. Evacuation, **sheltering in place**, and protecting/defending in place are discussed in greater detail in Chapter 6.

Emergency Response Section

The third section, *Emergency Response*, describes emergency response topics, including precautions for incidents involving fire, spills or leaks, and first aid. Several recommendations are listed under each of these areas to further assist in the decision-making process. The information on first aid is general guidance before seeking medical care.

Fire Section. This section includes information on what type of extinguishing agent to use on large fires, small fires, and fires involving bulk containers **(Figure 3.15)**. Examples might include foam or water, or a specific type of fire extinguisher for small fires. If foam is recommended, it will specify the type of foam to be used. Recommendations vary by Guide, but additional information may include such things as using unmanned hose holders or cooling containers with flooding quantities of water.

Spill or Leak Section. This section provides actions to take in regards to spills and leaks **(Figure 3.16)**. If a flammable liquid is involved, for example, it would recommend eliminating all ignition sources (see information box). It will also provide basic information needed to mitigate a spill, such as what materials to use to absorb the spill.

Self-Contained Breathing Apparatus (SCBA) — Respirator worn by the user that supplies a breathable atmosphere that is either carried in or generated by the apparatus and is independent of the ambient atmosphere. Respiratory protection is worn in all atmospheres that are considered to be Immediately Dangerous to Life and Health (IDLH). Also called Air Mask or Air Pack.

Chemical Protective Clothing (CPC) — Clothing designed to shield or isolate individuals from the chemical, physical, and biological hazards that may be encountered during operations involving hazardous materials.

Evacuation — Controlled process of leaving or being removed from a potentially hazardous location, typically involving relocating people from an area of danger or potential risk to a safer place.

Shelter in Place — Having occupants remain in a structure or vehicle in order to provide protection from a rapidly approaching hazard (fire, hazardous gas cloud, etc.). Also called *Sheltering, Protection-in-Place,* and *Taking Refuge.*

Figure 3.15 (left) The *Fire* section provides information for firefighters, including appropriate extinguishing agents, types of foam to use, and actions to take or avoid.

Figure 3.16 (right) The *Spill or Leak* section provides actions to take to mitigate a spill, such as what materials to use to absorb the material.

CAUTION

First responders must be properly trained to conduct the actions recommended by the *ERG* before attempting to perform them. They must also have the proper equipment to do so.

WARNING!

Awareness-Level personnel should not handle or touch contaminated or potentially contaminated victims!

Figure 3.17 The *First Aid* section provides basic steps to help victims, such as calling for medical assistance, moving them to fresh air, and flushing contaminated skin and eyes with running water.

Decontamination (Decon) — Process of removing a hazardous, foreign substance from a person, clothing, or area.

Ignition Sources at an Incident Scene

Many potential ignition sources may exist at the scene of a hazardous materials incident. Emergency responders need to be aware of the following items:

- Open flames
- Static electricity
- Existing pilot lights
- Electrical sources including non-explosion-proof electrical equipment
- Internal combustion engines in vehicles and generators
- Heated surfaces
- Cutting and welding operations
- Radiant heat
- Heat caused by friction or chemical reactions
- Cigarettes and other smoking materials
- Cameras
- Road flares

Explosive atmospheres can be ignited by several simple actions:

- Opening or closing a switch or electrical circuit (for example, a light switch)
- Turning on a flashlight
- Operating a radio
- Activating a cell phone

First Aid Section. This section provides basic steps to help victims affected by the hazardous material involved **(Figure 3.17)**. Common recommendations include calling for emergency medical service assistance, moving victims to fresh air, and flushing contaminated skin and eyes with running water (**decontamination**). Avoiding direct contact with the hazardous material is also emphasized. Many of the recommendations provided in this section will be beyond the scope of Awareness-Level personnel due to the need for specialized training and personal protective equipment, the dangers of **cross contamination**, and the necessity of decontaminating victims before first aid is provided (see Chapter 9, Decontamination, and Chapter 13, Victim Rescue and Recovery). For example, Awareness-Level personnel should never enter hazardous atmospheres or potentially contaminated areas. Victims at haz mat incidents may present serious hazards to rescuers because they may be contaminated with the hazardous material! Only first responders with appropriate training, personal protective clothing, and equipment should contact these victims directly. Awareness-Level personnel should not handle or touch contaminated or potentially contaminated victims at haz mat incidents, even to provide basic first aid.

ERG Table of Initial Isolation and Protective Action Distances (Green Pages)

This section of the *ERG* contains a table that lists (by ID number) TIH materials — including certain chemical warfare agents and water-reactive materials that produce toxic gases upon contact with water. The table provides two different types of recommended safe distances: initial isolation distances and **protective action distances (Figure 3.18).** These materials are highlighted for easy identification in both numeric (yellow-bordered) and alphabetic (blue-bordered) *ERG* indexes.

Cross Contamination — Contamination of people, equipment, or the environment outside the hot zone without contacting the primary source of contamination; sometimes called *secondary contamination.*

Protective Action Distance — Downwind distance from a hazardous materials incident within which protective actions should be implemented.

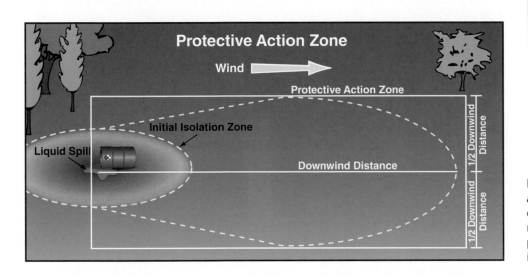

Figure 3.18 The *protective action distance* is the downwind distance from a hazardous materials incident within which protective actions should be implemented.

The table provides isolation and protective action distances for both small (approximately 53 gallons [200 L] or less) and large spills (more than 53 gallons [200 L]) (**Figure 3.19**). A *small spill* is one that involves a single, small package (up to a 55-gallon [208 L] drum), small cylinder, or small leak from a large package. A *large spill* is one that involves a spill from a large package or multiple spills from many small packages.

		SMALL SPILLS				LARGE SPILLS			
		(From a small package or small leak from a large package)				(From a large package or from many small packages)			
		First ISOLATE in all Directions		Then PROTECT persons Downwind during-		First ISOLATE in all Directions		Then PROTECT persons Downwind during-	
ID No.	NAME OF MATERIAL			DAY	NIGHT			DAY	NIGHT
		Meters (Feet)		Kilometers (Miles)	Kilometers (Miles)	Meters (Feet)		Kilometers (Miles)	Kilometers (Miles)
1005 1005	Ammonia, anhydrous Anhydrous ammonia	30 m (100 ft)		0.1 km (0.1 mi)	0.2 km (0.1 mi)	150 m (500 ft)		0.8 km (0.5 mi)	2.3 km (1.4 mi)
1008 1008	Boron trifluoride Boron trifluoride, compressed	30 m (100 ft)		0.1 km (0.1 mi)	0.6 km (0.4 mi)	300 m (1000 ft)		1.9 km (1.2 mi)	4.8 km (3.0 mi)
1016 1016	Carbon monoxide Carbon monoxide, compressed	30 m (100 ft)		0.1 km (0.1 mi)	0.1 km (0.1 mi)	150 m (500 ft)		0.7 km (0.5 mi)	2.7 km (1.7 mi)
1017	Chlorine	60 m (200 ft)		0.4 km (0.3 mi)	1.6 km (1.0 mi)	600 m (2000 ft)		3.5 km (2.2 mi)	8.0 km (5.0 mi)
1023 1023	Coal gas Coal gas, compressed	30 m (100 ft)		0.1 km (0.1 mi)	0.1 km (0.1 mi)	60 m (200 ft)		0.3 km (0.2 mi)	0.4 km (0.3 mi)
1026 1026	Cyanogen Cyanogen gas	30 m (100 ft)		0.2 km (0.1 mi)	0.9 km (0.5 mi)	150 m (500 ft)		1.0 km (0.7 mi)	3.5 km (2.2 mi)
1040 1040	Ethylene oxide Ethylene oxide with Nitrogen	30 m (100 ft)		0.1 km (0.1 mi)	0.2 km (0.1 mi)	150 m (500 ft)		0.8 km (0.5 mi)	2.5 km (1.6 mi)
1045 1045	Fluorine Fluorine, compressed	30 m (100 ft)		0.1 km (0.1 mi)	0.3 km (0.2 mi)	150 m (500 ft)		0.8 km (0.5 mi)	3.1 km (1.9 mi)
1048	Hydrogen bromide, anhydrous	30 m (100 ft)		0.1 km (0.1 mi)	0.4 km (0.3 mi)	300 m (1000 ft)		1.5 km (1.0 mi)	4.5 km (2.8 mi)
1050	Hydrogen chloride, anhydrous	30 m (100 ft)		0.1 km (0.1 mi)	0.4 km (0.2 mi)	60 m (200 ft)		0.3 km (0.2 mi)	1.4 km (0.9 mi)
1051	AC (when used as a weapon)	100 m (300 ft)		0.3 km (0.2 mi)	1.1 km (0.7 mi)	1000 m (3000 ft)		3.8 km (2.4 mi)	7.2 km (4.5 mi)
1051 1051 1051	Hydrocyanic acid, aqueous solutions, with more than 20% Hydrogen cyanide Hydrogen cyanide, anhydrous, stabilized Hydrogen cyanide, stabilized	60 m (200 ft)		0.2 km (0.1 mi)	0.6 km (0.4 mi)	400 m (1250 ft)		1.6 km (1.0 mi)	4.1 km (2.5 mi)

Page 300

TABLE 1 - INITIAL ISOLATION AND PROTECTIVE ACTION DISTANCES

Figure 3.19 The *Table of Initial Isolation and Protective Action Distances* provides isolation and protective action distances for several variables, including small and large spills at night and during the day.

What This Means To You

It's very important to understand what TIH materials are and how to use the *ERG* when confronted with them at an incident. A *toxic inhalation hazard (TIH)* material is a liquid or a gas known (or presumed) to be so toxic to humans as to pose a severe hazard to health during transportation. Small amounts of these products can kill you.

In the *ERG*, isolation or evacuation distances are shown in the Guides (orange-bordered pages) and in the Table of Initial Isolation and Protective Action Distances (green-bordered pages). These distances may be confusing if you aren't thoroughly familiar with the *ERG*.

Some Guides refer to non-TIH materials only (40 Guides) and some refer to both TIH and non-TIH materials (22 Guides). A Guide refers to both TIH and non-TIH materials only when the following sentences appear under the title *EVACUATION* and then *Spill: See the Table of Initial Isolation and Protective Action Distances for highlighted substances. For non-highlighted substances, increase, in the downwind direction, as necessary, the isolation distance shown under "PUBLIC SAFETY."* If these sentences do not appear, then this particular guide refers to non-TIH materials only.

When dealing with a TIH material (highlighted entries in the index lists), the isolation and evacuation distances are found directly in the green-bordered pages. The orange-bordered Guide pages also remind the user to refer to the green-bordered pages for evacuation-specific information involving highlighted materials.

When dealing with a non-TIH material and the Guide refers to both TIH and non-TIH materials, an immediate isolation distance is provided under the heading *PUBLIC SAFETY.* It applies to the non-TIH materials only. In addition, for evacuation purposes, the Guide informs you under the title *EVACUATION* and then *Spill* to increase (for nonhighlighted substances) in the downwind direction, if necessary, the immediate isolation distance listed under *PUBLIC SAFETY.*

For example, Guide 123, Gases -Toxic and/or Corrosive, the *PUBLIC SAFETY* section instructs the user as follows: *Isolate spill or leak area immediately for at least 100 meters (330 feet) in all directions.* In case of a large spill, the isolation area could be expanded from 100 meters (330 feet) to a distance deemed safe by the on-scene incident commander (IC) and emergency responders.

If you are dealing with a non-TIH material and the Guide refers only to non-TIH material, the immediate isolation and evacuation distances are specified as actual distances in the Guide (orange-bordered pages) and are not referenced in the green-bordered pages.

The list is further subdivided into daytime and nighttime situations. This division is necessary because atmospheric conditions significantly affect the size of a chemically hazardous area, and differences can be generally associated with typical daytime and nighttime conditions. The warmer, more active atmosphere normal during the day disperses chemical contaminants more readily than the cooler, calmer conditions common at night. Therefore, during the day, lower toxic concentrations may be spread over a larger area than at night, when higher concentrations may exist in a smaller area. The quantity of material spilled or released and the area affected are both important, but the single most critical factor is the concentration of the contaminant in the air.

As with the isolation distances provided in the orange-bordered pages, the initial isolation distances provided in the green-bordered pages are the distance within which all persons should be considered for evacuation in all directions from an actual hazardous materials spill/leak source. This distance will always be at least 100 feet (30 m). The **initial isolation zone** is a circular zone (with a radius equivalent to the initial isolation distance) within which persons may be exposed to dangerous concentrations upwind of the source and life-threatening concentrations downwind of the source (see Figure 3.11).

Protective actions are those steps taken to preserve the health and safety of emergency responders and the public. People in this area could be evacuated and/or sheltered in-place.

NOTE: If hazardous materials are on fire or have been leaking for longer than 30 minutes, this *ERG* table does not apply. Seek more detailed information on the involved material on the appropriate orange-bordered page in the *ERG*. Also, the orange-bordered pages in the *ERG* provide recommended isolation and evacuation distances for non-highlighted chemicals with poisonous vapors and situations where the containers are exposed to fire.

Emergency Response Centers

The *ERG* provides contact information for emergency response centers that can provide valuable assistance to first responders at haz mat incidents. Canada, Mexico, and the U.S. have government-operated emergency response centers. Contact numbers are provided in the white pages in both the front and the back of the *ERG*.

The Canadian Transport Emergency Centre (CANUTEC) is operated by Transport Canada. This national, bilingual (English and French) advisory center is part of the Transportation of Dangerous Goods Directorate. CANUTEC has a scientific data bank on chemicals manufactured, stored, and transported in Canada and is staffed by professional scientists who specialize in emergency response and are experienced in interpreting technical information and providing advice. Mexico has two emergency response centers: (1) National Center for Communications of the Civil Protection Agency (CENACOM) and (2) Emergency Transportation System for the Chemical Industry (SETIQ), which is operated by the National Association of Chemical Industries.

In the U.S., several emergency response centers, such as the Chemical Transportation Emergency Center (CHEMTREC®), are not government operated. CHEMTREC®, for example, was established by the chemical industry as a public service hotline for firefighters, law enforcement responders, and other emergency service responders to obtain information and assistance for emergency incidents involving chemicals and hazardous materials **(Figure 3.20, p. 180)**. These centers are staffed with experts who can provide 24-hour assistance to emergency responders dealing with haz mat emergencies.

A list of emergency response centers and their telephone numbers is provided in the *ERG*. The responder should collect and provide to the center as much of the following information as safely possible:

- Caller's name, callback telephone number, and FAX number
- Location and nature of problem (spill, fire, etc.)
- Name and identification number of material(s) involved

Initial Isolation Zone
— Circular zone (with a radius equivalent to the initial isolation distance) within which persons may be exposed to dangerous concentrations upwind of the source and may be exposed to life-threatening concentrations downwind of the source.

FOR HAZARDOUS MATERIALS EMERGENCY
Spill, Leak, Fire, Exposure or Accident
CALL CHEMTREC® — Day or Night
800-424-9300
Outside the United States,
Call 703-527-3887
Collect Calls Accepted
www.chemtrec.com
See reverse for more instructions

In a Hazardous Materials Emergency:
Isolate the area and contact CHEMTREC® immediately with as much of the following information as possible.

IDENTIFY:
☐ Your name/organization
☐ Location you are calling from
☐ Call-back number

INCIDENT:
☐ Location of incident
☐ Time of incident
☐ Weather/environment
☐ Product(s) involved
☐ Quantity
☐ Container type
☐ Any injuries/deaths
☐ Assistance on site/en route/requested

OTHER INFO:
☐ UN, NA, or STCC Code
☐ Origin of shipment and shipper
☐ Carrier
☐ Destination/consignee
☐ Truck/car/trailer/flight #
☐ Bill of lading #

CHEMTREC® • 800-424-9300

Figure 3.20 CHEMTREC® can provide first responders with a variety of information and assistance at haz mat incidents.

- Shipper/consignee/point of origin

- Carrier name, railcar reporting marks (letters and numbers), or truck number

- Container type and size

- Quantity of material transported/released

- Local conditions (weather, terrain, proximity to schools, hospitals, waterways, etc.)

- Injuries, exposures, current conditions involving spills, leaks, fires, explosions, and vapor clouds, etc.

- Local emergency services that have been notified

The emergency response center will do the following:

- Confirm that a chemical emergency exists.

- Record details electronically and in written form.

- Provide immediate technical assistance to the caller.

- Contact the shipper of the material or other experts.

- Provide the shipper/manufacturer with the caller's name and callback number so that the shipper/manufacturer can deal directly with the party involved.

United States Centers

- CHEMTREC®: 800-424-9300
- CHEM-TEL, INC.: 888-255-3924
- INFOTRAC: 800-535-5053
- 3ECOMPANY: 800-451-8346
- National Response Center: 800-424-8802
- U.S. Army Operations Center (DOD shipments involving explosives and ammunition): 703-697-0218 (call collect)
- Defense Logistics Agency (DOD shipments involving other dangerous goods): 800-851-8061

Canada Centers
- CANUTEC: 613-996-6666 collect (Emergency only)
- Alberta: 800-272-9600
- British Columbia: 800-663-3456
- Manitoba: 204-945-4888
- New Brunswick: 800-565-1633
- Newfoundland: 709-772-2083
- Northwest Territories: 867-920-8130
- Nova Scotia: 800-565-1633
- Nunavut: 800-693-1666
- Ontario, Quebec: Local police
- Prince Edward Island: 800-565-1633
- Saskatchewan: 800-667-7525
- Yukon Territory: 867-667-7244

Mexico Centers
- SETIQ: 01-800-00-214-00
- CENACOM: 01-800-00-413-00

Brazil Center
PRÓ-QUÍMICA: 0-800-118270

Argentina Center
CIQUIME: 0-800-222-2933

Isolation and Denial of Entry

Isolation and scene control is one of the most important means by which Awareness-Level personnel and Ops-Level responders can ensure their own safety and the safety of others. Separating people from the potential source of harm is necessary to protect the life safety of everyone involved. It is also necessary to prevent the spread of hazardous materials through cross contamination (see Chapter 9, Decontamination). Isolation involves physically securing and maintaining the emergency scene by establishing isolation perimeters (or cordons) and denying entry to unauthorized persons **(Figure 3.21, p. 182).** It also includes preventing contaminated or potentially contaminated individuals (or animals) from leaving the scene in order to stop the spread of hazardous materials. The process may continue with either evacuation, defending in place, or shelter in place of people located within protective-action zones.

The **isolation perimeter** (sometimes called the *outer perimeter* or *outer cordon*) is the boundary established to prevent access by the public and unauthorized persons. If an incident is inside a building, posting personnel at entrances can set the isolation perimeter by denying entry and exit from the building. If the incident is outside, the perimeter might be set at the surrounding intersections with response vehicles or law enforcement officers diverting

Isolation Perimeter — Outer boundary of an incident that is controlled to prevent entrance by the public or unauthorized persons.

Figure 3.21 Keeping others from entering the hazardous area is one of the most important things Awareness-Level personnel can do. Using barrier tape is an effective way to accomplish this. *Courtesy of the U.S. Air Force.*

Figure 3.22 Streets and intersections can be blocked by vehicles and barricades.

Figure 3.23 If a crime or terrorist attack is suspected at a haz mat incident, immediately notify law enforcement. *Courtesy of August Vernon.*

traffic and pedestrians **(Figure 3.22)**. Ropes, cones, and barrier tape can also be used. In some cases, a traffic cordon may be established beyond the outer cordon to prevent unauthorized vehicle access only, whereas pedestrian traffic is still allowed.

The isolation perimeter can be expanded or reduced as needed, and it is used to control both access and egress from the scene. For example, in some cases, the initial isolation perimeter established by first responders is expanded outward as additional help arrives. Law enforcement officers are often used to establish and maintain isolation perimeters. For more information, see Chapter 6, Strategic Goals and Tactical Objectives.

Terrorist Incidents

Because terrorist and criminal incidents may differ from ordinary haz mat incidents, there are some specific, unique actions that need to be taken, such as notifying law enforcement immediately **(Figure 3.23)**. It is also important to be alert for secondary devices and booby traps since terrorists and criminals may deliberately target first responders or crowds.

As true at all haz mat incidents, Awareness-Level personnel and Ops-Level responders should do all of the following:

• Protect themselves and others by isolating the incident and denying entry.

• Prevent contaminated persons and animals from leaving the scene, if possible, and direct them to a safe area to wait for help.

- Avoid contacting contaminants or contaminated surfaces.

- Remember that WMD agents may be deadly in very small amounts, and biological agents may not cause symptoms for several days.

Finally, Awareness-Level personnel are likely to be on or near the scene when an incident or attack occurs, and therefore they make important witnesses. Law enforcement will want to know what they saw and when. In addition personnel should do the following:

- Document their observations.

- Take pictures, if possible.

- Make note of other witnesses and observers at the scene.

- Protect evidence at the crime scene as best they are able.

Summary

Awareness-Level personnel need to understand their agency's predetermined procedures in regard to hazardous materials incidents and know which notifications to make (for example, who they need to call for help and how). They should also be able to use the *Emergency Response Guidebook* which may assist them identify the hazardous material involved. The *ERG* will also help them determine isolation distances for purposes of scene control and taking protective actions. They must be familiar with safety procedures to protect themselves and others. They also need to understand certain actions they need to take if terrorism or criminal activity is suspected at an incident.

Review Questions

1. What are the responsibilities of Awareness-Level personnel when faced with an incident involving hazardous materials?

2. What is the purpose of predetermined procedures and emergency response plans?

3. What is provided on the yellow-shaded pages of the *ERG*?

4. What is provided on the blue-shaded pages of the *ERG*?

5. What is provided on the orange-shaded pages of the *ERG*?

6. What are several potential ignition sources at a hazardous materials incident?

7. What is provided on the green-shaded pages of the *ERG*?

8. What is a TIH material?

9. What types of information should be provided to an emergency response center?

10. What is an isolation perimeter?

Using the U.N. Identification Number

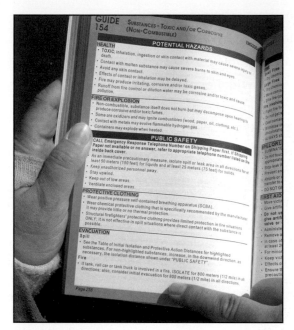

Step 1: Identify the four-digit U.N. identification number.

Step 3: Refer to the orange-bordered page with the appropriate guide number for information on managing the incident.

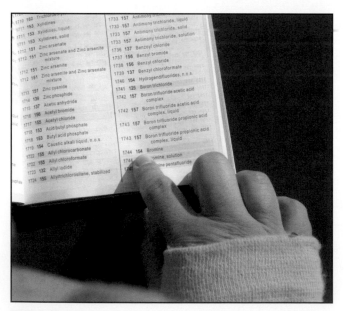

Step 2: Refer to the appropriate yellow-bordered pages reference guide number.

Step 4: For highlighted chemicals refer to the green-bordered pages for initial isolation by looking up the identification number.

Note: The green-bordered pages provide initial isolation and protective action distances. The action taken will vary according to whether or not the spill is large/small and whether it is night/day.

Using the Material Name

Step 1: Identify the name of the material.

Step 2: Refer to the name of the material in the blue-bordered pages to locate the correct guide number.

Step 3: Refer to the orange-bordered page with the appropriate guide number for information on managing the incident.

Step 4: For highlighted chemicals refer to the green-bordered pages for initial isolation by looking up the identification number.

Using the Container Profile

Step 2: Refer to the appropriate guide number in the circle and go to the appropriate orange-bordered page.

Note: Using this method does not identify whether or not the chemical is a highlighted chemical.

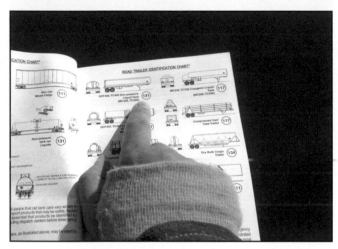

Step 1: Identify the profile of the container and locate the profile in the white pages of the *ERG*.

Using the Placard

Step 2: Refer to the appropriate guide number in the circle and go to the appropriate orange-bordered page.

Note: Using this method does not identify whether or not the chemical is a highlighted chemical.

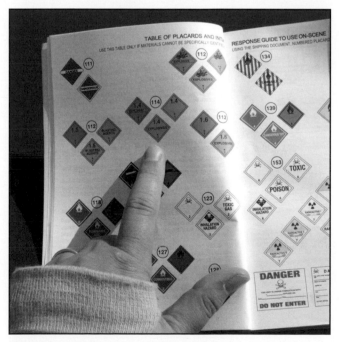

Step 1: Identify the placard and locate it in the white pages of the *ERG*.

Chemical Properties and Hazardous Materials Behavior

Chapter Contents

Divider page photo courtesy of Rich Mahaney.

Key Terms

Competencies

NFPA® 472:	5.2.3(1)(a)(v)	5.2.3(1)(a)(xi)	5.2.3(2)	5.2.4(1)
5.2.1.4	5.2.3(1)(a)(vi)	5.2.3(1)(a)(xii)	5.2.3(3)	5.2.4(2)
5.2.3(1)(a)(i)	5.2.3(1)(a)(vii)	5.2.3(1)(a)(xiii)	5.2.3(4)	
5.2.3(1)(a)(ii)	5.2.3(1)(a)(viii)	5.2.3(1)(a)(xiv)	5.2.3(5)	
5.2.3(1)(a)(iv)	5.2.3(1)(a)(ix)	5.2.3(1)(a)(xv)	5.2.3(6)	

Chemicial Properties and Hazardous Materials Behavior

Learning Objectives

1. Discuss the three states of matter. [NFPA® 472, 5.2.3(1)(a)(vii),5.2.3(1)(a)(ix)]

2. Discuss the flammability of various hazardous materials. [NFPA® 472, 5.2.3(1)(a)(iv-vi), 5.2.3(1)(a)(xii)]

3. Describe vapor pressure. [NFPA® 472, 5.2.3(1)(a)(xiv)]

4. Explain boiling point. [NFPA® 472, 5.2.3(1)(a)(i)]

5. Define melting point, freezing point, and sublimation.

6. Describe vapor density. [NFPA® 472, 5.2.3(1)(a)(xiii)]

7. Define solubility and miscibility. [NFPA® 472, 5.2.3(1)(a)(xv)]

8. Discuss specific gravity. [NFPA® 472, 5.2.3(1)(a)(xi)]

9. Define persistence. [NFPA® 472, 5.2.3(1)(a)(viii)]

10. Define reactivity and describe the reactivity triangle. [NFPA® 472, 5.2.3(1)(a)(il)]

11. Describe the General Hazardous Materials Behavior Model. [NFPA® 472, 5.2.1.4, 5.2.3(2-6), 5.2.4(2), 5.2.4(4)]

Chapter 4
Chemical Properties and Hazardous Materials Behavior

Case History

On April 7, 2000 units from a volunteer fire department responded to a reported grass fire around a large fuel tank. Two engines, a command unit, and one patrol car were dispatched to the incident. Arriving on the scene, the Incident Commander (IC) conducted a size-up and noticed a small grass fire around a fuel tank. The IC radioed for the other arriving units to pull a handline. The tank was lying on the ground, in an east-west orientation, with the east end of the tank slightly elevated. The grass fire was comprised of burning grass and rubbish located on the north, east, and south sides of the fuel tank.

Two firefighters were on the north side of the tank using water from a hose to knock down the fire and to cool the tank. Two other fire fighters were on the south side of the tank doing the same thing. A civilian approached the IC and informed him that he wanted to cut a hole in the end of the tank. The IC acknowledged him and agreed without finding out how the civilian planned to cut the hole in the tank. The civilian then began using a cutting torch to cut a hole near the threaded opening of the 2-inch (51 mm) fill pipe while another civilian stood nearby and watched. When the cutting began, there was a loud noise (reported as sounding like a jet engine), the tank swelled, and then exploded. The east end of the tank separated at the seam and was blown 114 feet (35 m) in an easterly direction. The tank turned 180 degrees, with the opened east end of the tank now facing west.

At the time of the explosion, one of the firefighters was in the direct path of the east end of the tank and was killed instantly. The IC, Assistant Chief, firefighters, and a city police officer began administering medical treatment to both civilians and another injured firefighter. This firefighter had been knocked to the ground and received numerous injuries to his legs. One of the civilians had been knocked to the ground, receiving severe burns, and was transported to the State's burn treatment center. The second civilian had both legs severed at the knees from the flying debris. He was transported by helicopter to the regional hospital where he was later pronounced dead as a result of the traumatic amputation of his lower extremities.

After investigating the incident, NIOSH investigators concluded that, to minimize the risk of similar occurrences, fire departments should:

- Ensure that, for fires involving potentially dangerous substances, fire fighters utilize and follow the guidelines set forth in the U.S. Department of Transportation's *Emergency Response Guidebook.*

- Develop, implement, and enforce standard operating procedures (SOPs) that address fire fighter safety regarding emergency operations for hazardous substance releases.

- Ensure that emergency response personnel adhere to the procedures outlined in 29 CFR 1910.120(q)2 - Emergency response to hazardous substance releases.

Source: NIOSH

An uncontrolled release of a hazardous material from a container can create a variety of problems. Operations-Level responders need to know how hazardous materials behave in addition to the symptoms and effects of an exposure. The material's physical properties often determine the behavior of hazardous materials. The material's physical state, flammability, boiling point, persistence, chemical reactivity, and other properties affect how it behaves, determine the harm it can cause, and influence the effect it may have on its container (as well as the people, other living organisms, other chemicals, and the environment it contacts) **(Figure 4.1)**. A material's physical properties will also determine how it behaves once it has been released from its container.

This chapter discusses some of the physical properties of hazardous materials that emergency responders need to know including the following:

- State of matter

- Flammability

- Vapor pressure

- Boiling point

- Melting point/freezing point/sublimation

- Persistence

- Particle size

- Vapor density

- Solubility/Miscibility

- Specific gravity

- Reactivity

Figure 4.1 A material's chemical and physical properties determine how it behaves once it is released from its container and enters the environment. *Courtesy of Phil Linder.*

Effects of radiological materials were discussed in Chapter 1. This chapter will also examine the ways in which hazardous materials may be released from their containers and how they will behave afterwards. The explanations of the terms found in this book may vary slightly from those used in scientific circles, but they will aid first responders in applying technical data to real-world incidents.

States of Matter

Matter is found in three states: *gas, liquid,* and *solid.* First responders must understand that different hazardous materials may be found any one of these states and that the material's state of matter influences its behavior **(Figure 4.2)**. This behavior in turn can influence the nature of the hazards that the material presents. The three states of matter are as follows:

- *Gas* — Fluid that has neither independent shape nor volume; gases tend to expand indefinitely.
- *Liquid* — Fluid that has no independent shape but does have a specific volume; liquids flow in accordance with the laws of gravity.
- *Solid* — Substance that has both a specific shape (without a container) and volume.

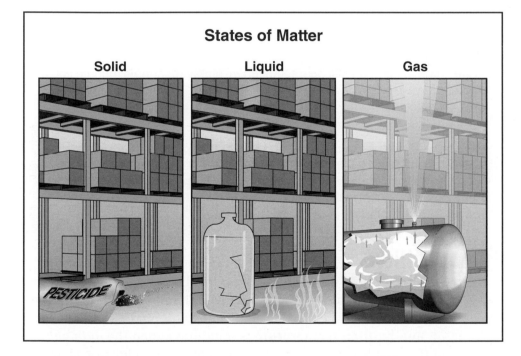

Figure 4.2 The state of a material (solid, liquid, or gas) influences the way it behaves. First responders must be prepared to deal with hazardous materials in all three states.

Simplistically, if a hazardous material is a gas, it will be present in the air, potentially presenting a breathing/inhalation hazard. Some may also present a contact hazard. Gases are difficult (if not impossible) to contain for mitigation purposes and will move according to prevailing wind and air movement. Compressed gases will expand rapidly when released, potentially threatening large areas.

Liquids will flow or pool according to surface contours and topography, permitting opportunities for containment or confinement. While many liquids may give off vapors that become inhalation hazards, the liquid itself is primarily a splash or contact hazard.

Solids may be moved by exterior forces (wind, water, gravity, etc.) but will typically remain in place unless acted upon. Particle size of solids such as dusts, fumes, or powders may influence their behavior, for example, how long particles may remain suspended in air. Larger particles will settle out more quickly.

Is It a Solid, Liquid, or Gas? Industry Terms Used to Describe Air Contaminants

Air contaminants are commonly classified as either particulate or gas and vapor contaminants. The most common particulate contaminants include dusts, fumes, mists, aerosols, and fibers. Definitions are as follow:

- *Dust* — Solid particle that is formed or generated from solid organic or inorganic materials by reducing its size through mechanical processes such as crushing, grinding, drilling, abrading, or blasting.

- *Fume* — Suspension of particles that form when material from a volatilized (vapor state) solid condenses in cool air. In most cases, the solid, smokelike particles resulting from the condensation react with air to form an oxide.

- *Mist* — Finely divided liquid suspended in the atmosphere. Mists are generated by liquids condensing from a vapor back to a liquid or by breaking up a liquid into a dispersed state by splashing, foaming, or atomizing.

- *Aerosol* — Form of mist characterized by highly respirable, minute liquid particles.

- *Fiber* — Solid particle whose length is several times greater than its diameter.

- *Vapor* — Gaseous form of a substance that is normally in a solid or liquid state at room temperature and pressure. It is formed by evaporation from a liquid or sublimation from a solid. Examples can be found where parts cleaning and painting takes place and solvents are used. Vapors are the volatile forms of these substances.

Autoignition Temperature — Same as ignition temperature except that no external ignition source is required for ignition because the material itself has been heated to ignition temperature; temperature at which autoignition occurs through the spontaneous ignition of the gases or vapors emitted by a heated material.

Flammability — Fuel's susceptibility to ignition.

Flash Point — Minimum temperature at which a liquid gives off enough vapors to form an ignitable mixture with air near the surface of the liquid.

Flammability

The majority of hazardous materials incidents involve materials that are flammable. Flammable materials can cause damage to life and property when they ignite, burn, or explode. A flammable hazard depends on properties such as flash point, **autoignition temperature** (sometimes called the *autoignition point*), and flammable (explosive or combustible) range **(Figure 4.3)**. Other important properties include boiling point, vapor pressure, vapor density, specific gravity, and solubility (see following sections). All of this information is vital for determining incident strategies and tactics.

Flash Point

Flash point is the minimum temperature at which a liquid or volatile solid gives off sufficient vapors to form an ignitable mixture with air near its surface **(Figure 4.4)**. At this temperature, the vapors will flash (in the presence of an ignition source) but will not continue to burn. Do not confuse flash point with

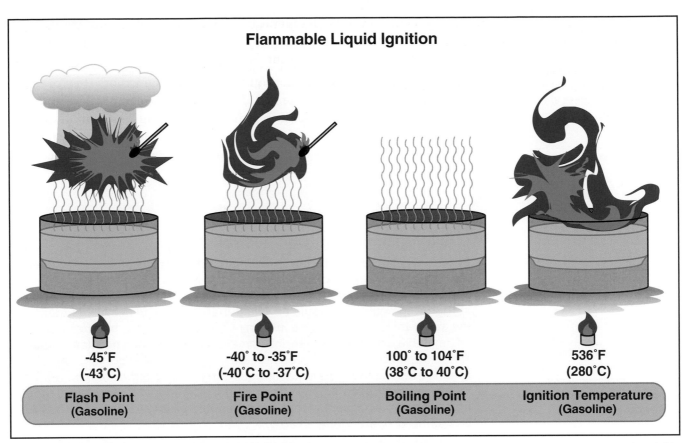

Flammable Liquid Ignition

-45°F (-43°C)	-40° to -35°F (-40°C to -37°C)	100° to 104°F (38°C to 40°C)	536°F (280°C)
Flash Point (Gasoline)	**Fire Point** (Gasoline)	**Boiling Point** (Gasoline)	**Ignition Temperature** (Gasoline)

Figure 4.3 Flammable and combustible liquids produce varying amounts of vapors depending on their temperature. In addition to flash point, there is the *fire point*, the temperature at which the liquid gives off sufficient vapors to sustain combustion. It will boil at the *boiling point* and automatically ignite at the *autoignition temperature*.

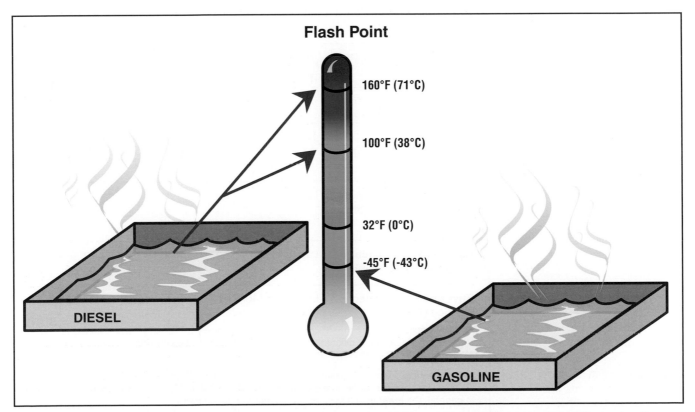

Flash Point

160°F (71°C)

100°F (38°C)

32°F (0°C)

-45°F (-43°C)

DIESEL

GASOLINE

Figure 4.4 *Flash point* is the temperature at which a liquid gives off sufficient vapors to ignite, but not sustain combustion. Liquids with low flash points are easily ignited.

Fire Point — Temperature at which a liquid fuel produces sufficient vapors to support combustion once the fuel is ignited. The fire point is usually a few degrees above the flash point.

fire point. **Fire point** is the temperature at which enough vapors are given off to support continuous burning. The fire point temperature is usually only slightly higher than the flash point.

The liquids themselves do not burn but rather the vapors they produce burn. As the temperature of the liquid increases, more vapors are emitted. Vapors are emitted below the flash point but not in sufficient quantities to ignite. Therefore, if a substance is not at its flash-point temperature, it will not burn. Flammable gases have no flash points because they are already in the gaseous state.

Nonflammable — Incapable of combustion under normal circumstances; normally used when referring to liquids or gases.

Is it Flammable or Combustible?

In everyday language, the terms *flammable* and *combustible* can be used interchangeably to denote a substance that will burn. In the world of hazardous materials, these terms take on very specific meanings, particularly in regard to liquids. It should be noted that the term *inflammable* means the same thing as *flammable* in many parts of the world. In other words, inflammable materials will burn; inflammable does not mean **nonflammable**. A nonflammable material does not ignite easily. In Mexico, for example, a tank truck carrying flammable liquids may read either *flammable* or *inflammable* **(Figure 4.5)**. While Transport Canada (TC) allows only the term *flammable*, Canadian responders should be aware that *inflammable* is the French word for *flammable*.

The flash point is commonly used to determine how flammable a liquid is. Liquids that have low flash points and burn very easily are designated as *flammable liquids*, whereas liquids with higher flash points which do not burn as easily are called *combustible liquids*. However, different U.S. agencies reference different flash points in their designations of *flammable* and *combustible* substances.

Flash Points Used to Determine Whether a Liquid Is Flammable or Combustible		
Agency	**Flammable Liquid**	**Combustible Liquid**
Department of Transportation (DOT)	140°F (60°C) or less	Greater than 141°F (60.5°C) and below 200°F (93°C)
Occupational Safety and Health Administration (OSHA)	Less than 100°F (38°C)	100° F (38°C) or greater
National Fire Protection Association (NFPA®)	Less than 100°F (38°C)	100°F (38°C) or greater
Environmental Protection Agency (EPA)	Less than 140°F (60°C) (Ignitable Waste)	

NOTE: Canada does not differentiate between flammable and combustible liquids for purposes of transportation. It uses only the term flammable.

Flammable liquids such as gasoline and acetone have flash points well below 100°F (38°C). Combustible liquids such as fuel oils and lubricating oils have flash points above 100°F (38°C). In general, the lower the flash point, the greater the fire hazard.

Figure 4.5 In many countries, the term *inflammable* means *flammable. Courtesy of Rich Mahaney.*

Autoignition Temperature

The *autoignition temperature* of a substance is the minimum temperature to which the fuel in air must be heated to initiate self-sustained combustion without initiation from an independent ignition source. This temperature is the point at which a fuel spontaneously ignites. All flammable materials have autoignition temperatures, and these are considerably higher than the flash and fire points. For example, the autoignition temperature of gasoline is about 536°F (280°C), but the flash point of gasoline is -45°F (-43°C). This difference means that at -45°F (-43°C), gasoline will ignite if a match is waved through its vapors, whereas at 536°F (280°C) it ignites all by itself.

NOTE: The term *ignition temperature* is often used synonymously with *autoignition temperature* or *autoignition point*. They are always the same temperature. However, the NFPA® defines *ignition temperature* as "the minimum temperature required to initiate or cause self-sustained combustion, independent of the heating or heated element." It defines *autoignition temperature* as "the temperature at which a mixture will spontaneously ignite."

Toxic Products of Combustion

While the heat energy from a fire is a danger to anyone directly exposed to it, toxic smoke causes most fire deaths. Smoke is an aerosol comprised of gases, vapor, and solid particulates. Fire gases, such as carbon monoxide, are generally colorless, while vapor and particulates give smoke its varied colors. Most components of smoke are toxic and present a significant threat to human life. The materials that make up smoke vary from fuel to fuel, but generally all smoke is toxic. **Table 4.1, p. 199** lists some of the more common products of combustion and their toxic effects.

Carbon monoxide (CO) is a byproduct of the incomplete combustion of organic (carbon-containing) materials. This gas is probably the most common product of combustion encountered in structure fires and exposure

Carbon Monoxide (CO) — Colorless, odorless, dangerous gas (both toxic and flammable) formed by the incomplete combustion of carbon. It combines more than 200 times as quickly with hemoglobin as oxygen, thus decreases the blood's ability to carry oxygen.

Hydrogen Cyanide (HCN) — Colorless, toxic gas with a faint odor similar to bitter almonds; produced by the combustion of nitrogen-bearing substances.

Carbon Dioxide (CO₂) — Colorless, odorless, heavier than air gas that neither supports combustion nor burns. CO₂ is used in portable fire extinguishers as an extinguishing agent to extinguish Class B or C fires by smothering or displacing the oxygen.

Lower Explosive Limit (LEL) — Lowest percentage of fuel/oxygen mixture required to support combustion. Any mixture with a lower percentage would be considered "too lean" to burn. Also called Lower Flammable Limit (LFL).

Upper Explosive Limit (UEL) — Maximum concentration of vapor or gas in air that will allow combustion to occur. Concentrations above this are called "too rich" to burn. Also called Upper Flammable Limit (UFL).

to it is frequently identified as the cause of death for civilian fire fatalities and firefighters who have run out of air in their SCBAs. Carbon monoxide acts as a chemical asphyxiant by binding with hemoglobin in the blood that transports oxygen throughout the body.

Hydrogen cyanide (HCN), produced in the combustion of materials containing nitrogen, is also commonly encountered in smoke, although at lower concentrations than CO. HCN also acts as a chemical asphyxiant but with a different mechanism of action. HCN acts to prevent the body from using oxygen at the cellular level. HCN is a significant byproduct of the combustion of polyurethane foam, which is commonly used in furniture and bedding.

Carbon dioxide (CO₂) is a product of complete combustion of organic materials. It is not toxic in the same manner as CO or HCN, but it acts as a simple asphyxiant by displacing oxygen. Carbon dioxide also acts as a respiratory stimulant, increasing respiratory rate.

These are only three of the more common products of combustion that can be hazardous to first responders. It is important to remember that the toxic effects of smoke inhalation are not the result of any one gas; it is the interrelated effect of all the toxic products present.

Irritants in smoke are those substances that cause breathing discomfort and inflammation of the eyes, respiratory tract, and skin. Depending on the fuels involved, smoke will contain a wide range of irritating substances.

Flammable, Explosive, or Combustible Range

The *flammable, explosive*, or *combustible range* is the percentage of the gas or vapor concentration in air that will burn or explode if ignited. The **lower explosive limit (LEL)** or *lower flammable limit (LFL)* of a vapor or gas is the lowest concentration (or lowest percentage of the substance in air) that will produce a flash of fire when an ignition source is present. At concentrations lower than the LEL, the mixture is too *lean* to burn.

The **upper explosive limit (UEL)** or *upper flammable limit (UFL)* of a vapor or gas is the highest concentration (or highest percentage of the substance in air) that will produce a flash of fire when an ignition source is present. At higher concentrations, the mixture is too *rich* to burn **(Figure 4.6, p. 200)**. Within the upper and lower limits, the gas or vapor concentration will burn rapidly if ignited. Atmospheres within the flammable range are particularly dangerous. **Table 4.2, p. 200** provides the flammable ranges for selected materials.

What This Means To You

Products with a very low LEL and products with a wide range between the LEL and UEL are the most dangerous. Also, just because the concentration is above the UEL does not mean that you are safe. If the concentration drops for any reason (for example, more fresh air is introduced, which dilutes the concentration, or the concentration is less than the UEL in places where you did not monitor), you could be in an explosive atmosphere anyway.

Table 4.1
Common Products of Combustion and Their Toxic Effects

Acetaldehyde	Colorless liquid with a pungent choking odor, which is irritating to the mucous membranes and especially the eyes. Breathing vapors will cause nausea, vomiting, headache and unconsciousness.
Acrolein	Colorless to yellow volatile liquid with a disagreeable choking odor, this material is irritating to the eyes and mucous membranes. This substance is extremely toxic; inhalation of concentrations as little as 10 ppm may be fatal within a few minutes.
Asbestos	A magnesium silicate mineral that occurs as slender, strong flexible fibers. Breathing of asbestos dust causes asbestosis and lung cancer.
Benzene	Colorless liquid with a petroleum-like odor. Acute exposure to benzene can result in dizziness, excitation, headache, difficulty breathing, nausea and vomiting. Benzene is also a carcinogen.
Benzaldehyde	Colorless to clear yellow liquid with a bitter almond odor. Inhalation of concentrated vapor is irritating to the eyes, nose, and throat.
Carbon Monoxide	Colorless, odorless gas. Inhalation of carbon monoxide causes headache, dizziness, weakness, confusion, nausea, unconsciousness, and death. Exposure to as little as 0.2% carbon monoxide can result in unconsciousness within 30 minutes. Inhalation of high concentration can result in immediate collapse and unconsciousness.
Formaldehyde	Colorless gas with a pungent irritating odor that is highly irritating to the nose. 50-100 ppm can cause severe irritation to the respiratory track and serious injury. Exposure to high concentrations can cause injury to the skin. Formaldehyde is a suspected carcinogen.
Glutaraldehyde	Light yellow liquid that causes severe irritation of the eyes and irritation of the skin.
Hydrogen Chloride	Colorless gas with a sharp, pungent odor. Mixes with water to form hydrochloric acid. Hydrogen chloride is corrosive to human tissue. Exposure to hydrogen chloride can result in irritation of skin and respiratory distress.
Isovaleraldehyde	Colorless liquid with a weak, suffocating odor. Inhalation causes respiratory distress, nausea, vomiting and headache.
Nitrogen Dioxide	Reddish brown gas or yellowish-brown liquid, which is highly toxic and corrosive.
Particulates	Small particles that can be inhaled and be deposited in the mouth, trachea, or the lungs. Exposure to particulates can cause eye irritation, respiratory distress (in addition to health hazards specifically related to the particular substances involved).
Polycyclic Aromatic Hydrocarbons (PAH)	PAH are a group of over 100 different chemicals that generally occur as complex mixtures as part of the combustion process. These materials are generally colorless, white, or pale yellow-green solids with pleasant odor. Some of these materials are human carcinogens.
Sulfur Dioxide	Colorless gas with a choking or suffocating odor. Sulfur dioxide is toxic and corrosive and can irritate the eyes and mucous membranes.

Source: *Computer Aided Management of Emergency Operations (CAMEO) and Toxicological Profile for Polycyclic Aromatic Hydrocarbons.*

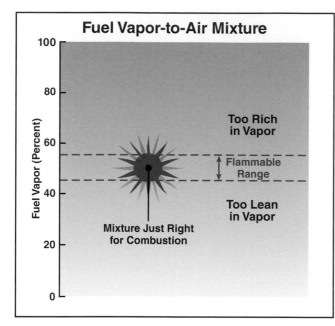

Fuel Vapor-to-Air Mixture

Figure 4.6 At concentrations lower than the LEL, the mixture is too lean to burn. At concentrations above the UEL, the mixture is too rich. The flammable range is the point at which the mixture will ignite and burn.

Table 4.2
Flammable Ranges for Selected Materials

Material	Lower Flammable Limit (LFL) (percent by volume)	Upper Flammable Limit (UFL) (percent by volume)
Acetylene	2.5	100.0
Carbon Monoxide	12.5	74.0
Ethyl Alcohol	3.3	19.0
Fuel Oil No. 1	0.7	5.0
Gasoline	1.4	7.6
Methane	5.0	15.0
Propane	2.1	9.5

Source: *NIOSH Pocket Guide to Chemical Hazards*

Vapor Pressure

Vapor Pressure — (1) Measure of the tendency of a substance to evaporate. (2) The pressure at which a vapor is in equilibrium with its liquid phase for a given temperature. Liquids that have a greater tendency to evaporate have higher vapor pressures for a given temperature.

Vapor pressure is the pressure exerted by a saturated vapor above its own liquid in a closed container, or more simply, it is the pressure produced or exerted by the vapors released by a liquid. Vapor pressure can be viewed as the measure of the tendency of a substance to evaporate.

Vapor pressure can be expressed in terms of pounds per square inch (psi), kilopascals (kPa), bars, or in *millimeters of mercury (mmHg)* or *atmospheres (atm)*. Vapor pressures reported in reference materials may use any of these units, but vapor pressures are usually reported in *mmHg* on safety data sheets (SDSs).

The following facts regarding vapor pressure are important to first-responders:

- The higher the temperature of a substance, the higher the vapor pressure will be **(Figure 4.7)**. In other words, the vapor pressure of a substance at 100°F (38°C) will always be higher than the vapor pressure of the same substance at 68°F (20°C). Higher temperatures provide more energy to a liquid, which in turn allows more liquid to escape into a gaseous form. The gas rises above the liquid and exerts a downward pressure.

- Vapor pressures reported on MSDSs in mmHg are usually very low; for comparison, 760 mmHg is equivalent to 14.7 psi or 1 atmosphere at standard temperature *(14.7 psi = 760 mmHg = 1 atm = 101 kPa = 1 bar)*.

- The lower the boiling point of a substance, the higher its vapor pressure will be. The low boiling point means less heat is required to change some of the liquid into a gas. For example, water requires a lot of heat to boil (212°F [100°C]), but some substances boil at room temperature (68°F [20°C]).

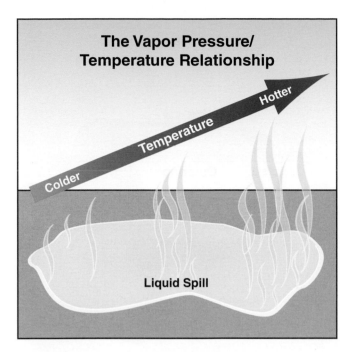

The Vapor Pressure/ Temperature Relationship

Colder

Temperature

Hotter

Liquid Spill

Figure 4.7 At a given temperature, materials with high vapor pressures produce more vapors than materials with lower vapor pressures. As the temperature increases, all materials will produce more vapors.

What This Means To You

Following are the vapor pressures of six substances at 68°F (20°C):

- *Motor oil:* less than 0.01 mmHg
- *Sarin:* 2.1 mmHg (at 70°F/21°C)
- *Water:* 25 mmHg
- *Acetone:* 180 mmHg
- *Isopropylamine:* 478 mmHg
- *Chlorine:* 5,168 mmHg

You can use vapor pressure as a general gauge to tell how fast a product will evaporate under normal circumstances. A product like acetone evaporates much more quickly at room temperature and normal atmospheric pressure than something like water with a much lower vapor pressure — or even motor oil, which does not evaporate easily at all.

Knowledge of this fact is important to you for many reasons. For example, under most normal conditions, a spill of a liquid with a high vapor pressure (such as isopropylamine) will produce more vapors in greater concentration than a substance with a low vapor pressure (such as sarin). These fumes or vapors could then be carried by the wind or travel distances on air currents and cause problems far from the spill itself (such as toxic or flammable vapors being blown into a residential neighborhood).

The vapor pressure may also be an indication of what state of matter a product is likely to be. For example, chlorine, with its extremely high vapor pressure, is likely to be released as a gas rather than a liquid because at normal atmospheric pressure and temperatures it will instantly evaporate **(Figure 4.8, p. 202)**. Also, the lower the vapor pressure, the less likely that the chemical will produce fumes or vapors that you might inhale (and vice versa).

Boiling Point — Temperature of a substance when the vapor pressure equals or exceeds atmospheric pressure. At this temperature, the rate of evaporation exceeds the rate of condensation. At this point, more liquid is turning into gas than gas is turning back into a liquid.

Boiling Liquid Expanding Vapor Explosion (BLEVE) — Rapid vaporization of a liquid stored under pressure upon release to the atmosphere following major failure of its containing vessel; failure is the result of over-pressurization caused by an external heat source, which causes the vessel to explode into two or more pieces when the temperature of the liquid is well above its boiling point at normal atmospheric pressure.

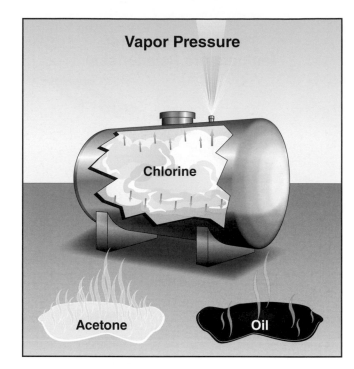

Figure 4.8 Chlorine has an extremely high vapor pressure (4,800 mm Hg at 68°F [20°C]). Because of this, if it is released from its container, it will escape as a gas.

Boiling Point

Boiling point is the temperature at which the vapor pressure of a liquid is equal to or greater than atmospheric pressure. In other words, it is the temperature at which a liquid changes to a gas at a given pressure. The boiling point is usually expressed in degrees Fahrenheit (Celsius) at sea level air pressure. For mixtures, the initial boiling point or boiling-point range may be given. Flammable materials with low boiling points generally present special fire hazards.

A **boiling liquid expanding vapor explosion (BLEVE)** (also called *violent rupture*) can occur when a liquid within a container is heated, causing the material inside to boil or vaporize (such as in the case of a liquefied petroleum gas tank exposed to a fire). If the resulting increase in internal vapor pressure exceeds the vessel's ability to relieve the excess pressure, it can cause the container to fail catastrophically **(Figure 4.9)**. As the vapor is released, it expands rapidly and ignites, sending flames and pieces of tank flying in a tremendous explosion. BLEVEs most commonly occur when flames contact a tank shell above the liquid level or when insufficient water is applied to keep a tank shell cool.

BLEVE!!
A more practical way of explaining the results of a BLEVE is a "<u>B</u>last <u>L</u>eveling <u>E</u>verything <u>V</u>ery <u>E</u>venly." Don't let one happen on your shift.

Melting Point/Freezing Point/Sublimation

Melting point is the temperature at which a solid substance changes to a liquid state at normal atmospheric pressure. For example, an ice cube melts at just above 32°F (0°C) — its melting point.

Figure 4.9 A BLEVE can occur when a liquid in a container is heated to its boiling point. The expanding vapor inside the tank can increase the pressure to such a degree that the tank will fail catastrophically. Any compressed liquified gas can BLEVE, so there may not be a fireball.

Freezing point is the temperature at which a liquid becomes a solid at normal atmospheric pressure. For example, water freezes at 32°F (0°C) — its freezing point. Some substances will actually *sublime* or change directly from a solid into a gas without going into a liquid state in between **(Figure 4.10)**. Dry ice (carbon dioxide) and mothballs both sublime rather than melt.

Figure 4.10 Carbon dioxide (dry ice) sublimes directly from a solid into a gas without changing into a liquid in between.

Vapor Density

Vapor Density — Weight of a given volume of pure vapor or gas compared to the weight of an equal volume of dry air at the same temperature and pressure. Vapor density less than 1 indicates a vapor lighter than air; vapor density greater than 1 indicates a vapor heavier than air.

Vapor density is the weight of a given volume of pure vapor or gas compared to the weight of an equal volume of dry air at the same temperature and pressure. A vapor density less than 1 indicates a vapor lighter than air, while a vapor density greater than 1 indicates a vapor heavier than air **(Figure 4.11)**. Examples of materials with a vapor density less than 1 include helium, neon, acetylene, and hydrogen (see info box). Examples of materials with a vapor density greater than 1 include propane, hydrogen sulfide, ethane, butane, chlorine, and sulfur dioxide. **The majority of gases have a vapor density greater than 1.**

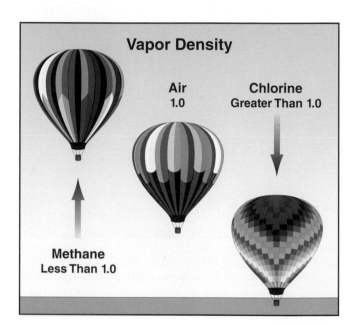

Figure 4.11 Methane weighs less than air, so its specific vapor density is less than 1. Chlorine weighs more than air, so its vapor density is greater than 1. Methane will rise under normal atmospheric conditions while chlorine will sink.

Lighter Than Air Gases

The 13 chemicals that have a vapor density lighter than air (normally presented using the acronym, 4H MEDIC ANNA) are:

Hydrogen (.07)	**M**ethane (.55)	**A**cetylene (.9)
Hydrogen Cyanide (1.0)	**E**thylene (.96)	**N**eon (.34)
Helium (.14)	**D**iborane (.96)	**N**itrogen (.96)
Hydrogen Fluoride (.34)	**I**lluminating Gases (.6)	**A**mmonia (.59)
	Carbon Monoxide (.96)	

All vapors and gasses will mix with air, but the lighter materials tend to rise and dissipate (unless confined). Heavier vapors and gases are likely to concentrate in low places along or under floors; in sumps, sewers, and manholes; and in trenches and ditches where they may create fire or health hazards. Unfortunately, the spread of vapors cannot be predicted exactly from the vapor density because topography, weather conditions, and the vapor mixture with air easily affect vapors. However, knowing the vapor density gives a general idea of what to expect from a specific gas.

Solubility/Miscibility

Solubility in water is a term expressing the percentage of a material (by weight) that will dissolve in water at ambient temperature. A substance's solubility affects whether it mixes in water. Solubility information can be useful in determining spill cleanup methods and extinguishing agents. When a non-water-soluble liquid such as a *hydrocarbon* (gasoline, diesel fuel, pentane) combines with water, the two liquids remain separate. When a water-soluble liquid such as a **polar solvent** (alcohol, methanol, methyl ethyl ketone [MEK]) combines with water, the two liquids mix easily.

Water solubility is also an important contributor for symptom development. Irritant agents that are water-soluble usually cause early upper respiratory tract irritation, resulting in coughing and throat irritation. Partially water-soluble chemicals penetrate into the lower respiratory system, causing delayed (12 to 24 hours) symptoms that include breathing difficulties, pulmonary edema, and coughing up blood.

Degrees of Solubility

The following are generally accepted terms for degrees of solubility:

- *Negligible (insoluble)* — Less than 0.1 percent dissolved in water
- *Slight (slightly soluble)* — Percents from 0.1 to 1 dissolved in water
- *Moderate (moderately soluble)* — Percents from 1 to 10 dissolved in water
- *Appreciable (partly soluble)* — More than 10 to 25 percent dissolved in water
- *Completely (soluble)* — Percents from 25 to 100 percent dissolved in all proportions in water

Miscibility is the degree or readiness to which two or more gases or liquids are able to mix with or dissolve into each other. Two liquids that dissolve into each other in any proportion are considered *miscible*. Typically, two materials that do **not** readily dissolve into each other are considered *immiscible*. For example, water and fuel oil are immiscible, which can create a hazard because the oil (which weighs less than water) will rise to the top and could ignite and burn on top of the water **(Figure 4.12, p. 206)**.

Specific Gravity

Specific gravity is the ratio of the density (heaviness) of a material to the density of some standard material at standard conditions of pressure and temperature. The weight of a substance compared to the weight of an equal volume of water is an expression of the density of a material. For example, if a volume of a material weighs 8 pounds (3.6 kg), and an equal volume of water weighs 10 pounds (4.5 kg), the material is said to have a specific gravity of 0.8. Materials with specific gravities less than 1 will float in (or on) water. Materials with specific gravities greater than 1 will sink in water.

Water Solubility — Ability of a liquid or solid to mix with or dissolve in water.

Polar Solvents — Flammable liquids that have an attraction for water, much like a positive magnetic pole attracts a negative pole; examples include alcohols, esters, ketones, and amines.

Miscibility — Two or more liquids' capability to mix together.

Specific Gravity — Weight of a substance compared to the weight of an equal volume of water at a given temperature. Specific gravity less than 1 indicates a substance lighter than water; specific gravity greater than 1 indicates a substance heavier than water.

Figure 4.12 Water and fuel oil are immiscible, so they don't mix. Since fuel oil weighs less than water, it rises to the surface of the water where it can ignite and burn. *Courtesy of the U.S. Coast Guard.*

Solubility plays an important role in specific gravity in that highly soluble materials will mix or dissolve more completely in water (distributing themselves more evenly throughout), rather than sinking or floating (without dissolving) according to their specific gravities. Most (but not all) flammable liquids have specific gravities less than 1 and, if not soluble, will float on water **(Figure 4.13)**. This fact is an important consideration for fire-suppression activities.

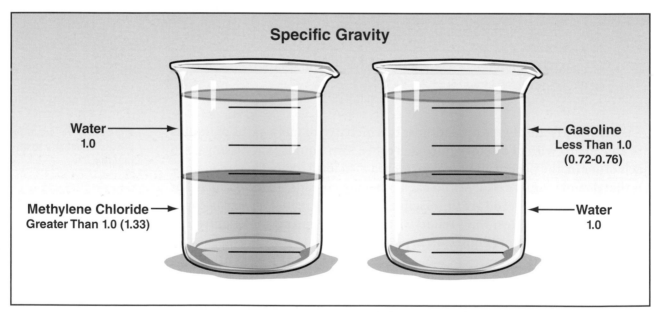

Figure 4.13 Gasoline, which has a specific gravity of less than one will float on water, while methylene chloride (with a specific gravity greater than one) will sink. In general, hydrocarbons float on water, and chlorinated solvents sink.

What This Means To You

Let's look at heptane, a major component of gasoline. Heptane has the following physical and chemical properties:

- *Vapor Pressure:* 45 mmHg
- *Flash Point:* 25°F (-4°C)
- *Boiling Point:* 210°F (98°C)
- *Vapor Density:* 3.5
- *Solubility in Water:* Negligible
- *Specific Gravity:* 0.7

By understanding how to interpret this information, you can predict how the material is likely to behave. If, for example, a significant amount of heptane is spilled into a pond or waterway, you might follow this chain of thought:

- First, because the heptane is spilled into the water, it is important to determine what the material is going to do relative to the water. Is it going to mix with the water? Is it going to sink or float? Since heptane's solubility in water is negligible, you can gather that it's going to stick together in the water rather than dissolve or mix in it. Because its specific gravity is less than 1, you know it is going float to the surface of water. So, you can deduce that the heptane is going to float on top of the water.

- Next, since you know that heptane will burn, you will want to know what it might be doing on top of the water in terms of emitting vapors or fumes that could be ignited accidentally. Its vapor pressure (higher than that of water) tells you that it will likely evaporate (emit some vapors) under most normal conditions. A flashpoint of 25°F (-4°C) tells you that those vapors will burn if exposed to most ignition sources, so keeping ignition sources away from the vapors (or visa versa) is an important priority.

- So what are the vapors likely to be doing and where are they going? Are they rising in the air or staying close to the surface of the water? A vapor density of 3.5 tells you (assuming no wind or other disturbances) that they will tend to stay low or close to the surface of the water.

Persistence

The **persistence** of a chemical is its ability to remain in the environment. Chemicals that remain in the environment for a long time are more persistent than chemicals that quickly dissipate or break down **(Figure 4.14, p. 208)**. For example, persistent nerve agents will remain effective at their point of **dispersion** for a much longer time than nonpersistent nerve agents.

Reactivity

The *reactivity* of a substance is its relative ability to undergo a chemical reaction with another material. Undesirable effects such as pressure buildup, temperature increase, and/or formation of noxious, toxic, or corrosive byproducts may occur as a result of a reaction. Substances referred to in the industrial world as **reactive materials** commonly react vigorously or violently with air, water, heat, light, each other, or other materials.

Many first responders are familiar with the fire tetrahedron or the four elements necessary to produce combustion: oxygen, fuel, heat, and a chemical chain reaction. Fire is just one type of chemical reaction, and a *reactivity*

Persistence — (1) Length of time a chemical agent remains effective without dispersing. (2) Length of time a chemical remains in the environment. (3) Length of time a chemical agent remains as a liquid; typically a liquid chemical agent is considered persistent if it remains for longer than 24 hours.

Dispersion — Act or process of being spread widely.

Reactive Material — Substance capable of or tending to react chemically with other substances; examples: materials that react violently and release energy when combined with air or water.

Figure 4.14 Persistent materials do not dissipate readily.

Activation Energy — Amount of energy that must be added to an atomic or molecular system to begin a reaction.

triangle can be used to explain the basic components of many (though not all) chemical reactions: an oxidizing agent (oxygen), a reducing agent (fuel), and an activation energy source (often heat, but not always so) **(Figure 4.15)**.

All reactions require some energy to get them started, which is commonly referred to as **activation energy**. How much energy is needed depends on the particular reaction. In some cases, the energy can be in the form of added heat from an external source (such as when starting a fire with a match). In

Figure 4.15 The reactivity triangle consists of an oxidizing agent, an activation energy source, and a reducing agent.

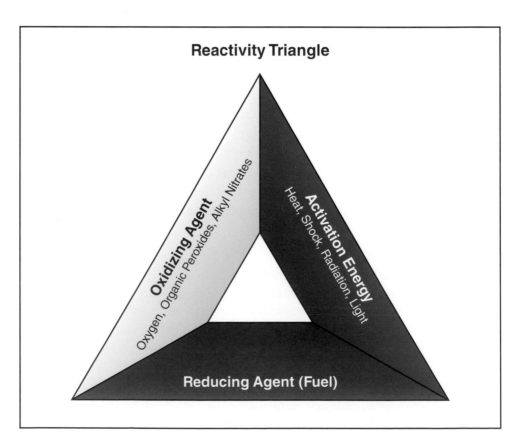

Reactivity Triangle

Oxidizing Agent — Oxygen, Organic Peroxides, Alkyl Nitrates

Activation Energy — Heat, Shock, Radiation, Light

Reducing Agent (Fuel)

some instances, radio waves, radiation, or another waveform of energy may provide the needed energy to the molecules (such as when food is heated in a microwave oven). In other reactions the energy could come from a shock or pressure change (such as might occur when nitroglycerin is jostled).

Reactions that have very low activation energies happen very easily or need very little help to begin the process. For example, materials that are generally classified as water-reactive typically react with water easily at room temperature simply because the heat being provided from the surroundings is sufficient to start the reaction. In addition to other major categories of chemical reactives, first responders may see terms like *light-sensitive, heat-sensitive,* or *shock-sensitive* on MSDSs and/or manufacturers' labels, indicating that those products have an increased susceptibility to those particular sources of activation energy. See **Table 4.3, p. 210** for a summary of the different ways in which chemicals can be reactive. This table supplies the definition and chemical examples of nine reactive hazard classes.

The oxidizing agent in the reactivity triangle provides the oxygen necessary for the chemical reaction. **Strong oxidizers** are materials that encourage a strong reaction (by readily accepting electrons) from reducing agents (fuels). The greater the concentrations of oxygen present in the atmosphere, the hotter, faster, and brighter a fire will burn. The same principle applies to oxidation reactions — generally speaking, the stronger the oxidizer, the stronger the reaction. For example, many hydrocarbons (such as petroleum products) ignite spontaneously when they come into contact with a strong oxidizer. If liquid oxygen (a cryogenic liquid) is spilled on an asphalt roadway and sufficient activation energy is supplied (from shock or friction such as someone stepping on it), the roadway could explode.

> **Strong Oxidizer** — Material that encourages a strong reaction (by readily accepting electrons) from a reducing agent (fuel).

The reducing agent in the fire tetrahedron acts as the fuel source for the reaction, which basically means that it is combining with the oxygen (or losing electrons to the oxidizer) in such a way that energy is being released. Oxidation-reduction reactions can be extremely violent and dangerous because they are releasing a tremendous amount of energy. Obviously, some reducing agents (fuels) are more volatile than others. Wood, for example, is not as prone to undergo rapid oxidation (that is, it will not burn as easily) as a highly flammable liquid like MEK.

Another term with which first responders should be familiar is polymerization, which is a chemical reaction in which a catalyst causes simple molecules to combine to form long chain molecules. Examples of catalysts include light, heat, water, acids, or other chemicals. If this reaction is uncontrolled, it often results in a tremendous release of energy. Materials that may undergo violent polymerization if subjected to heat or contamination are designated with a P in the blue and yellow sections of the *ERG* (**Figure 4.16, p. 211**).

Inhibitors are materials that are added to products that easily polymerize in order to control or prevent an undesired reaction. Inhibitors may be *time-sensitive* in that they may be *exhausted* over a period of time or when exposed to circumstances/unexpected contamination that cause them to be *consumed* more rapidly, such as being *overwhelmed* by exposure to heat or other reaction triggers. Shipments of polymerizing materials may become extremely unstable if delayed during transport or involved in accidents. For example, time-sensitive inhibitors are added to liquid styrene when it is shipped in

> **Inhibitor(s)** — Material that is added to products that easily polymerize in order to control or prevent an undesired reaction. Also called Stabilizer.

Table 4.3
Nine Reactive Hazard Classes

Reactive Hazard Class	Definition	Chemical Examples
Highly Flammable	Substances having flash points less than 100°F (38°C) and mixtures that include substances with flash points less than 100°F (38°C).	Gasoline, Acetone, Pentane, Ethyl Ether, Toluene, Methyl Ethyl Ketone (MEK), Turpentine
Explosive	A material synthesized or mixed deliberately to allow the very rapid release of chemical energy; also, a chemical substance that is intrinsically unstable and liable to detonate under conditions that might reasonably be encountered.	Dynamite, Nitroglycerin, Perchloric Acid, Picric Acid, Fulminates, Azide
Polymerizable	Capable of undergoing self-reactions that release energy; some polymerization reactions generate a great deal of heat. (The products of polymerization reactions are generally less reactive than the starting materials.)	Acrylic Acid, Butadiene, Ethylene, Styrene, Vinyl Chloride, Epoxies
Strong Oxidizing Agent	Oxidizing agents gain electrons from other substances and are themselves thereby chemically reduced, but strong oxidizing agents accept electrons particularly well from a large range of other substances. The ensuing oxidation-reduction reactions may be vigorous or violent and may release new substances that may take part in further additional reactions. Keep strong oxidizing agents well separated from strong reducing agents. In some cases, the presence of a strong oxidizing agent can greatly enhance the progress of a fire.	Hydrogen Peroxide, Fluorine, Bromine, Calcium Chlorate, Chromic Acid, Ammonium Perchlorate
Strong Reducing Agent	Reducing agents give up electrons to other substances and are thereby oxidized, but strong reducing agents donate electrons particularly well to a large range of other substances. The ensuing oxidation-reduction reactions may be vigorous or violent and may generate new substances that take part in further additional reactions.	Alkali metals (Sodium, Magnesium, Lithium, Potassium), Beryllium, Calcium, Barium, Phosphorus, Radium, Lithium Aluminum Hydride
Water-Reactive	Substances that may react rapidly or violently with liquid water and steam, producing heat (or fire) and often toxic reaction products.	Alkali metals (Sodium, Magnesium, Lithium, Potassium), Sodium Peroxide, Anhydrides, Carbides
Air-Reactive	Likely to react rapidly or violently with dry air or moist air; may generate toxic and corrosive fumes upon exposure to air or catch fire.	Finely divided metal dusts (Nickel, Zinc, Titanium), Alkali metals (Sodium, Magnesium, Lithium, Potassium), Hydrides (Diborane, Barium Hydrides, Diisobutyl Aluminum Hydride)
Peroxidizable Compound	Apt to undergo spontaneous reaction with oxygen at room temperature, to form peroxides and other products. Most such auto-oxidations are accelerated by light or trace impurities. Many peroxides are explosive, which makes peroxidizable compounds a particular hazard. Ethers and aldehydes are particularly subject to peroxide formation (the peroxides generally form slowly after evaporation of the solvent in which a peroxidizable material had been stored).	Isopropyl Ether, Furan, Acrylic Acid, Styrene, Vinyl Chloride, Methyl Isobutyl Ketone, Ethers, Aldehydes
Radioactive Material	Spontaneously and continuously emitting ions or ionizing radiation. Radioactivity is not a chemical property, but an additional hazard that exists in addition to the chemical properties of a material.	Radon, Uranium

Source: U.S. Environmental Protection Agency's CEPPO (Chemical Emergency Preparedness and Prevention Office) Computer-Aided Management of Emergency Operations (CAMEO) software was used to identify this information.

Sodium potassium alloys, solid	138	3404
Sodium selenite	151	2630
Sodium silicofluoride	154	2674
Sodium sulfide, anhydrous	135	1385
Sodium sulfide, hydrated, with not less than 30% water	153	1849
Sodium sulfide, with less than 30% water of crystallization	135	1385
Sodium sulphide, anhydrous	135	1385
Sodium sulphide, hydrated, with	153	1849

Strontium phosphide	139	2013
Strychnine	151	1692
Strychnine salts	151	1692
Styrene monomer, stabilized	(128P)	2055
Substituted nitrophenol pesticide, liquid, flammable, poisonous	131	2780
Substituted nitrophenol pesticide, liquid, flammable, toxic	131	2780

Figure 4.16 Materials designated with a *P* in the blue and yellow pages of the *ERG* may undergo rapid polymerization and become dangerously unstable if delayed during transport or involved in accidents.

order to prevent the styrene from polymerizing during transport. If containers holding the styrene rupture or emergency responders add water, the inhibitor becomes exhausted, and the polymerization reaction begins. A BLEVE may result because of the resulting release of heat and expanding gas.

Remember that some reactive materials are special concerns to emergency responders because they are oxidizers. Others are concerns because they are volatile reducing agents or because the reduction reaction is started in an unusual way (such as by exposure to water). Additionally, emergency responders must be concerned with the various ways in which activation energy may be supplied to certain chemicals in order to prevent a reaction from occurring. Under emergency conditions, reactive materials can be extremely destructive to life and property. Knowledge and extreme caution are vital factors in handling emergencies involving reactive materials. People and equipment should be kept upwind, uphill, and back a safe distance or in protected locations until pertinent facts are established and definite plans can be formulated. With advances in modern technology, more and more reactive and unstable materials are being used for various processes, and responders must be prepared to deal with them.

Hazardous Materials Properties Summary

Hazardous materials have a variety of properties that affect how they behave and how dangerous they may be. First responders need to be aware of how these properties may affect the behavior of containers (such as the potential for BLEVE) and/or their contents (such as the material's flammability). **Table 4.4, p. 212** (in both English and SI units) provides a list of common hazardous materials and summaries of their properties.

General Hazardous Materials Behavior Model

First responders need to have a basic understanding of how a hazardous material is likely to behave in any given situation. The general premise of the General Hazardous Materials Behavior Model (sometimes referred to as the **General Emergency Behavior Model (GEBMO)** is based on a definition of hazardous materials developed by Ludwig Benner Jr. His definition states that, *"Hazardous materials are things that can escape from their containers and hurt or harm the things that they touch."* Hazardous materials incidents have the following common elements:

General Emergency Behavior Model (GEBMO) — Model used to describe how hazardous materials are accidentally released from their containers and how they behave after the release.

Table 4.4
Hazardous Materials Properties Summary
(English Units)

Chemical Name	Lbs Per Gallon	Solid/Liquid/ Gas	Vapor Pressure (at 68°F)	Vapor Density (air=1)	Soluble (degree of solubility)	Specific Gravity (water=1)	Boiling Point (°F)	Melting Point (°F)	Freezing Point (°F)	Flash Point (°F)	LEL/ UEL	Ignition Temperature (°F)
Acetone	6.6	L	3.5 psi	2.0	Yes	0.791	133	-138	-138	0	2.5/13.0	869
Acrolein	7.0	L	4.1 psi	1.94	Yes (40%)	0.843	127	-125	-125	-15	2.8/31	428
Acrylonitrile	6.7	L	1.6 psi	1.83	Yes (7%)	0.800	171	-117	-117	32	3/17	898
Ammonia Anhydrous	6.0	G/L	14.7 psi	0.6	Yes (34%)	0.68	-28	-108	-108	N/A	15.5/27	1204
Chlorine	13	G/L	92.6 psi	2.5	Yes (0.7%)	1.424	-29	-150	-150	N/A	N/A	N/A
Ethylene Oxide	7.2	L/G	21 psi	1.5	Yes	0.974	51	-79	-79	-20	3.0/100	1058
Benzene	7.3	L	1.5 psi	2.7	No	0.879	176	42	42	12	1.3/7.9	928
Bromine	26.0	L	3.3 psi	5.5	Yes (4%)	3.12	138	19	19	N/A	N/A	N/A
Acetylene		G	587.8 psi	0.91	No	0.613 at 112°F	-118	-115.2	-115.2	N/A	2.5/100	581
Sulfuric Acid	15	L	0.019 psi at 295°F	3.3	Yes	1.84	554	37	37	N/A	N/A	N/A
Hydrochloric Acid	10.1	L	58.8 psi	1.3	Yes	1.19	123	-173.7	-173.7	N/A	N/A	N/A
Sodium Hydroxide	12.5	L/S	0 psi	1.4	Yes	2.13	2534	604	604	N/A	N/A	N/A
Ethyl Ether		L	8.5 psi	2.6	Yes (8%)	0.714	95	-177	-177	-49	1.9/36.0	356
Ethylene Diamine	7.5	L	0.2 psi	2.1	Yes	0.909	242	52	52	99	4.2/14.4	715
Hydrofluoric Acid	9.6 or 10.5	L	7.7 psi at 36.5°F	0.7	Yes	1.25	152	-117.6	-117.6	N/A	N/A	N/A
Isopropyl Alcohol		L	0.6 psi	2.1	Yes	0.8	181	-128	-128	53	2/12.7	750
Methyl Ethyl Ketone	6.7	L	1.4 psi	2.5	Yes	0.8	175.3	-123	-123	16	1.4/11.5	759
Carbon Dioxide		G/L	14.7 psi	1.53	Yes	1.56	Sublimes	-109	-109	N/A	N/A	N/A
Phenol	9.9	L/S	0.008 psi	3.2	Yes (9%)	1.058	358	106	106	175	1.7/8.6	1319

Continued

Table 4.4
Hazardous Materials Properties Summary
(International System Units)

Chemical Name	Kgs Per Gallon	Solid/Liquid/ Gas	Vapor Pressure (at 20°C)	Vapor Density (air=1)	Soluble (degree of solubility)	Specific Gravity (water=1)	Boiling Point (°C)	Melting Point (°C)	Freezing Point (°C)	Flash Point (°C)	LEL/ UEL	Ignition Temperature (°C)
Acetone	2.99	L	180 mmHg	2.0	Yes	0.791	56	-94	-94	-18	2.5/13.0	465
Acrolein	3.18	L	210 mmHg	1.94	Yes (40%)	0.843	53	-87	87	-26	2.8/31	220
Acrylonitrile	3.04	L	83 mmHg	1.83	Yes (7%)	0.800	77	-82.8	-82.8	0	3/17	481
Ammonia Anhydrous	2.7	G/L	760 mmHg	0.6	Yes (34%)	0.68	-33	-78	-78	N/A	15.5/27	651
Chlorine	6.0	G/L	6.3 atm or 4 788 mmHg	2.5	Yes (0.7%)	1.424	-34	-101	-101	N/A	N/A	N/A
Ethylene Oxide	3.27	L/G	1.43 atm or 1 086.8 mmHg	1.5	Yes	0.974	11	-62	-62	-29	3.0/100	570
Benzene	3.31	L	75 mmHg	2.7	No	0.879	80	6	6	-11	1.3/7.9	498
Bromine	12	L	172 mmHg	5.5	Yes (4%)	3.12	59	-7	-7	N/A	N/A	N/A
Acetylene		G	40 atm	0.91	No	0.613 at 44°C	-83	-82	-82	N/A	2.5/100	305
Sulfuric Acid	6.8	L	1 mmHg or 146°C	3.3	Yes	1.84	290	3	3	N/A	N/A	N/A
Hydrochloric Acid	4.6	L	4 atm 3 040 mmHg	1.3	Yes	1.19	51	-114	-114	N/A	N/A	N/A
Sodium Hydroxide	5.7	L/S	0 mmHg	1.4	Yes	2.13	1 390	318	318	N/A	N/A	N/A
Ethyl Ether		L	440 mmHg	2.6	Yes (8%)	0.714	35	-116	-116	-45	1.9/36.0	180
Ethylene Diamine	3.4	L	11 mmHg	2.1	Yes	0.909	117	11	11	37	4.2/14.4	379
Hydrofluoric Acid	4.4 or 4.8	L	400 mmHg at 2.5°C	0.7	Yes	1.25	67	-83.1	-83.1	N/A	N/A	N/A
Isopropyl Alcohol		L	33 mmHg	2.1	Yes	0.8	83	-89	-89	12	2/12.7	399
Methyl Ethyl Ketone	3.04	L	71.2 mmHg	2.5	Yes	0.8	80	-86	-86	-9	1.4/11.5	404
Carbon Dioxide		G/L	760 mmHg	1.53	Yes	1.56	Sublimes	-78	-78	N/A	N/A	N/A
Phenol	4.5	L/S	0.4 mmHg	3.2	Yes (9%)	1.058	181	41	41	79	1.7/8.6	715

- Material or materials presenting hazards to people, the environment, or property
- Container or containers that have failed or have the potential to fail
- Exposure or potential exposure to people, the environment, and/or property

GEBMO helps first responders predict the course of an incident, thereby enabling them to limit the effects of a hazardous material. GEBMO is basically a defensive-mode action that is concerned with potential haz mat emergencies involving containers. First responders must calmly assess the situation at hand and then identify the appropriate response depending on answers to the following questions:

- How long will the harmful exposure exist?
- What has stressed or is stressing the container?
- How will the stressed container and its material behave?
- What are the harmful effects of the container materials?

The application of information gained through answering these questions helps responders develop a strategy for mitigating the incident. If it is not possible to make all predictions required to analyze the incident, then further assistance is needed. This assistance may be sought from the manufacturer, shipper and consignee, carrier, or other agencies involved with hazardous materials.

The events in a hazardous materials incident follow a general pattern or model. As with many other types of events, prediction may be based on past experience. In hazardous materials incidents the following sequence generally occurs:

- **Stress** — If a container is stressed beyond its design strength, it fails or breaches.

↓

- **Breach** — The way in which a container breaches is based on the material of which it is constructed, type of stress that it is exposed to, and pressure inside the container at the time that it fails. A breach or failure of the container may be partial (as in a puncture) or total (as in disintegration). A breached container releases its contents.

↓

- **Release** — When a container is breached or fails, its contents, stored energy, and pieces of the container may release. A release always involves the product and may (depending on the product, container, and incident conditions) involve the release of energy and container parts. The released product disperses.

↓

- **Dispersion/Engulf** — When released, the product inside the container, any stored energy, and the container disperse. The patterns of dispersion are based on the laws of chemistry, physics, and the characteristics of the product.

↓

- *Exposure/Contact* — Anything (such as persons, the environment, or property) that is in the area of the release is exposed.

↓

- *Harm* — Depending on the container, product, and energy involved, exposures may be harmed.

This sequence is elaborated in the paragraphs that follow. The behavior of explosives, chemical and biological agents, and radiological materials used for purposes of terror attacks will be discussed in Chapter 7, Terrorist Attacks, Criminal Activities, and Disasters.

Stress

Container stress is classified as stimulus causing strain (excessive tension or compression), pressure (force applied at right angles to a surface), or deformity (distortion by torque or twisting). According to U.S. Department of Transportation (DOT) records, almost one-fourth of all reported haz mat incidents are caused by container failure. One or all three of the stressors listed may be encountered at each haz mat incident. For instance, heat (thermal stress) can initiate or speed a chemical reaction while weakening a container and increasing internal pressure. Similarly, a mechanical blow can initiate a violent chemical reaction in an unstable chemical while simultaneously damaging the container.

When evaluating container stress, consider the type of container, the type and amount of stress, and its potential duration. Container stress may involve a single factor or several stressors acting on the container simultaneously. The factors placing stress on a container may be readily visible such as a collision or a fire impinging on a container surface. In other cases, container stress cannot be directly observed and must be predicted based on conditions or other indirect indicators. If the container has already failed, think about other containers that may be exposed, and evaluate the impact of product contact. Preventing container failure may require reducing or eliminating the stress. Common stressors are as follows:

Figure 4.17 Thermal stress may simultaneously increase internal pressure and reduce container integrity, resulting in sudden failure. Responders must be very cautious around propane tanks of any size that are involved in fires. *Courtesy of Donny Howard, Agent, Oklahoma State Fire Marshal's Office.*

- *Thermal* — Excessive heat or cold causing intolerable expansion, contraction, weakening (loss of temper), or consumption of the container and its parts; thermal stress may also simultaneously increase internal pressure and reduce container shell integrity, resulting in sudden failure. Thermal stress may result from the heating or cooling of the container **(Figure 4.17)**. Clues may include observation of flame impingement on the container, operation of a relief device, or changing environmental conditions (such as increased temperature).

- *Chemical* — Uncontrolled reactions/interactions of contents in the container and the container itself, resulting in a sudden or long-term deterioration of the container. Reactions involving two chemicals placed into the same container can cause excessive heat and/or pressure, also resulting in container failure. Chemical stress may be the result of corrosive action or other chemical attack on an incompatible container material. Clues may include visible corrosion or other degradation of container surfaces. However, the interior of a container may experience chemical stress with no visible indication from the exterior.

- *Mechanical* — Physical application of energy resulting in container/attachment damage; mechanical application may change the shape of the container (crushing), reduce the thickness of the container surface (abrading or scoring), crack or produce gouges, unfasten or disengage valves and piping, or penetrate the container wall **(Figure 4.18)**. Common causes may include collision, impact, or internal over-pressure. Clues may include physical damage, the mechanism of injury (forces placed on the container), or operation of relief devices.

Figure 4.18 This damaged tank car shows evidence of mechanical stress. *Courtesy of Phil Linder.*

Figure 4.19 White residue surrounding this tank car is evidence of a breach. *Courtesy of Phil Linder.*

Breach

When a container is stressed beyond its limits of recovery (its design strength or ability to hold contents), it opens or breaches and releases its contents **(Figure 4.19)**. Different container types breach in different ways based on a variety of factors (including internal pressure). A breach is dependent upon the type of container and the stress applied. The extent of a breach or failure varies with container construction and the type of stress to which it is subjected. First responders should try to visualize how the container would be damaged by the stress that is being or has been applied. The nature of a breach is a major factor in planning offensive product control operations. Several types of breaches are as follows:

- *Disintegration* — Container suffers a general loss of integrity such as a glass bottle shattering or a grenade exploding. This type of breach occurs in containers that are made of a brittle material (or that have been made more brittle by some form of stress).

- *Runaway cracking* — Crack develops in a container as a result of some type of damage, which continues to grow rapidly, breaking the container into two or more relatively large pieces. This type of breach is associated with closed containers such as drums, tank cars, or cylinders. Runaway linear cracking is commonly associated with BLEVEs.

- *Attachments (closures) open or break* — Attachments (such as pressure-relief devices, discharge valves, or other related equipment) fail, open, or break off when subjected to stress, leading to a total failure of a container.

When evaluating an attachment failure, first responders should consider the entire system and the effect of failure at a given point.

- **Puncture** — Mechanical stress coming into contact with a container causes a puncture. Examples: forklifts puncturing drums and couplers puncturing a rail tank car.

- **Split or tear** — Welded seam on a tank or drum fails or a seam on a bag of fertilizer rips. Mechanical or thermal stressors may cause splits or tears **(Figure 4.20)**.

Figure 4.20 This intermodal tank has split along a seam. *Courtesy of Rich Mahaney.*

Figure 4.21 High explosives detonate faster than the speed of sound. *Courtesy of the U.S. Army, photo by Sgt. Jabob H. Smith.*

Release

When a container fails, three things may release: the product, energy, and the container (whole or in pieces). If a cylinder of pressurized, chlorine gas suffers an attachment failure at the valve due to mechanical stress, the product (a toxic, corrosive oxidizer) is released along with a substantial amount of energy (because of stored pressure) and rapid acceleration of the valve and/or cylinder in the opposite direction from the release. Depending on the situation, this release can occur quickly or over an extended time period. Generally, when a great amount of chemical/mechanical energy is stored, a more rapid release of the material (causing a greater risk to first responders) is possible. Releases are classified based on how fast they occur as follows:

- **Detonation** — Instantaneous and explosive release of stored chemical energy of a hazardous material. The results of this release include fragmentation, disintegration, or shattering of the container; extreme overpressure; and considerable heat release. The duration of a detonation can be measured in hundredths or thousandths of a second. An example would be detonation of a high explosive **(Figure 4.21)**.

- **Violent rupture** — Immediate release of chemical or mechanical energy caused by runaway cracks. The results are ballistic behavior of the container and its contents and/or localized projection of container pieces/parts and hazardous material such as with a BLEVE. Violent ruptures occur within a timeframe of one second or less.

- *Rapid relief* — Fast release of a pressurized hazardous material through properly operating safety devices caused by damaged valves, piping, or attachments or holes in the container. This action may occur in a period of several seconds to several minutes.

- *Spill/leak* — Slow release of a hazardous material under atmospheric or head pressure through holes, rips, tears, or usual openings/attachments can occur in a period of several minutes to several days **(Figure 4.22)**.

Figure 4.22 Leaking diesel fuel is an example of a spill or leak which might occur during routine accidents. *Courtesy of Rich Mahaney.*

Release Potential

When evaluating release potential, remember the total amount of product in the container. A valve blowout in a pressurized container causes a rapid release. If this size breach occurs in a 150-pound (68 kg) cylinder, the product remaining quickly releases. If this same type of release occurs in a ton container, cargo tank, or tank car, the release occurs over a longer period of time and may have a substantially greater effect.

Dispersion Patterns/Engulfment

The dispersion of material is sometimes referred to as **engulf (Figure 4.23a and b)**. Dispersion of the product, energy, and container components depends on the type of release as well as physical and chemical laws. The product may release in the form of a solid, liquid, or gas/vapor. Mechanical, thermal, or chemical energy and ionizing radiation may also be released. The path of product, energy, and container travel depends on its form and physical characteristics. Based on product characteristics and environmental conditions (such as weather and terrain), the pattern of dispersion may be predicted. Once a container has been compromised, the hazardous material will be distributed over the surrounding area according to the following five factors:

Engulf — (1) Dispersion of material as defined in the General Hazardous Materials Behavior Model (GEBMO). An engulfing event is when matter and/or energy disperses and forms a danger zone. (2) To flow over and enclose; in fire service context, it refers to being enclosed in flames.

Figure 4.23a and b When Hurricane Katrina moved an open top floating roof tank off its foundation, oil spilled from the tank and engulfed the surrounding neighborhoods. *Courtesy of FEMA News Photos, photo by Bob McMillan.*

1. Physical/chemical properties

2. Prevailing weather conditions

3. Local topography

4. Duration of the release

5. Control efforts of responders

The shape and size of the pattern also depends on how the material emerges from its container — whether the release is an instantaneous "puff," a continuous plume, or a sporadic fluctuation. The outline of the dispersing hazardous material, sometimes called its *dispersion pattern*, can be described in a number of ways. Dispersion patterns are as follows:

● *Hemispheric* — Semicircular or dome-shaped pattern of airborne hazardous material that is still partially in contact with the ground or water **(Figure 4.24)**. A **hemispheric release** generally results from a rapid release of energy (detonation, deflagration, violent rupture, etc.). *Other dispersion elements:*

— Energy generally travels outward in all directions from the point of release.

— Dispersion of energy is affected by terrain and cloud cover. Solid cloud cover can reflect the detonation shock wave, increasing the explosion impact.

Hemispheric Release — Semicircular or dome-shaped pattern of airborne hazardous material that is still partially in contact with the ground or water.

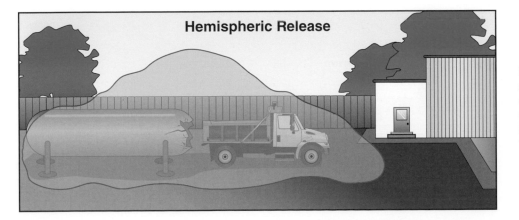

Figure 4.24 A *hemispheric release* is a semicircular or dome-shaped pattern of an airborne hazardous material that is still partially in contact with the ground or water.

— Energy release may propel the hazardous material and container parts; however, this dispersion may not be hemispherical. Large container parts generally (but **not always**) travel in line with the long axis of the container.

Cloud — Ball-shaped pattern of an airborne hazardous material where the material has collectively risen above the ground or water at a hazardous materials incident.

• *Cloud* — Ball-shaped pattern of the airborne hazardous material where the material has collectively risen above the ground or water **(Figure 4.25)**. Gases, vapors, and finely divided solids that are released quickly (puff release) can disperse in cloud form when wind conditions are minimal. Terrain and/or wind effects can transform a **cloud** into a plume.

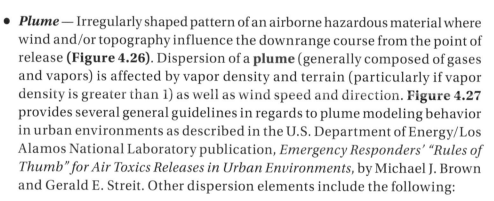

Figure 4.25 A *vapor cloud* is a ball-shaped patter of an airborne hazardous material where the material has collectively risen above the ground or water.

Plume — Irregularly shaped pattern of an airborne hazardous material where wind and/or topography influence the downrange course from the point of release.

• *Plume* — Irregularly shaped pattern of an airborne hazardous material where wind and/or topography influence the downrange course from the point of release **(Figure 4.26)**. Dispersion of a **plume** (generally composed of gases and vapors) is affected by vapor density and terrain (particularly if vapor density is greater than 1) as well as wind speed and direction. **Figure 4.27** provides several general guidelines in regards to plume modeling behavior in urban environments as described in the U.S. Department of Energy/Los Alamos National Laboratory publication, *Emergency Responders' "Rules of Thumb" for Air Toxics Releases in Urban Environments*, by Michael J. Brown and Gerald E. Streit. Other dispersion elements include the following:

— When all of the material is released at one time (puff release), the concentration of gas or vapor in the cloud or plume decreases over time.

— In an ongoing release, concentration increases over time until the leak is stopped or all of the product has been exhausted; then it decreases.

Figure 4.26 A *plume* is an irregularly-shaped pattern of an airborne hazardous material influenced by wind and/or topography in its downrange course.

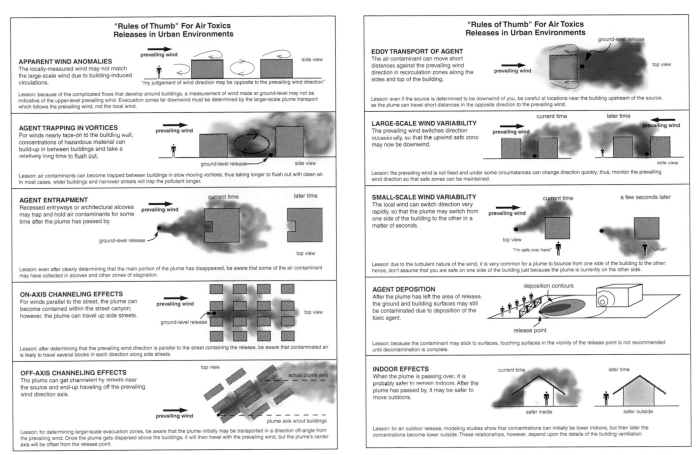

"Rules of Thumb" For Air Toxics
Releases in Urban Environments

APPARENT WIND ANOMALIES
The locally-measured wind may not match the large-scale wind due to building-induced circulations.

prevailing wind

side view

"my judgement of wind direction may be opposite to the prevailing wind direction"

Lesson: because of the complicated flows that develop around buildings, a measurement of wind made at ground-level may not be indicative of the upper-level prevailing wind. Evacuation zones far downwind must be determined by the larger-scale plume transport which follows the prevailing wind, not the local wind.

AGENT TRAPPING IN VORTICES
For winds nearly face-on to the building wall, concentrations of hazardous material can build-up in between buildings and take a relatively long time to flush out.

prevailing wind

ground-level release

side view

Lesson: air contaminants can become trapped between buildings in slow moving vortices, thus taking longer to flush out with clean air. In most cases, wider buildings and narrower streets will trap the pollutant longer.

AGENT ENTRAPMENT
Recessed entryways or architectural alcoves may trap and hold air contaminants for some time after the plume has passed by.

prevailing wind

current time

later time

ground-level release

top view

Lesson: even after clearly determining that the main portion of the plume has disappeared, be aware that some of the air contaminant may have collected in alcoves and other zones of stagnation.

ON-AXIS CHANNELING EFFECTS
For winds parallel to the street, the plume can become contained within the street canyon; however, the plume can travel up side streets.

prevailing wind

ground-level release

top view

Lesson: after determining that the prevailing wind direction is parallel to the street containing the release, be aware that contaminated air is likely to travel several blocks in each direction along side streets.

OFF-AXIS CHANNELING EFFECTS
The plume can get channeled by streets near the source and end-up traveling off the prevailing wind direction axis.

top view

actual plume axis

prevailing wind

plume axis w/out buildings

Lesson: for determining larger-scale evacuation zones, be aware that the plume initially may be transported in a direction off-angle from the prevailing wind. Once the plume gets dispersed above the buildings, it will then travel with the prevailing wind, but the plume's center axis will be offset from the release point.

EDDY TRANSPORT OF AGENT
The air contaminant can move short distances against the prevailing wind direction in recirculation zones along the sides and top of the building.

ground-level release

prevailing wind

top view

Lesson: even if the source is determined to be downwind of you, be careful at locations near the building upstream of the source, as the plume can travel short distances in the opposite direction to the prevailing wind.

LARGE-SCALE WIND VARIABILITY
The prevailing wind switches direction occasionally, so that the upwind safe zone may now be downwind.

current time

later time

prevailing wind

prevailing wind

side view

Lesson: the prevailing wind is not fixed and under some circumstances can change direction quickly; thus, monitor the prevailing wind direction so that safe zones can be maintained.

SMALL-SCALE WIND VARIABILITY
The local wind can switch direction very rapidly, so that the plume may switch from one side of the building to the other in a matter of seconds.

current time

a few seconds later

prevailing wind

top view

"I'm safe over here"

"uh-oh"

Lesson: due to the turbulent nature of the wind, it is very common for a plume to bounce from one side of the building to the other; hence, don't assume that you are safe on one side of the building just because the plume is currently on the other side.

AGENT DEPOSITION
After the plume has left the area of release, the ground and building surfaces may still be contaminated due to deposition of the toxic agent.

deposition contours

release point

Lesson: because the contaminant may stick to surfaces, touching surfaces in the vicinity of the release point is not recommended until decontamination is complete.

INDOOR EFFECTS
When the plume is passing over, it is probably safer to remain indoors. After the plume has passed by, it may be safer to move outdoors.

current time

later time

safer inside

safer outside

Lesson: for an outdoor release, modeling studies show that concentrations can initially be lower indoors, but then later the concentrations become lower outside. These relationships, however, depend upon the details of the building ventilation.

Figure 4.27 Responders should be aware of these rules of thumb regarding plume modeling behavior in urban environments. *Courtesy of Los Alamos National Laboratory.*

- *Cone* — Triangular-shaped pattern of a hazardous material with a point source at the breach and a wide base downrange **(Figure 4.28)**. An energy release may be directed (based on the nature of the breach) and may project solid, liquid, or gaseous material in a three-dimensional **cone**-shaped dispersion. Examples: container failures in a BLEVE or a pressurized liquid or gas release.

Cone — Triangular-shaped pattern of an airborne hazardous material release with a point source at the breach and a wide base downrange.

Cone

Figure 4.28 A *cone* is a triangular-shaped pattern of a hazardous material with a point source at the breach and a wide base downrange.

- **Stream** — Surface-following pattern of liquid hazardous material that is affected by gravity and topographical contours **(Figure 4.29)**. Liquid releases flow downslope whenever there is a gradient away from the point of release.

Figure 4.29 A *stream* is pulled by gravity, following the topographical contours of the surface.

- **Pool** — Three-dimensional (including depth), slow-flowing liquid dispersion. Liquids assume the shape of their container and pool in low areas **(Figure 4.30)**. As the liquid level rises above the confinement provided by the terrain, the substance flows outward from the point of release. If there is a significant gradient or confinement due to terrain, this flow forms a stream.

Figure 4.30 In a *pool*, liquids assume the shape of their container, typically accumulating in low areas.

- **Irregular** — Irregular or indiscriminate deposit of a hazardous material (such as that carried by contaminated responders) **(Figure 4.31)**.

Figure 4.31 *Irregular dispersion* results from indiscriminate deposit of a hazardous material such as that caused by contaminated vehicles or responders.

Some facilities' pre-incident surveys may contain plume dispersion models that can be used to estimate the size of an endangered area in the event of a release. Computer software such as *CAMEO* (Computer-Aided Management of Emergency Operations), *ALOHA* (Area Locations of Hazardous Atmospheres), and *HPAC* (Hazard Prediction and Assessment Capability) can also assist in the prediction of plume dispersion patterns. Of course, first responders should always consult the *ERG* for isolation and evacuation distances.

Dispersion of Solids

Solids in the form of dusts, powders, or small particles may also have dispersion patterns. In many cases, the simplest dispersion pattern of a released solid could best be described as a *pile* **(Figure 4.32)**. When the material is spilled or released from its container, it forms a pile. This pile can then be dispersed by wind, a moving liquid, or contact with a moving object (such as a forklift driving through the spill).

For example, a release of a pesticide powder could be picked up by the wind and dispersed in the form of a plume. Depending on the properties (such as solubility) of the solid, if it spills into a stream or sewer system (or it is washed away by a hose stream), it could be carried along and dispersed by the movement of the liquid in the *stream* dispersion pattern. Some dusts, powders, and particles can remain suspended in the air for hours or days such as in a *cloud*. An example of the latter would be microscopic asbestos fibers that can remain airborne for very long periods of time.

While *pile* is not normally included in the *GEBMO*, it is important for first responders to remember that hazardous materials also come in solid form, and that potential dispersion of the solid(s) is an important consideration in planning an appropriate response.

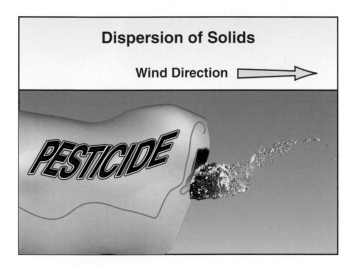

Figure 4.32 Dispersion of a released solid can sometimes be described as a *pile*.

Exposure/Contact

As a container is breached, it releases its materials and impinges upon exposures: people, the environment, and property. In some cases, the *ERG* can be used to estimate the size of an endangered area. Some hazardous materials may present a threat to one specific exposure type (such as marine pollutants that threaten fish and other marine plants and animals) and others present a threat to all types. In evaluating the severity of exposures, consider the hazards presented, concentration of the material, and duration of contact. Several types of exposures to consider in hazard and risk assessment are as follows:

- *People* — Includes responders and others in the path of a hazardous material.

- *Environment* — Includes the air, water, ground, and life forms other than humans. The potential effect on the environment varies with the location in which the product is released as well as its characteristics **(Figure 4.33)**.

- *Property* — Things threatened directly by the product or the energy liberated at the time of release (as with exposure to the heat from a burning product).

Contacts (impingements) are associated with the following general timeframes:

- *Immediate* — Milliseconds, seconds; examples: deflagration, explosion, or detonation

- *Short-term* — Minutes, hours; example: gas or vapor cloud

- *Medium-term* — Days, weeks, months; example: lingering pesticide

- *Long-term* — Years, generations; example: permanent radioactive source

Figure 4.33 Potential exposures include people, the environment, and property. *Courtesy of the U.S. Coast Guard, photo by PA2 Angel Deilmer.*

Harm

Injury or damage caused by exposure to a hazardous material. The health and physical hazards that could cause harm in a hazardous materials incident are thermal, mechanical, poisonous, corrosive, asphyxiation, radiation, and biological.

Summary

First responders need the ability to predict how a hazardous material will behave when it escapes its container. This behavior is often determined by the material's physical properties such as its flammability, boiling point, vapor pressure, chemical reactivity, and others. The physical properties of a material determine such things as whether or not it will burn, how fast it will evaporate, at what temperature it will freeze and/or boil, whether or not it is likely to react with other materials, and other important factors.

Review Questions

1. Describe the three states of matter.

2. What is flash point?

3. Define lower explosive limit (LEL) and upper explosive limit (UEL).

4. What is a BLEVE?

5. What are some of the chemicals that have a vapor density lighter than air?

6. What are the six portions of the GEBMO?

7. Describe several types of breaches.

8. What are the classifications of releases?

9. Describe each of the dispersion patterns.

10. What types of exposures should be considered in hazard and risk assessment?

Incident Management

chapter 5

Key Terms

Competencies

NFPA® 472:	5.4.1(5)(b)	5.4.3(4)(a)	5.4.3(5)	5.5.2(1)
5.4.1(5)(a)	5.4.3(3)	5.4.3(4)(b)	5.4.3(7)	

Incident Management

Learning Objectives

1. Describe incident priorities.

2. Discuss various incident management systems. [NFPA® 472, 5.2.2(6), 5.4.1(5)(a-b), 5.4.3(3), 5.4.3(4)(a-b), 5.4.3(5), 5.4.3(7)]

3. Identify communication procedures and guidelines for use at hazardous materials incidents. [NFPA® 472, 5.5.2(1)]

Chapter 5
Incident Managment

Case History

At 9:38 p.m. on Thursday, October 5, 2006 the Apex, North Carolina, Fire Department (AFD) was dispatched to a report of chlorine odor near the intersection of Schiefflein Road and Investment Boulevard. The Environmental Quality Company (EQ) operated a business that handled commercial hazardous waste at 1005 Investment Road. The company's business involved collecting, processing, and repackaging industrial waste for transport and proper disposal. The daily inventory of chemicals at the facility varied.

AFD dispatched its standard response of two engines and a chief officer (shift commander). Because the regularly assigned shift commander was on leave, the chief of the department responded in his place.

Upon arrival 4 minutes later, at 9:42 p.m., Engine Company 3 reported a large vapor cloud and requested a second-alarm assignment. Crews then began an initial reconnaissance to determine the source of the cloud. The chief of the department arrived at 9:52 p.m., took command, and established the initial Incident Command Post (ICP) near the intersection of Schiefflein Road and Investment Boulevard. He began the process of evacuating the community, which ultimately involved 17,000 people.

The town's reverse 9-1-1 system was activated as part of the notification and warning. This system sends prerecorded messages from the 9-1-1 center to threatened neighborhoods to provide emergency information such as evacuation or shelter-in-place instructions. Apex police officers began warning people downwind of the facility. They went door-to-door to inform people that they needed to leave. Regional haz mat resources and the town's Emergency Operations Plan (EOP) were activated. The haz mat resources included the Raleigh and Fayetteville Fire Departments' haz mat response teams. By activating the plan, the town's Emergency Operations Center (EOC) at Fire Station 3 automatically was opened, and town officials had a location to coordinate the serial evacuations and shelter operations.

The initial fire companies conducted reconnaissance and were unable to determine the source of the release. As a second team approached the area, the fire burned through the roof of the building. A plume of unknown chemical vapors was venting from the structure, and the wind was carrying the plume away from the site into the surrounding community.

At 10:12 p.m., a general alarm was sounded to recall the entire staff of the fire department. This recall included the remainder of the 27 full-time employees, 12 part-time employees, and 15 volunteers that staff the department. The ICP was relocated to a second location about a quarter of a mile away on Industrial Boulevard. This decision was based on changes in the wind direction that, in turn, caused the plume to move. The Apex Police Department (APD) continued with the door-to-door notice to evacuate. Over the course of the evening and next day the remainder of the police department also was recalled for extended duty.

In accordance with Apex's EOP, shelters were opened, and the EOC at Station 3 was activated. The first shelter was opened at the Community Center at 73 Hunter Street. Some residents who lived closest to the EQ facility were directed to shelter-in-place, that is, to stay indoors and shut down the heating, ventilation, and air conditioning systems to prevent outside air from coming indoors. That directive was given because those residents would have been evacuating *through* a toxic plume. It was determined to be safer to hold people indoors until the plume had passed.

Fire conditions prevented the reconnaissance teams from locating the daily manifest inside the structure. Since it was not possible to ascertain what chemicals were burning and what the plume might contain, the Incident Commander (IC) decided not to fight the fire. Instead, companies constructed a berm or dike to contain the liquid runoff that was starting to present near the edge of the EQ property. Companies arriving on subsequent alarms assisted with the evacuations. The fire department contacted the EQ Company and requested that a representative come to the scene. The EQ plant manager arrived on scene and reported that the fire involved pesticides, oxidizers, contaminated metals, flammable and combustible materials, lead, and sulfur. EQ also reported that they had requested a private firm specializing in chemical fires to respond. That response, however, did not arrive for 12 hours, because the firm was located in Arkansas.

Due to changing winds and the threat of explosions, at 10:35 p.m. the ICP again was moved several blocks away from the fire to the Crowne Plaza Hotel parking lot. As the winds continued to change, the ICP was moved two more times. The final move, at 5:30 a.m. on October 6, was to a shopping center parking lot to accommodate the space and logistical needs of the growing operation. As the plume moved, and the ICP moved, so did the shelter locations. The EOC itself had to be evacuated, and city government was essentially shut down while the elected officials relocated.

The evacuation of police headquarters caused a problem, as police equipment, including some vehicles that were assigned to off-duty personnel, was left behind and had to be decontaminated after the operation. A mobile command vehicle was requested from Raleigh through mutual-aid agreements. CSX freight rail and Amtrak passenger service were shut down, as the rail line was affected by the plume. Requests to close the airspace over the fire were made to the Federal Aviation Administration (FAA).

A joint information center and media site was established near the ICP. Media briefings initially were held on the hour, and the media was credited with assisting in providing essential emergency public information regarding hazards, evacuation orders, and evacuation routes.

Throughout the night of October 5, evacuations continued with the cooperation of the school board which provided school buses and shelters for the evacuees. In a remarkable operation, Apex EMS coordinated with public transportation, schools, and the area EMS to evacuate 100 nursing home residents. None of the residents were ambulatory and all needed wheelchairs to be moved. The patients were transported to three area hospitals: Western Wake, Wake Medical, and Rex Hospital, all in Wake County. This evacuation was completed in 4 hours with no injuries.

A Medical Branch was established that eventually included 16 EMS units, 2 buses, and 2 engine companies. Mutual-aid fire departments set up decontamination stations at the three area receiving hospitals. Three schools were used as shelters. Schools were closed Friday, as the town remained shut down. During the day on Friday, October 6, State and Federal support arrived. State law enforcement arrived to assist the Apex police with perimeter and traffic control. The North Carolina Department of Environment and Natural

Resources (NCDENR) and the U.S. Environmental Protection Agency (EPA) assisted with air and water sampling to determine if there had been a runoff problem. None was reported as the NCDENR and EPA took more than 250,000 air and water samples.

By about 9 a.m. on Friday, October 6, 2006, the fires had subsided enough to permit the Apex and the contract firefighters from EQ to begin offensive operations. By 5 p.m. Apex Fire Command was terminated, and the site was turned over to the EQ contract firefighters. The last of the fires was extinguished by 1 a.m. on Saturday, October 7.

By the time the incident demobilized, approximately 17,000 people had been evacuated from their homes due to the threat posed by the chemical plume. There were no fatalities. Thirty civilians sought medical treatment for respiratory distress and skin irritation. Twelve police officers and one firefighter were treated for respiratory difficulties that were consistent with exposure to tear gas.

At this point, on October 6 at 5 p.m., the Apex police took responsibility for coordinating traffic control to allow reentry to the evacuated areas. The reentry was conducted in phases, with traffic controls that allowed only people with proper identification to enter their neighborhoods.

From every account and after-action report, including the town's report, this potentially devastating situation was handled with the highest levels of skill and expertise. The multiagency cooperation was virtually a textbook application of Unified Command and the National Incident Management System (NIMS).

The key element contributing to the success of operations was that Apex had a very well-defined plan that was practiced routinely. They made a commitment to train to the plan, and when they had an incident they used the plan as a foundation for the response. Apex was prepared to evacuate the town in large part because police, fire, EMS, public works, and elected officials all participate in federally required biannual exercises for the Shearon Harris Nuclear Power Plant, located 10 miles (16 km) from the town. Communities within 50 miles (80 km) of a commercial nuclear reactor are required to plan for evacuation, emergency public information, sheltering, and other protective actions. Instructions on what to do in case of an emergency are routinely sent to the public via mailings with water and tax bills.

The entire town of Apex is within the Emergency Planning Zone (EPZ) for the Harris facility. Under the plans for the power plant, the town is designated for evacuation in the event of an emergency at the Harris facility, since Apex is so close to the reactor. This is the typical response if a plant declares a Site Area Emergency or General Emergency, (types of emergency designation which are unique to the nuclear power industry). It was due to the frequent and ongoing planning and exercise program for the nuclear power plant that a climate and culture of cooperation had been firmly established among the town and surrounding agencies. All of the key players knew each other and knew each others' capabilities.

Apex's 20-year emergency plan was originally designed for the purpose of meeting Federal requirements related to the nuclear power plant. However, as a result of a January 2002 ice storm, the EOP was revised to become an all-hazards plan, which included provisions for shelters, as part of the overall plans annex for evacuation. The shelters are located in schools and staffed by the American Red Cross. By taking a hazard-specific plan and expanding it to an all-hazards plan, the base of cooperation and coordination was expanded.

Another factor in Apex's success was that they applied the NIMS during the incident. Apex aggressively applied NIMS training and made it available to all city agencies. The police and fire departments required all personnel to complete the Federal Emergency Management Agency (FEMA) independent study programs IS-100, *Introduction to the Incident Command System*, IS-200, *ICS for Single Resources and Initial Action Incidents,* IS-700,

National Incident Management System, An Introduction and IS-800, *The National Response Plan, An Introduction.* The fire department provided instructors to the other city agencies for their ICS programs. Public works and other city agency members also completed the IS-700 and IS-800 courses.

The fire department also does something the study team found unique. The fire chief requires the shift commander to prepare and complete an I-204 form from NIMS at the beginning of every shift. This form is an assignment listing and is a fundamental part of a written Incident Action Plan (IAP). Large incidents such as major chemical fires require written IAPs. By requiring the shift commander to have an I-204 completed in advance, the formal written IAP process had already been set in motion at the time of the fire. While every IC has a plan, and those plans are given out verbally on emergencies, the transitional time between oral and written plans can be reduced if some of the documentation is begun ahead of time. Certainly this practice helped the Apex Fire Department move from a local response to one that involved many mutual-aid agencies.

Source: U.S. Fire Administration/Technical Report Series, Chemical Fire in Apex, North Carolina, USFA-TR-163/April 2008, reported by Daryl Sensenig and Patrick Simpson

Haz mat incidents are challenging enough without adding confusion, indecisiveness, and a lack of coordination and communication on the part of first responders. First responders must bring order to the incident, and it is their responsibility to begin managing the incident. ***Mistakes made in the initial response to the incident can make the difference between solving the problem and becoming part of it*** **(Figure 5.1).**

Figure 5.1 If responders make mistakes in the initial stages of the incident, the consequences can be severe. *Courtesy of the U.S. Air Force.*

All fire and emergency services organizations must have predetermined guidelines or procedures detailing how to manage incidents involving hazardous materials. These guidelines include incident management elements that, if put into use properly, will enable first responders to mitigate haz mat emergencies effectively and safely. In those cases where haz mat incidents progress very slowly, first responders may be tempted to take shortcuts to speed the process, or local businesses or other officials may pressure them into moving quickly. It is very important that first responders continue to follow their predetermined incident management elements to avoid taking shortcuts that may result in adverse consequences.

In general, predetermined guidelines or procedures for how to manage a haz mat incident contain the following incident management elements:

- *Priorities*
  - Life safety
  - Incident stabilization
  - Protection of property and the environment
- *Management structure*
  - Command system (in the U.S., NIMS)
  - Predetermined procedures and guidelines such as emergency response plans, standard operating procedures (SOPs), standard operating guidelines (SOGs), or operating instructions (OIs [military]) that outline operational procedures regarding communications, equipment, personnel, resources, mutual aid, and other needs
- *Problem-solving process*
  - Analyzing the incident through scene analysis, information gathering, and/or size-up to try to understand and identify the problem(s) and assess hazards and risks
  - Planning the response by determining specific strategic goals
  - Implementing the response using tactical objectives and assignment of tasks
  - Evaluating progress

This chapter focuses on incident priorities and management structure, including communication. In-depth details relating to problem-solving (including strategies and tactics) are covered in Chapter 6, Strategic Goals and Tactical Objectives.

Incident Priorities

There are three incident priorities for all haz mat incidents (which are the same for all emergency services organizations — law enforcement, fire, emergency medical services [EMS], or other). The three priorities for haz mat incidents (in order) are as follows:

1. Life safety
2. Incident stabilization
3. Protection of property and the environment

All decisions during the problem-solving process must be made with these priorities in mind. The first priority is the safety of emergency responders and civilians because if responders do not protect themselves first, they cannot protect the public **(Figure 5.2)**. Life safety must be a consideration from the moment an incident is reported until its termination — from the response to the scene until the ride back to the station. A dead, injured, or unexpectedly contaminated first responder becomes part of the problem, *not* the solution.

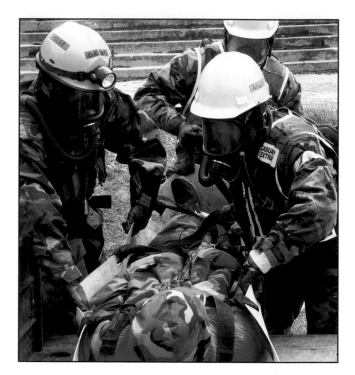

Figure 5.2 Injured or incapacitated responders become part of the problem, rather than the solution. The life safety of responders must be a top priority at all incidents. *Courtesy of the U.S. Marines, photo by Sgt. Christopher D. Reed.*

If there is no immediate threat to either responders or civilians, the next consideration is stabilizing the incident. When the first two priorities are satisfied, conservation (or protection) of property and the environment can be addressed. Stabilizing the incident can minimize environmental and property damage. If the situation calls for it, these priorities can be changed, but generally, first responders need to consider them in the order presented.

Decisions weighing the life safety of responders versus the life safety of the public must be based upon a careful risk/benefit analysis. Adopting a policy of cautious assessment before taking action is vital. A risk/benefit analysis should consider the following variables:

- Risk to rescuers
- Ability of rescuers to protect themselves
- Probability of rescue
- Difficulty of rescue
- Capabilities and resources of on-scene forces
- Possibilities of explosions or sudden material releases
- Available escape routes and safe havens
- Constraints of time and distance

Incident Management Systems

Incident management systems form the framework for a control and coordination structure that enables emergency personnel to turn chaos into order and effectively manage any emergency incident **(Figure 5.3)**. Incident management systems coupled with standard operating procedures (or the equivalent) provide a predetermined set of procedures to follow at every haz mat incident. When used correctly, these management tools enable first responders to quickly establish control of haz mat incidents.

Figure 5.3 The incident management system must be implemented at all incidents.

An **incident management system (IMS)** is a management framework used to organize emergency incidents and provides the following advantages:

- *Modular organization* — The management system is designed to provide direction from the top down (via the Incident Commander) and is expanded on an as needed basis according to the size and complexity of the incident.

- *Manageable span of control* — The management system enables an effective supervisory span of control at each level of the organization, which is determined by the ability of supervisors to monitor and effectively communicate with the personnel assigned to them. A reasonable span of control is usually 3 to 7, optimally 5.

- *Organizational facilities* — Incident facilities will be established to support the incident depending upon the type and complexity of the incident. Common terminology is used by the Incident Management System to define organizational facilities such as the incident command post, staging areas, base, and camp.

- *Position titles* — Incident Management System positions have distinct titles that use standard terminology to identify individuals and their assignment within the incident command organization. Standard titles help avoid confusion between the individual's day-to-day job position and the position assigned at an incident.

Incident Management System (IMS) — System described in NFPA® 1561, Standard on Fire Department Incident Management System, that defines the roles, responsibilities, and standard operating procedures used to manage emergency operations. Such systems may also be referred to as Incident Command Systems (ICS). U.S. emergency response agencies comply with Homeland Security Presidential Directive 5, which establishes the National Incident Management System (NIMS) as a standardized approach to incident management.

Incident Action Plan (IAP)
— Written or unwritten plan for the disposition of an incident. The IAP contains the overall strategic goals, tactical objectives, and support requirements for a given operational period during an incident. All incidents require an action plan. On relatively small incidents, the IAP is usually not in writing. On larger, more complex incidents, a written IAP is created for each operational period, and is disseminated to units assigned to the incident. When written, the plan may have a number of forms as attachments (for example forms ICS-202, 203, 204, 205, and 206 from NIMS.)

National Incident Management System - Incident Command System
— The U.S. mandated system that defines the roles, responsibilities, and standard operating procedures used to manage emergency operations. Also called NIMS- ICS.

Incident Commander (IC) — Person in charge of the incident management system and responsible for the management of all incident operations during an emergency, may also be called the On-Scene Incident Commander as defined in 29 CFR 1910.120q.

- *Incident Action Plan (IAP)* — **Incident Action Plans** provide the means to communicate overall incident strategies, tactics, and tasks. They are developed with input from all organizations/agencies involved. IAPs are used to issue assignments and direct efforts to attain the strategic objectives.

- *Organizational resources* — Execution of an IAP requires the assignment of personnel, equipment, and facilities. These resources are classified by type (ie. engine, truck, or bulldozer) by the Incident Management System according to their function and capability.

- *Integrated communication* — In many areas, law enforcement, fire, and EMS personnel are on different frequencies and in some cases have different dispatch/telecommunication centers. For the management system to function correctly, all of the emergency service organizations involved in the incident must be able to communicate with each other.

- *Accountability* — Accountability for resources operating on an incident is essential for maintaining a safe work environment. The Incident Management System requires an orderly chain of command and a system for tracking resources from check-in to demobilization.

In 2003, the U.S. implemented and mandated use of the **National Incident Management System (NIMS)** (see U.S. NIMS section, following). Use of incident management systems is also required by NFPA® 1561, *Standard on Emergency Services Incident Management System* (2002). Per this standard, all emergency services organizations (including law enforcement organizations, fire departments, fire brigades, EMS, and other support organizations such as the American Red Cross) must adopt an incident management system to manage all emergency incidents. Details of the incident management system must be incorporated into the organization's SOPs. Also, in the U.S., the Occupational Safety and Health Administration (OSHA) regulations (29 *CFR* 1910.120, *Hazardous Waste Site Operations and Emergency Response [HAZWOPER]*) mandate implementation of an incident command system as part of the requirements for emergency response to hazardous materials incidents.

What This Means To You

Regardless of what type of incident management system your organization uses, you must understand your role and responsibilities (as a first responder) within the command structure. Freelancing or taking action on your own without consent or knowledge of the **Incident Commander (IC)** is unacceptable and potentially very dangerous.

IMS and NIMS Terms

IMS has several terms that should be understood by all response personnel including the following:

Branch — That organizational level having functional/geographic responsibility for major segments of incident operations. The Branch level is organizationally between Section and Division or Group.

Command — That function of NIMS that determines the overall strategy of the incident, with input from throughout the IMS structure.

Division — That organizational level having responsibility for operations within a defined geographic area. The division level is organizational between single resources, Task Force, or the Strike Team and the Branch. Divisions are assigned clockwise around an outdoor incident with Division A at the front of the incident. In buildings or areas, Divisions are usually identified by the floor, area, or level to which they are assigned: First floor is Division 1, second floor is Division 2, and so on. In a one-story building, the entire interior may be assigned as a Division (Interior Division). All functional resources operating within that specific geographic area report to that division supervisor.

Group — That organizational level having responsibility for a specified functional assignment at an incident. When its assigned function has been completed, it is available for reassignment. Examples include Ventilation Group, Rescue Group, Evacuation Group, etc.

Incident Action Plan (IAP) — The strategic goals, tactical objectives, and support requirements for the incident. A plan should be formulated for every incident. For simple incidents, the incident action plan is usually verbal and not in written form. Large or complex incidents require that the incident action plan be documented in writing.

Incident Commander (IC) — The individual responsible for the management of all incident operations. The Incident Commander is primarily responsible for formulating the incident action plan and for coordinating and directing all incident resources to implement the plan and meet its goals and objectives.

Command Post (CP) — An established location from which the command function operates. The **Command Post** should be easily identified and accessible (either directly or indirectly). A CP can be a predetermined location such as at a facility, a conveniently located building, or a radio-equipped vehicle located in a safe area; also called *Incident Command Post* or *ICP.*

Resources — All personnel, tools and equipment, and extinguishing agents (water, foam, dry chemical, etc.) available or potentially available for assignment to incident tasks on which status is maintained. Resources may be individual companies, task forces, strike teams, or other specialized units. Resources are considered available when they have checked in at the incident and are not currently committed to an assignment. It is imperative that the status of these resources be tracked so that they may be assigned when and where they are needed without delay.

Supervisor — Individual responsible for the command of a Division or Group. The supervisor may be assigned to an area initially to evaluate and report conditions and advise Command of the needed task and resources.

Command Post (CP) — The designated physical location of the command and control point where the incident commander and command staff function during an incident and where those in charge of emergency units report to be briefed on their respective assignments.

U.S. NIMS and the National Response Framework

After the terror incidents of September 11, 2001, the United States government decided that all U.S. emergency services organizations needed to have common terminology and command structures. Homeland Security Presidential Directive/HSPD-5 formalized this idea by mandating that all state governments, local governments, and tribal entities must adopt NIMS. The lessons learned from September 11, 2001, and other major incidents clearly demonstrated the need for a single incident management system.

NIMS is designed to be applicable to small, single-unit incidents that may last only a few minutes as well as complex, large-scale incidents involving several agencies and many mutual aid units that could last for days or weeks

(Figure 5.4). NIMS builds from the ground up and is the basic operating system for all incidents within each facility or agency. By design, the NIMS can grow from a small-scale organization to a large-scale organization depending on the needs of the incident.

Figure 5.4 The flexibility of IMS allows management and command of the incident to expand as needed. This could range from a few initial response units to multiple jurisdictions and hundreds of emergency workers at a major, long-term incident. *Courtesy of the U.S. Department of Defense.*

Depending on incident complexity, the Incident Commander (IC) may delegate responsibilities and assign personnel to subordinate management roles. If the functions are not delegated, the IC retains these responsibilities and must ensure that all requisite functions are completed as part of the IAP.

In addition to the five usual functions (see sections following) NIMS adds another position, Intelligence, which is responsible gathering of information relating to the incident **(Figure 5.5)**. Depending on the type and complexity of the incident, the Intelligence function may be staffed as a staff officer within the Command Staff, as part of the Operations Section or the Planning Section, or assigned to a Section level position on its own. Because terrorist incidents are criminal acts and fall under the jurisdiction of the FBI in the U.S., it is likely the Intelligence position will be staffed at the Section level for these types of incidents.

In addition to NIMS, emergency responders in the U.S. should be familiar with the **National Response Framework (NRF)**. The NRF explains how, at all levels, the U.S. effectively manages all-hazards response. The framework incorporates best practices and procedures from the following incident management disciplines:

National Response Framework (NRF) — Document that provides guidance on how communities, States, the Federal Government, and private-sector and nongovernmental partners conduct all-hazards emergency response.

- Homeland security
- Emergency management
- Law enforcement
- Fire fighting
- Public works

Figure 5.5 The Intelligence Section is responsible for gathering information at the incident and is typically staffed by law enforcement. *Courtesy of FEMA News Photos, photo by Michael Rieger.*

- Public health
- Responder and recovery workers
- Health and safety
- Emergency medical services
- Private sector organizations

The NRF integrates these disciplines into a unified structure. It forms the basis of how the federal government coordinates with state, local, and tribal governments and the private sector during incidents.

The NRF, using NIMS, can be partially or fully implemented in the context of a threat, anticipation of a significant event, or the response to a significant event. Selective implementation through the activation of one or more of the system's components allows maximum flexibility in meeting the unique operational and information-sharing requirements of the situation at hand and enabling effective interaction between various federal and non-federal entities.

A basic premise of the NRF is that incidents are generally handled at the lowest jurisdictional level possible. Police, fire, public health and medical, emergency management, and other personnel are responsible for incident management at the local level. In some instances, a federal agency in the local area may act as a first responder and may provide direction or assistance consistent with its specific statutory authorities and responsibilities. In the vast majority of incidents, state and local resources and interstate mutual aid normally provide the first line of emergency response and incident management support. When an incident or potential incident is of such severity, magnitude, and/or complexity that additional resources are needed, states can request federal assistance **(Figure 5.6)**.

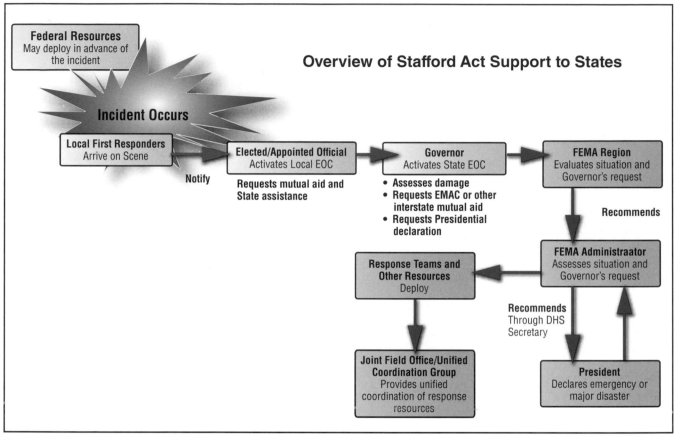

Figure 5.6 The U.S. Stafford Act (Public Law 100-707) details how local and state governments request federal assistance during disasters. This process is reflected in the National Response Framework.

Reflecting the NIMS construct, the NRF includes the following command and coordination structures:

- Incident Command Posts (ICPs) on-scene using the Incident Command System (ICS)/**Unified Command**

- Area Command (if needed)

- State, local, tribal, and private-sector Emergency Operations Centers (EOCs)

- Unified Coordination Group (formerly called the Joint Field Office), which is responsible for coordinating Federal assistance and supporting incident management activities locally

- Regional Response Coordination Center (RRCC) and Homeland Security Operations Center (HSOC), which serve as regional and national-level multiagency situational awareness and operational coordination centers

- Interagency Incident Management Group (IIMG), which serves as the national headquarters-level multiagency coordination entity for domestic incident management

Unified Command — In the Incident Management System, a shared command role that allows all agencies with responsibility for the incident, either geographical or functional, to manage the incident by establishing a common set of incident objectives and strategies. In unified command there is a single incident command post and a single operations chief at any given time.

- Homeland Security Council (HSC) and other White House organizations, which serve as the national-level multiagency coordination entities to advise and assist the President on homeland security and other policy issues

In the case of a catastrophic terrorist attack, the NRF's primary mission is to:

- Save lives
- Protect critical infrastructure, property, and the environment
- Contain the event
- Preserve national security

Additionally, it allows standard assistance-request procedures to be expedited or suspended, and, while notification and full coordination with State will occur, the coordination will not delay rapid deployment. Selected Federal response resources will mobilize and deploy and begin necessary operations as quickly as possible.

The following teams are resources that have been established by the NRF:

- **Weapons of Mass Destruction-Civil Support Teams (WMD-CST)** — Teams that support civil authorities at a domestic chemical, biological, radiological, nuclear, or high-yield explosive incident site by identifying CBRNE agents/substances. The National Guard Bureau fosters the development of WMD-CSTs. There are plans for at least one CST in each state. *Duties:*

  — Assess current and projected consequences

  — Advise on response measures

  — Assist with appropriate requests for state support

  — Provide an extensive communications capability

- **Disaster Medical Assistance Teams (DMAT)** — Groups of professional and paraprofessional medical personnel (supported by a cadre of logistical and administrative staff) designed to provide emergency medical care during a disaster or other event **(Figure 5.7)**. The National Disaster Medical System (NDMS), through the U.S. Public Health Service (PHS), fosters the development of DMATs.

Figure 5.7 DMAT teams provide emergency medical care during disasters or other events. *Courtesy of FEMA News Photos, photo by Andrea Booher.*

- *Disaster Mortuary Operational Response Teams (DMORT)* — Teams that work under the guidance of local authorities by providing technical assistance and personnel to recover, identify, and process deceased victims. The teams are composed of private citizens, each with a particular field of expertise, who are activated in the event of a disaster. The NDMS, through the PHS and the National Association for Search and Rescue (NASAR) fosters the development of DMORTs.

- *National Medical Response Team-Weapons of Mass Destruction (NMRT-WMD)* — Specialized response forces designed to provide medical care following a nuclear, biological, and/or chemical incident. Four teams are geographically dispersed throughout the U.S. The NDMS, through the PHS, fosters the development of NMRT-WMD. These units are capable of providing the following services:

  — Mass casualty decontamination

  — Medical triage

  — Primary and secondary medical care to stabilize victims for transportation to tertiary-care facilities in a hazardous material environment

- *Urban Search and Rescue (US&R) Task Forces* — Highly trained teams that provide search-and-rescue operations in damaged or collapsed structures and stabilization of damaged structures. They can also provide emergency medical care to the injured. Currently there are 28 federal US&R teams and numerous state teams that follow the DHS-FEMA US&R model regarding training, equipment, and personnel. The task forces are a partnership among the following entities:

  — Local fire departments

  — Law enforcement agencies

  — Federal and local governmental agencies

  — Private companies

- *Incident Management Teams (IMT)* — Teams of highly trained, experienced individuals who are organized to manage large and/or complex incidents. They provide full logistical support for receiving and distribution centers. National IMTs are hosted and managed by Geographic Area Coordination Centers. The teams are hosted by the U.S. Forest Service (USFS) during wildland fires. Both states and regions can have IMTs. *Factors and examples:*

  — Many fire and emergency services want to develop local and regional/ metropolitan IMTs, which would be based on USFS models.

  — IMTs would train to support command and general staff functions of the Incident Command System (ICS).

  — Many states have organized IMTs at the state and local level. For example, Tualatin Valley Fire and Rescue, Oregon, maintains five IMTs, rotating on-call status on a weekly basis. They provide strategic incident management and support for incidents involving large areas, long durations, technical or political complexities, or any other aspects extending beyond routine response capabilities.

IMS Operational Functions

Under NIMS and many other incident management systems, there are five major operational positions or functions:

- Command
- Operations
- Planning
- Logistics
- Finance/Administration

Command

Command is the function of directing, ordering, and controlling resources by virtue of explicit legal, agency, or delegated authority **(Figure 5.8)**. It is important that lines of authority be clear to all involved. Lawful commands by those in authority need to be followed immediately and without question. The basic configuration of the Command organization includes the following three levels:

- *Strategic level* — Entails the overall direction and goals of the incident
- *Tactical level* — Identifies the objectives that the tactical level supervisor/officer must achieve to meet the strategic goals
- *Task level* — Describes the specific tasks needed to meet tactical-level requirements, and assigns these tasks to operational units, companies, or individuals

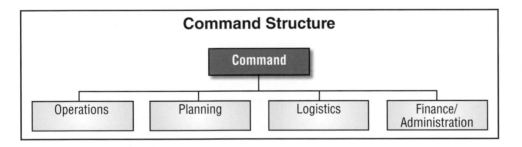

Figure 5.8 The basic NIMS command structure.

The person in overall command of an incident is the IC. The IC establishes the Command Post and is ultimately responsible for all incident activities, including the development and implementation of a strategic plan. This process may include making a number of critical decisions and being responsible for the results of those decisions. The IC has the authority both to call resources to the incident and release them from it. Additional duties and responsibilities of the IC are discussed in the next section.

If the size and complexity of the incident requires, the IC may delegate authority to the following Command Staff positions:

- *Safety Officer* (OSHA requires the appointment of a safety officer at haz mat incidents)
- *Liaison Officer*
- *Public Information Officer*

Incident commander. The IC is the officer at the top of an incident chain of command and in overall charge of the incident **(Figure 5.9)**. The IC is ultimately responsible for everything that takes place at the emergency scene and primarily responsible for formulating the IAP and coordinating and directing all incident resources to implement the plan and meet its goals and objectives. No aggressive plan should be undertaken unless sufficient information is available to make logical decisions and the safe coordination of operations can be accomplished. The IC must make it known to the telecommunicator and other responders when Command is assumed or transferred. ICs at haz mat incidents have specific responsibilities in addition to standard IC functions. The IC does not have to actually perform or supervise each function but may choose to delegate them to others. The IC is required to perform the following functions at haz mat incidents:

● Establish the site safety (also called *scene safety*) plan

● Implement a site security and control plan to limit the number of personnel operating in the control zones (see Chapter 6, Strategic Goals and Tactical Objectives)

● Designate a safety officer

● Identify the materials or conditions involved in the incident

● Implement appropriate emergency operations

● Ensure that all emergency responders (not just those of their own organizations) wear appropriate personal protective equipment (PPE) in restricted zones

● Establish a decontamination plan and operation

● Implement post-incident emergency response procedures (incident termination)

Figure 5.9 The Incident Commander (IC) has overall responsibility for managing the incident.

HAZWOPER IC Requirements

According to OSHA regulations (29 *CFR* 1910.120, *HAZWOPER*), any emergency responder expected to perform as an IC must be trained to fulfill the obligations of the position at the level of response the responder is providing (Operations Level, minimum), which includes the ability to perform the following duties:

● Analyze a hazardous substance incident to determine the magnitude of the response problem.

● Plan and implement an appropriate response plan within the capabilities of available personnel and equipment.

● Implement a response to favorably change the outcome of an incident in a manner consistent with the local emergency response plan and the organization's predetermined procedures.

● Evaluate the progress of the emergency response to ensure that the response objectives are being met safely, effectively, and efficiently.

● Adjust the response plan to the conditions of the response and notify higher levels of response when required by changes to the response plan.

Appendix E, Hazardous Materials Incident Commander Checklist, provides a sample checklist designed to ensure compliance with *HAZWOPER*.

Safety officer. The **Safety Officer** is responsible for monitoring and identifying hazardous and unsafe situations and developing measures for ensuring operational and personnel safety **(Figure 5.10)**. Although the Safety Officer may exercise emergency authority to stop or prevent unsafe acts when immediate action is required, the officer generally chooses to correct them through the regular line of authority. Many unsafe acts or conditions are addressed in the

Safety Officer — (1) Fire officer whose primary function is to administrate safety within the entire scope of fire department operations. Also referred to as the Health and Safety Officer. (2) Member of the IMS Command Staff responsible to the incident commander for monitoring and assessing hazardous and unsafe conditions and developing measures for assessing personnel safety on an incident. Also referred to as the Incident Safety Officer.

Figure 5.10 The safety officer monitors the scene for unsafe conditions.

IAP. The Safety Officer must be trained to the level of operations conducted at the incident and is required to perform the following duties:

- Obtain a briefing from the IC.
- Review IAPs for safety issues.
- Identify hazardous situations at the incident scene.
- Participate in the preparation and monitoring of incident safety considerations, including medical monitoring of entry team personnel before and after entry.
- Maintain communications with the IC, and advise the IC of deviations from the incident safety considerations and of any dangerous situations.
- Alter, suspend, or terminate any activity that is judged to be unsafe.
- Conduct safety briefings.

At hazardous waste cleanup sites and other nonemergency operations involving hazardous materials, OSHA *HAZWOPER* and the Canadian Ministry of Labour require appointment of a *Site Safety and Health Supervisor* whose duty is to oversee onsite safety and health issues. These individuals must have training specific to the safety of haz mat incidents.

The Safety Officer also needs to ensure that safety briefings are conducted for entry team personnel before entry **(Figure 5.11)**. Safety briefings include the following information about the status of the incident based upon the preliminary evaluation and subsequent updates:

- Identification of hazards
- Description of the site
- Tasks to be performed
- Anticipated duration of the tasks
- PPE requirements
- Monitoring requirements

Figure 5.11 The safety officer conducts safety briefings with entry team members before they enter hazardous areas.

- Notification of identified risks
- Additional, pertinent information

At incidents involving potential criminal or terrorist activities, the safety briefing should also cover the following items:

- Being alert for secondary devices
- Not touching or moving any suspicious-looking articles (bags, boxes, briefcases, soda cans, and the like)
- Not touching or entering any damp, wet, or oily areas
- Wearing full protective clothing, including self-contained breathing apparatus (SCBA)
- Limiting the number of personnel entering the crime scene
- Documenting all actions
- Not picking up or taking any souvenirs
- Photographing or videotaping anything suspicious
- Not destroying any possible evidence
- Seeking professional crime-scene assistance

Command post (CP). Establishing a CP to which information flows and from which orders are issued is vital to a smooth operation. The IC must be accessible (either directly or indirectly) and a CP ensures this accessibility. A CP can be a predetermined location at a facility, a conveniently located building, or a radio-equipped vehicle located in a safe area **(Figure 5.12)**. Ideally, the CP is located where the IC can observe the scene, although such a location is not absolutely necessary. The location of the CP is relayed to the telecommunicator/dispatcher and emergency responders. A CP needs to be readily identifiable. One common identifier is a green flashing light. Other methods include pennants, signs, and flags.

Figure 5.12 Many departments have mobile command posts, however, a command post can be a predetermined location at a facility, a conveniently located building, or a radio-equipped vehicle located in a safe area. *Courtesy of Ron Jeffers.*

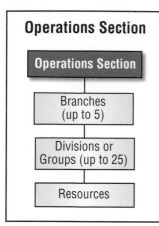

Figure 5.13 The Operations Section can be divided into Branches, Divisions, or Groups.

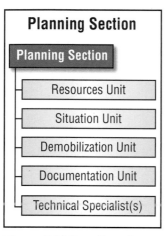

Figure 5.14 Specific units under the Planning Section include the Resource Unit, Situation Unit, Demobilization Unit, Documentation Unit, and Technical Specialists.

Operations Section

The *Operations Section* is responsible for the direct management of all incident tactical activities, the tactical priorities, and the safety and welfare of personnel working in the Operations Section **(Figure 5.13)**. The *Operations Section Chief* reports directly to the IC and is responsible for managing all operations that directly affect the primary mission of eliminating a problem incident. The Operations Section Chief directs the tactical operations to meet the strategic goals developed by the IC.

Planning Section

The Planning Section is responsible for gathering, assimilating, analyzing, and processing information needed for effective decision-making **(Figure 5.14)**. Information management is a full-time task at large incidents. The Planning Section serves as the IC's clearinghouse for incidents, which allows the IC's staff to provide information instead of having to deal with dozens of information sources. Command uses the information compiled by the Planning position to develop strategic goals and contingency plans. Specific units under Planning include the *Resources Unit, Situation Unit, Documentation Unit, Demobilization Unit*, and any technical specialists whose services are required.

Logistics Section

The *Logistics Section* is the support mechanism for the organization. It provides services and support systems to all the organizational components involved in the incident including the following:

- Facilities
- Transportation needs
- Supplies
- Equipment
- Maintenance
- Fueling supplies
- Meals
- Communications
- Responder medical services **(Figure 5.15)**

Support Branch and *Service Branch* are two branches within the logistics section. The Service Branch includes medical, communications, and food services. The Support Branch includes supplies, facilities, and ground support (vehicle services).

Finance/Administration Section

The *Finance/Administration Section* is established on incidents when agencies involved have a specific need for financial services **(Figure 5.16)**. Not all agencies require the establishment of a separate Finance/Administration Section. In some cases, such as cost analysis, that position could be established as a Technical Specialist in the Planning Section. Specific units under the Finance/Administration Section include the *Time Unit, Procurement Unit, Compensation Claims Unit*, and *Cost Unit*.

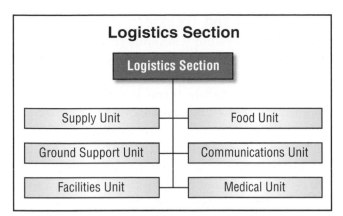

Logistics Section

Logistics Section

Supply Unit	Food Unit
Ground Support Unit	Communications Unit
Facilities Unit	Medical Unit

Figure 5.15 There are two branches within Logistics: the support branch (left) and the service branch (right).

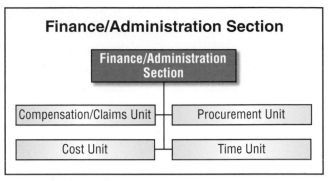

Finance/Administration Section

Finance/Administration Section

| Compensation/Claims Unit | Procurement Unit |
| Cost Unit | Time Unit |

Figure 5.16 The Finance/Administration Section is often only activated at large-scale, long-term incidents.

Staging Area

The *staging area* is where personnel and equipment awaiting assignment to the incident are held. This practice keeps the responders and their equipment a short distance from the scene until they are needed and minimizes confusion at the scene. Staging is discussed in greater detail in Chapter 6.

Resources

Resources are all personnel, equipment, and major pieces of apparatus on scene or en route on which status is maintained. Resources may be individual companies, task forces, strike teams, or other specialized units **(Figure 5.17)**. Resources are considered to be *available* when they have checked in at the incident and are not currently committed to an assignment. The status of these resources must be tracked so that they can be assigned when and where they are needed without delay.

Figure 5.17 This decon trailer would be considered a resource at a haz mat incident.

Incident Command Establishment and Transfer

When using the Incident Management System (IMS) one directive is clear: The first person on the scene or the ranking individual of the first company on the scene assumes command of the incident. That individual maintains command until a higher ranking or more extensively trained responder arrives on the scene and assumes command. *The IC must have IMS training and be at the hazardous materials Operations Level.* Before command is transferred, the person accepting it must be capable of assuming command (that is, have the necessary qualifications) and be willing to accept it. If the transfer cannot take place face to face, it can be accomplished over the radio, but *command can only be transferred to someone who is on scene.* As an incident grows larger, command may be transferred several times before the situation is brought under control. A smooth and efficient transfer of command contributes greatly to bringing an incident to a timely and successful conclusion.

The person relinquishing command must provide the person assuming command with as clear a picture of the situation as possible. This update can be accomplished by giving a briefing or situation status report, an updated version of the incident evaluation performed on arrival. The person assuming command acknowledges receipt of the information by repeating it back to the other person. If the reiteration is accurate, the recipient is ready to accept control of and responsibility for the management of the incident. The former IC can then be reassigned to an operating unit or retained at the CP as an aide or as a member of the Command Staff.

When a complex emergency occurs, command may be transferred several times as the organization grows. It is important that transitions are as smooth and as efficient as possible. When command is transferred, the former IC must announce the change to avoid any possible confusion caused by others hearing a different voice acknowledging messages and issuing orders. If all involved follow the chain of command and use correct radio protocols, they will not call anyone by name, rank, or job title, so it does not matter who answers their radio messages. Because the early stages of an emergency can be chaotic, anything done to reduce confusion is desirable. Announcing a transfer of command is one way of accomplishing that objective.

There is only *ONE* IC. There may be many involved in decision making, but there is only one IC. A multijurisdictional incident involves services (fire, law enforcement, EMS, and the like) from one city but beyond the jurisdiction of one organization/agency. The chain of command must be clearly defined, especially when a unified command is used. One person issues all orders through the chain of command to avoid the confusion caused by conflicting orders.

NIMS Incident Command Transfer Steps

Per NIMS, the following steps should be taken to transfer command:

Step 1: The incoming Incident Commander should, if at all possible, personally perform an assessment of the incident situation with the existing Incident Commander.

Step 2: The incoming Incident Commander must be adequately briefed. This briefing must be by the current Incident Commander, and take place face-to-face if possible. The briefing must cover the following:

- Incident history (what has happened)
- Priorities and objectives
- Current plan
- Resource assignments
- Incident organization
- Resources ordered/needed
- Facilities established
- Status of communications
- Any constraints or limitations
- Incident potential
- Delegation of Authority

Step 3: After the incident briefing, the incoming Incident Commander should determine an appropriate time for transfer of command.

Step 4: At the appropriate time, notice of a change in incident command should be made to:

- Agency headquarters (through dispatch)
- General Staff members (if designated)
- Command Staff members (if designated)
- All incident personnel

Step 5: The incoming Incident Commander may give the previous Incident Commander another assignment on the incident. There are several advantages of this:

- The initial Incident Commander retains first-hand knowledge at the incident site.
- This strategy allows the initial Incident Commander to observe the progress of the incident and to gain experience.

The ICS Form 201 is especially designed to assist in incident briefings (Step 2). It should be used whenever possible because it provides a written record of the incident as of the time prepared. The ICS Form 201 contains:

- Incident objectives
- A place for a sketch map
- Summary of current actions
- Organizational framework
- Resources summary

Source: NIMS

Unified Command

Control of an incident involving multiple agencies with overlapping authority and responsibility is accomplished through the use of unified command **(Figure 5.18, p. 252)**. The concept of *unified command* simply means that all agencies that have a jurisdictional responsibility at a multijurisdictional incident contribute to the process by taking the following actions:

- Determine overall incident objectives.
- Select strategies.

Figure 5.18 Many different organizations may be part of a unified command. *Courtesy of the U.S. Air Force.*

- Accomplish joint planning for tactical activities.

- Ensure integrated tactical operations.

- Use all assigned resources effectively.

Proactive organizations identify target hazards in their areas of jurisdiction and also identify any other agencies with authority and responsibility for those target hazards. Ideally, those agencies meet, identify differences in agency IMS practices, and establish a *memorandum of understanding for unified command:* a written agreement defining roles and responsibilities within a unified command structure. It is signed by the lead officials of the agencies and becomes policy governing the personnel within those agencies.

Controlling hazardous material incidents may require the coordinated efforts of several agencies/organizations such as the following:

- Fire service

- Law enforcement

- EMS

- Private concerns
  — Material's manufacturer
  — Material's shipper
  — Facility manager

- Government agencies (local, state/provincial, federal) with mandated interests in health and environmental issues

- Privately contracted cleanup and salvage companies

- Specialized emergency response groups, organizations, and technical support groups

- Utilities and public works

To avoid jurisdictional and command disputes, the specific agency/organization responsible for handling and coordinating response activities should be identified before an incident happens. It is important to know what mutual aid contracts do and do not cover. Pre-incident coordination should be done at the local level so that jurisdictional disputes can be avoided. The responsible or *lead* agency can then begin documenting the identities and capabilities of nearby support sources. Proper planning and preparation lead to safe and successful responses to a hazardous material incident. The occurrence of a serious haz mat incident is not the time to discover that a neighboring fire department or industry cannot provide desperately needed equipment, personnel, or technical expertise. When emergency services organizations work together to develop their haz mat pre-incident surveys, they can meet the following objectives:

- Share vital resource information
- Develop rapport among participating emergency services organizations
- Identify and pool needed resources

IMS Haz Mat Positions

IMS provides an organizational structure for necessary supervision and control of the essential functions required at virtually all hazardous materials incidents. This organizational structure may incorporate resources called to the scene such as hazardous materials response teams.

Incident Organization

The IMS organization will be determined by the complexity of the incident and the resources needed for mitigation. A hazardous material branch or group may be added to the IMS structure when resources are needed to specifically address the problems created by the hazardous material(s) **(Figure 5.19)**. For example, if a hazardous materials response team is called in for purposes of mitigation of a spill or leak, they would be assigned in the IMS structure within the Haz Mat Branch or Group.

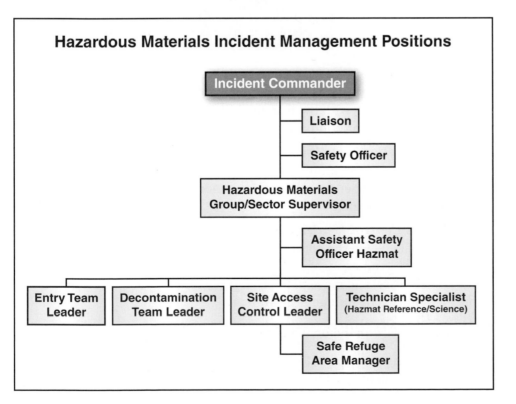

Figure 5.19 At haz mat incidents, the IMS structure may need to add a haz mat branch or group.

These positions are at the Technician Level, but Awareness- and Operations-Level responders need to be familiar with them. There are standard IMS positions particular to hazardous materials incidents:

- *Hazardous Materials Branch Director / Group Supervisor* — Manages the resources assigned to the branch or group and directs the primary tactical functions (called the Haz Mat Group Supervisor hereafter) **(Figure 5.20)**

Figure 5.20 The Haz Mat Group Supervisor directs the primary tactical functions of the haz mat group.

- *Entry Team Leader* — Supervises all companies and personnel operating in the hazardous area, with the responsibility to direct all tactics and control the positions and functions of all personnel in the hazardous area

- *Decontamination Team Leader* — Supervises operations in the scene control zone where decontamination is conducted, and ensures that all rescued citizens, response personnel, and equipment have been decontaminated before leaving the incident

- *Site Access Control Leader* — Controls all movement of personnel and equipment between the control zones, and is responsible for isolating the control zones and ensuring proper routes; also has the responsibility for the control, care, and movement of people before they are decontaminated; may appoint a Safe Refuge Area Manager

- *Assistant Safety Officer (Hazardous Materials)* — Is responsible for the overall safety of assigned personnel within the Hazardous Materials Group, and reports directly to the Safety Officer; must be appointed at hazardous materials incidents and have the requisite knowledge to function as the Assistant Safety Officer at a haz mat incident

- *Technical Specialist (Hazardous Materials Reference/Science Technical Specialist)* — Is responsible for providing technical information and assistance to the Hazardous Materials Group and the Planning Section using various sources such as computer databases, technical journals, public and private technical information agencies, facility representatives, and product specialists

- *Safe Refuge Area Manager* — Is responsible for evaluating and prioritizing victims for treatment, collecting information from the victims, and preventing the spread of contamination by these victims; also it is recommended that this person have an EMS background

The functional positions of the Hazardous Materials Branch/Group (Entry Leader, Decontamination Leader, and Site Access Control Leader) require a high degree of control and close supervision. The Haz Mat Group Supervisor manages the functional responsibilities, which includes all tactical operations carried out in the hazardous area.

All rescue operations come under the direction of the Haz Mat Group Supervisor. In addition to the primary functions, the Haz Mat Group Supervisor works with an Assistant Safety Officer who is trained in hazardous materials and must be present at the hazardous site. The Haz Mat Group Supervisor may also supervise one or more Technical Specialists. Evacuation and all other tactical objectives that are outside the scene control zones are not responsibilities of the Haz Mat Group Supervisor. These tactical operations as well as many other hazardous materials related functions are managed by regular IMS positions.

Communications

Effective communication is vital for incident management and mitigation efforts. Departmental/organizational predetermined procedures usually cover methods of communication (both externally and internally) at incidents, whether by radio, cell phone, hand-light signals, or hand signals. Traditionally, two-way radios are the primary means of communication; however, advances in technology have made a variety of communication tools available to emergency responders **(Figure 5.21, p. 256)**. First responders must be able to communicate the need for assistance through their department/organization's communications equipment. Some of these communications might be requests for additional personnel or special equipment or to notify others at the incident of any apparent hazards. Several procedures and guidelines for using communications equipment correctly are as follows:

- *Radio procedures* — Communication protocols, priority transmission methods, and terminology:
  - Use *plain language*.
  - Transmit *only essential information* when sending information and orders.
  - Use appropriate channels to communicate with both the IC and the telecommunicator.
  - Use appropriate number of channels for the size and complexity of the incident:
    - Routine, day-to-day incidents are usually handled on a single channel.

Plain Language — Communication that can be understood by the intended audience and meets the purpose of the communicator. For the purpose of the *National Incident Management System*, plain language is designed to eliminate or limit the use of codes and acronyms, as appropriate, during incident response involving more than a single agency. Also called *plain text*.

Figure 5.21 Advances in technology have improved the number of ways emergency responders can communicate and gather information at incidents.

Figure 5.22 Plain language should be used for all radio transmissions.

⊚ Large incidents may require using several channels to allow for clear and timely exchanges of information.

⊚ Separate channels may be needed for Command, entry, decontamination, research support, tactical, and support functions.

● *Requesting additional help* — Additional units, specialized equipment, alarm signals and what to do when they are received:

— Know what types of additional help are available.

— Identify these items in the local emergency response plan.

● *Emergency radio traffic* — Distress messages sent to telecommunicators at telecommunications centers (who are better equipped than on-scene personnel to hear weak signals from portable and mobile radios) and urgent messages for additional resources or to relay detailed instructions:

— Make the urgency clear to the telecommunicator.

— Wait while the telecommunicator gives an attention tone (if used in that system), advises all other units to stand by, and advises the caller to proceed with the emergency message.

— Resume normal or routine radio traffic when the emergency communication is complete after the telecommunicator notifies all units to do so.

- *Evacuation signals* — Broadcasts pulling all emergency personnel from a hazardous area because conditions have deteriorated beyond the point of safety; similar to emergency traffic procedures:

  — Broadcast a radio message ordering evacuation.

  — Sound audible warning devices (sirens and air horns) on the apparatus at the incident scene for an extended time period (works well outside small structures).

  — Broadcast the message several times.

Cell Phone Usage

Many departments have written cell phone usage into their emergency communications plan. However, immediately after a large emergency or disaster, there is such a great demand on the cell phone system from the public that efficient cell phone communication may be impossible.

The radio communications system should reflect the size and complexity of the incident. Routine, day-to-day incidents can usually be handled on a single channel, but larger incidents may require using several channels to allow for clear and timely exchanges of information. Separate channels may be needed for command, tactical, and support functions.

Consider, also, communication difficulties that may be associated with the location of the incident. For example, radio communications may be ineffective in underground subway systems, or cell phone coverage may be inadequate in rural areas. Plans and adjustments must be made to address these issues before an emergency occurs.

Under NIMS, all radio communications should be transmitted in plain language **(Figure 5.22)**. Codes, abbreviations, and acronyms should not be used because they may not be universally understood. Any misunderstanding can add to the confusion that is often a part of emergency operations — especially those involving more than one agency.

Personnel should exercise proper radio discipline and follow all organizational protocols. They should also confine transmissions to essential information and keep them as brief as possible. Long messages should be interrupted at frequent intervals to allow others to break in with high-priority traffic. Emergency transmissions always have priority over other traffic, and others should avoid transmitting whenever anyone declares they have emergency traffic. Local protocols dictate how messages are phrased, but everyone should be called by the position they occupy. For example, regardless of who the IC is, that individual is always called the Incident Commander or IC. The incident name may be included (Warehouse IC) if more than one incident is in progress. Radio communications should be established between the IC and any mutual or automatic aid agencies that would respond to the incident.

An external communication system facilitates communication between onsite and offsite personnel, which may be necessary to coordinate an emergency response while maintaining contact with essential offsite personnel and technical experts. The primary means of external communication are cell phone, telephone, and radio, but other technologies may enable use of computers (e-mail, digital messaging for pagers, etc.) and other devices. Predetermined procedures may also address external communications with the media and public.

Information regarding the incident received and shared among emergency responders once they arrive on the scene is considered to be internal communication. Examples of internal communication are as follows:

- Alerting team members to emergencies
- Passing along safety information
- Communicating changes in the action plan
- Maintaining site control

Verbal communication at a site can be impeded by onsite background noise and the use of PPE. For example, speech transmission through a respirator can be poor, and protective hoods and respirator airflow can impair hearing. Often for effective internal communication, commands must be prearranged. In addition, audio or visual cues can help convey messages. The most important thing is that signals must be agreed upon in advance.

Both primary and backup systems of communication are recommended, particularly when developing a set of signals for use only during emergencies. All communication devices used in a potentially explosive atmosphere must be intrinsically safe and not capable of sparking; check them daily to ensure that they are operating. Several examples of internal communication devices are as follows:

- Radios
- Cell phones
- Pagers
- Noisemakers
  - Bells
  - Compressed-air horns
  - Megaphones
  - Sirens
  - Whistles
- Visual signals
  - Flags
  - Flare or smoke bombs
  - Hand signals
  - Lights
  - Signal boards
  - Whole body movements

Summary

Emergency response to haz mat incidents must be conducted within a certain management framework and structure to ensure successful mitigation of the incident. Every response must be conducted with these priorities in mind: life safety, incident stabilization, and protection of property and the environment. Furthermore, IMS must be implemented at all incidents, and operations must be guided by emergency response plans and predetermined procedures that specify communication procedures and other important predetermined actions. While not all responders are responsible for every aspect of managing the incident (for example, the IC will decide the specific strategic goals and objectives), everyone must be conscious of their role or roles in the greater scheme of things.

Review Questions

1. What are the priorities for haz mat incidents?

2. What are the advantages of an incident management system?

3. What incident management system is used in the U.S?

4. What structures are included in the National Response Framework (NRF)?

5. What are the five major operational functions of most incident management systems?

6. What is the function of the Incident Commander?

7. What are the duties of a safety officer?

8. Discuss the steps in transferring command.

9. What are the standard IMS positions particular to hazardous materials incidents?

10. What are some guidelines for using communications equipment correctly?

Strategic Goals and Tactical Objectives

Chapter Contents

Divider page photo courtesy of Rich Mahaney.

chapter 6

Competencies

NFPA® 472:	5.2.4(4)	5.3.2(1)	5.4.1(3)(a)	5.4.3(6)	5.5.1(1)
5.2.1.4	5.3.1(1)	5.3.2(2)	5.4.1(3)(b)	5.4.4(1)	5.5.1(2)
5.2.2(7)	5.3.1(2)	5.4.1(1)	5.4.1(6)	5.4.4(2)	5.5.2(2)
5.2.4(2)	5.3.1(3)	5.4.1(2)	5.4.3(2)	5.4.4(3)	

Strategic Goals and Tactical Objectives

Learning Objectives

1. Describe each of the steps of the basic problem-solving formula. [NFPA® 472, 5.3.1(1-3), 5.3.2(1), 5.4.3(2), 5.5.1(1)]

2. Discuss isolation and scene control. [NFPA® 472, 5.2.1.4, 5.2.4(2), 5.2.4(4), 5.4.1(1-2), 5.4.1(3)(a-b), 5.4.1(6)]

3. Explain the notification process. [NFPA® 472, 5.2.2(7), 5.4.3(6), 5.5.2(2)]

4. Discuss protection of responders, the public, the environment, and property. [NFPA® 472, 5.3.2(2), 5.4.4(1-3)]

5. Describe recovery and termination. [NFPA® 472, 5.5.1(2)]

Chapter 6
Strategic Goals and Tactical Objectives

Case History

On January 6, 2005 two Norfolk Southern trains collided in Graniteville, South Carolina near an Avondale Mills textile plant, resulting in the deaths of nine people, the treatment of 250 people for chlorine exposure, and the 14-day evacuation of 5,400 residents within a 1-mile (1.6 k) radius during rescue and decontamination efforts **(Figure 6.1 a and b)**. The accident occurred when Train No. 192 was diverted onto an improperly lined siding (a lighter rail track meant for lower speed or less heavy traffic) already in use by unoccupied Train No. P22, which consisted of a locomotive and two freight cars. Train No. 192 consisted of two locomotives and 42 freight cars and was transporting chlorine gas, sodium hydroxide, and cresol.

The resulting collision derailed both locomotives and 16 of the 42 freight cars on No. 192 and the locomotive and one of the two freight cars on No. P22. More than 200 emergency 9-1-1 calls were placed within the four hours after the collision, and another 200 calls were made to the nonemergency line of the railway call center. These calls mobilized local, state, and federal emergency response crews which had to work together to evacuate, stabilize, and decontaminate the accident site and immediate vicinity.

The first fire response personnel were on the scene within three minutes of receiving the 9-1-1 emergency call. Minutes after arriving at the wreckage, the senior officer on scene, the fire chief, had difficulty breathing and other response teams were asked to stand by until a better size-up could be conducted. Haz mat response teams were requested and emergency personnel established a command center close to the incident. According to the

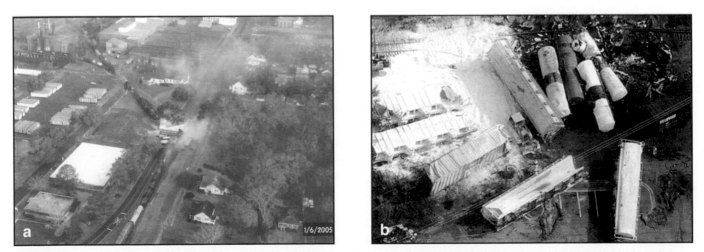

Figure 6.1a and b Nine people died as a result of this train derailment in Graniteville, South Carolina. *Courtesy of the U.S. EPA.*

2005 NTSB incident report, due to light winds in the area, the escaping chlorine gas cloud extended at least 2,500 feet (762 m) to the north, 1,000 feet (305 m) to the east, 900 feet (274 m) to the south, and 1,000 feet (305 m) to the west. The sudden release and expansion of the escaping gas caused the product remaining in the tank to auto-refrigerate and remain in the liquid state, slowing the release of additional gas.

After establishing a 1 mile (1.6 k) perimeter and receiving a consist list of materials on the train, fire personnel began to receive reports of a steady stream of people leaving the area. Authorities set up the first of four decontamination stations to treat those exposed to the chlorine. Reports of people down inside the Avondale Mills facility made it necessary for a firefighter and mill supervisor to enter the steam plant to prevent a possible boiler explosion, and the decision was made to evacuate the plant.

After moving the command station farther back for safety, a firefighter with proper PPE was able to get close enough to the accident site to determine which car had been breached. During this process, he discovered two individuals close to the accident site. One was suffering chemical inhalation symptoms, and the other was trapped under a train car (they were both later successfully rescued).

The main evacuation process consisted of entry teams wearing PPE riding in privately owned pickup trucks that were organized and dispatched to search for individuals or groups affected by the chlorine gas. They then transported them to one of the decontamination sites. This evacuation cycle was repeated for several hours. During this process one entry team reported downed electrical power lines near the wreckage. The utility company responded to a request to disconnect the power feed, and then a report of bright orange and green smoke emanating from two separate cars was received.

About 3 hours into the incident an entry team entered the steam plant to shut it down and had to revise the mission when five or six individuals were reported to be trapped in a room at the plant. One person was found and rescued, after which a complete shut down and sweep of the plant was performed. The immediate area around the accident site was relatively stable until later that day when a fire was reported at the steam plant. An entry team entered the plant to discover that a fire had ignited in coal chutes feeding the boilers. Using a pumper truck supplying water to an unmanned, waterline-fed monitor nozzle, the fire was brought under control, but was not extinguished. Entry teams were then used to monitor the fire while clean up of the accident site occurred. Nine days later fire crews were able to enter the steam plant to extinguish the remaining fire in the coal feeders.

While evacuation efforts were occurring, haz mat response teams worked to stabilize and decontaminate the wreckage and surrounding area. Within three minutes of fire personnel reaching the scene, a haz mat team was requested, and the chief who had difficulty breathing withdrew from the scene. Four minutes after the emergency response personnel's arrival, a Reverse 9-1-1 Emergency Notification was initiated advising residents to shelter indoors. While evacuation efforts continued, haz mat responders inserted a temporary polymer patch into the puncture on the ninth tank car. Two days later they began to unload the sodium hydroxide from the eighth car. During this process, chlorine vapors were released again when the temporary patch failed. This release caused unloading to stop. The next day, a second temporary polymer patch was inserted into the ninth car. The chlorine vapor was drawn from the car to create a vacuum to reduce the amount of chlorine escaping. This vapor was transferred to the sodium hydroxide solution in the eighth car to neutralize it.

When the unloading of the eighth car was completed, construction for a permanent lead patch for the ninth car began. The efforts to unload other cars were delayed because of rotating the ninth car to place the puncture at the highest elevation point on the car. These efforts recommenced with the unloading of chlorine from the sixth train car. During this process responders rejected the plan for a lead patch and decided to use a steel one instead. Unloading of the chlorine in the seventh car was completed and the steel patch was placed on the punctured ninth tank car. Because of the extensive damage to the car, the chlorine in this tank car could not be removed as it had been from the undamaged cars. The process of transferring the gas in order to bubble it through a sodium hydroxide solution in a separate tank took several days, but this converted the chlorine into a relatively safe and transportable bleach, salt solution.

The stabilization and removal of the chlorine took nine days to accomplish. The evacuation efforts were complicated because the closest rescue station was only 100 yards (91 m) from the accident site. Therefore, all personnel and equipment were contaminated, requiring the mobilization of other departments and adding time to the rescue/evacuation process. Post-incident reports emphasized two lessons to be learned from this episode. The first is that response crews should not rush in without proper PPE. Doing so creates extra risk to fire personnel, and the response teams who are there to help can become victims themselves without appropriate protection. The second lesson is to establish a unified command early. In a large incident with multiple crews working together it is vital that rescue and decontamination efforts be coordinated in order to achieve the greatest level of safety for rescue crews and those in the affected area.

In order to effectively mitigate a hazardous materials/WMD incident, Incident Commanders must develop **strategic goals** and **tactical objectives** that will help bring the situation under control and lessen the danger(s) presented. It is a problem-solving process aimed at resolving the incident in accordance with the priorities of life safety, incident stabilization, and protection of the environment and property.

> **Strategic Goal** — Broad statements of desired achievements to control an incident; achieved by the completion of tactical objectives.

This chapter will discuss the problem solving process with a focus on the following elements:

- Analyzing the incident through scene analysis, information gathering, and/or size-up to try to understand and identify the problem(s) and assess hazards and risks
- Planning the response by determining specific strategic goals
- Implementing the response using tactical objectives and assignment of tasks
- Evaluating progress (feedback loop)

> **Tactical Objectives** — Specific operations that must be accomplished to achieve strategic goals.

This chapter will also address some of the strategic goals and tactical objectives commonly used at haz mat incidents as part of the problem-solving process. Strategic goals are broad statements of what must be done to resolve an incident. Tactical objectives are specific operations that must be done in order to accomplish those goals.

Some of the standard strategic goals of hazardous materials incidents are as follows:

- Isolation
- Notification
- Identification
- Protection (life safety)

- Rescue
- Spill control/confinement
- Leak control/containment
- Crime scene and evidence preservation
- Fire control
- Recovery/termination

Some of these strategic goals have already been discussed in this manual. For example, identification of hazardous materials was covered in Chapter 2, Hazardous Materials Identification. Spill, leak, and fire control will be addressed in Chapter 10, Product Control.

While these are some of the common strategic goals, ICs can set whatever goals they deem appropriate, using whatever terms they prefer. For example, *rescue* might be considered an important strategic goal at one incident but not at another. If conditions at an incident change suddenly, rapid evacuation might become a strategic goal that springs to the top of the priority list.

Strategic goals are prioritized depending on available resources and the particulars of the incident. Some of these goals may not be needed if the hazard is not present at the incident. For example, if the material involved is nonflammable, fire control may not be an issue. Some goals may require the use of specialized resources (such as chemical protective clothing or specific absorbent materials) that are not yet available and therefore must be postponed or eliminated. Others may require the use of so many of the available resources that the ability to complete other goals in an expedient timeframe might be compromised. The goals listed in this chapter are very broad strategic categories, and mitigation of an actual incident may require a variety of specific strategies based on the problems presented at the scene. **Appendix F,** Typical Haz Mat Problems with Potential Mitigating Strategies and Tactics, provides a table with examples of common problems presented at haz mat incidents with more narrowly defined strategies and tactics.

Problem-Solving Process

Incident priorities, IMS, and predetermined procedures provide a management structure for first responders. But the incident still needs mitigation (actions taken to lessen the harm or hostile nature of an incident). The problem must be solved through a process of problem-solving and decision-making. While not all first responders are in a command position responsible for planning an appropriate response to a haz mat emergency, it is important for them to understand this process. The most common haz mat management processes contain the elements of the basic four-step problem-solving formula developed by George Polya, a former professor of mathematics at Stanford University:

Step 1: Understand the problem.

Step 2: Devise a plan.

Step 3: Carry out the plan.

Step 4: Look back.

While there are many problem-solving and decision-making models, this particular four-step problem-solving process has been widely adapted for use in any problem-solving or decision-making situation (see information box). Other models may have more steps, but most will contain the following four common elements:

1. Information gathering or input stage

2. Processing, analysis, and/or planning stage

3. Implementation or output stage

4. Review or evaluation stage

In the case of haz mat incidents, it must be understood that problem-solving and decision-making are fluid processes. A first responder's understanding of a problem (and consequent plans to address it) may change as more information becomes available and/or conditions change.

Incident Problem-Solving Process Models

Many incident management process models are in use to provide first responders a series of steps or actions to take at haz mat incidents and terrorist attacks. Some of the better known systems are as follows:

- *GEDAPER* by David Lesak (taught by the National Fire Academy):

  G — Gather information

  E — Estimate potential course and harm

  D — Determine strategic goals

  A — Assess tactical options and resources

  P — Plan of action implementation

  E — Evaluate operations

  R — Review the process

- *DECIDE* by Ludwig Benner

  D — Detect the presence of hazardous materials

  E — Estimate likely harm without intervention

  C — Choose response objectives

  I — Identify action options

  D — Do best option

  E — Evaluate progress

- *APIE* by the International Association of Fire Fighters (IAFF)

  A — Analyze

  P — Plan

  I — Implement

  E — Evaluate (and repeat)

Incident Problem-Solving Process Models (continued)

- **SISIACMRD** – Source: New South Wales Fire Brigades (NSWFB) Australia. This acronym is easily remembered by the using the following mnemonic: **S**ick **I**n **S**ide **I A**lways **C**all **M**y **R**egular **D**octor

  S — Safe approach

  I — Incident command/control

  S — Secure the scene

  I — Identify risk/hazmat

  A — Assess potential harm

  C — Call in resources

  M — Monitor information

  R — Render safe

  D — Decontaminate

- *OODA* – U.S. Military

  O — Observe

  O — Orient

  D — Decide

  A — Act

Gregory G. Noll, Michael S. Hildebrand, and James G. Yvorra developed the Eight Step Incident Management Process©, which is a tactical decision-making model that focuses on haz mat/WMD incident safe operating practices.* The eight steps are as follows:

1. Site management and control
2. Identify the problem
3. Hazards and risk evaluation
4. Select personal protective clothing and equipment
5. Information management and resource coordination
6. Implement response objectives
7. Decontamination
8. Terminate the incident

- The **RAIN** concept can be used to simplify objectives at WMD incidents:

R — **R**ecognize characteristics of WMDs

A — **A**void, by protection, the hazards of WMDs

I — **I**solate the hazards of WMDs

N — **N**otify the appropriate resources and authorities when responding to an event possible involving WMDs.

---

* See *Hazardous Materials: Managing the Incident,* 3rd edition, by Gregory G. Noll, Michael S. Hildebrand, and James G. Yvorra, 2005, distributed by Fire Protection Publications.

The basic four-step formula for problem solving is roughly analogous to the division of required first responder competency tasks in NFPA® 472, *Standard for Professional Competence of Responders to Hazardous Materials Incidents* (2008), and the *APIE* haz mat incident management **(Table 6.1)**. For that reason, when discussing haz mat incident management, this manual references the four-step process rather than *GEDAPER, DECIDE*, or the Eight Step Incident Management Process©.

Table 6.1 Four-Step Problem Solving Process
Problem-Solving Stages at Haz Mat and Terrorist Incidents
Analysis Stage and Information Gathering. • Recognize the type of incident (Is it a haz mat incident or terrorist attack? If so, what materials are involved?) • Identify all the hazards presented by the incident (chemical, mechanical, electrical, etc.) • Predict the likely behavior of chemical, explosive, biological, and radiological materials • Estimate potential harm
Planning Stage. • Determine if additional help is needed • Identify actions to protect emergency responders and others from hazards • If appropriate, consult the *ERG* for instructions and guidelines regarding hazardous materials • Determine strategies and tactics to stabilize the incident • Determine appropriate personal protective equipment • Determine appropriate decontamination methods • Devise the Incident Action Plan
Implementation Stage. • Implement the incident management system • Transmit information to an appropriate authority and call for appropriate assistance • Establish and enforce scene control perimeters • Implement the Incident Action Plan • Implement strategies and tactics appropriate to training such as isolating the area and denying entry, conducting rescues, conducting mass decontamination, etc. • Identify and preserve evidence
Evaluation and Review Stage. • Evaluate effectiveness of approach (is the incident stabilizing?) • Process and provide feedback to IC

Analyzing the Incident

It is impossible to solve a problem without enough information to understand it. For example, first responders cannot fully mitigate a haz mat incident if they don't know what material is involved. For that matter, they cannot take appropriate steps to protect themselves or others if they do not know that hazardous materials are present.

Within the framework of incident priorities, IMS, and predetermined procedures, understanding the problem becomes the first step in mitigating (or solving the problem of) any emergency incident. Understanding the problem (including all of its subproblems or multiple elements) enables first responders to form an overall plan of action.

What This Means To You

Understanding what is happening (and what has happened) at an emergency incident is not a linear process. Unfortunately, there is not a checklist of *All the Things You Need to Know Right Now* that you can follow, check off all that apply, and subsequently act upon them. Every incident is going to be different. You may be bombarded with visual, audible, and sometimes conflicting information; yet you must be able to make sense of what you are given in order understand the situation. Skillful ICs are able to quickly identify relevant information and analyze it in order to form a clear picture of the incident.

Figure 6.2 First responders must survey the scene to detect the presence of hazardous materials.

In the case of hazardous materials incidents, one of the key pieces of information needed to begin understanding the problem is the identity of the hazardous material. Responders must first survey the scene to detect the presence of hazardous materials, and they must then correctly identify them **(Figure 6.2).** This detection/identification process can be done by paying attention to the seven clues to the presence of hazardous materials as discussed in Chapter 2. The information gathered must then be correctly interpreted and verified by emergency response agencies, manufactures, shippers, and/or other resources that can confirm handling procedures, product identity, and appropriate response information. These resources assist the first responder in estimating the size of the endangered area (by using the *ERG's* recommended isolation distances, for example) and the potential harm posed by the material.

However, the identity of the hazardous material is just one piece of the information needed. Many factors may have an effect on the situation — everything from wind direction, topography, land use, and the presence of victims to concerns such as equipment access and available response personnel. The initial survey should include the answers to the following questions:

- Where is the incident scene in relation to population and environmental and property exposures?

- Is the incident scene inside a building or outside?

- What are the hazardous materials?

- What hazard classes are involved?

- What quantities are involved?

- What concentrations are involved?

- How could the material react?

- How is the material likely to behave?

- Is it a liquid or solid spill or a gas release?

- Is something burning?

- What kind of container holds the material?

- What is the condition of the container?

- How much time has elapsed since the incident began?

- What personnel, equipment, and extinguishing agents are available?

- Is there private fire protection or other help available?

- What effect can the weather have?

- Are there nearby lakes, ponds, streams or other bodies of water?

- Are there overhead wires, underground pipelines, or other utilities?

- Where are the nearest storm and sewer drains?

- What has already been done?

All of these factors and others can affect the problem. The first responder must also be able to make reasonable determinations as to the amount or level of hazard present and the risks associated with dealing with the incident. This information is gathered during the incident scene analysis or **size-up** and then analyzed through a hazard and risk assessment model such as the General Hazardous Materials Behavior Model.

Scene Analysis: Size-up and Hazard and Risk Assessment

Size-up is the assessment of incident conditions and recognition of cues indicating problems and potential problems presented by an incident. It is the mental process of considering all available factors that will affect an incident during the course of the operation. The information gained from the size-up is used to determine the strategies and tactics that are applied to the incident during the planning and implementation stages. **Hazard and risk assessment** is part of the size-up process, focusing particularly on the dangers, hazards, and risks presented by the incident. (See the information box for the distinction between hazard and risk.)

Size-Up — Ongoing mental evaluation process performed by the operational officer in charge of an incident that enables him or her to determine and evaluate all existing influencing factors that are used to develop objectives, strategy, and tactics for fire suppression before committing personnel and equipment to a course of action. Size-up results in a plan of action that may be adjusted as the situation changes. It includes such factors as time, location, nature of occupancy, life hazard, exposures, property involved, nature and extent of fire, weather, and fire fighting facilities.

Hazard and Risk Assessment — Formal review of the hazards and risk that may be encountered while performing the functions of a firefighter or emergency responder; used to determine the appropriate level and type of personal and respiratory protection that must be worn.

What Is the Difference Between Hazard and Risk?

An important part of the hazard assessment is determining which physical and health hazards associated with hazardous materials are present at the haz mat scene. Of course, there may be other hazards present that have nothing to do with the hazardous material itself. Traffic hazards or electrical hazards are examples of dangers that need to be considered during the hazard assessment as well.

Risk, on the other hand, deals more with probabilities — the probability of getting hurt or injured or suffering damage, harm, or loss because of the hazards present. Assessing the risk at a haz mat incident is a matter of determining the ifs of a situation: *If I do this, then this might happen. If this happens, then this will follow. If I don't do this, then this could occur.* Once you know the hazards (or potential hazards), it is a matter of estimating how likely it is that harm or loss will actually occur. Assessing risk is often more difficult than assessing the hazards themselves. Experience and knowledge are valuable assets in the ability to quickly predict future events. ICs who are skillful at estimating the course of an incident (and its potential harm) are better able to conduct a risk/benefit analysis and choose a wise plan of action in which the benefits outweigh the potential risks.

Risk — Likelihood of suffering harm from a hazard; exposure to a hazard; the potential for failure or loss.

As with a structure fire, haz mat size-up must consider all six sides of the incident, often referred to as *Alpha, Bravo, Charlie, Delta*, and the *top* and *bottom* **(Figure 6.3)**. Haz mat size-up is frequently complicated by limited information. The IC's view of the incident may be limited by the size of the hazard area or location of the release (for example, inside a vehicle or structure). In addition, limited or conflicting information regarding the product or products involved is possible. Initial assessment is based on anticipated conditions and updated as additional information becomes available.

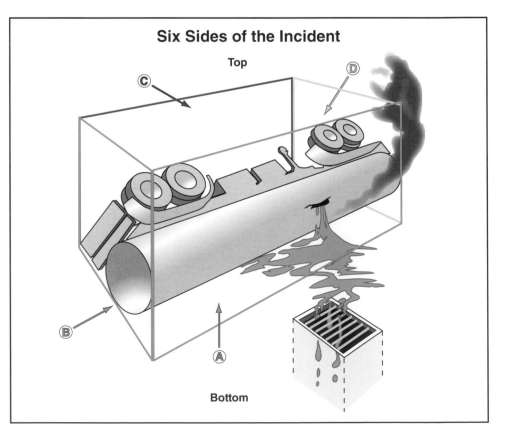

Figure 6.3 Size-up must consider all six sides of an incident. Always remember that hazardous vapors and gases rise or sink depending on their vapor density.

What This Means To You

It's important to realize that size-up/hazard and risk assessment isn't something that's done solely by the IC upon arrival at the scene — it's a continuing process. You always need to be aware of the situation around you. Furthermore, you are responsible for reporting this information to the IC through appropriate channels. Conditions can change rapidly and you must be continually alert. When a cloud of green gas starts drifting in your direction because the wind direction has changed unexpectedly, you need to notice it, react to it accordingly, and then report it!

The following information needed for hazard and risk assessment can be obtained at the time the incident is reported:

- Number and type of injuries
- Occupancy type
- Type of incident
- Product and container information if available
- Location of the incident
- Equipment and resources responding
- Time of day
- Weather

Additional Size-Up Actions

To process more information relevant to size-up, IC's can do the following:

- Review pre-incident surveys and sketches.
- Review topography maps and utility plans (for drains/sewers, rivers, streams, and the like).
- Note arrival time of other responding units.
- Note exposure types and distances.
- Review hydrant and water supply conditions.
- Consider scene access conditions.
- Consider resource staging areas.
- Make preliminary plans for apparatus placement at the scene.
- Secure any additional information from the telecommunicator (dispatcher).
- Decide what additional units are needed, if any.

Once on the scene, additional pieces of the hazard and risk assessment are added to the information made available before arrival. The following conditions should be evaluated:

- Unusual signs (smoke, fire, explosions, leaking material, vapor clouds, and the like)

- Life hazards
- Product(s) involved
- Container types
- Amount of product involved
- Product travel or path of fire
- Actions already taken by people on the scene

After the material has been identified, responders can use references such as safety data sheets (SDSs) and the generic information provided by the *ERG* to determine health and physical hazards presented by the material. This will assist in determining the level of risk presented by the hazardous material itself.

Using the *ERG* and other sources (for example plume-modeling software if available), first responders should also be able to predict (or attempt to predict) where the hazardous material may be going given its physical state of matter (liquid, gas, or solid) and the environmental conditions present (night or day, wind or no wind, indoors or outdoors) **(Figure 6.4)**. Monitoring and detection devices can be employed to determine concentration and spread of the material. Given this information, responders can estimate the size of the endangered area and predict potential exposures (including the number of people, buildings, property, and the environmental concerns such as sewer drains, streams, lakes, ponds, and wells in the area) **(Figure 6.5)**.

Figure 6.4 Responders should be able to predict the behavior of hazardous materials based on the material's properties and conditions at the scene, for example, topography and wind direction. *Courtesy of Steve Irby, Owasso Fire Department.*

Figure 6.5 Using available resources, responders must be able to estimate the size of the area endangered by a haz mat incident. *Courtesy of Rich Mahaney.*

Situational Awareness

Effective mitigation of any hazardous material incident requires that emergency responders establish and maintain **situational awareness** of the event. Situational awareness is more than just size-up of the incident, but a continuous process that includes size-up, interpreting signs and clues to assess what is happening over the life of the incident and predicting outcomes based on a plan of action **(Figure 6.6)**. Maintaining situational awareness is one of the greatest challenges to emergency responders as the process is also met with barriers such as competing priorities, distractions, and information overload. Failure to establish and maintain situational awareness of the incident is likely to result in a failure to achieve the desired outcome.

Situational Awareness – An individual's perception and comprehension of the details of their surrounding environment, and the understanding of how events occurring in the moment may affect the future.

Figure 6.6 It is important to maintain situational awareness, continually assessing what is happening at the incident. *Courtesy of MSA.*

Hazard and risk assessment is a continual evaluation. It starts with pre-incident planning and continues throughout the course of an incident. The first IC who arrives on the scene conducts an extensive size-up, then continues assessing hazards throughout the incident, altering the mitigation process to minimize risk and maximize benefit as appropriate.

Incident Levels

After the initial size-up has determined the scope of an incident, the level of the incident can be determined in accordance with the definitions in the Local Emergency Response Plan (LERP). Most incident level models define three levels of response graduating from Level I (least serious) to Level III (most serious). By defining the levels of response, an increasing level of involvement and necessary resources can be identified based on the severity of the incident. These levels are described as follows:

- *Level I* — This type of incident is within the capabilities of the fire or emergency services organization or other first responders having jurisdiction. A Level I incident is the least serious and the easiest to handle. It may pose a serious threat to life or property, although this situation is not usually the case. Evacuation (if required) is limited to the immediate area of the incident. The following are examples of Level I incidents:

— Small amount of gasoline or diesel fuel spilled from an automobile **(Figure 6.7)**

Figure 6.7 Level I incidents include small fuel spills. *Courtesy of Rich Mahaney.*

— Leak from domestic natural gas line on the consumer side of the meter

— Broken containers of consumer commodities such as paint, thinners, bleach, swimming pool chemicals, and fertilizers (owner or proprietor is responsible for cleanup and disposal)

- *Level II* — This type of incident is beyond the capabilities of the first responders on the scene and may be beyond the capabilities of the first response agency/organization having jurisdiction. Level II incidents may require the services of a formal haz mat response team. A properly trained and equipped response team could be expected to perform the following tasks:

— Use chemical protective clothing.

— Dike and confine within the contaminated areas.

— Perform plugging, patching, and basic leak control activities **(Figure 6.8)**.

— Sample and test unknown substances **(Figure 6.9)**.

— Perform various levels of decontamination.

The following are examples of Level II incidents:

— Spill or leak requiring limited-scale evacuation

— Any major accident, spillage, or overflow of flammable liquids

— Spill or leak of unfamiliar or unknown chemicals

— Accident involving extremely hazardous substances

— Rupture of an underground pipeline

— Fire that is posing a boiling liquid expanding valor explosion (BLEVE) threat in a storage tank

Figure 6.8 A haz mat team might be needed to conduct offensive operations at a Level II incident. *Courtesy of New South Wales Fire Brigades.*

Figure 6.9 Sampling and testing of unknown substances must be conducted by individuals with appropriate training.

Figure 6.10 Level III incidents will likely involve unified command, multiple agencies, and a lengthy mitigation period. *Courtesy of Chris Mickal.*

- *Level III* — This type of incident requires resources from state/provincial agencies, federal agencies, and/or private industry and also requires unified command. A Level III incident is the most serious of all hazardous material incidents. A large-scale evacuation may be required. Most likely, the incident will not be concluded by any one agency **(Figure 6.10)**. Successful handling of the incident requires a collective effort from several of the following resources/procedures:

  — Specialists from industry and governmental agencies

  — Sophisticated sampling and monitoring equipment

  — Specialized leak and spill control techniques

— Decontamination on a large scale

The following are examples of Level III incidents:

— Incidents that require an evacuation extending across jurisdictional boundaries

— Incidents beyond the capabilities of the local hazardous material response team

— Incidents that activate (in part or in whole) the federal response plan

U.S. NIMS Incident Types

Per U.S. NIMS, incidents may be typed in order to make decisions about resource requirements. Incident types are based on the following five levels of complexity:

Type 5 – Details:

- The incident can be handled with one or two single resources with up to six personnel.

- Command and General Staff positions (other than the Incident Commander [IC]) are not activated.

- No written Incident Action Plan (IAP) is required.

- The incident is contained within the first operational period and often within an hour to a few hours after resources arrive on scene.

- Examples include a vehicle fire, an injured person, or a police traffic stop.

Type 4 – Details:

- Command staff and general staff functions are activated only if needed.

- Several resources are required to mitigate the incident.

- The incident is usually limited to one operational period in the control phase.

- The agency administrator may have briefings, and ensure the complexity analysis and delegation of authority are updated.

- No written IAP is required but a documented operational briefing will be completed for all incoming resources.

- The role of the agency administrator includes operational plans including objectives and priorities.

Type 3 – Details:

- When capabilities exceed initial attack, the appropriate ICS positions should be added to match the complexity of the incident.

- Some or all of the Command and General Staff positions may be activated, as well as Division/Group Supervisor and/or Unit Leader level positions.

- A Type 3 Incident Management Team (IMT) or incident command organization manages initial action incidents with a significant number of resources, an extended attack incident until containment/control is achieved, or an expanding incident until transition to a Type 1 or 2 team.

- The incident may extend into multiple operational periods.

- A written IAP may be required for each operational period.

Type 2 – Details:

- This type of incident extends beyond the capabilities for local control and is expected to go into multiple operational periods. A Type 2 incident may require the response of resources out of area, including regional and/or national resources, to effectively manage the operations, command, and general staffing.

- Most or all of the Command and General Staff positions are filled.

- A written IAP is required for each operational period.

- Many of the functional units are needed and staffed.

- Operations personnel normally do not exceed 200 per operational period and total incident personnel do not exceed 500 (guidelines only).

- The agency administrator is responsible for the incident complexity analysis, agency administrator briefings, and the written delegation of authority.

Type 1 – Details:

- This type of incident is the most complex, requiring national resources to safely and effectively manage and operate.

- All Command and General Staff positions are activated.

- Operations personnel often exceed 500 per operational period and total personnel will usually exceed 1,000.

- Branches need to be established.

- The agency administrator will have briefings, and ensure that the complexity analysis and delegation of authority are updated.

- Use of resource advisors at the incident base is recommended.

- There is a high impact on the local jurisdiction, requiring additional staff for office administrative and support functions.

Source: U.S. Fire Administration

Planning the Appropriate Response

Once first responders have a basic understanding of the problem, they can begin to plan their solution to the problem by establishing strategic goals and tactical objectives (sometimes referred to as **response objectives**). Strategic goals must be selected based on the following criteria:

- Their ability to be achieved

- Their ability to prevent further injuries and/or deaths

- Their ability to minimize environmental and property damage within the constraints of safety, time, equipment, and personnel

When response objectives are determined by the hazards and risks present at the haz mat incident, it is sometimes called **risk-based response**. Risk-based response strategies are based upon the hazards present at the incident. For example, if materials with higher levels of toxicity are involved, a more cautious response using higher levels of personal protective equipment might be used.

Response Objectives — Statements based on realistic expectations of what can be accomplished when all allocated resources have been effectively deployed that provide guidance and direction for selecting appropriate strategies and the tactical direction of resources.

Risk-Based Response — Method using hazard and risk assessment to determine an appropriate mitigation effort based on the circumstances of the incident.

An incident involving a hazardous material in a gaseous or vapor form might dictate a different strategy for control than an incident involving a hazardous material in a solid or liquid form that is far easier (and safer) to contain.

Some additional risk-based response principles are as follows:

- Activities that present a significant risk to the safety of members shall be limited to situations where there is a potential to save endangered lives.

- Activities that are routinely employed to protect property shall be recognized as inherent risks to the safety of members, and actions shall be taken to reduce or avoid these risks.

- No risk to the safety of members shall be acceptable when there is no possibility to save lives or property.

Making the right strategic decision at a haz mat incident is critical because of the variety of things that can occur. Poorly developed decision-making processes can lead to greater problems. The sections that follow detail two elements of setting strategic goals: determining modes of operation and developing IAPs.

Determining Modes of Operation

Strategies are divided into three options that relate to modes of operation: defensive, offensive, and nonintervention. A **nonintervention strategy** allows the incident to run its course on its own. A **defensive strategy** provides confinement of the hazard to a given area by performing diking, damming, or diverting actions. An **offensive strategy** includes actions to control the incident such as plugging a leak **(Figure 6.11)**.

Nonintervention Strategy — Overall plan for incident control established by the Incident Commander in which responders take no direct actions on the actual problem. Sometimes referred to as Natural Stabilization.

Defensive Strategy — Overall plan for incident control established by the Incident Commander (IC) that involves protection of individuals and exposures as opposed to aggressive, offensive intervention.

Offensive Strategy — Overall plan for incident control established by the Incident Commander (IC) in which responders take aggressive, direct action on the material, container, or process equipment involved in an incident.

Figure 6.11 Plugging a leak is an offensive strategy.

Selection of the strategic mode is based on the risk to responders, their level of training, and the balance between the resources required and those that are available. The safety of first responders is the uppermost consideration in selecting a mode of operation. The mode of operation may change during the course of an incident. For example, first-arriving responders may be restricted to nonintervention or defensive mode. After the arrival of the haz mat team, the IC may switch to offensive mode and initiate offensive tactics. The following three incident-based elements affect the selection of strategic mode, and the IC must always have a clear picture of them as the IAP is developed:

1. *Value* — Related directly to the incident priorities of life safety, incident stabilization, and property conservation. Value is stated in terms of *yes* or *no* — either there is value *(yes)* or there is no value *(no)*. Once value has been determined with a yes, the degree of value can be assessed. If a civilian life hazard (savable victim or victims) exists, the value is high **(Figure 6.12)**. If no civilian life hazard exists but environmental harm may be prevented or property may be saved, the value is somewhat less. If no civilian life hazard exists and responder actions will have little affect on environmental or property protection, there is no value. In the absence of value, nonintervention or defensive strategy is indicated.

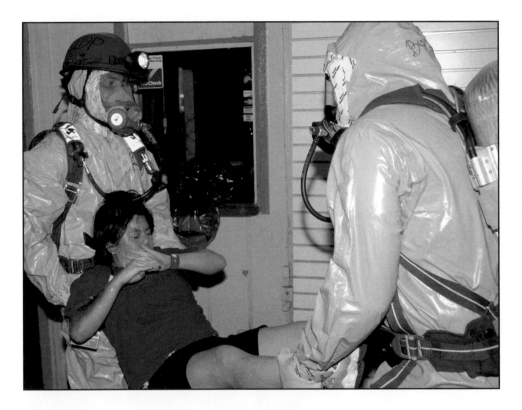

Figure 6.12 It is acceptable to assume a higher level of risk in order to save a life. *Courtesy of the U.S. Marine Corps, photo by Sgt. Christopher D. Reed.*

2. *Time* — Possible limited window of opportunity exists to intervene before an incident escalates dramatically (such as cooling of a liquefied gas container exposed to direct flame impingement on its vapor space); estimated time during which offensive operations may be initiated. In other cases, the reaction and response times of Technician-Level responders may be the driving factor in selecting the strategic mode for incident operations.

3. *Size* — Most frequently driven by the need to conduct protective action (evacuation or protection in place) concurrently with incident control operations. Resource requirements are driven by tactical requirements. In most cases (with the exception of fires involving hazardous materials), fire flow is not the descriptor of size.

The sections that follow describe operations designed to meet each of the three overall strategies that may be applied to an incident.

Nonintervention Operations
— Operations in which responders take no direct actions on the actual problem.

Figure 6.13 Nonintervention is an acceptable strategy at some incidents. *Courtesy of the U.S. Army Corps of Engineers*.

Nonintervention operations. **Nonintervention operations** are operations in which the responders take no direct actions on the actual problem **(Figure 6.13)**. Not taking any action is the only safe strategy in many types of incidents and the best strategy in certain types of incidents when mitigation is failing or otherwise impossible. An example of a situation for nonintervention is a pressure vessel that cannot be adequately cooled because it is exposed to fire. In such incidents, responders should evacuate personnel in the area and withdraw to a safe distance. The nonintervention mode is selected when one or more of the following circumstances exist:

- The facility or LERP calls for it based on a pre-incident evaluation of the hazards present at the site.

- The situation is clearly beyond the capabilities of responders.

- Explosions are imminent.

- Serious container damage threatens a massive release.

In such nonintervention situations, first responders should take the following actions:

- Withdraw to a safe distance.

- Report scene conditions to telecommunications center.

- Initiate an incident management system.

- Call for additional resources as needed.

- Isolate the hazard area and deny entry.

- Commence evacuation where needed.

Figure 6.14 Defensive operations aim to confine the emergency without directly contacting the hazardous material involved. In this case, responders have constructed a dam to contain the spill.

Defensive Operations
— Operations in which responders seek to confine the emergency to a given area without directly contacting the hazardous materials involved.

Defensive operations. **Defensive operations** are those in which responders seek to confine the emergency to a given area without directly contacting the hazardous materials involved **(Figure 6.14)**. The defensive mode is selected when one of the following two circumstances exists:

1. The facility or LERP calls for it based on a pre-incident evaluation of the hazards present at the site.

2. Responders have the training and equipment necessary to confine the incident to the area of origin.

In defensive operations, Ops-Level first responders should take the following actions:

- Report scene conditions to telecommunications center.
- Initiate an incident management system.
- Call for additional resources as needed.
- Isolate the hazard area and deny entry.
- Establish and indicate zone boundaries.
- Commence evacuation where needed.
- Control ignition sources.
- Use appropriate defensive control tactics.
- Protect exposures.
- Perform rescues when safe and appropriate.
- Evaluate and report incident progress.
- Perform emergency decontamination procedures.

Offensive operations. **Offensive operations** are those where responders take aggressive, direct action on the material, container, or process equipment involved in the incident **(Figure 6.15)**. These operations may result in contact with the material and therefore require responders to wear appropriate chemical-protective clothing and respiratory protection. Some offensive operations are beyond the scope of responsibilities for first responders and are conducted by more highly trained hazardous materials personnel.

Offensive Operations — Operations in which responders take aggressive, direct action on the material, container, or process equipment involved in an incident.

Figure 6.15 Offensive operations typically involve a higher degree of risk. *Courtesy of the U.S. Navy, photo by Photographer's Mate 1st Class Aaron Ansarov.*

Developing Incident Action Plans (IAPs)

IAPs are critical to the rapid, effective control of emergency operations. An IAP is a well-thought-out, organized course of events developed to address all phases of incident control within a specified time. The timeframe specified is one that allows the least negative action to continue. Written IAPs may not be necessary for short-term, routine operations; however, large-scale or complex incidents require the creation and maintenance of a written plan for each operational period **(Figure 6.16, p. 284)**.

Figure 6.16 Large scale, complex incidents require a written IAP. *Courtesy of Phil Linder.*

Action planning starts with identifying the strategy to achieve a solution to the confronted problems. Strategy is broad in nature and defines *what* has to be done. Once the strategy has been defined, the Command Staff needs to select the tactics (the *how, where,* and *when*) to achieve the strategy. Tactics are measurable in both time and performance. An IAP also provides for necessary support resources such as water supply, utility control, or SCBA cylinder filling.

The IAP essentially ties the entire problem-solving process together by stating what the analysis has found, what the plan is, and how it shall be implemented. Once the plan is established and resources are committed, it is necessary to assess its effectiveness. Information must be gathered and analyzed so that necessary modifications may be made to improve the plan if necessary. This step is part of a continuous size-up process. Elements of an IAP include the following:

- Strategies/incident objectives
- Current situation summary
- Resource assignment and needs
- Accomplishments
- Hazard statement
- Risk assessment
- Safety plan and message
- Protective measures
- Current and projected weather conditions
- Status of injuries

- Communications plan
- Medical plan

All incident personnel must function according to the IAP. Company officers or sector officers should follow predetermined procedures, and every action should be directed toward achieving the goals and objectives specified in the plan.

How Many Plans Do I Need?

At various times throughout this manual, we refer to different types of plans: *pre-incident survey* (sometimes called *pre-incident plan* or *preplan*), *emergency action plan*, *site safety and health plan*, *emergency response plan*, *site safety and control plan*, *integrated contingency plan*, *incident action plan (IAP)*, *local emergency response plan (LERP)*, and others. Numerous U.S. state and federal laws require the development of both facility response plans and community emergency plans, and the different names for these plans can be confusing. **Appendix G**, Emergency Plans, summarizes many of these plans.

For practical purposes, all first responders should be familiar with the concept of IAPs and site safety plans because they have a direct effect on actions taken at the emergency incident scene. A first responder assuming the role of IC will need to develop and implement an IAP.

Implementing the Incident Action Plan

After strategic goals have been selected and the IAP formulated, the IC can begin to implement the plan. Strategic goals are achieved through tactics (or tactical objectives). Strategies and tactics are accomplished or conducted by performing specific tasks. **Figure 6.17** illustrates the relationship between priorities, strategic goals, tactics, tasks, and the IMS level normally associated with each.

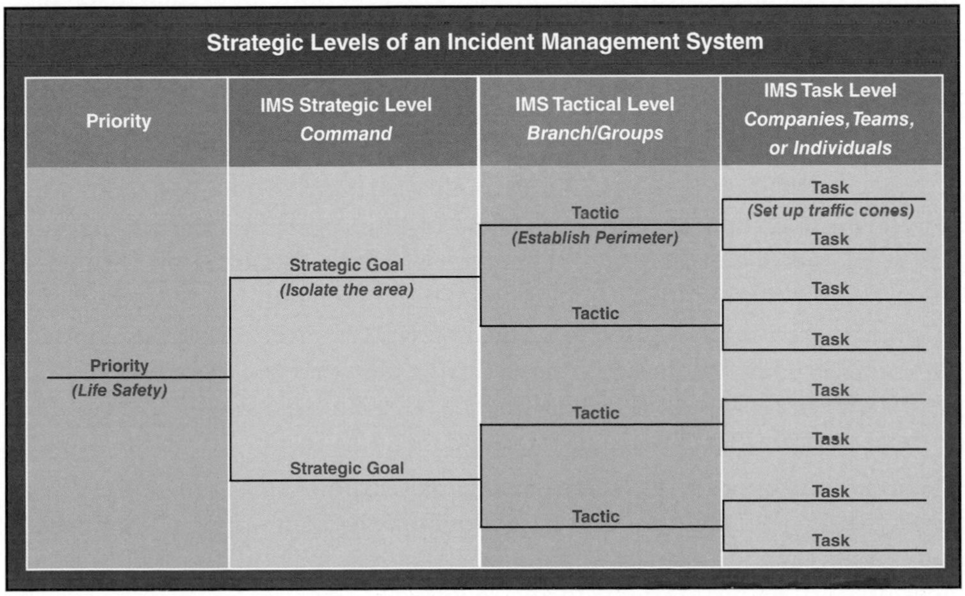

Figure 6.17 The relationship between priorities, strategic goals, tactics, tasks, and the IMS level normally associated with each.

Tactics follow their respective strategies in that they can be nonintervening, offensive, or defensive in nature. Tactics related to controlling chemical releases basically fall into two categories: confinement (spill control) and containment (leak control), with the majority of defensive control options being related to confinement. Other tactics may include such things as establishing scene control zones, calling for additional resources, wearing appropriate PPE, decontamination, and fire extinguishment.

When implementing the response, the IC must maintain contact with the officers and crew and constantly reevaluate the situation. If crews cannot achieve the objective, the IC must reassess the situation and prepare an alternate plan.

Evaluating Progress

The final aspect of the problem-solving process is *looking back* or evaluating progress. If an IAP is effective, the IC should receive favorable progress reports from tactical and/or task supervisors and the incident should begin to stabilize. If, on the other hand, mitigation efforts are failing or the situation is getting worse (or more intense), the plan must be reevaluated and very possibly revised. The plan must also be reevaluated as new information becomes available and circumstances change. If the initial plan isn't working, it must be changed either by selecting new strategies or by changing the tactics used to achieve them. In accordance with predetermined communication procedures, it is important for first responders to communicate the status of the planned response and the progress of their actions to the IC.

Isolation and Scene Control

As discussed in Chapter 3, Isolation and Denial of Entry, the isolation perimeter may be comprised of an inner and outer perimeter, and it may be expanded upon or reduced in size as needed. In most cases, the initial isolation perimeter established by first-arriving responders is determined by the outcomes of an on-site risk assessment.

Once resources have been committed to an incident, it is easier to reduce the isolation perimeter in size than it is to extend it. If resources have arrived and have been tasked at an incident, it may be very difficult to then disengage and relocate those resources should the initial perimeter be inadequate. One example of this situation might be when the perimeter fails to ensure emergency responder safety, crowd control, crime scene preservation or general safety for other hazards associated with toxic plumes or hazardous materials.

The IC must undertake a risk assessment or size-up of the incident in order to determine an appropriate size for the isolation perimeter. Where appropriate, the perimeter size should be determined in consultation with other onsite agency commanders to ensure that the spatial requirements and tactical objectives of other agencies can be met. **NOTE:** From a risk management perspective, it is better to encompass a larger area that can be reduced in size once incident site conditions have been assessed for risks such as secondary devices, unidentified hazardous materials, and atmospheric monitoring.

The isolation perimeter is also used to control both access and egress from the incident site. Unauthorized personnel may be kept out, while witnesses and persons with information about the incident may be directed to a safe location until being interviewed and released.

Another important aspect of scene control at haz mat/WMD incidents is the establishment of hazard-control zones and staging areas. These topics will be discussed in sections that follow.

Hazard-Control Zones

Hazard-control zones provide for the scene control required at hazmat and terrorist incidents to protect responders from interference by unauthorized persons, help regulate movement of first responders within the zones, and minimize contamination (including secondary contamination from exposed or potentially exposed victims). At large, multiagency response incidents, establishing hard perimeters for hazard-control zones can assist in the challenge of ensuring accountability for all personnel involved in the response.

Hazard-control zones divide the levels of hazard of an incident, and what a zone is called generally depicts this level. These zones are often referred to as hot, warm, and cold **(Figure 6.18)**. Traditional hazardous materials response models have taught emergency responders the use of concentric circles to delineate the hot, warm, and cold zones. U.S. Occupational Safety and Health Administration (OSHA) and the U.S. Environmental Protection Agency (EPA) refer to these zones collectively as *site work zones*. They are sometimes called *scene-control zones* as well. Other countries may use different terminology for these zones with which responders must be familiar.

Hazard-Control Zones — System of barriers surrounding designated areas at emergency scenes intended to limit the number of persons exposed to a hazard and to facilitate its mitigation; major incident has three zones: restricted (hot), limited access (warm), and support (cold). U.S. EPA/OSHA term: Site Work Zones. Also called Scene-Control Zones and Control Zones.

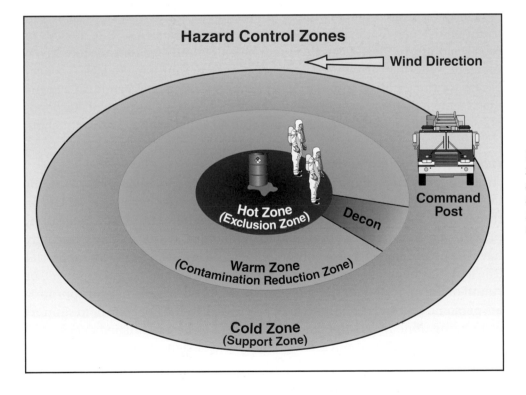

Figure 6.18 Hazard-control zones divide the levels of hazard at an incident into hot, warm, and cold zones, with the hot zone indicating the highest degree of danger.

Included within these control zones, certain tasks are performed: rescue and leak control in the hot zone; the establishment of the *Safe Refuge Area* and decontamination corridor in the warm zone; and triage, treatment, transportation, and incident support functions in the cold zone.

These control zones are not necessarily static and can be adjusted as the incident changes. It should be noted that different agencies may have different needs in terms of establishing control zones. At incidents involving crimes, law enforcement may designate a zone to incorporate the entire crime scene which may not correspond to traditional fire service activities. For example, at terrorist incidents in the U.S., the FBI establishes an evidence search perimeter 1.5 times the distance of the farthest known piece of evidence **(Figures 6.19)**. These law enforcement zones might change as evidence is processed and the crime scene is released **(Figure 6.20)**. These dynamics create an extreme need for flexibility on the part of all agencies in the unified command when establishing these zones.

Incidents involving bombs are a good example of where traditional control zones and the operations that are usually conducted within those zones may be different. Because of the blast effects, there may be multiple buildings that are in danger of collapse requiring the designation of a much larger hot zone. Another example is that in order to preserve evidence in a bombing incident, law enforcement may require that the hot zone be extended out as far as the perimeter of the debris field. This was the case in Oklahoma City. In these cases, there will likely be very tight perimeter control as well as a very large hot zone. Due to the logistics of the event, it may be necessary to conduct operations such as triage, treatment, and transportation in an area designated as the hot zone.

Another example that is unique to terror events and may require a non-traditional scene management plan is an event that has multiple devices or release points. In these cases there may be more than one hot zone for a given incident. Whether the incident involves bombs or multiple devices, it is necessary for the incident commander to remain flexible and establish a scene management plan and control zones to meet the needs of the incident. Responders at the incident must be made aware control zones as they are established.

Hot Zone

Hot Zone — Potentially hazardous area immediately surrounding the incident site requiring appropriate protective clothing and equipment and other safety precautions for entry; typically limited to technician-level personnel; also called the Exclusion Zone.

Traditionally, the **hot zone** (also called *exclusion zone*) is an area surrounding an incident that is potentially very dangerous either because it presents a threat in the form of a hazardous material or the effects thereof **(Figure 6.21, p. 290)**. The area may be contaminated by chemical warfare agents, or it may have the potential to become contaminated by a released hazardous material (the area has been or could be exposed to the gases, vapors, mists, dusts, or runoff of the material). Responders must have proper training and appropriate personal protective equipment (PPE) to work in the hot zone or to support work being done inside the hot zone. There will be established access and egress points to ensure both accountability and designated PPE prior to entry.

The hot zone extends far enough to prevent people outside the zone from suffering ill effects from the released material, explosion, or other threat. Work performed inside the hot zone is often limited to highly trained personnel such as SWAT teams, US&R teams, hazardous materials technicians, Joint Hazard Assessment Teams (JHAT), and bomb technicians.

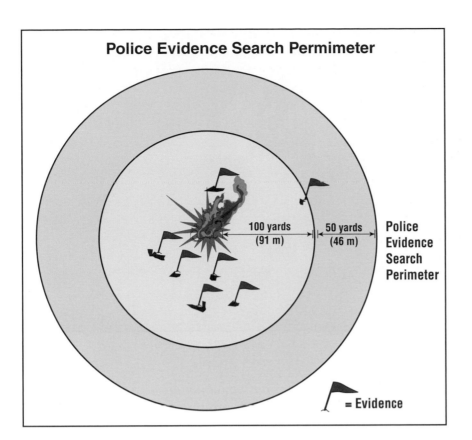

Police Evidence Search Permimeter

100 yards (91 m) | 50 yards (46 m) | Police Evidence Search Perimeter

🚩 = Evidence

Figure 6.19 In the U.S., the FBI will establish a control perimeter at 1.5 times the distance of the farthest known piece of evidence.

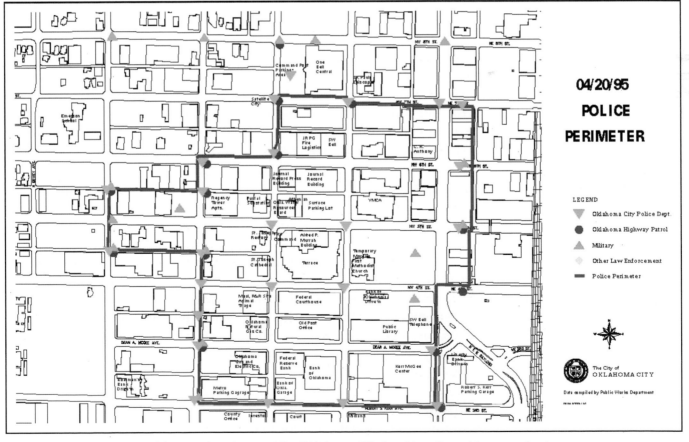

04/20/95
POLICE
PERIMETER

LEGEND
🔻 Oklahoma City Police Dept.
⬤ Oklahoma Highway Patrol
🔺 Military
◇ Other Law Enforcement
▬ Police Perimeter

The City of OKLAHOMA CITY

Data compiled by Public Works Department

Figure 6.20 If evidence is widespread, as it was at the Oklahoma City bombing, the evidence perimeter may encompass a very large area.

WARNING!
Responders must have proper training and appropriate personal protective equipment (PPE) to work in the hot zone or to support work being done inside the hot zone.

Figure 6.21 In most cases, personal protective equipment will be required for entry into the hot zone. *Courtesy of the U.S. Coast Guard, photo by PA3 Brent Erb.*

Warm Zone — Area between the hot zone and the cold zone, usually containing the decontamination corridor and typically requiring a lesser degree of personnel protection than the Hot Zone; also called the contamination reduction zone.

Cold Zone — Safe zone outside of the warm zone where equipment and personnel are not expected to become contaminated and special protective clothing is not required; the incident command post and other support functions are typically located in this zone; also called the support zone.

Staging Area — Location where incident personnel and equipment are assigned on an immediately available status.

Warm Zone

The **warm zone** (also called *contamination reduction zone* or *corridor*) is an area adjoining the hot zone and extending to the cold zone (see following section). The warm zone is used as a buffer between the hot and cold zones and is the place to decontaminate personnel and equipment exiting the hot zone. Decontamination usually takes place within a corridor (decon corridor) located in the warm zone **(Figure 6.22)**. At incidents involving crimes, parts of the warm zone may also be considered to be part of the crime scene with emphasis on minimal disturbance. PPE will normally be required in this zone, although in some circumstances it may be at a reduced level from the hot zone. Monitoring and detection may be conducted around the perimeter of the warm zone to determine the extent of the hazards. The level of PPE to be required for work within this zone will be approved by Unified Command or the Incident Commander after input from others.

Cold Zone

The **cold zone** (also called *support zone*) surrounds the warm zone and is used to carry out all logistical support functions of the incident. Workers in the cold zone are not required to wear personal protective clothing because the zone is considered safe, although some personnel may still be wearing PPE to ensure safe evacuation in the case of rapid expansion of the hot zone (for example, body armor in case of secondary devices and/or attacks). The multi-agency command post (CP), staging area, donning/doffing area, backup teams, research teams, logistical support, criminal investigation teams, triage/treatment/rehabilitation (rehab), and transportation areas are located within the cold zone.

Staging

The **staging area** needs to be located at an isolated spot in a safe area where occupants cannot interfere with ongoing operations. Staging minimizes confusion and freelancing at the scene, and staging areas should be located at

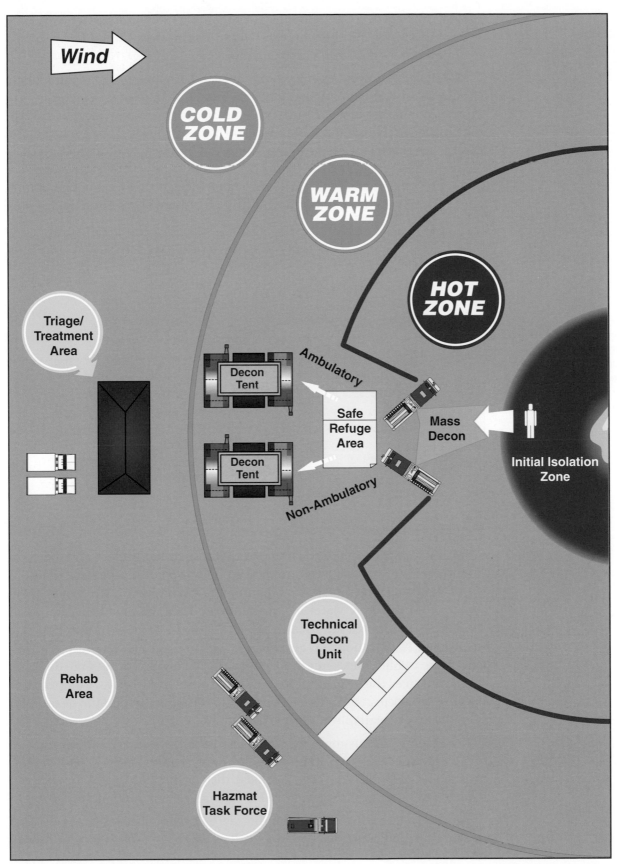

Figure 6.22 A sample incident layout showing several areas and zones. Rarely will an incident be divided into hazard-control zones arranged in neat concentric circles, however, the area where decontamination is conducted is typically designated as the warm zone.

isolated spots in the cold zone where occupants cannot interfere with ongoing operations. A safe direction of travel to the staging area should be broadcast to all resources responding to the incident.

Ideally, staging of emergency responders and equipment at terror incidents should be spread out between multiple locations in case staging areas are attacked. Some departments use the concept of a cornering/quartering staging procedure **(Figure 6.23)**. This has two basic purposes. The first is that the emergency response personnel are spread out from one another to limit their exposure as a target. Having them separated minimizes the effects of a secondary type of attack/device. The second purpose is that it allows personnel to envelop the scene and provide mulitple treatment or points of operation.

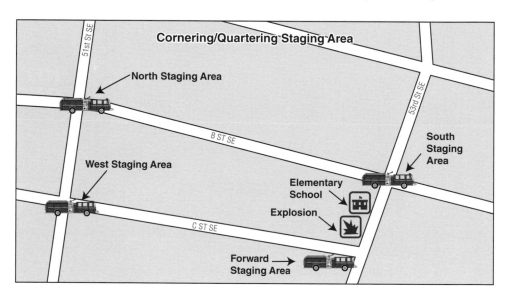

Figure 6.23 Some departments use a cornering/quartering staging procedure to spread their resources between multiple points.

Notification

Emergency response plans must ensure that responders understand their role in notification processes and predetermined procedures such as standard operating procedures (SOPs). As mentioned in Chapter 3, Awareness Level Actions at Hazardous Materials Incidents, notification may be as simple as dialing 9-1-1 (in North America) to report an incident and get additional help dispatched. Notification may also include such items as incident-level identification and public emergency information/notification. It is better to dispatch more resources than necessary in an initial response to ensure appropriate weight of attack to combat incident conditions. Responders should be familiar with the assets available in their jurisdictions.

Because hazardous materials incidents have the potential to overwhelm local responders, it is important to know how to request additional resources. This process should be described in local, district, regional, state and national emergency response plans as well as through mutual/automatic aid agreements.

Notification also involves contacting law enforcement whenever a terrorist or criminal incident is suspected, as well as notifying other agencies that an incident has occurred (public works, the local emergency operations center, etc.). Procedures will differ between military and civilian agencies as well as from location to location. Always follow SOPs/SOGs/OIs and emergency plans for notification procedures. Methods to notify the public in the event of an emergency are listed later in this chapter.

In the U.S. the notification process is described in the National Response Framework, and all local, state, and federal emergency response plans must comply with these provisions. While all incidents are handled at the lowest geographic, organizational, and jurisdictional level, when local agencies need additional assistance, they must contact their state authorities to request help. State governors have the authority to mobilize state resources and may include activation of mutual aid agreements with other states. If state resources are overwhelmed, state governors have the authority to request federal aid.

The local emergency response plan (LERP) should be the first resource a responder in the U.S. should turn to if they need to request outside assistance for an incident. Per the NRF, the local response agency should be closely tied with the community's **Emergency Operations Center (EOC)**. If local assets are insufficient to manage the emergency, requests for additional assistance (such as activation of National Guard units) will be made to the state EOC. States may then request federal assistance through the Department of Homeland Security. Even if additional assistance is not required for an incident, the proper authorities (local, state, and federal) must be informed that an incident has occurred.

Emergency Operations Center (EOC) — Facility that houses communications equipment, plans, contact/notification list, and staff that are used to coordinate the response to an emergency.

Notification Example: National Response Center

When the amount of a hazardous substance release or oil spill exceeds established reporting quantities in the U.S., by law the federal government's National Response Center (NRC) must be notified **(Figure 6.24, p. 294)**. Once a report is made, the NRC immediately notifies a predesignated EPA or U.S. Coast Guard (USCG) On-Scene Coordinator (OSC), based on the location of the spill. The OSC may then determine that local action is sufficient and no additional federal action is required. If the incident is large or complex, the federal OSC may remain on the scene to monitor the response and advise on the deployment of personnel and equipment. However, the federal OSC takes command of the response in any one of the following situations:

● When the party responsible for the chemical release or oil spill is unknown or not cooperative

● When the OSC determines that the spill or release is beyond the capacity of company, local, or state responders to manage

● For oil spills, when the incident is determined to present a substantial threat to public health or welfare due to the size or character of the spill.

Protection

Protection is the overall goal of ensuring safety of responders and the public. Protection goals also include measures taken to protect property and the environment. Protection goals are accomplished through such tactics as the following:

● Identifying and controlling materials and hazards

● Avoiding contact with hazardous materials

● Maximizing distance between people and hazardous areas (hazard control zones and incident perimeters)

Figure 6.24 In the U.S., when an oil spill exceeds established reporting quantities, the National Response Center (NRC) must be notified. *Courtesy of the U.S. Coast Guard.*

- Using and wearing appropriate PPE (see Chapter 8, Personal Protective Equipment)

- Using time, distance, and shielding, when appropriate

- Conducting rescues (see Rescue section and Chapter 12, Victim Rescue and Recovery)

- Implementing shoring and stabilization at incidents involving structural collapses **(Figure 6.25)**

- Providing decontamination (see Chapter 9, Decontamination)

- Providing emergency medical care and first aid

- Ensuring victims and responders stay upwind, upstream, and uphill of hazardous materials

- Taking any other measures to protect responders and the public, including conducting evacuations and sheltering in place

Figure 6.25 Implementing shoring and stabilization at an incident involving structural collapse is a protection tactic. *Courtesy of FEMA News Photos, photo by Jocelyn Augustino.*

Protection of Responders

The protection and safety of emergency responders is the first priority at any incident. Injured or incapacitated responders are unable to assist in mitigation efforts or protection of the public.

Measures to protect responders include the following:

- Staying uphill, upstream, and upwind of hazardous materials
- Wearing appropriate PPE
- Using time, distance, and shielding for protection
- Decontaminating responders when necessary **(Figure 6.26)**
- Ensuring accountability of all personnel
- Tracking and identifying all personnel working at an incident
- Working as part of a team or buddy system
- Assigning safety officers
- Putting evacuation and escape procedures in place

The sections that follow discuss some of these measures in greater detail.

Figure 6.26 Protection of responders includes conducting decontamination when necessary. *Courtesy of the U.S. Air Force.*

Figure 6.27 All personnel assigned to the incident must check in and out via the established accountability system.

Accountability Systems

One of the most important functions of an incident management system is to provide a means of tracking all personnel and equipment assigned to the incident **(Figure 6.27)**. Primarily this responsibility falls on the IC. Most units responding to an incident arrive fully staffed and ready to be assigned an operational objective; other personnel may have to be formed into units at the scene. To handle these and other differences in the resources available, the IAP must contain a tracking and accountability system that has the following elements:

- Procedure for checking in at the scene
- Way of identifying and tracking the location of each unit and all personnel on scene
- Procedure for releasing people, equipment, and apparatus that are no longer needed

Accountability systems are especially important for terrorist incidents where multiple agencies/organizations may be responding, all of which may have different levels of PPE and training. The agency/organization in command is responsible for tracking all responders (not just their own). Methods for tracking accountability should be determined in preplans and implemented as soon as possible at the incident scene. Types of accountability systems include traditional systems such as the fire service passport system, T-card systems for wildland incidents, as well as systems that utilize newer technologies such as GPS and GIS systems. NFPA® 1500 and 1561 addresses the requirements of accountability systems.

Buddy Systems and Backup Personnel

Use of buddy systems and backup personnel are mandated by NFPA® and OSHA at haz mat incidents. A *buddy system* is a system of organizing personnel into workgroups in such a manner that each member has a *buddy* or partner, so that nobody is working alone **(Figure 6.28)**. The purpose of the buddy system is to provide rapid help in the event of an emergency.

In addition to using the buddy system, backup personnel shall be standing by with equipment ready to provide assistance or rescue if needed. Qualified basic life support personnel (as a minimum) shall also be standing by with medical equipment and transportation capability. Any haz mat team working within the hazardous area must have at least two members. The minimum number of personnel necessary for performing tasks in the hazardous area is four — two working in the area itself and two standing by as backup. Backup personnel must be dressed in the same level of personal protective clothing as entry personnel.

Figure 6.28 Responders working in the hot zone must always work with a buddy, never alone.

Time, Distance, and Shielding

Using time, distance, and shielding is an effective protection strategy for first responders at hazardous materials incidents **(Figure 6.29)**. Responders can protect themselves by utilizing the following:

- **Time** — Limiting the time to which they are exposed (or potentially exposed) to hazards and hazardous materials reduces the likelihood of serious harm. This can be accomplished by restricting work times in the hot zone and frequently rotating personnel on work groups.

- **Distance** — Maximizing distance from potential hazards will often prevent or reduce harm. For example, the closer a responder is to the source of an explosion, the greater the harmful effects. Staying well away from hazardous areas will also prevent harmful exposures. Distance may be controlled by implementing hazard-control zones.

- **Shielding** — Shielding places a physical barrier between a responder and the hazard. Shielding may consist of wearing PPE or positioning personnel so that another object such as a wall, building, or apparatus is between them and the hazard, thereby minimizing the chance of contact or effect.

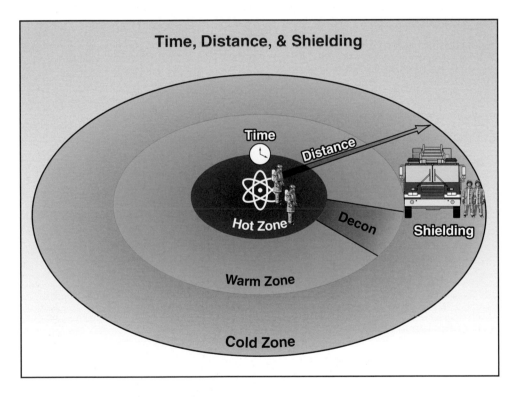

Figure 6.29 Use the concepts of time, distance, and shielding to protect from hazards.

Evacuation/Escape Procedures

Because haz mat/WMD incidents have a high potential for hazardous conditions, it is important to adopt and use a signaling system that will advise personnel inside the danger area when to evacuate. For example, the FEMA US&R Task Force program has developed a system for evacuating rescuers from dangerous areas. Notification can be made using devices such as handheld CO_2 boat air horns, air horns on fire apparatus, or vehicle horns. Other communication methods in the event of an emergency can include portable radios, voice, hand signals, or the use of predetermined signals. The US&R designated signals and their meanings are as follows:

- Cease Operations/All Quiet: one long blast (three seconds)
- Evacuate the Area: three short blasts (one second each)
- Resume Operations: one long and one short blast

Responders must also plan escape procedures. If the primary means of egress becomes blocked, rescuers should determine the feasibility of using the alternate route.

Protection of the Public

Measures to protect the public include such things as conducting rescues, performing mass decontamination, and providing emergency medical care and first aid. Additional measures include evacuation, sheltering in place, and protecting/defending in place. The options for protective action are evacuation, shelter in place, or a combination of both. The IC selects the best option (or combination of options) based on factors that include but are not limited to the following:

- Material considerations
  - Toxicity
  - Quantity
  - Rate of release
  - Type of release and dispersion
  - Possibility of controlling and/or stopping the release
  - Direction of spread
- Environmental conditions
  - Wind direction
  - Wind velocity
  - Temperature
  - Humidity
  - Precipitation
  - Tides and currents
  - Topography
- Population at risk
  - Population density
  - Proximity
  - Warning/notification systems
  - Method of transport
  - Ability to control and/or stop the release
  - Special needs

Evacuation

Evacuate means to move all people from a threatened area to a safer place. To perform an evacuation, there must be enough time to warn people, for them to get ready, and for them to leave the area by a safe route (uphill, upwind, etc.). Generally, if there is enough time for evacuation, it is the best protective action. Emergency responders should begin evacuating people who are most threatened by the incident in accordance with distances recommended by the *ERG*, pre-incident surveys, or other sources. Even after people move these recommended distances, they are not necessarily completely safe from harm. Evacuees should not be permitted to congregate at the scene. Instead, they should be sent to a designated place (or area of safe refuge) along a specific route.

The number of responders needed to perform an evacuation varies with the size of the area and number of people to evacuate. Evacuation can be an expensive, labor-intensive operation, so it is important to assign enough personnel resources to an incident to conduct it.

Evacuation and traffic-control activities on the downwind side could cause responders and evacuees to become contaminated and, consequently, need decontamination. Responders may also need to wear PPE to safely conduct the evacuation. Evacuation plans (including casualties) for likely terrorist targets such as stadiums and other public gathering places should be made in advance as part of the local emergency response plan.

Large-scale evacuations present many factors that must be addressed by the IC including:

- *Notification* — The public must be alerted of the need to evacuate, plus they must understand where to go. Notification methods to be used should be spelled out in the local emergency response plan. When notifications are made, clear and concise information must be relayed to avoid confusion or additional panic. Notification methods include the following:
  — Knocking on doors
  — Public address systems
  — Radio
  — TV
  — Sirens
  — Building alarms
  — Short message service (SMS) through cell phones (text messages)
  — Reverse 9-1-1
  — Emergency Alerting System (EAS, in the United States)
  — Loudspeakers mounted on helicopters or emergency vehicles
  — Electronic billboards

- *Transportation* — As evidenced in the evacuation of New Orleans prior to hurricane Katrina, some individuals will not be able to self evacuate (not everyone owns a car). Alternate means of transportation must be planned in advance using school buses, public transit systems, or other means (planes, trains, boats, barges, ferries, etc.) **(Figure 6.30, p. 300)**.

- *Relocation Facilities and Temporary Shelters* — Evacuees must have some place to go. These places must be able to provide and care for the people relocated to them, including providing food, water, medicine, bathroom and shower facilities, and places to sleep (for evacuations of long duration) **(Figure 6.31, p. 300)**. Many evacuees may bring pets. Appropriate evacuation

Figure 6.30 Not all people will be able to self-evacuate. Plans must be made in advance to provide transportation to individuals without a means to leave the hazard area. *Courtesy of FEMA News Photos, photo by Win Henderson.*

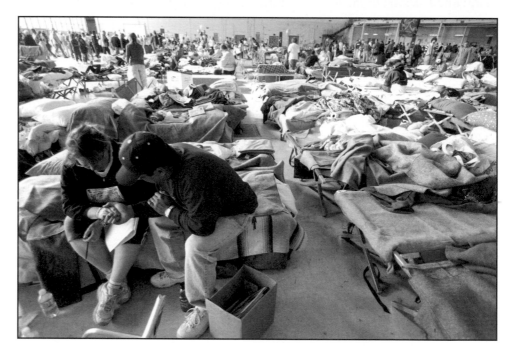

Figure 6.31 Evacuees must have some place to stay. *Courtesy of FEMA News Photos, photo by Andrea Booher.*

shelters must be designated in the local emergency response plan, and staffing arrangements should be determined in advance. An information/registration system should also be established to track the whereabouts of evacuees so their friends and relatives can find them.

- *Prevention of Looting* — Protection of property, while not as high a priority as life safety, must be taken into consideration. When home and business owners are no longer able to protect their property, the IC must ensure that steps are taken to prevent looting by unscrupulous individuals who are willing to risk staying in the evacuation area in order to take advantage of the situation.

- *Reentry* — Consideration must be given as to how people will be allowed to return to evacuated areas.

Evacuation of Contaminated Victims

Individuals who have been exposed or potentially exposed to chemical, biological, or radiological agents must be decontaminated (see Chapter 9, Decontamination). While it may be impossible to keep these individuals at the scene, efforts to keep them in place should be made in order to prevent the spread of harmful or potentially deadly materials to other locations. Evacuate contaminated or potentially contaminated individuals to an area of safe refuge (or a triage and treatment area as appropriate) within the isolation perimeter to await decontamination. Because victims may leave the scene before emergency responders arrive (or ignore requests to stay in order to undergo decon), shelters, hospitals, and other public health care facilities must be prepared to conduct decon of **self-presenters** at their facilities.

Self-Presenters — Individuals who seek medical assistance and have not been treated or undergone decontamination at the incident scene.

Sheltering in Place

Sheltering in place means to direct people to go quickly inside a building and remain inside until danger passes. It may be determined that shelter in place is the preferred option over evacuation. The decision to shelter in place may be guided by the following factors:

- The population is unable to initiate evacuation because of health care, detention, or educational occupancies.

- The material is spreading too rapidly to allow time for evacuation.

- The material is too toxic to risk any exposure.

- Vapors are heavier than air, and people are in a high-rise structure.

When protecting people inside a structure, directions should be given to close all doors, windows, heating, ventilation and air-conditioning systems. Shelter in place may not be the best option if vapors or gases are explosive as it may take a long time for vapor or gases to dissipate. Additionally, vapor or gases may permeate into any building that cannot be sealed from the outside atmosphere. Vehicles are not as effective as buildings for shelter in place, but they can offer temporary protection if windows are closed and the ventilation system is turned off. Whether using evacuation or shelter in place, the public needs to be informed as early as possible and receive additional instructions and information throughout the course of an emergency. Shelter in place may be more effective if public education has been done ahead of time through emergency planning.

First responders should also pay attention to the condition of surrounding buildings before ordering sheltering in place. For example, some areas may have old and dilapidated structures without air-conditioning or with openings between floorboards. Sheltering in place might not provide sufficient protection in such cases, making evacuation the better option.

Protecting/Defending in Place

Protecting/defending in place is an active (offensive) role or aggressive posture to physically protect those in harm's way. For example, using hose streams to diffuse a plume, or sending in law enforcement to secure a neighborhood or area. When appropriate and safe to do so, defending in place eliminates the need for unnecessary evacuations which, if initiated, will require additional logistical support to ensure the health and safety of evacuees.

Rescue

Due to the defensive nature of most Operations-Level actions, rescue can be a difficult strategy to implement for first responders, particularly in the initial stages of a response. Search and rescue attempts should be made within the framework of the IAP with appropriate training, PPE, backup personnel, and other safety considerations in place.

The safety of emergency personnel is the IC's first priority. When a rescue is too dangerous, the proper decision may be to protect the victims in place. This rule may directly conflict with the fire-fighting strategic priority of rescue first as well as with many responders' natural desire to help victims as quickly as possible. However, because of the dangers presented by hazardous materials, responders who rush to the rescue often require the need for rescue themselves.

Fight the Urge!

Never rush to conduct a rescue without proper PPE, planning, and coordination under the direction of the IC!

The IC makes decisions about rescue based on a variety of factors at the incident, including a risk-benefit analysis that weighs the possible benefits of taking certain actions versus the potential negative outcomes that might result. The following factors affect the ability of personnel to perform a rescue:

- Nature of the hazardous material and incident severity
- Training
- Availability of appropriate PPE
- Availability of monitoring equipment
- Number of victims and their conditions
- Time needed (including a safety margin) to complete a rescue
- Tools, equipment, and other devices needed to affect the rescue

Chapter 12, Victim Rescue and Recovery, addresses information needed by Ops-Level responders who will enter the hot zone to conduct rescues. Chapter 2, Hazardous Materials Identification, provides information about the hazards associated with each DOT hazard class so that first responders can assess potential risks at incidents involving these materials (for example, at an incident involving corrosives, they can determine that chemical burns are probably one of the major hazards). Without this additional training, first responders should avoid contact with hazardous materials, and they should not physically touch or move a victim who is either contaminated (or potentially contaminated) or located within the initial-isolation zone, warm zone, or hot zone. These responders' rescue actions are limited to telling people what to do and/or where to go. Thus, these responders can take the following actions from a distance:

- Direct people to an area of safe refuge or evacuation point located in a safe place within the hot zone that is upwind and uphill of the hazard area.
- Instruct victims to move to an area that is less dangerous before moving them to an area that offers complete safety.

- Direct contaminated or potentially contaminated victims to an isolation point, safe refuge area, safety shower, eyewash facility, or decontamination area **(Figure 6.32).**

- Give directions to a large number of people for mass decontamination **(Figure 6.33).**

- Conduct searches during reconnaissance or defensive activities.

- Conduct searches on the edge of the hot zone.

If there are injured victims at the scene, first responders must also be aware of the potential dangers of contamination and the need to decontaminate as part of the treatment process (see Chapter 9, Decontamination). They must follow local procedures for determining prioritization of emergency medical care and decontamination.

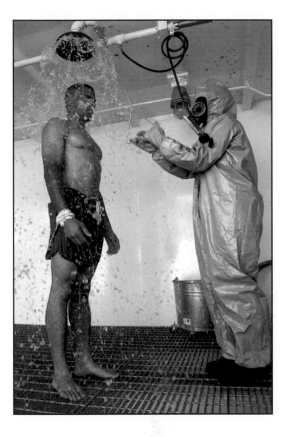

Figure 6.33 First responders can provide directions and instructions for mass decon operations.

Figure 6.32 First responders can direct contaminated or potentially contaminated victims to safety showers or areas of safe refuge. *Courtesy of the U.S. Marine Corps photo by Sgt J.A. Lee II.*

Protection of the Environment and Property

Exposure protection is a defensive control tactic. Most firefighters should be familiar with the concept of protecting exposures in fire situations, usually in terms of protecting property that is exposed to a fire in order to keep it from spreading. However, at the haz mat scene, the same concept is expanded to include protecting the environment and protecting property that is threatened by an expanding incident (including closed containers and piping).

Protecting environment and property also includes protecting exposures from fires involving hazardous materials but also protecting the environment from the harmful effects of hazardous materials that are not burning. For example, diking a storm drain is a tactic that protects the environment from being exposed to (and harmed by) potentially toxic materials **(Figure 6.34, p. 304)**.

Figure 6.34 Diking a storm drain can protect the environment from harm.

Protecting the Environment

Environmental damage is also an important concern. The air, surface water, wildlife, water table, and land surrounding an incident may be seriously affected by released materials. Water used during fire-control activities may potentially become contaminated. The nonbiodegradable nature of many materials means that the consequences of contamination may take years for the full effect to be realized. The result of contamination may also require large sums of money to repair. All released materials and runoff need to be confined and held until their effect on the environment can be determined.

Protecting Property

The property risk is similar to that created by other fire hazards except that the threatening material may not always be readily evident. Flammable and toxic gases, mists, and vapors can contaminate and pose an ignition threat with no visible signs. Protective actions must be tailored to the material, its properties, and any reactions to the proposed protective medium. ICs have appropriately decided not to save property when operations were potentially risky. Lives or the environment must not be unduly compromised to save property.

Recovery and Termination

Normally, the last strategic goals for the proper management of a hazardous materials emergency are the recovery and termination efforts. It is important to remember that there is a distinct difference between these functions. Recovery deals with returning the incident scene and responders to a pre-incident level of readiness. Termination involves documenting the incident and using this information to evaluate the response. This evaluation leads to an improvement of future response capabilities based upon problems that were identified during the original incident.

Recovery

The major goals of the recovery phase are as follows:

- Return the operational area to a safe condition.
- Debrief personnel before they leave the scene.
- Return the equipment and personnel of all involved agencies to the condition they were in before the incident.

On-scene Recovery

On-scene recovery efforts are directed toward returning the scene to a safe condition. These activities may require the coordinated effort of numerous agencies, technical experts, and contractors. Generally, fire and emergency services organizations do not conduct remedial cleanup actions unless those actions are absolutely necessary to eliminate conditions that present an imminent threat to public health and safety. If such imminent threats do not exist, contracted remediation firms under the oversight of local, state/provincial and federal environmental regulators generally provide for these cleanup activities. In these situations, the fire and emergency services organization may also provide control and safety oversight according to local SOPs.

On-scene Debriefing

On-scene debriefing, conducted in the form of a group discussion, gathers information from all operating personnel, including law enforcement, public works, and EMS responders. During the debriefing stage, obtain the following information from responders:

- Important observations
- Actions taken
- Timeline of those actions

In addition, one very important step in this process is to provide information to personnel concerning the signs and symptoms of overexposure to the hazardous materials, which is referred to as the hazardous communication briefing (required by OSHA in the U.S.). It is extremely important that this debriefing process be thoroughly documented. Each person attending must receive and understand the instructions and sign a document stating those facts. The information provided to responders before they leave the scene includes the following:

- Identity of material involved
- Potential adverse effects of exposure to the material
- Actions to be taken for further decontamination
- Signs and symptoms of an exposure
- Mechanism by which a responder can obtain medical evaluation and treatment
- Exposure documentation procedures

Operational Recovery

Operational recovery involves those actions necessary to return the resource forces to a level of pre-incident readiness. These actions involve the release of units, resupply of materials and equipment, decontamination of equipment and PPE, and preliminary actions necessary for obtaining financial restitution.

The financial effect of hazardous materials emergencies can be far greater than any other activity conducted by the fire and emergency services. Normally, a fire and emergency services organization's revenues obtained from taxes or subscriber fees are calculated based upon the equipment and personnel needs necessary to conduct fire-suppression and other emergency activities. It is recommended that communities have in place the necessary ordinances to allow for the recovery of costs incurred from such emergencies. In addition, the proper documentation of costs through the use of forms such as the *Unit Log* and other tracking mechanisms is a vital part of this process.

Termination

In order to conclude an incident, the IC must ensure that all strategic goals have been accomplished and the requirements of laws have been met. Documentation, analysis, and evaluation must be completed. The termination phase involves two procedural actions: critiques and after-action analysis (also known as an *After Action Report* or *AAR*). Analysis includes study of all postincident reports and critiques.

Critiques

OSHA Title 29 *CFR* 1910.120 mandates that incidents be critiqued for the purposes of identifying operational deficiencies and learning from mistakes. As with all critiques performed by the fire and emergency services, hazardous materials incident critiques need to occur as soon as possible after the incident and involve all responders, including law enforcement, public works, and EMS responders. As with other administrative and emergency-response functions, the critique is documented to identify those in attendance as well as any operational deficiencies that were identified.

After-Action Analysis

The after-action analysis process compiles the information obtained from the debriefings, postincident reports, and critiques to identify trends regarding operational strengths and weaknesses. Once trends have been identified, recommendations for improvements are made. These recommendations may be made in the following categories:

- Operational weaknesses
- Training needs
- Necessary procedural changes
- Required additional resources
- Plan updates and/or required changes

Also included in the after-action analysis is the completion of necessary reporting procedures required to document personal exposures, equipment exposures, incident reports, and staff analysis reports. After-action analysis forms the basis for improved response. Therefore, any recommendations for change or improvement are benchmarked for further consideration. Schedule follow-up activities to ensure successful implementation.

Summary

With the incident management system in place, responders can focus on the problem-solving process itself. First, they must analyze the incident to understand the problem by successfully identifying the material, conducting size-up and hazard/risk assessment, and predicting the behavior of the hazardous material and the containers involved. Second, based on the understanding they have gained from analyzing the incident, they must plan an appropriate response by setting strategic goals, determining the mode(s) of operation, and developing the IAP. Third, once a plan is in place, it must be implemented by choosing tactics and assigning tasks. Finally, the progress of the response must be evaluated by determining if mitigation efforts are succeeding and the incident is stabilizing. If they are not, the plan must be reevaluated and changed or altered accordingly.

The IC must determine the strategic goals and tactical objectives that will begin to stabilize the incident and bring it to a successful conclusion with the least amount of harm and damage. The resources available, a risk/benefit analysis, and other factors that are specific to the individual problems presented by the incident largely determine these goals and objectives. Strategic goals range from isolation to recovery and termination. Tactics dictate how these goals will be accomplished.

Review Questions

1. What are several strategic goals commonly used at hazardous materials incidents?

2. What are the four steps of the basic problem-solving formula?

3. What are some questions that should be answered during an initial survey?

4. What information is needed for a hazard and risk assessment?

5. What is the difference between hazard and risk?

6. Describe each of the incident levels.

7. What elements affect the selection of strategic mode?

8. What are the elements of an incident action plan (IAP)?

9. Describe the hazard-control zones.

10. What tactics are used to accomplish protection goals?

Terrorist Attacks, Criminal Activities, and Disasters

Chapter Contents

Divider page photo courtesy of FEMA News Photos, photo by Mike Rieger.

Key Terms

Competencies

NFPA® 472:	5.2.3(9)(a)	5.2.3(9)(c)	5.2.3(9)(e)	5.2.3(9)(g)	5.4.2
5.2.2(6)	5.2.3(9)(b)	5.2.3(9)(d)	5.2.3(9)(f)	5.2.4(5)	

Terrorist Attacks, Criminal Activities, and Disasters

Learning Objectives

1. Define terrorism.

2. Distinguish between a terrorist attack and a routine emergency.

3. Discuss terrorist tactics and types of attacks.

4. Discuss explosive attacks.

5. Discuss chemical attacks. [NFPA® 472, 5.2.3(9)(a-g)]

6. Discuss biological attacks.

7. Discuss radiological and nuclear attacks. [NFPA® 472, 5.2.4(5)]

8. Identify hazards of illegal haz mat dumps.

9. Describe proper evidence preservation. [NFPA® 472, 5.4.2]

10. Discuss hazardous materials during and after disasters.

Chapter 7
Terrorist Attacks, Criminal Activities, and Disasters

Case History

On March 11, 2004, ten bombs detonated on four commuter trains during the morning rush hour in Madrid, Spain. All the trains left Alcalá de Henares station between 7:01 and 7:14 a.m. local time, and they were traveling on the same line, going the same direction. They all detonated at approximately the same time, between 7:37 and 7:40 a.m., killing 191 people and wounding over 600 others. The bombs were composed of approximately 22 pounds (10 kg) of explosives packed into backpacks. They were detonated remotely by cell phones.

Three bombs detonated in a train that had just arrived at Atocha Station; two blasts that exploded approximately 4 seconds apart followed a minute later. Almost simultaneously, four bombs went off in another train approaching the station. The second train was running slightly late, so investigators believe it was the bombers' intention that all the bombs would explode inside Atocha Station. Three unexploded bombs were also found at the station.

Elsewhere, but at approximately the same time as the Atocha Station bombs, two bombs exploded in separate carriages of a train as it left El Pozo del Tío Raimundo Station. Another single bomb exploded on a train at Santa Eugenia Station.

Train station video footage widely available on the internet shows how the Atocha Station bombs may have been timed to herd frightened commuters away from the first blast, toward the station exit, and into the path of additional explosions. Assuming it was not merely coincidence, it vividly demonstrates the high level of sophisticated planning that was taken in advance of the attacks in order to maximize casualties.

The attacks took place three days before Spain's general elections. Initially, it was speculated that the ETA, who have carried out many bombings in Spain, were responsible for the attacks. However, investigators concluded the perpetrators were Islamic extremists, some with connections to al-Qaeda although direct al-Qaeda influence has not been proven.

It is believed by many political analysts that the attacks may have played a role in the defeat of the ruling party in the general Spanish elections three days later. The ruling party supported Spain's involvement in the Iraq war, a stance which was extremely unpopular with the general population. The attacks may have been perceived as a direct result of Spain's involvement in the war, turning voters against the incumbent government.

Incidents involving terrorism are an increasing concern for communities across the globe. Terrorists now have the knowledge and the capability to strike anywhere in the world, and they deliberately target locations where civilians are present. Recent examples of terrorist attacks include bombings in Iraq and Israel; the Mumbai, India attacks; and the 9/11 attacks in the U.S. **(Figure 7.1)**. The list will continue to grow. All societies, especially those that are free, are vulnerable to incidents involving terrorism.

Figure 7.1 Terrorists have the knowledge, will, and capability to strike anywhere in the world at any time. *Courtesy of the U.S. Navy, photo by Journalist 1st Class Mark D. Faram.*

Emergency responders must understand the role they play at incidents involving terrorism because these incidents differ from other emergency incidents in critical ways. These differences often present additional dangers to first responders, ranging from exposure to biological agents to encountering violent and armed individuals. Response personnel must know their limitations and realize when they cannot proceed or must do things differently than they normally would. Because terrorist acts can have such devastating consequences, response to any act of terrorism requires coordination and cooperation among many different agencies. Terrorist attacks are crimes, and the proper law-enforcement agencies must be involved. Law enforcement will provide scene security and force protection, and they will direct operations involving crime scene processing such as evidence collection and preservation.

Terrorists are difficult to stop, even when security precautions are taken and attacks are expected. An act of terrorism can occur anywhere, at any time, when least expected. Terrorist will attack targets on land, sea, or air **(Figures 7.2)**. No jurisdiction — urban, suburban, or rural — is immune from terrorist acts.

This chapter defines terrorism and explores the different types of terrorist attacks. It includes brief information on operations at terrorist and criminal incidents. The chapter also discusses illegal chemical dumps and haz mat issues during and after disasters.

Figure 7.2 Terrorists will attack anywhere they detect vulnerability, on land, in the air, or at sea. In October 2002, terrorists attacked the USS Cole while it was refueling in the port of Aden, Yemen. *Courtesy of the U.S. Department of Defense.*

What Is Terrorism?

While there are many different definitions of terrorism, the U.S. Federal Bureau of Investigation (FBI, the lead federal law enforcement agency in the U.S. at terrorist attacks) defines terrorism as *the unlawful use of force against persons or property to intimidate or coerce a government, the civilian population, or any segment thereof, in the furtherance of political or social objectives.* This definition includes the following three elements:

1. Terrorist activities are illegal and involve the use of force.

2. Actions intend to intimidate or coerce.

3. Actions are committed in support of political or social objectives.

Other popular definitions of terrorism include the threat of terrorism, for example, *the unlawful or **threatened** use of force or violence against individuals or property to coerce and intimidate governments or societies, often to achieve political, religious, or ideological objectives.* The criminal component is the most important element separating a terrorist organization and its actions from a legitimate organization. However, any organization, legitimate or not, can resort to terrorist means to achieve its political or social agenda **(Figure 7.3)**. Terrorists can operate as a group or act alone.

Figure 7.3 Terrorist organizations use terrorism to publicize their political, religious, or ideological agendas. *Courtesy of the U.S. Navy.*

Terrorist organizations plan activities that will have an emotional effect on the target population. They desire the target population to react to their attacks and demands in a manner that furthers their goals. For the most part, terrorism is designed to cause disruption, fear, and panic in order to draw attention to their cause; coerce or intimidate governments into granting their demands; or provoke governments into repressive actions thereby inspiring an uprising on the part of the oppressed masses.

Terrorism and Emergency Response

In some ways, it makes no difference to emergency responders whether the incident is a terrorist act or not. The same priorities of life safety, incident stabilization, and protection of property and the environment will apply. Operations will be managed under the same incident command/management system. Emergency responders will still respond and be among the first on the scene, and they will consider traditional strategies and tactics to manage the incident. They will use the same procedures to ensure safety and protection of responders and the public. The size and type of incident are key factors in how a response is managed. There may even be an unfortunate delay in identifying an incident as a terrorist attack.

However, there are some very important differences between terrorist incidents and other emergencies, and because of these differences it is vital for first responders to recognize that an incident might be a terrorist attack as soon as possible. Some of the key differences between more routine emergencies and a terrorist attack include:

- *Intent* — An act of terrorism is essentially different from normal emergencies in that it is intended to cause damage, inflict harm, and kill. The fire that starts in someone's home as a result of careless smoking did not occur with the intention of damaging something, hurting someone, or killing someone. Exceptions, of course, include cases of arson, but most emergency incidents are not criminal in nature.

- *Severity and Complexity* — Because of the deliberate intent to do damage at terrorist incidents, first responders may have to deal with circumstances far different from the usual structural fire, vehicle accident, or even hazardous materials incident **(Figure 7.4)**. For example, terrorist events may involve large numbers of casualties or materials (such as radioactive materials) with which first responders have little experience. Secondary contamination from handling patients may present a threat. Issues such as securing the scene and managing the incident will be more complex and difficult.

Figure 7.4 Terrorist attacks can be far more devastating and severe than many emergencies. *Courtesy of the U.S. Department of Defense.*

- *Crime Scene Management* — Terrorist attacks are crimes, and preservation of evidence becomes an extremely important consideration during a response to a terrorist attack **(Figure 7.5)**. If an incident is not recognized as a terrorist attack quickly, valuable information may be lost or accidentally destroyed (see Evidence Preservation section). For this reason, notification of federal law enforcement is very important.

Figure 7.5 Preservation of evidence is an important element of emergency operations at terrorist incidents. It is important for responders to recognize potential evidence and avoid disturbing it if possible. *Courtesy of the U.S. Navy, photo by Journalist 1st Class Mark D. Faram.*

- *Command Structure* — Most terrorist incidents require some form of unified command structure. Law enforcement will have jurisdiction over all incidents involving terrorism.

- *Secondary Devices/Attacks and Armed Resistance* — Terrorists may specifically target emergency responders in an attempt to incapacitate or kill them **(Figure 7.6, p. 316)**. Some additional hazards in these situations include the following:

  — Secondary events intended to incapacitate or delay emergency responders

  — Armed resistance and assault

  — Use of weapons

  — Booby traps

Many emergency response organizations may have supplies, equipment, and emergency response plans for other disasters that can be applied to terrorist incidents. For example, personal protective equipment used for clandestine drug lab responses and other haz mat incidents may also provide protection at a terrorist incident, depending on the materials involved (see Chapter 8, Personal Protective Equipment, for more information). Decontamination tents, trailers, and equipment can be used at terrorist incidents

Figure 7.6 This car bomb was aimed at U.S. and Iraqi forces arriving to inspect a car bomb that detonated in the same area an hour earlier, in southern Baghdad, Iraq, April 14, 2005. The attack was aimed at the Iraqi police force, and two police officers were among 18 casualties. *Courtesy of the U.S. Army, photo by Spc. Ronald Shaw Jr.*

and hazardous materials incidents (see Chapter 9, Decontamination). Evacuation plans used for natural disasters or other emergencies can be adapted for use during a terrorist incident. For more in-depth information about emergency operations at incidents involving terrorism, see the IFSTA's manual, **Emergency Response to Terrorist Attacks**.

Terrorist Tactics and Types of Attacks

Traditionally, terrorists have used conventional weapons such as firearms and explosives to achieve their goals and objectives. Traditional tactics of terrorism are assassination, armed assault, and bombings (including suicide bombings). Some conventional attacks may produce devastating effects equal to or exceeding those produced by the use of weapons of mass destruction. For example, assassination of a political leader could affect regime stability, or the use of conventional weapons could produce mass casualties and destruction exceeding the response capability of the community **(Figure 7.7)**. New tactics such as **cyber terrorism** and **agroterrorism** (also called *agricultural terrorism*) present threats to computer/network security and food supplies.

Experts now fear that terrorists have the means to broaden their tactics to include the use of weapons of mass destruction. According to the U.S. Government (*United States Code*, Title 50, Chapter 40, Section 2302, and Title 18, Part I, Chapter 113B, Section 2332a), the term *weapon of mass destruction* (WMD) means any weapon or device that is intended or has the capability to cause death or serious bodily injury to a significant number of people through the release, dissemination, or impact of one of the following means:

- Toxic or poisonous chemicals or their precursors
- A disease organism
- Radiation or radioactivity

Other parts of the *U.S. Code* also include certain explosive and incendiary devices in the definition. Weapons of mass destruction can be divided into the five categories discussed earlier in this manual: chemical, biological, radiological, nuclear, and explosive (CBRNE).

Agroterrorism – A terrorist attack directed against agriculture, for example, food supplies or livestock.

Cyber Terrorism — The premeditated, politically motivated attack against information, computer systems, computer programs, and data which result in violence against noncombatant targets by sub-national groups or clandestine agents.

Figure 7.7 The use of conventional weapons can produce mass casualties. In April, 1995, a truck bomb destroyed the Alfred P. Murrah Federal Building in Oklahoma City, killing 168 people and injuring over 600 others.

Other Acronyms for WMDs

While this chapter discusses types of attacks based on the *CBRNE* acronym, other terms are sometimes used to indicate essentially the same thing. These terms may include *COBRA* (chemical, ordinance, biological, radiological agents), *B-NICE* (biological, nuclear, incendiary, chemical, explosive) and *NBC* (nuclear, biological, chemical) as well as others. This manual will often refer to *CBR* materials when discussing chemical, biological, and radiological attacks.

Despite the media hype, it is not easy to manufacture most WMDs in a basement laboratory. Nerve agents, sophisticated biological agents, and nuclear bombs require high levels of technical expertise to make, as well as hard-to-acquire materials and expensive equipment in order to manufacture. Scientific personnel, high funding, and good infrastructure are needed to produce high-quality chemical and biological agents. For example, from 1983-1990, Iraq spent hundreds of millions of dollars to develop and use sarin, and it took thousands of scientists to do so. Large-scale efforts are difficult to conceal.

Many WMDs present significant hurdles that must be overcome before they can be successfully deployed. Some are difficult to store, so they must be used very quickly. For example, Iraqi VX nerve agent had to be used almost immediately after production because it was of a very poor quality and degraded quickly. Other materials such as botulism toxin may have to be kept in temperature-controlled environments. The greatest threat of mass-produced WMD agents comes from nations with the infrastructure, finances, and scientific knowledge to produce them, not from isolated terrorist groups. Even with these resources, it is not easily done.

However, some WMDs are more readily produced and/or acquired than others. The biological toxin, ricin, for example, can be made from castor beans, and recipes for making the toxin are available on the internet. TATP, an explosive, can be made from common household products without expensive laboratory equipment. Biological agricultural threats such as foot-and-mouth disease have natural reservoirs in nature. Radiological materials can be stolen and/or acquired from a variety of accessible medical and construction sources. Toxic industrial chemicals are available in every jurisdiction.

So what type of WMD incidents are first responders most likely to face? While there isn't universal consensus among experts, it is generally believed that the WMD threat reality is something close to this:

1. Explosives (IEDs, vehicle bombs, suicide bombers; also explosives potentially combined with other materials such as industrial chemicals, biological materials, or radiological materials) **(Figure 7.8)**

2. Biological toxins (ricin) **(Figure 7.9)**

3. Industrial chemicals (chlorine, phosgene, or others)

4. Biological pathogens (such as contagious diseases)

5. Radiological materials (such as those used in a radiological dispersal device)

6. Military-grade chemical weapons (nerve agents)

7. Nuclear weapons (nuclear bomb)

NOTE: Conventional attacks such as hijackings, sniper attacks, and/or shootings are also highly likely, but not considered a WMD threat for purposes of this list.

Explosives and conventional attacks have been the weapons of choice for terrorists throughout history, and most experts agree that explosives are the greatest WMD threat today. It is for this reason that explosives are discussed first in the sections that follow. Chemical attacks, biological attacks, and nuclear/radiological attacks will also be described.

Figure 7.8 First responders are more likely to face explosive attacks than any other type of WMD incident. Suicide bombers, IEDs, car and truck bombs, and attacks utilizing military ordnance or homemade rockets (as shown in the picture) are all possibilities. *Courtesy of Adi Moncaz.*

Figure 7.9 Attacks using biological toxins such as ricin are also possible. This photograph shows the leaves and seed pods of a castor oil plant. Ricin is made from castor beans. *Courtesy of the USDA.*

Figure 7.10 A replica of a suicide bomb vest used in Israel.

Explosive Attacks

Explosive devices can be anything from homemade pipe bombs to sophisticated military ordinance, but nonmilitary first responders are more likely to encounter improvised explosive devices (IEDs) than military weapons **(Figure 7.10)**. While most bombs made by criminals or terrorists are likely to be homemade or constructed in an improvised manner, they usually have one thing in common: they are designed to kill, maim, or destroy property. The truck bomb that exploded April 19, 1995, outside the Murrah Federal Building in Oklahoma City, killing 168 people and injuring many others, is testimony to the potential destructive power of such devices.

Anatomy of an Explosion

An explosive (or energetic) material is any material or mixture that will undergo an extremely fast, self-propagating reaction when subjected to some form of energy. Explosive materials combine an oxidizing component with a fuel component. This is done either by mixing an oxidizer with a fuel (for example, forming a mixture such as black powder) or combining the two at a molecular level (for example, forming a compound such as TNT) **(Figure 7.11a and b, p. 320)**.

An **explosion** results when a material undergoes a physical or chemical reaction that releases rapidly expanding gases. These gases are formed almost instantaneously, in approximately 1/10,000th of a second. The expanding gases

Explosion — A physical or chemical process that results in the rapid release of high pressure gas into the environment.

Figure 7.11a and b Explosive materials may be mixtures like black powder **(a)**, or they can be compounds like TNT **(b)**.

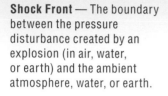

Shock Front — The boundary between the pressure disturbance created by an explosion (in air, water, or earth) and the ambient atmosphere, water, or earth.

Blast-Pressure Wave — Shock wave created by rapidly expanding gases in an explosion.

move outward in a circle from the point of detonation in a wave at speeds up to 13,000 miles (20,921 km) per hour, compressing the surrounding atmosphere into a **shock front** that is sometimes visible expanding outward from the point of detonation. The pressure wave formed by the expanding gases is sometimes called the **blast-pressure wave**, and it can demolish virtually anything in its way.

There are actually two phases to the blast-pressure wave: the positive-pressure phase and the negative-pressure phase, sometimes called the suction phase. Both phases can be destructive.

In the positive-pressure phase, the shock front leads the positive-pressure wave, striking anything in its path with destructive, hammering force **(Figure 7.12)**. The positive-pressure wave will continue outwards in an expanding radius until its energy is diminished by distance or transferred to objects standing in its path (such as buildings).

After the initial expansion energy has been dissipated, a negative pressure or suction phase is created when displaced atmosphere rushes into fill the vacuum left at the center of the explosion. This rush of air also has destructive power although not to the same degree as the positive-pressure wave. However, structures damaged in the initial blast can be further damaged in the negative-pressure phase **(Figure 7.13)**. The negative-pressure phase of an explosion lasts about three times longer than the positive-pressure phase. **Figure 7.14** shows all the effects of an explosion including the shock front, blast-pressure effect, fragmentation effect, and incendiary thermal effect.

The size of an incident involving explosives is determined by the amount and type of explosives used. The crime scene may be limited to a small area such as a single business establishment or it could extend for blocks. For example, it is estimated that in a city center or downtown area surrounded by tall buildings (5 to 15 stories), a 2,205 pound (1,000 kg) device, while unlikely to cause total structural collapse, could blow in walls in a one block radius and cause injuries over nine square blocks from breaking glass. A 15,432 pound (7,000 kg) blast could cause severe injury and death from building collapse out to 328 feet (100 m).

Figure 7.12 The blast pressure of an explosion compresses the surrounding atmosphere into a rapidly expanding shock front. Depending on its force, this positive pressure wave can destroy virtually anything standing in its way.

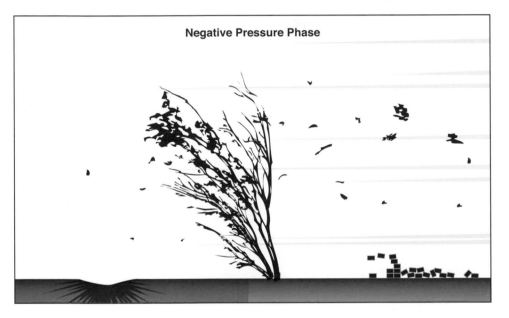

Figure 7.13 While typically less destructive than the positive pressure phase, additional damage can be done during the negative pressure phase, particularly to buildings and structures damaged in the initial blast.

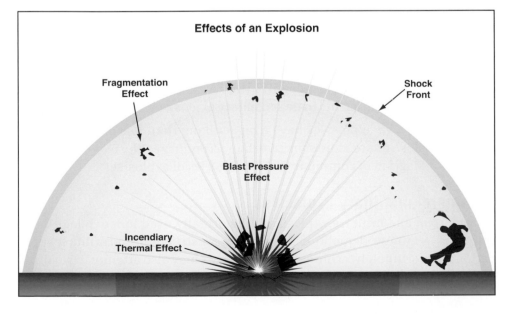

Figure 7.14 Effects of an explosion include the blast pressure effect, incendiary/thermal effects, the shock front, and the fragmentation effect.

Classification of Explosives

Most commonly, explosives are categorized by chemical reaction or rate of decomposition:

- *High explosives* — Decompose extremely rapidly (almost instantaneously), which is normally called **detonation**. In other words, **high explosives** detonate (explode). Detonation velocities for high explosives range from 3,300 feet per second (fps) to 29,900 fps (1 006 mps to 9 114 mps). Simply put, high explosives have detonation velocities that are faster than the speed of sound. Examples of commercially available high explosives include:

  — Plastic explosives

  — Nitroglycerin

  — TNT

  — Blasting caps

  — Dynamite

  — Ammonium nitrate and fuel oil (ANFO) and other blasting agents

- *Low explosives* — Decompose rapidly but do not produce an explosive effect unless they are confined; rather, they *deflagrate* (burn rapidly). For example, black powder undergoes **deflagration** rather than detonation. **Low explosives** confined in small spaces or containers are commonly used as propellants (bullets and fireworks). Pyrotechnic substances are used in fireworks and road flares. Unconfined low explosives may also be considered incendiary materials. Many experts do not separate incendiary devices/materials from explosives.

Emergency responders should also be familiar with the following classifications based on high explosives' susceptibility to initiation (or sensitivity):

- *Primary explosives* — Easily initiated and highly sensitive to heat; usually used as **detonators**. Small amounts such as a single grain or crystal can detonate. Examples of **primary explosives** are lead azide, mercury fulminate, and lead styphnate.

- *Secondary explosives* — Designed to detonate only under specific circumstances; usually by activation energy from a primary explosive. **Secondary explosives** are less sensitive to initiating stimuli such as heat or flame. Example: TNT.

- *Tertiary explosives (blasting agents)* — Very insensitive materials based on ammonium nitrate (AN), usually requiring initiation from a secondary explosive. Not all experts recognize this category and would consider AN and other blasting agents to be secondary explosives.

In general, *high* explosives have a bigger bang than *low* explosives, and *primary* explosives are more sensitive than *secondary* explosives. **Figure 7.15** provides a diagram of commonly used categories of explosives.

Commercial/Military Explosives

Commercial and military explosives are normally used for such legitimate purposes as mining, demolition, excavation, construction, and military applications. Unfortunately, criminals and terrorists may also attempt to steal and use explosives to inflict injury and death to persons and damage to property

---

Detonation — (1) Supersonic thermal decomposition, which is accompanied by a shock wave in the decomposing material. (2) Explosion with an energy front that travels faster than the speed of sound.

High Explosive — Explosive material that detonates at a velocity faster than the speed of sound.

Deflagration — (1) Chemical reaction producing vigorous heat and sparks or flame and moving through the material (as black or smokeless powder) at less than the speed of sound. A major difference among explosives is the speed of the reaction. (2) Can also refer to intense burning; a characteristic of Class B explosives. (3) An explosion involving a chemical reaction in which the reaction (energy front) proceeds at less than the speed of sound.

Low Explosive — Explosive material that deflagrates, producing a reaction slower than the speed of sound.

Detonator — A device used to trigger less sensitive explosives, usually composed of a primary explosive. Detonators may be initiated mechanically, electrically, or chemically. A blasting cap is an example of a detonator. Also called an Initiator.

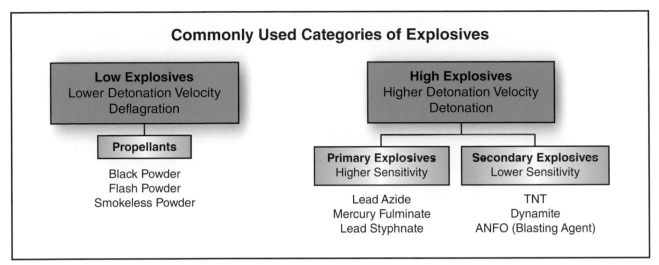

Commonly Used Categories of Explosives

Low Explosives
Lower Detonation Velocity
Deflagration

Propellants

Black Powder
Flash Powder
Smokeless Powder

High Explosives
Higher Detonation Velocity
Detonation

Primary Explosives
Higher Sensitivity

Lead Azide
Mercury Fulminate
Lead Styphnate

Secondary Explosives
Lower Sensitivity

TNT
Dynamite
ANFO (Blasting Agent)

Figure 7.15 Commonly used categories of explosives.

and the environment. **Table 7.1, p. 324** provides pictures of many commercial and military explosive materials.

Military munitions may also be used. These may include mortars, grenades, anti-personnel mines, surface-to-air missiles, rocket propelled grenades, and other types of military explosives to which terrorists can get access **(Table 7.2, p. 325)**.

Homemade/Improvised Explosive Materials

First responders are more likely to encounter homemade or improvised explosive materials rather than military weapons in their day-to-day response activities. Improvised explosive materials are typically made by combining an oxidizer with a fuel **(Figure 7.16, p. 326)**. Many of these materials are fairly simple to make and require very little technical expertise or specialized equipment. However, the explosive materials created are often highly unstable, and more than one would-be bomb maker has been killed trying to make these homemade explosives.

The sections that follow will discuss peroxide-based explosives, potassium chlorate, and urea nitrate. These four do not represent a comprehensive list, and many oxidizers and fuels can be combined to form improvised explosive materials. **Appendix H** contains a comprehensive list of explosive materials as identified by the U.S. Bureau of Alcohol, Tobacco, Firearms and Explosives (ATF).

Peroxide-Based Explosives

Peroxide-based explosives such as acetone peroxide (triacetonetriperoxide or TATP) and hexamethylene triperoxide diamine (HMTD) are a growing concern. Used in several recent terrorist attacks including the 2005 London mass transit bombings, peroxide-based explosives can be made by mixing concentrated hydrogen peroxide, acetone, and either hydrochloric or sulfuric acid. Both TATP and HMTD are very dangerous to make and handle because they are very unstable, both during the manufacturing process and as a finished product. Both are normally made in **illicit laboratories** located almost anywhere because highly specialized equipment is not needed for the

Primary Explosive — High explosive that is easily initiated and highly sensitive to heat; often used as detonators.

Secondary Explosive — High explosive that is designed to detonate only under specific circumstances.

WARNING!
Never attempt to handle commercial or military explosives!

Illicit Laboratory — Laboratory established to produce or manufacture illegal or controlled substance such as drugs, chemical warfare agents, explosives, or biological agents.

Table 7.1
Commercial Explosives

Ammonium Nitrate

Binary Explosives

Black Powder

Blasting Caps

C-4

C-3 Sheet Explosive

DET Cord–RDX

DET Cord–PETN

Dynamite

Semtex

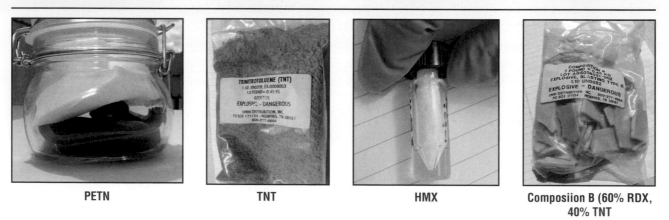

PETN

TNT

HMX

Composiion B (60% RDX, 40% TNT)

Table 7.2
Military Ordnance

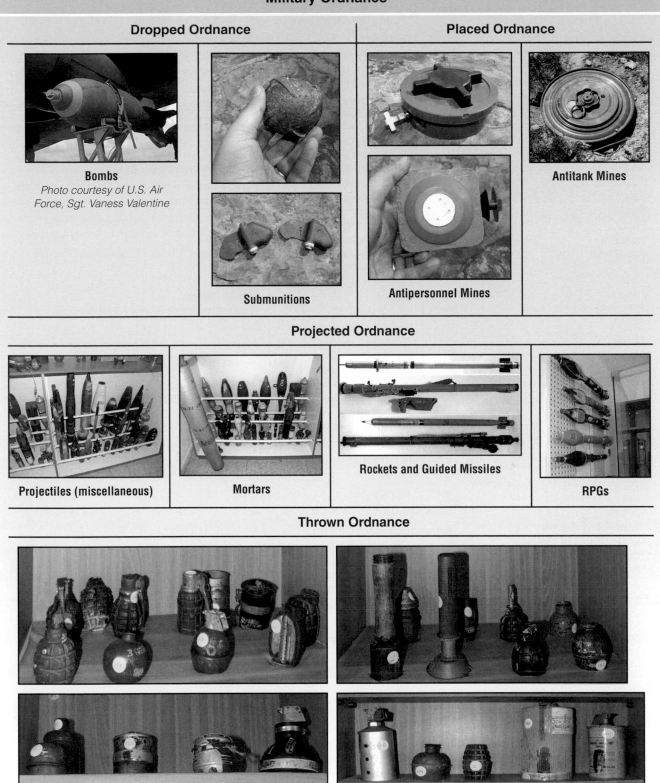

Dropped Ordnance

Bombs
Photo courtesy of U.S. Air Force, Sgt. Vaness Valentine

Submunitions

Placed Ordnance

Antitank Mines

Antipersonnel Mines

Projected Ordnance

Projectiles (miscellaneous)

Mortars

Rockets and Guided Missiles

RPGs

Thrown Ordnance

Fragmentation Grenades, Offensive Grenades, Antitank Grenades, Smoke Grenades, and Illumination Gremades

Components of Improvised Explosives

🔥 Potential Fuels + 🔥 Potential Oxidizers = 💥 Explosive Blends (Oxidizer + Fuel)

Hydrocarbons:
Alcohol
Carbon Black
Charcoal
Dextrin
Diesel
Ethylene Glycol
Gas
Kerosene
Naphtha
Rosin
Sawdust
Shellac
Sugar
Vaseline
Wax/Parfin

Energetic Hydrocarbons:
Nitrobenzene
Nitromethane
Nitrocellulose

Elemental "Hot" Fuels:
Powdered Metals
- Aluminum
- Magnesium
- Zirconium
- Copper
Phosphorus
Sulfur
Antimony Trisulfide

Oxidizers:
Perchlorate
Chlorate
Hypochlorite
Nitrate
Peroxide
Iodate
Chromate
Dichromate
Permaganate
Sodium Chlorate
Potassium Chlorate
Ammonium Nitrate
Potassium Nitrate
Hydrogen Peroxide
Barium Peroxide
Ammonium Perchlorate
Calcium Hypochlorite
Nitric Acid
Lead Iodate
Sodium Chlorate
Potassium Permanganate
Lithium Chromate
Potassium Dichromate

Nitrate Blends:
ANFO (Ammonium Nitrate + Diesel Fuel)

ANAl (Ammonium Nitrate + Aluminum Powder)

ANS (Ammonium Nitrate + Sulfur Powder

ANIS (Ammonium Nitrate + Icing Sugar)

Black Powder (Potassium Nitrate + Charcoal + Sulfur)

Chlorate/Perchlorate Blends:
Flash Powder (Potassium Chlorate/Perchlorate + Aluminum Powder + Magnesium Powder + Sulfur)

Poor Man's C-4 (Potassium Chlorate + Vaseline)

Armstrong's Mixture (Potassium Chlorate + Red Phosphorus)

Liquid Blend:
Hellhoffite (Nitric Acid + Nitrobenzene)

Common Precursors Used To Make Explosives

🪓 Precursors:
Hydrogen Peroxide
Sulfuric Acid (battery acid)
Nitric Acid
Hydrochloric Acid (muriatic acid)
Urea
Acetone
Methyl Ethyl Ketone
Alcohol (Ethyl or Methyl)
Ethylene Glycol (antifreeze)
Glycerin(e)
Hexamine (camp stove tablets)
Citric Acid (sour salt)

💥 Nitrated Explosives:
Nitroglycerine (Glycerine + Mixed Acid [Nitric Acid + Sulfuric Acid])

Ethylene Glycol Dinitrate (EGDN) (Ethylene Glycol + Mixed Acid [Nitric Acid + Sulfuric Acid])

Methyl Nitrate (Methyl Alcohol [methanol] + Mixed Acid [Nitric Acid + Sulfuric Acid])

Urea Nitrate (Urea + Nitric Acid)

Nitrocotton (Gun Cotton) (Cotton + Mixed Acid [Nitric Acid + Sulfuric Acid])

Peroxide Explosives:
Triacetone Triperoxide (TATP) (Acetone + Hydrogen Peroxide + Strong Acid [Sulfuric, Nitric, or Hydrochloric])

Hexamethylene Triperoxide Diamine (HMDT) (Hexamine + Hydrogen Peroxide + Citric Acid)

Methyl Ethyl Ketone Peroxide (MEKP) (Methyl Ethyl Ketone + Hydrogen Peroxide + Strong Acid [Sulfuric, Nitric, or Hydrochloric])

Figure 7.16 Most homemade explosives are made by combining an oxidizer with a fuel.

manufacturing process. Many youths experiment in making these explosives as recipes are readily available. TATP is typically a white crystalline powder with a distinctive acrid smell. Dependant on the quality of the manufacturing process and ingredients, TATP can range in color from a yellowish to white color **(Figure 7.17a and b)**.

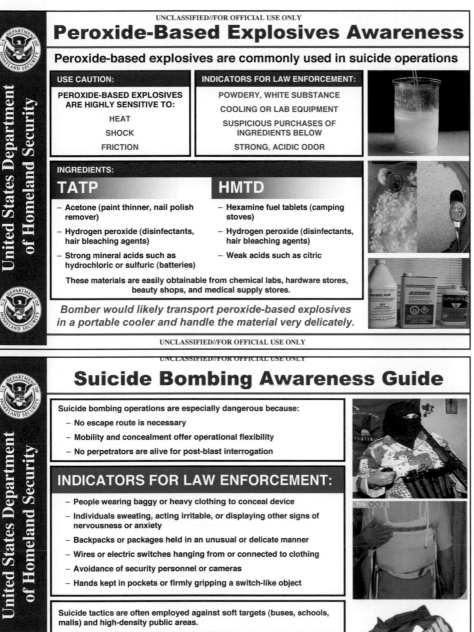

UNCLASSIFIED//FOR OFFICIAL USE ONLY

Peroxide-Based Explosives Awareness

Peroxide-based explosives are commonly used in suicide operations

USE CAUTION:	INDICATORS FOR LAW ENFORCEMENT:
PEROXIDE-BASED EXPLOSIVES ARE HIGHLY SENSITIVE TO:	POWDERY, WHITE SUBSTANCE
HEAT	COOLING OR LAB EQUIPMENT
SHOCK	SUSPICIOUS PURCHASES OF INGREDIENTS BELOW
FRICTION	STRONG, ACIDIC ODOR

INGREDIENTS:

TATP

– Acetone (paint thinner, nail polish remover)

– Hydrogen peroxide (disinfectants, hair bleaching agents)

– Strong mineral acids such as hydrochloric or sulfuric (batteries)

HMTD

– Hexamine fuel tablets (camping stoves)

– Hydrogen peroxide (disinfectants, hair bleaching agents)

– Weak acids such as citric

These materials are easily obtainable from chemical labs, hardware stores, beauty shops, and medical supply stores.

Bomber would likely transport peroxide-based explosives in a portable cooler and handle the material very delicately.

UNCLASSIFIED//FOR OFFICIAL USE ONLY

UNCLASSIFIED//FOR OFFICIAL USE ONLY

Suicide Bombing Awareness Guide

Suicide bombing operations are especially dangerous because:

– No escape route is necessary

– Mobility and concealment offer operational flexibility

– No perpetrators are alive for post-blast interrogation

INDICATORS FOR LAW ENFORCEMENT:

– People wearing baggy or heavy clothing to conceal device

– Individuals sweating, acting irritable, or displaying other signs of nervousness or anxiety

– Backpacks or packages held in an unusual or delicate manner

– Wires or electric switches hanging from or connected to clothing

– Avoidance of security personnel or cameras

– Hands kept in pockets or firmly gripping a switch-like object

Suicide tactics are often employed against soft targets (buses, schools, malls) and high-density public areas.

DO NOT rely solely on conventional perceptions of suicide bombers for identification. Focus on behavior and situation context.

UNCLASSIFIED//FOR OFFICIAL USE ONLY

Figure 7.17a and b (a) Emergency responders need to recognize peroxide-based explosives and the labs that make them. *Courtesy of the U.S. Department of Homeland Security.* **(b)** *TATP.*

⚠️ **WARNING!**
You should be very cautious of any items, materials, or locations that arouse your curiosity!

Potassium Chlorate

Potassium chlorate is another white crystal or powder form explosive that may be used in IEDs. It has approximately 83% of the power of TNT. Potassium chlorate is a common ingredient in some fireworks and can be purchased in bulk form fireworks/chemical supply houses. Potassium Chlorate is used in printing, dying, steel, weed killer, matches and the explosive industry.

Urea Nitrate

Urea nitrate is considered a type of fertilizer-based explosive composed of nitric acid and urea. The prill used for de-icing sidewalks is composed of urea, and urea can also be derived from concentrated urine. Often, sulfuric acid is added to assist with catalyzing the constituents. Urea nitrate has a destructive power similar to ammonium nitrate.

Improvised Explosive Devices

As their name implies, IEDs are not commercially manufactured. They are homemade, are usually constructed for a specific target, and can be contained within almost any object (**Figures 7.18 a and b**). Depending on the sophistication of the device, they are relatively easily to make and can be constructed in virtually any location or setting. Bomb makers who specialize in IED manufacture often make more sophisticated varieties.

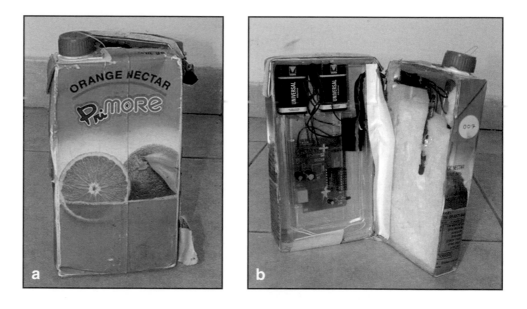

Figure 7.18a and b IEDs can be contained in virtually anything. This is a reproduction of an IED hidden in a juice box, which detonated in Israel.

Explosive devices made by inexperienced designers (or with unsafe materials) may fail to detonate, and in some cases will detonate during the building process or when being moved or placed. Some groups are known to produce sophisticated devices that are constructed with components scavenged from conventional munitions and standard consumer electronics components, such as speaker wire, cellular phones, or garage door openers (**Figure 7.19**). The sophistication of an IED depends on the training of the designer and the tools and materials available on hand.

The majority of IEDs use conventional explosives. However, there is a growing concern that improvised explosives like TATP and HMTD will be used more

Figure 7.19 U.S. troops discover ordinary objects scavenged in Iraq that are used to build IEDs. *Courtesy of the U.S. Army, photo by Spc. Ben Brody.*

Figure 7.20 IEDs can be placed anywhere. One threat is vehicle bombs driven and detonated on busy streets. *Courtesy of the U.S. Department of Defense, photo by MSGT Michael E. Best.*

in the future. IEDs often include nails, tacks, broken glass, bolts, and other items that will cause additional shrapnel damage and fragmentation injuries.

IEDs may be placed anywhere **(Figure 7.20)**. Usually, bombers want to succeed without being detected or caught. The level of security and awareness of the public, security forces, employees, and the like determine where and how an IED is placed.

Types of IEDs

IEDs are typically categorized by their container (such as pipe, backpack or vehicle bomb) and the way in which they are initiated. Bomb types, based on the outer container, can include the following:

- *Vehicle bombs* — **Vehicle-Borne Improvised Explosives Devices (VBIEDs)** may contain many thousands of pounds (kilograms) of explosives that can cause massive destruction and are perhaps the most devastating of all IEDs **(Figure 7.21, p. 330)**. They can be easy to conceal, and the explosives can be placed anywhere in a vehicle but are often located in the trunk when small vehicles (such as passenger cars) are used. These bombs are described in greater detail later in this chapter.

- *Pipe bombs* — The most common type of IED found in the United States **(Figure 7.22, p. 330)**. Can be anywhere from 4 to 14 inches (102 to 356 mm) in length. Usually made of steel or polyvinyl chloride (PVC) pipe sections that are filled with explosives and capped or sealed on the ends. These can be filled with easily obtained materials such as black powder or match heads. Pipe bombs are often filled or wrapped with nails or other materials so that when detonated, they can throw shrapnel up to 300 feet (91 m) with lethal

Vehicle-Borne Improvised Explosives Device (VBIED) — An improvised explosive device placed in a car, truck, or other vehicle, typically creating a large explosion. Also called a Car Bomb or Vehicle Bomb.

Figure 7.21 Vehicle bombs can be extremely devastating. *Courtesy of the U.S. Department of Defense.*

Figure 7.22 Pipe bombs are the most common type of IED in the U.S. *Courtesy of the U.S. Department of Defense, photo by 2LT William Powell.*

force. Pipe bombs may be detonated with a homemade fuse or commercially available fuses. Explosive filler can get into the threads of the pipe making the device extremely sensitive to shock or friction.

- *Satchel, backpack, knapsack, duffle bag, briefcase, or box bombs* — Filled with explosives or an explosive device **(Figures 7.23)**. Terrorists have and continue to extensively use these devices because it is very common in today's world to see people carrying backpacks or other types of bags. They may include electronic timers or radio-controlled triggers so there may be no external wires or other items visible. These bombs come in any style, color, or size of carrying container (even as small as a cigarette pack). These types of devices were used in the Eric Rudolph United States attacks and Spain train bombings by being left in the target areas. They were also carried by suicide bombers in the London, England mass transit bombings.

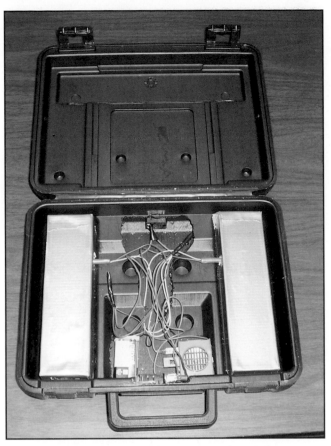

Figure 7.23 Backpack and briefcase bombs can be detonated by suicide bombers or left in crowded locations to be detonated by timer or remote signal. *Courtesy of August Vernon.*

Figure 7.24 Person-borne bombs include suicide bombers and individuals coerced into carrying explosives. *Courtesy of August Vernon.*

- *Person-borne bombs* — Person-Borne Improvised Explosives Devices (PBIEDs) typically consist of belts or bombs worn or carried by suicide bombers, often vests with many pockets sewn into them to hold explosive materials **(Figures 7.24)**. This category also includes bombs attached to coerced or unwilling victims. These bombs are described in greater detail later in this chapter.

- *Mail, package, or letter bombs* — Explosive device or material is concealed in a package or letter. Opening the package or letter usually triggers the bomb. A large number of potentially suspicious letters and packages continue to be reported to state, federal, and local emergency response agencies. The list of possible indicators of package or letter bombs is long **(Figure 7.25, p. 332)**. Some of the most common indicators are as follows:

— Package or letter has no postage, noncancelled postage, or excessive postage. Normally a bomber does not want to mail a parcel over the postal counter and communicate with a window clerk face to face.

— Parcels may be unprofessionally wrapped with several combinations of tape to secure them and endorsed *Fragile — Handle With Care* or *Rush — Do Not Delay.*

— Sender is unknown, no return address is available, or the return address is fictitious.

Person-Borne Improvised Explosives Device (PBIED) — An improvised explosive device carried by a person; includes suicide bombers as well as individuals coerced into carrying the bomb against their will.

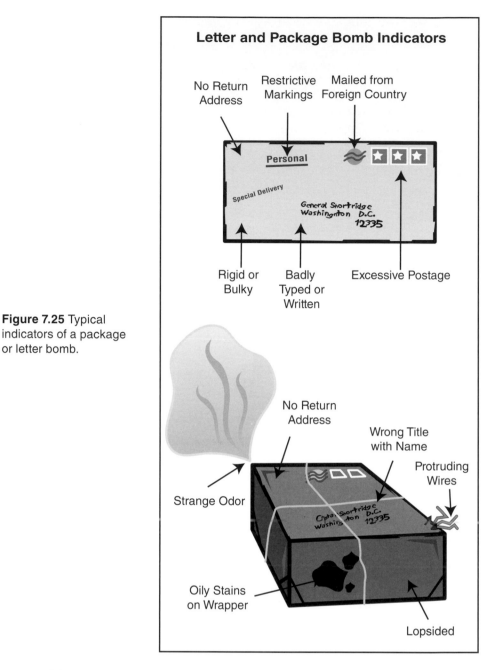

Figure 7.25 Typical indicators of a package or letter bomb.

— Addressee does not normally receive mail at that address.

— Mail may bear restricted endorsements such as *Personal* or *Private*. These endorsements are particularly important when addressees do not usually receive personal mail at their work locations.

— Postmarks may show different locations than return addresses.

— Common words are misspelled on mail.

— Mail may display distorted handwriting, or the name and address may be prepared with homemade labels or cut-and-paste lettering.

— Package emits a peculiar or suspicious odor.

— Mail shows oily stains or discoloration.

— Letter or package seems heavy or bulky for its size and may have an irregular shape, soft spots, or bulges.

— Letter envelopes may feel rigid or appear uneven or lopsided.

— Mail may have protruding wires or aluminum foil.

— Package makes ticking, buzzing, or whirring noises. Unidentified person calls to ask if a letter or package was received.

NOTE: An additional mail threat is a letter or package containing unidentified powders or chemicals.

- *Plastic bottle bombs* — Plastic pop bottles are filled with a material (such as dry ice) or combination of reactive materials that expand rapidly, causing the container to explode. Devices can be made with plastic soda or drink bottles of any size from small to large. There are dozens of variations of bottle bombs and a large amount of information on bottle bombs is available on the Internet. Responders should be careful around plastic containers containing multi-layered liquids and containers with white or gray liquids with possible cloudy appearance. Materials such as pool chemicals, dry ice, alcohol, acid, aluminum, toilet bowl, drain and driveway cleaners, etc., may be used. Responders must not attempt to move or open these types of containers because once they are initiated there is no accurate way of telling when they may detonate. These are commonly found being built by teenagers and others experimenting with the materials. These devices can cause injury. These may also include homemade chemical bombs (IICBs, also referred to as bottle bombs), acid bombs, and MacGyver bombs.

- *Fireworks* — Legally obtained fireworks modified and/or combined to form more dangerous explosive devices. Responders must be cautious of home-based labs where some of these materials may be produced.

- *M-devices* — Devices constructed of cardboard tubes (often red) filled with flash powder and sealed at both ends; are ignited by fuses. The most common are M-80s, which measure $5/8 \times 1\frac{1}{2}$ inches (16 mm by 25 mm). M-devices, while generally more dangerous than most fireworks, present more of a hazard as a potential incendiary than a destructive explosive (shock and fragmentation hazards are less than those from a pipe or vehicle bomb). M-80s were available in the U.S. as commercial fireworks but were made illegal for safety reasons in 1966.

- *Carbon dioxide (CO_2) grenades (also called crickets)* — Devices made by drilling a hole in and filling used CO_2 containers (such as those used to power pellet pistols) with an explosive powder; usually initiated by a fuse and can produce deadly shrapnel. Shrapnel can also be added to the outside of the container. The range of these crickets is very small but can be very dangerous. These were used extensively during the 1999 Columbine High School attack.

- *Tennis ball bombs* — Devices made by filling a tennis ball with an explosive mixture that ignites by a simple fuse.

- *Other existing objects* — Items that seem to have an ordinary purpose can be substituted or used as the bomb container. It is up to the imagination and materials available to the builder. Examples: fire extinguishers, propane bottles, trash cans, gasoline cans, and books have all been utilized in the past.

WARNING!
Responders should not move, handle, or disturb an IED when found!

Identification of IEDs

IEDs can be found during an explosives incident response or will be discovered when conducting normal, routine activities. IEDs can be designed to be concealed or made to look like ordinary items **(Figures 7.26a and b)**. The exterior inspection of a suspected device does not ensure its safety. The design and implementation of IEDs are limited only by the imagination of the bomber. Responders should be very cautious of any item(s) that attract attention because they seem out of place, anomalous, out of the ordinary, curious, suspicious, out of context, or unusual, including the following:

WARNING!
The design and implementation of IEDs are limited only by the imagination of the bomber. An IED can look like ANYTHING!

- Containers with unknown liquids or materials; unusual devices or containers with electronic components such as wires, circuit boards, cellular phones, antennas and other items attached or exposed

- Devices containing quantities of fuses, fireworks, match heads, black powder, smokeless powder, incendiary materials or other unusual materials

- Materials attached to, or surrounding, an item such as nails, bolts, drill bits, marbles, etc. that could be used for shrapnel

- Ordnance such as blasting caps, detcord, military explosives, commercial explosives, grenades, etc.

- Any combination of the above described items

Person-Borne Improvised Devices

Modern suicide bombing was introduced in Lebanon in the 1980s. Suicide bombers are individuals who carry IEDs on their person to detonate in a location with the intention of taking the lives of bystanders as well as their own. Suicide bombings have proven to be one of the most effective ways to success-

Figure 7.26 IEDs can be designed to look like anything, as demonstrated by these replicas of actual bombs used in Israel. In the case of the melon, a real fruit was used.

fully penetrate a target and create injuries and havoc. The suicide bomber is also a very difficult threat to deter as the bomber is able to shift targets, avoid security forces, etc. There is no complete defense against suicide bombers.

There are several indicators that an individual may be a suicide bomber, including following:

- Fear and nervousness; suicide bombers are often nervous or anxious about their impending mission; fear or over-enthusiasm may give them away. Other indicators may include:

  — Profuse sweating

  — Keeping one's hands in one's pockets

  — Repeated or nervous handling or patting of clothing

  — Slow-paced walking while constantly shifting eyes to the left and right

  — Major attempts to stay away from security personnel

- Bulky suicide vests or belts; contours of suicide vests or belts may be visible to observant security forces prior to detonation **(Figure 7.27)**.

- Unseasonable attire, for example, wearing a coat (which may conceal a suicide belt) during warm weather.

- Wires or other materials exposed on or around the body

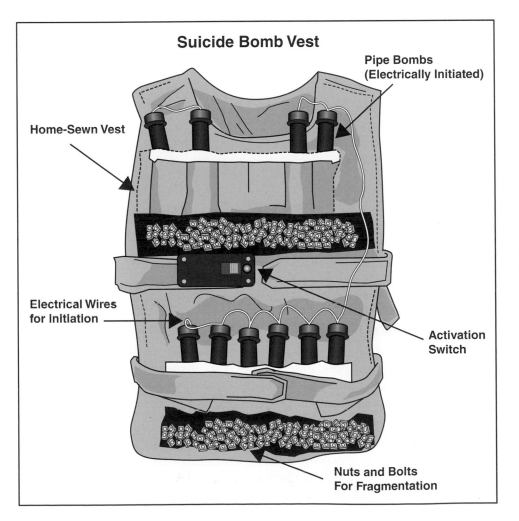

Figure 7.27 Components of a typical suicide bomb vest.

- Carrying or wearing items that could conceal bomb; such items include:
  — Briefcases
  — Luggage
  — Shopping bags
  — Backpacks

  This is especially true at high-risk locations. Individuals carrying briefcases or packages are inherently suspicious to security forces, particularly in locations where personnel have to pass through a security checkpoint.

- Obvious or awkward attempts to blend in with a crowd

- Obvious disguising of appearance

- Dyed or short cut hair

- Actions indicating a strong determination to get to a target

- Repeated visits to a location; bombers may scout locations during the recon/target acquisition phase. It is very suspicious if an individual, vehicle, or truck repeatedly leaves and comes back to a high-risk location.

- Anything that seems out of place, unusual, abnormal, or arouses curiosity

- Any combination of the above

None of the above indicators are *proof* of the existence of a suicide bomber. However, if someone is showing the clues or indicators of being a possible suicide bomber, you need to pay more attention to your situation and circumstances. Be suspicious and take appropriate precautions!

ALERT

The FBI has designated indicators of a possible suicide bomber using the acronym, ALERT:

Alone and nervous

Loose and/or bulky clothing

Exposed wires (possibly through a sleeve)

Rigid mid-section (explosives device or a rifle)

Tightened hands (may hold a detonation device)

Bomb Squad — Crew of emergency responders specially trained and equipped to deal with explosive devices. Also called Hazardous Devices Units or Explosive Ordnance Disposal [EOD] Personnel.

Never approach a suspected or confirmed suicide bomber who is injured or deceased. If there are several strong indicators that there is a suicide bomber, the first priority is to clear and isolate the area and observe the bomber with binoculars or spotting scopes. The first approach must be conducted by trained personnel from an equipped **bomb squad** (sometimes called a *hazardous devices unit* or *explosive ordnance disposal [EOD] personnel*) or by using a bomb disposal robot **(Figure 7.28)**.

Vehicle Bombs (VBIEDs)

Two notable vehicle bomb events that have occurred in the U.S. were the first World Trade Center bombing in 1993 and the Oklahoma City bombing in 1995. In each case, a rental truck was used to deliver the device to the scene, and the

Figure 7.28 Never approach a suspected suicide bomber who is injured, deceased, or has surrendered. Let bomb technicians orchestrate the first approach. *Courtesy of the U.S. Marine Corps, photo by Sgt. Lukas M. Atwell.*

bomb was assembled from commercially available materials. The growing use and frequency of lethal VBIEDs is cause for continuing concern worldwide. Responders should be very cautious of any vehicles that attract attention or curiosity. Never approach a suspicious vehicle once an indicator of possible VBIED has been noticed. Clear and isolate the surrounding area and observe the vehicle with binoculars or a spotting scope. First approach tactics must utilize deployment of specially trained bomb personnel and/or robots.

Indicators of a possible VBIED include:

- Pre-incident intelligence or 911 calls leading to the suspected vehicle
- A vehicle parked suspiciously for a prolonged amount of time in a central location, choke point, or other strategic location
- Vehicle abandoned in a public assembly, tourist area, pedestrian area, retail area, or transit facility
- Vehicle parked between, against, or very close to the columns of a multi-story building
- Vehicle which appears to be weighted down or sits unusually low on its suspension
- Vehicle with stolen, non-matching plates, or no plates at all
- Wires, bundles, electronic components, packages, unusual containers, liquids or materials visible in the vehicle
- Unknown liquids or materials leaking under vehicle
- Anything that seems out of place, unusual, abnormal, or arouses curiosity
- Unusually screwed, riveted, or welded sections located on the vehicle's bodywork
- Unusually large battery or extra battery found under the hood or elsewhere in the vehicle
- Blackened windows or covered windows
- The hollows of front or rear bumpers have been sealed, taped, or otherwise made inaccessible
- Tires that seem solid
- Bright chemical stains or unusual rust patches on a new vehicle

WARNING!
Never approach a suspicious vehicle once an indicator of possible VBIED has been noticed.

- Chemical odor present or unusual chemical leak beneath vehicle
- Wiring protruding from the vehicle, especially from trunk or engine compartment
- Wires or cables running from the engine compartment, through passenger compartment, to the rear of vehicle
- Wires or cables leading to a switch behind sun visor
- The appearance or character of the driver does not match the use or type of vehicle
- The driver seems agitated, lost, and unfamiliar with vehicle controls
- **Any combination of the above described items**

Response to Explosive/IED Events

All operations must be conducted within an incident command system and determined by the risk/benefit analysis. In addition, do the following:

- **ALWAYS** proceed with caution, especially if an explosion has occurred or it is suspected that explosives may be involved in an incident.
- Understand that secondary devices may be involved.
- Request bomb squads, haz mat, and other specialized personnel as needed.
- Establish control zones **(Figure 7.29)**.
- Treat the incident scene as a crime scene until proven otherwise.
- Attempt to detect the presence of hazardous materials.
- **NEVER touch or handle a suspected device, even if someone else already has. Only certified, trained bomb technicians should touch, move, defuse, or otherwise handle explosive devices (Figure 7.30).**
- Do not use two-way radios, cell phones, mobile data terminals (MDT) within a minimum of 300 feet (91 m) of the device or suspected device. The bigger the suspicious device - the bigger the standoff distance.
- Follow designated SOPs.
- Use intrinsically safe communications equipment within the isolation zone.
- Note unusual activities or persons at the scene and report observations to law enforcement.
- Limit the exposure of personnel until the risk of secondary devices is eliminated.

Train with Specialized Bomb Personnel

If there are bomb squads in your area, ask for their assistance with your training and planning. Most bomb technicians will be glad to provide your agency with training on their procedures and equipment, since they will require your support during an incident. One key issue for fire and EMS departments is to become familiar with your local bomb squad operations and entry suits so you will know how to remove one from an injured bomb technician in case of emergency.

UNCLASSIFIED

Bomb Threat Stand-Off Distances

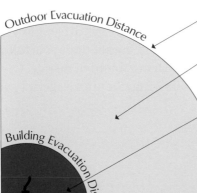

Threat Description		Explosives Capacity[1](TNT Equivalent)	Building Evacuation Distance[2]	Outdoor Evacuation Distance[3]
	Pipe Bomb	5 LBS/ 2.3 KG	70 FT/ 21 M	850 FT/ 259 M
	Briefcase/ Suitcase Bomb	50 LBS/ 23 KG	150 FT/ 46 M	1,850 FT/ 564 M
	Compact Sedan	500 LBS/ 227 KG	320 FT/ 98 M	1,500 FT/ 457 M
	Sedan	1,000 LBS/ 454 KG	400 FT/ 122 M	1,750 FT/ 533 M
	Passenger/ Cargo Van	4,000 LBS/ 1,814 KG	600 FT/ 183 M	2,750 FT/ 838 M
	Small Moving Van/ Delivery Truck	10,000 LBS/ 4,536 KG	860 FT/ 262 M	3,750 FT/ 1,143 M
	Moving Van/ Water Truck	30,000 LBS/ 13,608 KG	1,240 FT/ 378 M	6,500 FT/ 1,981 M
	Semi-Trailer	60,000 LBS/ 27,216 KG	1,500 FT/ 457 M	7,000 FT/ 2,134 M

This table is for general emergency planning only. A given building's vulnerability to explosions depends on its construction and composition. The data in these tables may not accurately reflect these variables. Some risk will remain for any persons closer than the Outdoor Evacuation Distance.

Outdoor Evacuation Distance

Building Evacuation Distance

Preferred area (beyond this line) for evacuation of people in buildings and mandatory for people outdoors.

All personnel in this area should seek shelter immediately inside a building **away from windows and exterior walls**. Avoid having anyone outside—including those evacuating—in this area.[4]

All personnel must evacuate (both inside of buildings and out).

1: Based on maximum volume or weight of explosive (TNT equivalent) that could reasonably fit in a suitcase or vehicle.

2: Governed by the ability of typical US commercial construction to resist severe damage or collapse following a blast. Performances can vary significantly, however, and buildings should be analyzed by qualified parties when possible.

3: Governed by the greater of fragment throw distance or glass breakage/ falling glass hazard distance. Note that pipe and briefcase bombs assume cased charges that throw fragments farther than vehicle bombs.

4: A known terrorist tactic is to attract bystanders to windows, doorways, and the outside with gunfire, small bombs, or other methods and then detonate a larger, more destructive device, significantly increasing human casualties.

NCTC

UNCLASSIFIED

000841ID 10-05

Figure 7.29 Recommended evacuation distances for VIEDs, briefcase bombs, and pipe bombs. *Source: National Counterterrorism Center.*

Figure 7.30 Only certified, trained bomb technicians should touch, move, defuse, or otherwise handle explosive devices. *Courtesy of the U.S. Air Force, photo by Airman Matthew Flynn.*

WARNING!

Secondary devices may be of fragmentation design or contain anti-personnel features such as nails or shrapnel. Resources should be staged in clear areas away from gardens, garbage bins, or other vehicles that could conceal explosive or incendiary devices. Limit the exposure of personnel until the risk of secondary devices is eliminated.

Chemical Attack — Deliberate release of a toxic gas, liquid, or solid that can poison people and the environment.

Nerve Agent — Toxic agent that attacks the nervous system by affecting the transmission of impulses.

When responding to a bombing incident, a primary search will have to be completed. In these situations, local protocol must be followed. Regardless of who completes the primary search and subsequent rescue operations when required, the incident commander should limit the number of personnel in the blast area to the minimum number of personnel required to carry out critical lifesaving operations. In addition to the potential for additional explosions, structural stability must also be evaluated. For more information about emergency response to explosive attacks, see IFSTA's **Emergency Response to Terrorist Attacks** manual.

Chemical Attacks

A **chemical attack** is the deliberate release of a toxic gas, liquid, or solid that can poison people and the environment. Chemical agents (also called chemical warfare agents) or toxic industrial materials (TIMs) may be used in a chemical attack. Chemical agents are chemical substances that are intended for use in warfare or terrorist activities to kill, seriously injure, or seriously incapacitate people through their physiological effects. TIMs are particularly poisonous hazardous materials that are normally used for industrial purposes, but could be used by terrorists to deliberately kill, injure or incapacitate people (see Toxic Industrial Materials section).

As with most hazardous materials, there are different ways of categorizing chemical agents. Much of the information in this and the sections that follow was adapted from material provided by the U.S. Department of Justice, CDC, and other U.S. government resources. The following chemical-agent types are discussed:

- Nerve agents
- Blister agents (vesicants)
- Blood agents (cyanide agents)
- Choking agents (pulmonary or lung-damaging agents)
- Riot control agents (irritants)
- Toxic industrial materials (normal hazardous materials used for terrorist purposes)

Table 7.3 provides the UN/DOT identification number and hazard class for some of the common chemical warfare agents, as well as the *Emergency Response Guide (ERG)* where responders can obtain additional information to manage the initial response phases of the incident.

NOTE: Letters in parentheses next to the name represent military designations, not chemical formulas.

Nerve Agents

Nerve agents attack the nervous system by affecting the transmission of impulses and are the most toxic of the known chemical warfare agents. Exposure to even minute quantities can kill very quickly. Nerve agents are stable, easily dispersed, highly toxic, and have rapid effects when absorbed through the skin or respiratory system. Some of the raw materials used to make them may be difficult to acquire because of regulations restricting their sale and use.

Table 7.3
UN/DOT ID Number, Hazard Class, and *ERG* Guide Number
for Selected Chemical Agents

Agent	UN/DOT ID #	UN/DOT Class	*ERG* Guide U.S.	Military Symbol
Nerve Agents				
Tabun (GA)	2810	6.1	153	GA
Sarin (GB)	2810	6.1	153	GB
Soman (GD)	2810	6.1	153	GD
V Agent (VX)	2810	6.1	153	VX
Blister Agents/Vesicants				
Mustard (H)	2810	6.1	153	H
Distilled mustard (HD)	2810	6.1	153	HD
Nitrogen mustard (HN)	2810	6.1	153	
Lewisite (L)	2810	6.1	153	L
Phosgene Oxime (CX)	2811	6.1	154	
Blood Agents				
Hydrogen Cyanide (AC)	1051	6.1	117	
Cyanogen Chloride (CK)	1589	2.3	125	
Choking Agents				
Chlorine (CL)	1017	2.3	124	
Phosgene (CG)	1076	2.3	125	
Riot Control Agents / Irritants				
Tear Gas (CS)	1693	6.1	159	
Tear Gas (CR)	1693	6.1	159	
Mace (CN)	1697	6.1	153	
Pepper Spray (OC)		2.2 (6.1)*	159	
Adamsite (DM)	1698	6.1	154	DM

Hazard class can be 2.2 or 6.1, depending on how the pepper spray is packaged.

Although people will sometimes use the term *nerve gas*, nerve agents are liquids at ambient temperatures and dispersed as an aerosolized liquid (vapor). (The term *nerve gas* is a misnomer.) Colors and odors can vary with impurities. However, nerve agents are generally clear and colorless.

First responders should be familiar with the following nerve agents (military designations are provided in parentheses):

Volatility — A substance's ability to become a vapor at a relatively low temperature; volatile chemical agents have low boiling points at ordinary pressures and/or high vapor pressures at ordinary temperatures.

- *Tabun (GA)* — Usually low-**volatility** persistent chemical agent that is absorbed through skin contact or inhaled as a vapor
- *Sarin (GB)* — Usually volatile, nonpersistent chemical agent mainly inhaled
- *Soman (GD)* — Usually moderately volatile chemical agent that can be inhaled or absorbed through skin contact
- *Cyclohexyl sarin (GF)* — Low-volatility persistent chemical agent that is absorbed through skin contact and inhaled as a vapor
- *V-agent (VX)* — Low-volatility persistent chemical agent that can remain on material, equipment, and terrain for long periods; main route of entry is through the skin but also through inhalation of the substance as a vapor. First responders may also see reference to other *V*-agents including VE, VG, and VS, but the most common is VX. VX is primarily a contact exposure hazard.

Nerve agents in their pure states are colorless liquids. Their volatility varies widely. However, the *G*-agents tend to be nonpersistent (unless thickened with some other agent to increase their persistency), whereas the *V*-agents are persistent. For example, the consistency of VX is similar to motor oil. Its primary route of entry is through direct contact with the skin. GB is at the opposite extreme: an easily volatile liquid that is primarily an inhalation hazard. The volatilities of GD, GA, and GF are between those of GB and VX. The vapors are heavier than air.

Antidote — A substance that will counteract the effects of a poison or toxin.

Considering the very low vapor pressures, nerve agent vapors will not travel far under normal conditions. Therefore, the size of the endangered area may be relatively small. However, the vapor hazard can be significantly increased if the liquid is exposed to high temperatures, spread over a large area, or aerosolized. **Table 7.4** provides information about nerve agents including descriptions and symptoms of exposure.

Autoinjector — A spring-loaded syringe filled with a single dose of a life-saving drug.

Nerve Agent Antidotes

Speed is the most important factor in medical management of individuals who have been exposed to nerve agents because of their extremely rapid effects. Effective treatment is best achieved by immediate use of **autoinjectors** containing **antidotes.**

Autoinjectors can be self-injected or administered directly into a large muscle such as the thigh or buttocks of the victim **(Figure 7.31)**. Be aware that the needles in autoinjectors are not retractable.

NOTE: The provider must ensure that all antidote kits are maintained within the designated shelf life or expiration date.

While there are many types of autoinjectors available, a common nerve agent autoinjector kit, the Mark-1™ kit (recently replaced by the DuoDote™), contains 2mg of atropine in one intramuscular autoinjector and 600mg of pralidoxime chloride (2 PAM) in another. (The DuoDote™ delivers 2.1 mg atropine followed by 600 mg pralidoxime chloride in a single injector.) These are adult dosages only. Separate injectors are used for children. These may be followed with additional atropine and an anti-convulsant drug such as diazepam (better known as valium) or midazolam (also known as versed).

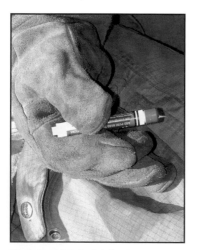

Figure 7.31 Autoinjectors are self-injected or administered directly into the thigh or buttocks.

Table 7.4
Nerve Agent Characteristics

Nerve Agent (Symbol)	Descriptions	Symptoms (All Listed Agents)
Tabun (GA)	• Clear, colorless, and tasteless liquid • May have a slight fruit odor, but this feature cannot be relied upon to provide sufficient warning against toxic exposure • ***Probable Dispersion Method:*** Aerosolized liquid	***Low or moderate dose by inhalation, ingestion (swallowing), or skin absorption:*** Persons may experience some or all of the following symptoms within seconds to hours of exposure:
Sarin (GB)	• Clear, colorless, tasteless, and odorless liquid in pure form • ***Probable Dispersion Method:*** Aerosolized liquid	• Runny nose • Diarrhea • Watery eyes • Increased urination • Small, pinpoint pupils • Confusion • Eye pain • Drowsiness • Blurred vision • Weakness
Soman (GD)	• Pure liquid is clear, colorless, and tasteless; discolors with aging to dark brown • May have a slight fruity or camphor odor, but this feature cannot be relied upon to provide sufficient warning against toxic exposure • ***Probable Dispersion Method:*** Aerosolized liquid	• Drooling and excessive sweating • Headache • Cough • Nausea, vomiting, and/or abdominal pain • Chest tightness • Slow or fast heart rate • Rapid breathing • Abnormally low or high blood pressure
Cyclohexyl sarin (GF)	• Clear, colorless, tasteless, and odorless liquid in pure form • Only slightly soluble in water • ***Probable Dispersion Method:*** Aerosolized liquid	***Skin contact:*** Even a tiny drop of nerve agent on the skin can cause sweating and muscle twitching where the agent touched the skin ***Large dose by any route:*** These additional health effects may result: • Loss of consciousness • Convulsions
V-Agent (VX)	• Clear, amber-colored odorless, oily liquid • Miscible with water and dissolves in all solvents • Least volatile nerve agent • Very slow to evaporate (about as slowly as motor oil) • Primarily a liquid exposure hazard, but if heated to very high temperatures, it can turn into small amounts of vapor (gas) • ***Probable Dispersion Method:*** Aerosolized liquid	• Paralysis • Respiratory failure possibly leading to death ***Recovery Expectations:*** • Mild or moderately exposed people usually recover completely • Severely exposed people are not likely to survive • Unlike some organophosphate pesticides, nerve agents have *not* been associated with neurological problems lasting more than 1 to 2 weeks after the exposure

Source: Information on symptoms provided by the Centers for Disease Control and Prevention (CDC).

Blister Agents

Blister agents (vesicants) burn and blister the skin or any other part of the body they contact. They act on the eyes, mucous membranes, lungs, skin and blood-forming organs. These agents damage the respiratory tract when inhaled and can cause vomiting and diarrhea when ingested. Blister agents are likely to produce more casualties rather than fatalities, although exposure to such agents can be fatal.

Blister agents are usually persistent and may be employed in the form of colorless gases and liquids. However, the oily liquids range from colorless to pale yellow to dark brown, depending on purity. It can take several days or

Blister Agent — Chemical warfare agent that burns and blisters the skin or any other part of the body it contacts. Also called Vesicant.

weeks for them to evaporate. The oily consistency also means that these agents will be more difficult to remove during decontamination than less viscous products. **Table 7.5** provides information about blister agents.

Blister agents can be categorized into the following groups:

- *Mustard agents*
    - Sulfur mustards (H, HD [also called distilled mustard], and HT)
    - Nitrogen mustards (HN, HN-1, HN-2, and HN-3)
- *Arsenical vesicants*
    - Lewisite (L, L-1, L-2, and L-3)
    - Mustard/lewisite mixture (HL) (a mixture of lewisite [L] and distilled mustard [HD])
    - Phenyldichloroarsine (PD)
- *Halogenated oximes* — Phosgene oxime (CX)

Blood Agents

Chemical Asphyxiant — Substance that reacts to keep the body from being able to use oxygen. Also called Blood Poison, Blood Agent, or Cyanogen Agent.

Blood agents are **chemical asphyxiants** that interfere with the body's ability to use oxygen either by preventing red blood cells from carrying oxygen to other cells in the body or by inhibiting the ability of cells to use oxygen for producing the energy required for metabolism. The terms *blood agents* are sometimes used synonymously with *cyanogen agents*, but not all blood agents are cyanogens (for example, arsine is not). Neither are all cyanogens necessarily blood agents. Blood agents are sometimes categorized as TIMs because they also have industrial applications.

First responders should be familiar with the following blood agents:

- *Arsine (SA)* — Arsine gas is formed when arsenic comes in contact with an acid. It is a colorless, nonirritating toxic gas that has a mild garlic odor that can only be detected at levels higher than those necessary to cause poisoning **(Table 7.6, p. 346)**. Details:
    - Arsine gas also has chronic health effects associated with exposure, namely kidney damage and neuropsychological problems such as memory loss and irritability.
    - Arsine gas is considered a nonpersistent hazard.
- *Hydrogen cyanide (AC)* — Hydrogen cyanide is a colorless, highly volatile liquid that is extremely flammable, highly soluble, and stable in water; gas/air mixtures may be explosive **(Table 7.7, p. 347)**. The vapor is less dense than air and has a faint odor, somewhat like bitter almonds; although about 25 percent of the population is unable to smell it. Details:
    - Because of its physical properties, the agent will not remain long in its liquid state, so decon may simply consist of allowing the material to evaporate.
    - The agent represents a nonpersistent hazard.
- *Cyanogen chloride (CK)* — Cyanogen chloride is a colorless, highly volatile liquid that dissolves readily in organic solvents but is only slightly soluble in water (also see Table 7.7). Its vapors are heavier than air. Cyanogen chloride has a pungent, biting odor. Normally, it is a nonpersistent hazard. The effects of exposure to cyanogen chloride are similar to hydrogen cyanide but with additional irritation to the eyes and mucous membranes.

Table 7.5
Common Blister Agent Characteristics

Blister Agent (Symbol)	Description	Symptoms
Sulfur Mustard (H/HD)	• Can be clear to yellow or brown when in liquid or solid form • Sometimes smells like garlic, onions, or mustard; sometimes has no odor • Can be a vapor, an oily-textured liquid, or a solid • Vapors are heavier than air • ***Probable Dispersion Method:*** Aerosolized liquid	Symptoms include: • ***Skin:*** Redness and itching of the skin may occur 2 to 48 hours after exposure and change eventually to yellow blistering of the skin. • ***Eyes:*** Irritation, pain, swelling, and tearing may occur within 3 to 12 hours of a mild to moderate exposure. A severe exposure may cause symptoms within 1 to 2 hours and may include the symptoms of a mild or moderate exposure plus light sensitivity, severe pain, or blindness (lasting up to 10 days). • ***Respiratory tract:*** Runny nose, sneezing, hoarseness, bloody nose, sinus pain, shortness of breath, and cough within 12 to 24 hours of a mild exposure and within 2 to 4 hours of a severe exposure. • ***Digestive tract:*** Abdominal pain, diarrhea, fever, nausea, and vomiting. Other factors include: • Typically, signs and symptoms do not occur immediately. • Depending on the severity of the exposure, symptoms may not occur for 2 to 24 hours. • Some people are more sensitive than others. • Exposure is usually not fatal.
Nitrogen Mustard (HN)	• Comes in different forms that can smell fishy, musty, soapy, or fruity • Can be in the form of an oily-textured liquid, a vapor (the gaseous form of a liquid), or a solid • Is liquid at normal room temperature (70°F or 21°C) • Can be clear, pale amber, or yellow colored when in liquid or solid form • Vapors are heavier than air • ***Probable Dispersion Method:*** Aerosolized liquid	Symptoms include: • ***Skin:*** Redness usually develops within a few hours after exposure followed by blistering within 6 to 12 hours. • ***Eyes:*** Irritation, pain, swelling, and tearing may occur. High concentrations can cause burns and blindness. • ***Respiratory tract:*** Nose and sinus pain, cough, sore throat, and shortness of breath may occur within hours. Fluid in the lungs is uncommon. • ***Digestive tract:*** Abdominal pain, diarrhea, nausea, and vomiting. • ***Brain:*** Tremors, incoordination, and seizures are possible following a large exposure. Other factors include: • Typically, signs and symptoms do not occur immediately. • Depending on the severity of the exposure, symptoms may not occur for several hours.

Lewisite (L)	• Colorless liquid in its pure form; can appear amber to black in its impure form
	• Has an odor like geraniums
	• Vapors are heavier than air
	• ***Probable Dispersion Method:*** Aerosolized liquid
		Signs and symptoms occurring immediately following exposure include:
		• ***Skin:*** Pain and irritation within seconds to minutes; redness within 15 to 30 minutes followed by blister formation within several hours.
		  — Blister begins small in the middle of red areas and then expands to cover the entire reddened area of skin.
		  — Lesions (sores) heal much faster than lesions caused by other blistering agents (sulfur mustard and nitrogen mustards).
		  — Discoloring of the skin that occurs later is much less noticeable.
		• ***Eyes:*** Irritation, pain, swelling, and tearing may occur on contact.
		• ***Respiratory tract:*** Runny nose, sneezing, hoarseness, bloody nose, sinus pain, shortness of breath, and cough.
		• ***Digestive tract:*** Diarrhea, nausea, and vomiting.
		• ***Cardiovascular:*** *Lewisite shock* or low blood pressure.

Table 7.5 (concluded)

Blister Agent (Symbol)	Description	Symptoms

Phosgene Oxime (CX)	• Colorless in its solid form and yellowish-brown when liquid
	• Has a disagreeable, irritating odor
	• Vapors are heavier than air
	• *Probable Dispersion Method:* Aerosolized liquid
		Signs and symptoms occur immediately following exposure:
		• *Skin:* Pain occurring within a few seconds, and blanching (whitening) of the skin surrounded by red rings occurring on the exposed areas within 30 seconds.
		  — Within about 15 minutes, the skin develops hives.
		  — After 24 hours, the whitened areas of skin become brown and die, and a scab is then formed.
		  — Itching and pain may continue throughout the healing process.
		• *Eyes:* Severe pain and irritation, tearing, and possibly temporary blindness.
		• *Respiratory tract:* Immediate irritation to the upper respiratory tract, causing runny nose, hoarseness, and sinus pain.
		• Absorption through the skin or inhalation may result in fluid in the lungs (pulmonary edema) with shortness of breath and cough.

Source: Information on symptoms provided by the Centers for Disease Control and Prevention (CDC).

**Table 7.6
Arsine (SA)**

Description	Symptoms

• Colorless, nonirritating toxic gas with a mild garlic odor that is detected only at levels higher than those necessary to cause poisoning
• Is formed when arsenic comes in contact with an acid
• **Probable Dispersion Method:** Vapor release
	Low or moderate dose by inhalation: Persons may experience some or all of the following symptoms within 2 to 24 hours of exposure:
	
	• Weakness • Fatigue
	• Headache • Drowsiness
	• Confusion • Shortness of breath
	• Rapid breathing • Nausea, vomiting, and/or abdominal pain
	• red or dark urine • Yellow skin and eyes (jaundice)
	• muscle cramps
	
	Large dose by any route: These additional health effects may result:
	• Loss of consciousness
	• Convulsions
	• Paralysis
	• Respiratory failure, possibly leading to death
	Other factors:
	• Showing these signs and symptoms does not necessarily mean that a person has been exposed.
	• If people survive the initial exposure, chronic effects may include:
	  — Kidney damage
	  — Numbness and pain in the extremities
	  — Neuropsychological symptoms such as memory loss, confusion, and irritability

Table 7.7
Blood Agent Characteristics for AC and CK

Blood Agent (Symbol)	Description	Symptoms
Hydrogen cyanide (AC)	• Colorless gas or liquid • Characteristic *bitter almond* odor • Slightly lighter than air • Miscible • Extremely flammable • Explosive gas/air mixtures • Reacts violently with oxidants and hydrogen chloride in alcoholic mixtures, causing fire and explosion hazard • ***Probable Dispersion Method:*** Aerosolized liquid	May be absorbed through skin and eyes. Symptoms include: • ***Inhalation:*** Headache, dizziness, confusion, nausea, shortness of breath, convulsions, vomiting, weakness, anxiety, irregular heartbeat, tightness in the chest, and unconsciousness. Effects may be delayed. • ***Skin:*** May be absorbed. See *Inhalation* for other symptoms. • ***Eyes:*** Redness; vapor is absorbed. See *Inhalation* for other symptoms. • ***Ingestion:*** Burning sensation. See *Inhalation* for other symptoms.
Cyanogen chloride (CK)	• Colorless gas • Pungent odor • Heavier than air • ***Probable Dispersion Method:*** Vapor release	Symptoms include: • ***Inhalation:*** Runny nose, sore throat, drowsiness, confusion, nausea, vomiting, cough, unconsciousness, edema with symptoms which may be delayed. • ***Skin:*** Readily absorbed through intact skin, causing systemic effects without irritant effects on the skin; frostbite may occur on contact with liquid; liquid may be absorbed; redness and pain. • ***Eyes:*** Frostbite on contact with liquid; redness, pain, and excess tears.

Source: *Information on symptoms provided by the Centers for Disease Control and Prevention (CDC).*

Choking Agents

Choking agents are chemicals that attack the lungs causing tissue damage. For this reason they are sometimes called *pulmonary* or *lung-damaging agents*. Like blood agents, these chemicals also have industrial applications, and first responders may encounter them during normal haz mat incidents (as opposed to terrorist attacks). This section discusses phosgene (CG) and chlorine (CL) because of their easy availability, but other chemicals such as diphosgene (DP), chloropicin (PS), ammonia, hydrogen chloride, phosphine, and elemental phosphorus may also be classified as choking agents.

Choking Agent — Chemical warfare agent that attacks the lungs causing tissue damage.

Chlorine

In 2007, insurgents in Iraq began to manufacture *chlorine bombs* by attaching explosives to chlorine tanks, including cargo tank trucks. While the first attempts to use chlorine in these attacks resulted in most of the gas being burned off, it is feared that future attacks may be far more sophisticated and potentially deadly **(Figure 7.32, p. 348)**. Chlorine was also used as a choking agent during World War I, but most people are more familiar with it because of its industrial uses.

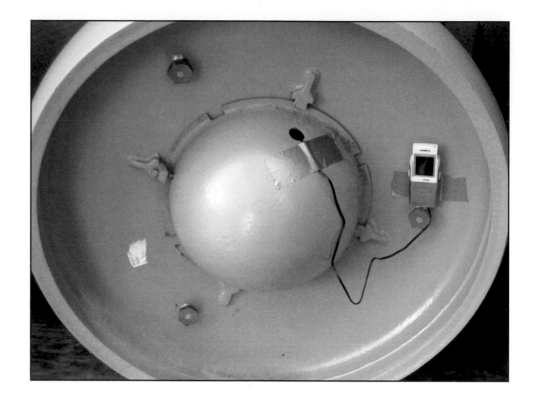

Figure 7.32 Attacks in Iraq have combined explosives and chlorine.

Chlorine gas is usually pressurized and cooled to a liquid state for storage and transportation. When liquid chlorine is released, it quickly turns into a gas that is heavier than air. Chlorine gas can be recognized by its pungent, irritating odor, which is like the odor of bleach. Chlorine gas is usually yellow-green in color. Chlorine itself is not flammable, but it can react explosively or form explosive compounds with other chemicals such as turpentine and ammonia. Because of its physical and chemical properties, the agent (like phosgene) does not remain in its liquid form very long. Thus decon is usually not required. **Table 7.8** provides additional information about chlorine including symptoms of exposure.

Phosgene

Phosgene is a colorless, nonflammable gas that has the odor of freshly cut hay. Its odor threshold is well above its permissible exposure limit, so it is already at a harmful concentration by the time someone smells it **(Table 7.9, p. 350)**. It is used in the manufacture of dyestuffs, pesticides, plastics, pharmaceuticals, and other chemicals products. Phosgene was used as a chemical weapon for the first time in World War I, and it accounted for the majority of all chemical fatalities in that war.

Phosgene is a gas at room temperature but is sometimes stored as a liquid under pressure or refrigeration. Its boiling point is 47°F (8.2°C), making it an extremely volatile and nonpersistent agent. Its vapor density is much heavier than air; therefore, it may remain for long periods of time in trenches and other low-lying areas. Because of the agent's physical and chemical properties, it will not remain in its liquid form very long. Thus decon is usually not required except when the agent is used in very cold climates or when it has soaked clothing or skin.

Table 7.8
Chlorine (CL)

Description	Symptoms

• Gas can be recognized by its pungent, irritating odor that smells like bleach.
 	
• Gas is yellow-green in color.
 	
• Is not flammable but can react explosively or form explosive compounds with other chemicals such as turpentine and ammonia.
 	
• ***Probable Dispersion Method:*** Vapor release
	Dangerous-level concentrations: During or immediately after exposure, the following signs and symptoms may develop:
	
	• Coughing
	• Chest tightness
	• Burning sensation in the nose, throat, and eyes
	• Watery eyes
	• Blurred vision
	• Nausea and vomiting
	• Burning pain, redness, and blisters on the skin if exposed to gas
	• Skin injury similar to frostbite if exposed to liquid
	• Difficulty breathing or shortness of breath
	  — May appear immediately if high concentrations of gas are inhaled
	  — Symptom may be delayed if low concentrations of gas are inhaled
	• Fluid in the lungs (pulmonary edema) within 2 to 4 hours
	
	Showing these signs or symptoms does not necessarily mean that a person has been exposed.
	
	Recovery Expectations:
	
	• Long-term complications from exposure are not found in people who survive a sudden exposure unless they suffer complications such as pneumonia during therapy.
	• Chronic bronchitis may develop in people who develop pneumonia during therapy.

Source: *Information on symptoms provided by the Centers for Disease Control and Prevention (CDC).*

Riot Control Agents

Riot control agents (sometimes called *tear gas* or *irritating agents*) are chemical compounds that temporarily make people unable to function by causing immediate irritation to the eyes, mouth, throat, lungs, and skin. Several different compounds are considered riot control agents.

All are solids and require dispersion in the form of aerosolized particles, usually released by pyrotechnics (such as with an exploding tear gas canister) or a propelled spray with the particles suspended in a liquid. Some are sold in small containers as personal defense devices containing either a single agent or a mixture. Some devices also contain a dye to visually mark a sprayed assailant. When dispersed, riot control agents are usually heavier than air.

Table 7.10, p. 351 provides characteristics of common riot control agents. Because the symptoms of exposure are very similar for all the agents, they are listed only once.

In addition to tear gas, mace, pepper spray, and other irritants, the following agents are sometimes categorized as riot control agents:

- ***Incapacitant*** — Chemical agent that produces a temporary disabling condition that persists for hours to days after exposure has occurred (unlike that produced by most riot control agents). *Examples:*

Riot Control Agent — Chemical compound that temporarily makes people unable to function by causing immediate irritation to the eyes, mouth, throat, lungs, and skin. Also called Tear Gas or Irritating Agent.

Table 7.9
Phosgene

Description	Symptoms
• Is a poisonous gas at room temperature. • May appear colorless or as a white to pale yellow cloud. • At low concentrations, has a pleasant odor of newly cut hay or green corn. • At high concentrations, odor may be strong and unpleasant. • Is nonflammable (not easily ignited and burned) but can cause flammable substances around it to burn. • Gas is heavier than air. • **Probable Dispersion Method:** Vapor release	***Exposure to dangerous-level concentrations:*** During or immediately after exposure, the following signs and symptoms may develop: • Coughing • Burning sensation in the throat and eyes • Watery eyes • Blurred vision • Difficulty breathing or shortness of breath • Nausea and vomiting • Skin contact can result in lesions similar to those from frostbite or burns ***Exposure to high concentrations:*** Person may develop fluid in the lungs (pulmonary edema) within 2 to 6 hours. ***Delayed effects:*** May not be apparent for up to 48 hours, even if a person feels better or appears well following removal from exposure. Monitor people who have been exposed for 48 hours. Delayed effects that can appear up to 48 hours include the following: • Difficulty breathing • Coughing up white to pink-tinged fluid (sign of pulmonary edema) • Low blood pressure • Heart failure Showing these signs or symptoms does not necessarily mean that a person has been exposed. ***Recovery Expectations:*** Most people who recover after an exposure make a complete recovery; however, chronic bronchitis and emphysema have been reported as an exposure result.

Source: Information on symptoms provided by the Centers for Disease Control and Prevention (CDC).

— Central nervous system (CNS) depressants (anticholinergics)

— CNS stimulants (lysergic acid diethylamide or LSD)

• *Vomiting agent* — Agent that causes violent, uncontrollable sneezing, cough, nausea, vomiting and a general feeling of bodily discomfort. It is dispersed as an aerosol and produces its effects by inhalation or direct action on the eyes. Principal agents:

— Diphenylchlorarsine (DA)

— Diphenylaminearsine chloride (Adamsite or DM)

— Diphenylcyanarsine (DC)

Toxic Industrial Materials (TIMs)

A TIM (also known as a toxic industrial chemical [TIC]) is an industrial chemical that is toxic at a certain concentration and is produced in quantities exceeding 30 tons (30.5 tonnes) per year at one production facility. TIMs are not as lethal as the highly toxic nerve agents, but because they are produced in very large quantities (multi-tons) and are readily available, they pose a far greater threat than chemical warfare agents. For example, sulfuric acid is not as lethal as a nerve agent, but it is easier to disseminate large quantities of sulfuric acid because large amounts of it are manufactured and transported every day.

Table 7.10
Riot Control Agent Characteristics

Riot Control Agent (Symbol)	Descriptions	Symptoms (All Listed Agents)
Chlorobenzylidene malononitrile (CS)	• White crystalline solid • Pepper-like smell	***Immediately after exposure:*** People exposed may experience some or all of the following symptoms: • ***Eyes:*** Excessive tearing, burning, blurred vision, and redness • ***Nose:*** Runny nose, burning, and swelling • ***Mouth:*** Burning, irritation, difficulty swallowing, and drooling • ***Lungs:*** Chest tightness, coughing, choking sensation, noisy breathing (wheezing), and shortness of breath • ***Skin:*** Burns and rash • ***Other:*** Nausea and vomiting
Chloroacetophenone (CN, mace)	• Clear yellowish brown solid • Poorly soluble in water, but dissolves in organic solvents • White smoke smells like apple blossoms	Long-lasting exposure or exposure to a large dose, especially in a closed setting, may cause severe effects such as the following: • Blindness • Glaucoma (serious eye condition that can lead to blindness)
Oleoresin Capsicum (OC, pepper spray)	• Oily liquid, typically sold as a spray mist • ***Probable Dispersion Method:*** Aerosol	• Immediate death due to severe chemical burns to the throat and lungs • Respiratory failure possibly resulting in death
Dibenzoxazepine (CR)	• Pale yellow crystalline solid • Pepper-like odor • ***Probale Dispersion Method:*** Propelled	Prolonged exposure, especially in an enclosed area, may lead to long-term effects such as the following: • Eye problems including scarring, glaucoma, and cataracts • May possibly cause breathing problems such as asthma
Chloropicrin (PS)	• Oily, colorless liquid • Intense odor • Violent decomposition when exposed to heat	***Recovery Expectations:*** If symptoms go away soon after a person is removed from exposure, long-term health effects are unlikely to occur.

Source: Information on symptoms provided by the Centers for Disease Control and Prevention (CDC).

What This Means To You

You are far more likely to have to deal with a toxic industrial chemical used as a weapon than chemical warfare agents in part because toxic industrial chemicals are much cheaper and easier to obtain. Chemical warfare agents like sarin are notoriously difficult (and expensive) to produce. For example, it is estimated that the cult, Aum Shinrikyo, spent over $30 million to produce the sarin used in their attacks in Japan. In contrast, chlorine cylinders could potentially be stolen from your local public swimming pool. Additionally, some TIMs are nearly as dangerous and deadly as warfare agents. Phosgene, for example, has industrial applications, but is also listed as a chemical warfare agent. Some commercially available pesticides disrupt nerve impulses in the same way nerve agents do, and they are very similar chemically.

TIMs may be well suited for use in terrorist attacks due to a variety of factors. Some may have poor warning properties. Others may be lethal in extremely low doses. Some may have ideal dispersal properties. You should pay attention to the TIMs in your community and determine what chemicals might be used for terrorist purposes in your jurisdiction. Then, be prepared to respond to incidents involving them.

TIMs are divided into three hazard categories. **Table 7.11** lists them with respect to their hazard index ranking (high, medium, or low hazard) as provided by OSHA. Definitions are as follows:

- *High hazard* — Indicates a widely produced, stored, or transported TIM that has high toxicity and is easily vaporized

- *Medium hazard* — Indicates a TIM that may rank high in some categories but is lower in others such as number of producers, physical state, or toxicity

- *Low hazard* — Indicates that this TIM is not likely to be a hazard unless specific operational factors indicate otherwise

Emergency responders should attempt to identify the material involved just as they would for any other hazardous materials incident. All predetermined procedures and the guidelines provided in the *ERG* and other sources should be followed when responding to emergencies involving TIMs.

Operations at Chemical Attack Incidents

The primary operational objective at a chemical attack is to do the greatest good for the greater number. Responders must be familiar with SOPs/SOGs for handling chemical terrorist attacks and hazardous materials incidents. Elements of a chemical attack that may differ from other incidents include the severity of hazards present (ie. deadly nerve agents), the possibility of secondary devices, mass casualties, the necessity of using appropriate PPE (see Chapter 8), the need for rapid decon (see Chapter 9, Decontamination), and administration of antidotes.

Biological Attacks

The Centers for Disease Control and Prevention (CDC) defines biological terrorism as an intentional release of viruses, bacteria, or their toxins for the purpose of harming or killing citizens. In addition to aerosolization, food, water, or insects must be considered as potential vehicles of transmission for biological weapons. First responders must be prepared to address various **biological agents**, including pathogens that are rarely seen in North America. They should also be aware of the possible indicators of biological attacks, the different types of biological agents, and the signs and symptoms of those biological agents most likely to be used as weapons. Four types of biological agents that first responders need to be concerned about are as follows:

1. *Viral agents* — Viruses are the simplest types of microorganisms that can only replicate themselves in the living cells of their hosts. Viruses do not respond to **antibiotics**, making them an attractive weapon.

2. *Bacterial agents* — Bacteria are microscopic, single-celled organisms. Most bacteria do not cause disease in people, but when they do, two different mechanisms are possible: invading the tissues or producing poisons (toxins).

3. *Rickettsias* — Rickettsias are specialized bacteria that live and multiply in the gastrointestinal tract of arthropod carriers (such as ticks and fleas). They are smaller than most bacteria, but larger than viruses. They have properties that are similar to both. Like bacteria, they are single-celled organisms with their own metabolisms, and they are susceptible to broad-

Biological Agent — Viruses, bacteria, or their toxins used for the purpose of harming or killing people, animals, or crops. Also called Biological Weapon.

Antibiotic — Type of antimicrobial agent made from a mold or a bacterium that kills, or slows the growth of other microbes, specifically bacteria. Examples include penicillin and streptomycin. Antibiotics are ineffective against viruses.

Table 7.11
Toxic Industrial Materials Listed by Hazard Index Ranking

High Hazard	Medium Hazard	Low Hazard
Ammonia	Acetonic cyanohydrin	Allyl isothiocyanate
Arsine	Acrolein	Arsenic trichloride
Boron trichloride	Acrylonitrile	Bromine
Boron trifluoride	Allyl alcohol	Bromine chloride
Carbon disulfide	Allylamine	Bromine pentafluoride
Chlorine	Allyl chlorocarbonate	Bromine trifluoride
Diborane	Boron tribromide	Carbonyl fluoride
Ethylene oxide	Carbon monoxide	Chlorine pentafluoride
Fluorine	Carbonyl sulfide	Chlorine trifluoride
Formaldehyde	Chloroacetone	Chloroacetaldehyde
Hydrogen bromide	Chloroacelonitrile	Chloroacetyl chloride
Hydrogen chloride	Chlorosulfonic acid	Crotonaldehyde
Hydrogen cyanide	Diketene	Cyanogen chloride
Hydrogen fluoride	1,2-Dimethylhydrazine	Dimethyl sulfate
Hydrogen sulfide	Ethylene dibromide	Diphenylmethane-4,4'-diisocyanate
Nitric acid, fuming	Hydrogen selenide	Ethyl chloroformate
Phosgene	Methanesulfonyl chloride	Ethyl chlorothioformate
Phosphorus trichloride	Methyl bromide	Ethyl phosphonothioic dichloride
Sulfur dioxide	Methyl chloroformate	Ethyl phosphonic dichloride
Sulfuric acid	Methyl chlorosilane	Ethyleneimine
Tungsten hexafluoride	Methyl hydrazine	Hexachlorocyclopentadiene
	Methyl isocyanate	Hydrogen iodide
	Methyl mercaptan	Iron pentacarbonyl
	Nitrogen dioxide	Isobutyl chloroformate
	Phosphine	Isopropyl chloroformate
	Phosphorus oxychloride	Isopropyl isocyanate
	Phosphorus pentafluoride	n-Butyl chloroformate
	Selenium hexafluoride	n-Butyl isocyanate
	Silicon tetrafluoride	Nitric oxide
	Stibine	n-Propyl chloroformate
	Sulfur trioxide	Parathion
	Sulfuryl chloride	Perchloromethyl mercaptan
	Sulfuryl fluoride	sec-Butyl chloroformate
	Tellurium hexafluoride	tert-Butyl isocyanate
	n-Octyl mercaptan	Tetraethyl lead
	Titanium tetrachloride	Tetraethyl pyrophosphate
	Trichloroacetyl chloride	Tetramethyl lead
	Trifluoroacetyl chloride	Toluene 2,4-diisocyanate
		Toluene 2,6-diisocyanate

Source: "Summary of the Final Report of the International Task Force 25: Hazard from Industrial Chemicals," April 15, 1999.

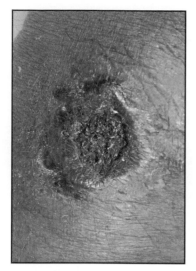
Figure 7.33 Natural anthrax is more of a skin hazard, causing lesions such as this. Weaponized anthrax is a more dangerous inhalation hazard. *Courtesy of the CDC.*

spectrum antibiotics. However, like viruses, they only grow in living cells. Most rickettsias are spread only through the bite of infected arthropods and not through human contact.

4. *Biological toxins* — Biological toxins are poisons produced by living organisms; however, the biological organism itself is usually not harmful to people. Some biological toxins have been manufactured synthetically and/ or genetically altered in laboratories. They are similar to chemical agents in the way they are disseminated (and in their effectiveness) as biological weapons.

As the 2001 anthrax attacks in the U.S. demonstrated, an attack using a biological weapon may not be as immediately obvious as one using a bomb or industrial chemical **(Figure 7.33)**. Generally, biological weapons agents do not cause immediate health effects. Most biological agents take hours, days, or weeks to make someone ill, depending on the incubation period of the agent. Because of this delay, the cause of illness may not be immediately evident, and the source of the attack may be difficult to trace.

In the beginning, patients may be few in number, with the number increasing when the disease continues to transmit from person to person (such as might happen with smallpox). The scope of the problem may not be evident for days or even weeks. However, certain biological toxins (such as saxitoxin, a neurotoxin produced by marine organisms) could potentially act more quickly (in minutes to hours).

Biological Agent Categories

Most biological agents and toxins fall under the United Nations/U.S. Department of Transportation (UN/DOT) Hazard Class 6.2. The CDC divides potential biological agents into three categories: A, B, and C. **Table 7.12** lists Category A agents that have the highest priority because they include organisms that pose a risk to national security in the following ways:

• Can be easily disseminated or transmitted person to person

• Cause high mortality and subsequently have a major public health effect

• Might cause public panic and social disruption

• Require special action for public-health preparedness

The CDC recommends that other less critical agents (Categories B and C) also receive attention for bioterrorism preparedness. These categories include new or emerging pathogens. **Table 7.13, p. 356** gives the CDC's list of critical biological agents for Category B. A subset of Category B agents includes pathogens that are foodborne or waterborne. Category B agents include those that pose a risk in the following ways:

• Are moderately easy to disseminate

• Cause moderate morbidity and low mortality

• Require specific enhancements of CDC's diagnostic capacity and enhanced disease surveillance

Category C agents include emerging pathogens that could be engineered for mass dissemination in the future because they pose risks in the following ways:

Table 7.12
Category A* Biological Agents

Name	Biological Agent	Common Signs and Symptoms**
Smallpox	Virus *(Variola major)*	• Acute rash, fever, fatigue, head and back aches • Rash lesions start red and flat, but fill with pus after first week
Anthrax	Bacteria *(Bacillus anthracis)*	• ***Inhalation anthrax:*** Acute respiratory distress with fever. Initial symptoms may resemble a common cold. • ***Intestinal anthrax:*** Nausea, loss of appetite, vomiting, and fever followed by abdominal pain, vomiting of blood, and sever diarrhea. • ***Cutaneous anthrax:*** Skin infection begins as a raised itchy bump that resembles an insect bite but within 1 to 2 days develops into a vesicle and then a painless ulcer, usually 0.4 to 1 inch (10 mm to 30 mm) in diameter, with a characteristic black necrotic (dying) area in the center.
Plague	Bacteria *(Yersinia pestis)*	Acute respiratory distress, fever, weakness and headache
Botulism	Toxin from the bacteria *Clostridium botulinum*	***Neurological syndromes:*** Double vision, blurred vision, drooping eyelids, slurred speech, difficulty swallowing, dry mouth, and muscle weakness
Tularemia	Bacteria *(Francisella tularensis)*	***Influenzalike illness:*** Sudden fever, chills, headaches, muscle aches, joint pain, dry cough, progressive weakness, and sometimes pneumonia
Hemorrhagic Fever	Virus (for example, Ebola, Marburg, and Lassa)	• Specific signs and symptoms vary by the type of viral hemorrhagic fever (VHF), but initial signs and symptoms often include marked fever, fatigue, dizziness, muscle aches, loss of strength, and exhaustion. • Patients with severe cases of VHF often show signs of bleeding under the skin, in internal organs, or from body orifices such as the mouth, eyes, or ears. • Rash and red eyes may be seen in some patients infected with the Ebola virus.

* The highest-priority biological agents according to the Centers for Disease Control and Prevention (CDC).

** List of symptoms is *not* all-inclusive.

- Availability
- Ease of production and dissemination
- Potential for high morbidity and mortality and major health effect

Category C Agents are as follows:

- Nipah virus
- Hantaviruses
- Tickborne hemorrhagic fever viruses
- Tickborne encephalitis viruses
- Yellow fever virus
- Multidrug-resistant tuberculosis (*Mycobacterium tuberculosis*)

Table 7.13
Category B * Biological Agents

Name	Biological Agent	Common Signs and Symptoms**
Brucellosis	Bacteria *(Brucella melitensis, abortus, suis, and canis)*	Intermittent or irregular fever of variable duration, headache, weakness, profuse sweating, chills, weight loss, and generalized aching
Epsilon Toxin of Clostridium perfringens	Toxin from the bacteria *Clostridium perfringens* (a common cause of foodborne illness)	Intense abdominal pain (cramping and bloating) and diarrhea
Glanders	Bacteria *(Burkholderia mallei)*	• Symptoms depend upon the route of infection with the organism. • Types of infection include localized, pus-forming cutaneous infections, pulmonary infections, bloodstream infections, and chronic suppurative infections of the skin. • Generalized symptoms include fever, muscle aches, chest pain, muscle tightness, and headache. • Additional symptoms include excessive tearing of the eyes, light sensitivity, and diarrheal.

Melioidosis	Bacteria *(Burkholderia pseudomallei)*	• ***Acute, localized infection:*** Generally localized as a nodule resulting from inoculation through a break in the skin
		  — Can produce fever and general muscle aches
		  — May progress rapidly to infect the bloodstream
		• ***Pulmonary infection:*** Can produce a clinical picture of mild bronchitis to severe pneumonia
		  — Onset is typically accompanied by high fever, headache, anorexia, and general muscle soreness.
		  — Chest pain is common, but with a nonproductive or productive cough with normal sputum.
		• ***Acute bloodstream infection:*** Symptoms vary depending on the site of original infection, but generally include respiratory distress, severe headache, fever, diarrhea, development of pus-filled lesions on the skin, muscle tenderness, and disorientation.
		• ***Chronic suppurative infection:*** Infection that involves the organs of the body, typically including the joints, viscera, lymph nodes, skin, brain, liver, lung, bones, and spleen.

Psittacosis	Bacteria *(Chlamydia psittaci)*	• Fever • Chills • Headache • Muscle aches • Dry cough

Table 7.13 (continued)

Name	Biological Agent	Common Signs and Symptoms**
Q Fever	Bacteria (Rickettsia) *(Coxiella burnetii)*	Only about one-half of those infected show signs of clinical illness; most acute cases begin with sudden onset of one or more of the following: • High fevers (up to 104–105°F) [40–40.5°C]; usually lasts for 1 to 2 weeks • Severe headache • General malaise • Myalgia • Confusion • Sore throat • Chills, sweats • Nonproductive cough • Nausea, vomiting • Diarrhea • Abdominal pain and chest pain • Weight loss can occur and persist for some time Thirty to fifty percent of patients with a symptomatic infection will develop pneumonia. Additionally, a majority of patients have abnormal results on liver function tests and some develop hepatitis.
Ricin Toxin	From *Ricinus communis* (castor beans)	• **Inhalation:** Acute onset of fever, chest pain, and cough, progressing to severe respiratory distress • **Ingestion:** Nausea, vomiting, abdominal pain and cramping, diarrhea, gastrointestinal bleeding, low or no urinary output, dilation of the pupils, fever, thirst, sore throat, headache, vascular collapse, and shock
Staphylococcal Enterotoxin B (SEB)	Toxin from the bacteria *Staphylococcus aureus* (common cause of foodborne illness)	• High fever (103–106°F) [39–41°C] and chills • Headache • Myalgia • Nonproductive cough
Typhus Fever	Bacteria (Rickettsia) *(Rickettsia prowazekii)*	In general presents with features similar to a bad cold with fever, chills, headache, and muscle pains as well as body rash
Viral Encephalitis	Alphaviruses such as: • Venezuelan equine encephalitis (VEE) • Eastern equine encephalitis (EEE) • Western equine encephalitis (WEE)	• Fever, chills • Severe headache • Rigors • Photophobia • Muscle pain (especially in the legs and lower back) • Cough, sore throat • Vomiting

Continued

Table 7.13 (concluded)

Name	Biological Agent	Common Signs and Symptoms**
Food-Safety Threats	Bacteria such as: • Salmonella • Escherichia coli 0157:H7 (sometimes called E. coli) • Shigella	• **Salmonellosis:** Diarrhea, fever, and abdominal cramps • **E. coli:** Often causes severe bloody diarrhea and abdominal cramps; sometimes causes nonbloody diarrhea or no symptoms • **Shigellosis:** Diarrhea (often bloody), fever, and stomach cramps
Water-Safety Threats	• **Bacteria:** Vibrio cholerae (causes Asiatic or epidemic cholera) • **Protozoa (single-celled organism):** Cryptosporidium parvum	• **Cholera:** Severe disease is characterized by profuse watery diarrhea, vomiting, and leg cramps; rapid loss of body fluids leads to dehydration and shock. • **Cryptosporidiosis:** If symptoms are present: — Intestinal cryptosporidiosis is characterized by severe watery diarrhea. — Pulmonary and tracheal cryptosporidiosis is associated with coughing and frequently a low-grade fever, often accompanied by severe intestinal distress.

* Less critical biological agents than Category A according to the Centers for Disease Control and Prevention (CDC).

** List of symptoms is *not* all-inclusive.

How Can You Tell the Difference Between Chemical and Biological Incidents?

Typically, a *chemical incident* is characterized by a rapid onset of medical symptoms (minutes to hours), and you may observe features such as colored residue, dead foliage, pungent odor, and dead insect and animal life.

With *biological incidents*, the onset of symptoms usually requires days to weeks. There may not be any characteristic features because biological agents are usually odorless and colorless. The area affected can be larger in biological incidents because of the movement of infected individuals between the time of infection and onset of symptoms. During that time, an infected person could transmit the disease to another person. EMS and public health personnel are more likely to notice unusual patterns of illness during a biological incident.

Disease Transmission

Specific methods of infectious disease transmission include (**Figure 7.34**):

• ***Airborne transmission (inhalation of airborne organisms or toxins)*** — Diseases transmitted in this way can remain suspended in air for a long time and when inhaled may penetrate deep into the respiratory tract. Airborne-transmitted diseases can typically survive outside the body for long periods of time. Examples include influenza, pneumonia, and polio.

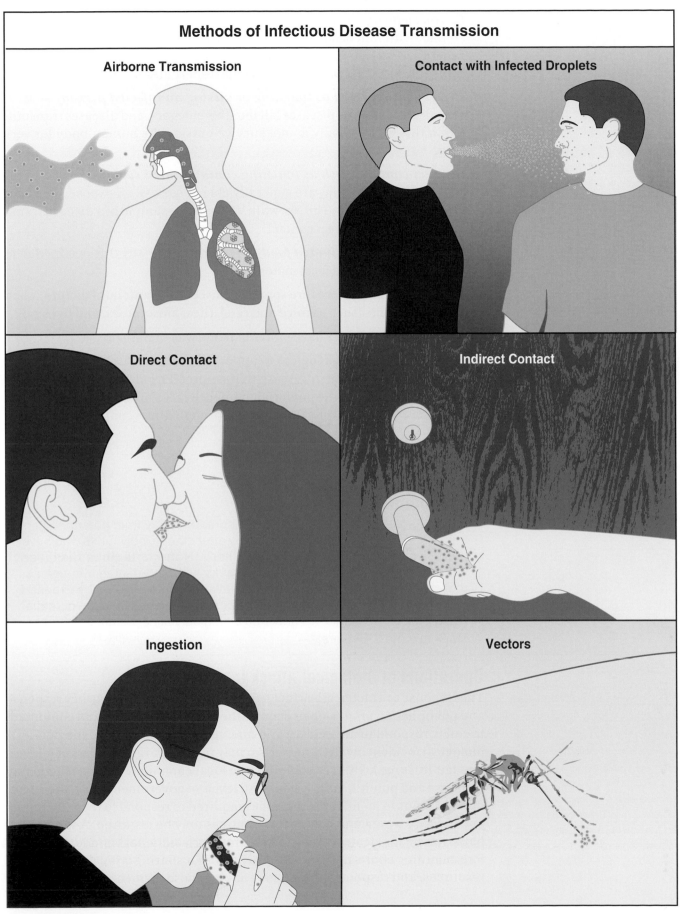

Figure 7.34 Methods of disease transmission.

- *Contact with infected droplets* — Infected droplets transmit disease through contact with mucous membranes of the eyes, nose, and mouth. Droplets generally do not stay airborne for long periods of time. Examples of diseases transmitted in this way include rubella, tuberculosis, and SARS.

- *Direct contact (such as touching or kissing an infected person)* — Most sexually transmitted diseases fall into this category, and diseases transmitted in this way typically do not survive outside the human body for very long.

- *Indirect contact (such as touching contaminated surfaces)* — Diseases transmitted in this way are generally able to survive on exposed surfaces for extended periods. The Norwalk Virus is an example of a disease transmitted through indirect contact.

- *Ingestion of contaminated food or water* — Normally this occurs due to contact with infected fecal material.

- *Vectors* — Some diseases are spread by insects and animals (**vectors**), such as fleas, rodents (such as mice and rats), flies, mosquitoes, and livestock.

  NOTE: Many diseases have more than one way of transmission.

When developing a biological weapon, the method of transmission is an important consideration. A disease that is spread by airborne transmission (such as smallpox) has the potential to infect a large number of people more quickly than one that is only transmitted through direct contact (such as HIV or Ebola).

Vector — An animate intermediary in the indirect transmission of an agent that carries the agent from a reservoir to a susceptible host.

Contagiousness

A contagious disease is one that can spread rapidly from person to person. Non-contagious diseases will only affect those individuals who have direct exposure to the disease agent itself. **Non-contagious diseases will not spread to other people.** Attacks with non-contagious agents, then, will be limited to the number of individuals who are directly exposed to the agent, whereas an attack with a contagious agent has the potential to become an epidemic. Anthrax and biological toxins are not contagious; smallpox and SARS are.

Operations at Biological Attack Incidents

The response to a bioterrorism incident requires that local emergency response and public health professionals acquire the necessary training and equipment to safely respond to an incident and manage the health care of the public. A biological incident involving terrorism may result in large numbers of injuries and fatalities, quickly overwhelming the local capabilities of the emergency response and public health systems. All levels of government must, therefore, develop plans to identify the resources required to respond to these incidents and identify where and how these resources may be acquired. Additionally, bioterrorism incidents will most likely cross jurisdictional boundaries; therefore, planning efforts must include provisions for sharing critical information, resources, and responsibilities for managing these incidents.

Before the recognition of and during a recognized disease outbreak caused by an act of biological terrorism, first responders may have had contact with patients who are infected by the biological agent. Most agents of bioterrorism are *not* transmitted from person to person; however, for agents such as smallpox or pneumonic plague, a first responder is at risk of acquiring infection from a patient.

When the presence or nature of a bioterrorism agent is not known, first responders should adhere to **universal precautions** whenever they have contact with broken or moist skin, blood, or body fluids. These precautions include using disposable gloves, washing the hands immediately after removing gloves, and using disposable PPE and a face shield if any splashing is anticipated **(Figure 7.35)**. Gloves should be changed between patients to prevent the worker from transmitting infection from patient to patient.

Universal Precautions — A set of precautions designed to prevent transmission of biological pathogens (specifically, bloodborne pathogens) when providing first aid or health care.

Figure 7.35 Common infection control procedures are important in protecting responders from biological agents. *Courtesy of the CDC, photo by Kimberly Smith.*

Once a specific agent is identified, additional precautions are applied based on the agent's mode of transmission, whether airborne, droplet, or contact. The local health department should be contacted for additional instructions for vaccinations, prophylactic antibiotic therapy, or other measures that may be appropriate for a given disease.

In the event of an overt attack or incident, isolation and containment of the biological agent is very important to prevent the spread of pathogens or toxins. An overt attack would include incidents involving dispersal of white powders (with a credible threat), discovery of a suspected biologic laboratory, witnessed use of spray devices, etc.

Indoor attacks may be contained by the following measures:

- Turning off ventilation systems
- Closing doors and windows
- Turning off elevators
- Restricting air flow by sealing ducts, windows, and doors using tape, plastic sheets, and expanding foams

Overt, outdoor attacks may be contained using the following actions:

- Covering the device or dispersed agent with tarps or other physical barriers to prevent spreading
- Decontaminating dispersed agent with a light spray of water and bleach
- Securing and placing the suspect item, package, object or substance in a sealed hazmat recovery bin or container to mitigate spread

If possible, individuals who have been exposed to biological agents should not be allowed to leave the scene until a thorough risk assessment has been conducted and appropriate measures taken (potentially in consultation with local health authorities). Decontamination is generally recommended for any credible threat involving aerosols or contact with potentially harmful substances even though victims are not showing signs or symptoms of illness (see Chapter 9, Decontamination). Important operations to ensure containment may include the following:

- Initial containment of persons that may be affected
- Decon of victims if circumstances indicate the need prior to treatment and transport to a medical facility
- Registration (record name and contact information) of all persons potentially exposed at the incident in case follow-up is required

In the event of a covert attack, isolation and containment issues will primarily deal with managing infected victims, and these will likely be managed by public health authorities. Local plans for handling a **pandemic** flu may translate to other contagious disease outbreaks.

Radiological and Nuclear Attacks

Although there is reason for concern about the potential for radiological terrorist attacks, history has seen relatively few actual attempts at radiological terrorism. Threats have been made, and plans to carry out nuclear attacks have been foiled, but we still have yet to experience the disastrous consequences from a radiological terrorism attack.

Response to a radiological incident is more likely to be similar to the response to other emergency incidents. For example, a response to an attack on a shipment of radioactive materials might follow *ERG* guidelines for radiological materials, with additional consideration given to secondary devices and evidence preservation. In the case of an attack with a dirty bomb, it may not be immediately evident that radiological materials are involved. **Emergency response agencies must include radiation monitoring as a normal part of response to any fire and/or explosion incident.** The use of radiological monitoring equipment is the only sure way to confirm the presence of a radiological hazard.

In the event of a nuclear attack, local first responders will probably be overwhelmed by the scale and scope of the disaster facing them. That fact does not mean that they cannot or should not take appropriate action. Outside assistance will undoubtedly be needed to successfully mitigate the incident. Communication, transportation, water supplies, and resources may be limited or nonexistent. The number of casualties and destruction may be overwhelming. When an organized response is possible, the same framework for any

Pandemic — An epidemic occurring over a very wide area (several countries or continents) and usually affecting a large proportion of the population.

emergency response should be applied with special consideration given to nuclear/radiological hazards.

Radiological Devices

There are several different devices designed to expose people to radiation or to disperse radiological material. They are sometimes referred to as *dirty bombs* because the contamination they spread could ruin property, crops, and livestock and cause large areas to become unusable. These devices include; radiation-emitting devices (REDs), radiological-dispersal devices (RDDs), and radiological-dispersal weapons (RDWs).

Radiation-Emitting Devices

A **radiation-emitting device (RED)** is a powerful gamma-emitting radiation source that can be placed in a high-profile location, such as a high-traffic urban area, entertainment arena, or a shopping complex which could expose a large number of people to the intense radiation source. REDs can also be used to target specific individuals and/or harm a limited number of people over a long period of time.

Radiation-Emitting Device (RED) — A powerful gamma-emitting radiation source used as a weapon.

Radiological-Dispersal Devices

Radiological-dispersal devices (RDDs) are defined by the U.S. Department of Defense as any device, including weapons or equipment, other than a nuclear explosive device, specifically designed to employ radioactive material by disseminating it to cause destruction, damage, or injury by means of the radiation produced by the decay of such material. More simply stated, an RDD is a device intended to disperse radioactive material over a large area, but it is not capable of producing a nuclear yield. The intent of an RDD is to create fear and panic by exposing people to radioactive material or to contaminate areas and buildings, making them unusable until decontaminated. An RDD typically uses the force of conventional explosives to scatter radioactive material.

Radiological-Dispersal Weapons

Radiological-dispersal weapons (RDWs) or Simple Radiological Dispersal Devices (SRDD) are nonexplosive RDDs. RDWs can use inexpensive and common items such as pressurized containers, building ventilation systems, fans, and mechanical devices to spread radioactive contamination. For example, radioactive material could be placed into a ventilation system and then dispersed throughout a building when the building's ventilation system is operated. Dispersal by these means would require putting the radioactive material into a dispersible form (powder or liquid) and would require large amounts of radioactive material to pose a hazard once dispersed.

Radiological Dispersal Weapons (RDW) — Device that spreads radioactive contamination without using explosives; instead, radioactive contamination is spread using pressurized containers, building ventilation systems, fans, and mechanical devices.

Nuclear Devices and Weapons

There is a distinction between a nuclear device and a nuclear weapon. A *nuclear device* is typically referred to as an **improvised nuclear device (IND)**. A *nuclear weapon* is a weapon typically owned by a country/state and is strictly controlled and highly secured.

According to the U.S. Department of Defense, an IND is a device incorporating radioactive materials designed to result in the dispersal of radioactive

Improvised Nuclear Device (IND) — An illicit nuclear weapon bought, stolen, or otherwise originating from a nuclear state; or a weapon fabricated by a terrorist group from illegally obtained fissile nuclear weapons material that produces a nuclear explosion.

material or in the formation of nuclear-yield reaction. An improvised nuclear device is a crude nuclear bomb fabricated using fissile material (acquired either by purchase, diversion, theft, or force). The fissile material is either plutonium or highly enriched uranium. It is generally assumed that successful INDs have yields in the 10-20 kiloton range, the equivalent to 10,000-20,000 tons of TNT **(Table 7.14)**.

Table 7.14
Blast Range and Significant Effects

Range in miles for significant effects.*
Significant effects are 50% mortality from shockwave and heat, and a radiation dose of 400 rads.

Yield (Kt)	Shockwave	Heat	Initial Radiation	Fallout Radiation (downwind)
1	0.2	0.4	0.5	up to 3.4
10	0.4	1.1	0.8	up to 6.0

* National Council on Radiation Protection and Measurements, Report No. 138, 2001

Source: The National Academies and the U.S. Department of Homeland Security.

Terrorists could also try to seize an intact nuclear weapon that is residing in a nuclear weapon country's arsenal. There is a high level of security at these arsenals and to date, no seizure attempts have been successful. Highly enriched uranium, plutonium, and stockpiled weapons are carefully inventoried and guarded. Nuclear attack is also impeded because of the following two reasons:

1. Building a nuclear weapon is difficult; general principles are available in open literature but constructing a workable device requires advanced technical knowledge in areas such as nuclear physics and material science.

2. Crude nuclear weapons are typically very heavy, ranging from a few hundred pounds to several tons, and are difficult to transport, especially by air. Suitcase bombs (see next section) are much lighter, but they are relatively difficult to acquire and very difficult to construct.

Suitcase Bomb — Small, suitcase-or backpack-sized nuclear weapon.

Suitcase Bombs

A **suitcase bomb** is described as a very compact and portable nuclear weapon that some estimate could have the dimensions of 24 by 16 by 8 inches (61 by 41 by 20 centimeters) **(Figure 7.36)**. The destructive power is described as having an expected yield of 0.5 to 2 kilotons, and it could radiate approximately 6 square miles (10 square kilometers).

Sabotage of Nuclear Facilities

Nuclear facility sabotage could target any of the following:

● Nuclear power plants

● Cooling pools for spent, nuclear fuel rods

Figure 7.36 Carrying case of a Special Atomic Demolition Munition (SADM), a very small nuclear device.

- Nuclear reactors used for research or other, nonelectricty generating purposes
- Nuclear reprocessing facilities
- Calibration laboratories
- Nuclear waste sites
- Trucks or railcars carrying nuclear weapons or spent nuclear fuel **(Figure 7.37)**

Because of the robust safeguards in place for nuclear facilities and nuclear shipments, the likelihood of a successful attack on a nuclear installation or nuclear shipment is minimal. While there are also significant obstacles terrorists would face in the development and deployment of a radiological device, it is possible that they could do so.

Figure 7.37 Railcars carrying radiological materials could be targeted by terrorists. *Courtesy of Rich Mahaney.*

Operations at Radiological and Nuclear Attack Incidents

Priorities for a radiological or nuclear terrorist incident are accomplished through the incident command system (ICS) and by following local/jurisdictional procedures. For most terrorism events, individual fire departments will eventually fold into a larger ICS structure. After multiple agencies with overlapping authority arrive, incident control will be accomplished through a unified command structure. Until those agencies arrive, an incident management system provides the necessary structure for managing virtually all incidents at the lowest level.

The Incident Commander, as well as all responders at the scene, has the responsibility to carefully and continually assess the incident scene. Identification and characterization of the hazards at the scene are extremely important. Responders should always be evaluating the area for new hazards, weather changes, or changing conditions that may affect the selected protective clothing.

While conducting scene size-up, responders need to look for unusual or out-of-place incident-scene indicators; size and shape of smoke plumes, odors, large debris fields, or craters from explosions. If responders suspect terrorism, they should proceed cautiously, evaluating the scene for radiation levels and noting areas where secondary devices might be placed. Responders should evaluate the geographic and environmental factors that can complicate a radiological terrorism incident, including prevailing winds that can carry airborne radioactive particulates, broken water mains, vehicle and/or pedestrian traffic flow, ventilation systems, air and rail corridors or other natural or man-made influences. Scene control zones should be designated and enforced.

Tactics for radiological incidents include the following:

- Position apparatus upwind of the incident.

- Secure the area and prevent entry of unauthorized or unprotected people and vehicles.

- Be alert to the possibility of small explosive devices designed to disseminate an agent.

- Use time, distance and shielding as protective measures.

- Use full PPE including SCBA (**Figure 7.38**).

Figure 7.38 Full PPE plus SCBA may be needed at radiological incidents. *Courtesy of U.S. Customs and Border Protection.*

- Avoid contact with the source and stay out of any visible smoke or fumes.

- Monitor radiation and contamination levels.

- Establish background levels of radiation outside the suspected contamination area.

- Detain or isolate uninjured people or equipment.

- Remove victims from high hazard areas.

- Assist the medical personnel as necessary to triage, treat, and decontaminate trauma victims.

- Call for expert guidance.

- Preserve possible evidence for subsequent criminal and forensic investigations.

- Do *not* conduct overhaul and clean-up operations, and disturb the incident scene as little as possible.

The *ERG* provides response information for general radiological incidents involving low to high levels of radiation in Guide No.163. Following sections will discuss phases of a nuclear response, protection strategies, PPE, radiation detection, protective action recommendations, mass decon, patient triage, and search and rescue. Radiological materials fall under UN/DOT Class 7.

Protection Strategies: Time, Distance, and Shielding

The same protection strategies of time, distance, and shielding that can be used at all haz mat incidents can also be applied to radiological incidents **(Figure 7.39)**:

- *Time* — Decrease the amount of time spent in areas where there is radiation. For nuclear fallout following a nuclear detonation, the "rule of seven" applies **(Table 7.15, p. 368)**. This rule states that for every sevenfold increase in time, the radioactivity level due to fallout decreases by a factor of 10. For example, if the level at 1 hour after detonation was 100 rem/hour, it will decline to 10 rem/hr in 7 hours and to 1 rem/hour in 49 hours (7x7).

- *Distance* — Increase the distance from a radiation source. Doubling the distance from a point source divides the dose by a factor of four. This calculation is sometimes referred to as the **inverse square law**. When the radius doubles, the radiation spreads over four times as much area, so the dose is only one-fourth as much **(Figure 7.40)**. If sheltered in a contaminated area, keep a distance from exterior walls and roofs.

- *Shielding* — Create a barrier between responders and the radiation source with a building, earthen mound, or vehicle. Buildings — especially those made of brick or concrete — provide considerable shielding from radiation. Exposure from fallout is reduced by about 50 percent inside a one-story building and by about 90 percent at a level belowground.

Using time, distance, and shielding to limit exposure to radiation is sometimes referred to as the ALARA (As Low As Reasonably Achievable) method or principle.

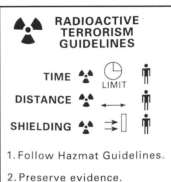

Figure 7.39 Time, distance, and shielding should be used to limit exposure to radiological hazards. *Courtesy of New South Wales Fire Brigades.*

Inverse Square Law — Physical law in which the amount of radiation present is inversely proportional to the square of the distance from the source of radiation.

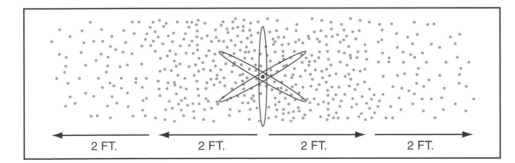

Figure 7.40 When the radius of the affected area doubles, the radiation spreads over four times as much area, so the dose is only one-fourth as much.

Table 7.15	
Rule of Seven for Radiation Dose Rates after Nuclear Explosion	
Elapsed Time	**Percent of Initial Radiation Dose Rates Remaining**
0 hours	100
7 hours	10
49 hours (2 days)	1.0
343 hours (14 days)	0.1

Illegal Haz Mat Dumps

Chemicals may be dumped illegally for a variety of reasons. In some cases, lawful disposal may be considered too expensive or complicated. In other cases, the materials may have been used in illegal clandestine labs or other illegal activities. Some chemical dumpsites may have been created years before any regulations prohibited such actions.

Discovery of an **illegal dump** may or may not be considered an emergency, depending on the chemicals involved and where the site is located. However, first responders are often the first persons called to the scene. Illegal dumpsites may be extremely expensive to clean up, and often require state/provincial and/or federal/national involvement. Frequently the following significant problems and hazards are associated with illegal dumps:

- *Unlabeled containers* — Chemicals may have been removed from their original containers or labels and identification information may have been deliberately removed.

- *Mixed chemicals* — Containers may have many different (and potentially incompatible chemicals) mixed together, making hazard and risk assessment extremely difficult.

- *Aged chemicals* — Many chemicals become unstable when subjected to age and weathering in outside climates.

- *Environmental contamination* — When chemicals are deliberately dumped in ponds, streams, rivers, and lakes, environmental contamination becomes a serious issue. Even if chemicals are not dumped in a body of water, leaking drums and other containers can pose a threat to groundwater sources.

Evidence Preservation

Incidents involving WMD or other illegal activities are crimes, and the locations where they occur are crime scenes. Law enforcement must be notified as soon as a crime is suspected. Fire service first responders should not try to collect **evidence**. Primarily they need to be aware of the need to identify and preserve evidence so that law enforcement personnel can collect and properly document it. Local emergency response plans should spell out responsibilities of individual agencies at such incidents as well as detail the acceptable procedures and techniques to be used.

Illegal Dump — Site where chemicals are disposed of illegally.

Evidence — (1) In law, something legally presented in court that bears on the point in question. (2) Information collected and analyzed by an investigator.

It is important for first responders to preserve evidence so that investigators can identify and successfully prosecute guilty parties. The more the scene is disturbed, the more difficult it becomes for investigators to develop a clear and accurate picture of what actually occurred. Law enforcement must gather accurate, acceptable information about the crime to be used in court. Even seemingly irrelevant things (footprints, wrapping paper, containers, debris placement, victim locations, vehicles in the vicinity, location of witnesses and bystanders, and the like) can have tremendous significance to forensic experts and other law enforcement investigators.

Evidence can take many forms; it is not limited to pieces of a bomb or an incendiary device (**Figure 7.41**). Evidence can include everything from body fluids to tire tracks to cigarette butts. The way debris is scattered can tell investigators about the force of an explosion (and consequently, how big the bomb was). Residue on debris can help identify what explosive materials were used. Clothing and jewelry removed from victims is considered evidence. At illegal clandestine labs, evidence may include such things as fingerprints, weapons, chemical containers, notes, letters, and papers. Evidence can be anything; therefore, responders must — to the degree possible — avoid disturbing a scene.

Figure 7.41 Piece of evidence from the Khobar Tower bombing. *Courtesy of U.S. Department of Defense.*

The preservation of life, of course, is more important than the preservation of evidence. Life-saving operations take precedence. However, even while conducting rescues and other vital operations, responders should do as much as possible to preserve the scene.

As soon as it is known or suspected that criminal or terrorist activity is involved at an incident, first responders should try to do the following to help preserve evidence and assist law enforcement:

- *DO NOT* touch anything unless it is necessary.

- Avoid disturbing areas not directly involved in rescue activities.

- Remember what the scene looked like upon first arrival as well as details about the progression of the incident. Note as many of the *W's* as possible: *who, what, when, where,* and *why.* If possible, pay attention to the following:

  — *Who* was present (including victims, people running from the scene, people acting suspiciously, bystanders, potential witnesses, and the like)

  — *What* happened

  — *When* important events occurred

- *Where* objects/people/animals were located
- *Why* events unfolded as they did

- Document observations as quickly as possible. While it may be quite some time before responders have the opportunity, the sooner information is written down, the more accurate it will probably be. This documentation may be used as evidence for legal proceedings.

- Take photographs and videos of the scene as soon as possible.

- Remember and document when something was touched or moved. Document in the report where it was and where it was placed. Photograph the item before doing anything if possible. **DO NOT** try to recreate the scene as it looked before something was touched or moved. In other words, don't move something back into the position you found it in after you have already moved it.

- Minimize the number of people working in the area if possible. Establish travel routes that minimize disturbance **(Figure 7.42)**.

Figure 7.42
Establishment of travel routes through the scene can minimize disturbance. *Courtesy of FEMA News Photos, photo by Jocelyn Augustino.*

- Leave fatalities and their surroundings undisturbed.

- Isolate and secure areas where evidence is found, and report findings to law enforcement authorities.

- Identify witnesses, victims, and the presence of evidence. Investigators will want to interview witnesses and victims as part of their investigations. These witnesses should be advised to remain at the scene in a safe location until they have been interviewed and released.

- Preserve potentially transient physical evidence (evidence present on victims or evidence that may be compromised by weather conditions such as chemical residue, body fluids, or footprints).

- Have evidence collection points (such as ground tarps) located near decontamination corridors and hot zone exit locations so that evidence can be gathered during decon or doffing operations.

- At chemical or biological incidents, secure and isolate restaurants or food vendors near the incident area in case contaminated food can be used as evidence.

- Follow predetermined procedures regarding operations at crime scenes.

Additional information about evidence preservation and sampling is provided in Chapter 10, Evidence Preservation and Sampling. The information found there is aimed at personnel who may be charged with evidence collection, sampling, and documenting chain of custody.

Haz Mat During and After Disasters

Natural disasters such as floods, hurricanes, tornadoes, and earthquakes can cause a variety of haz mat issues. For example, during floods, containers of all shapes, sizes, and contents can be moved about by water currents **(Figure 7.43)**. Tanks can be floated off foundations; entire chemical storage yards can be swept away. Some containers may release their contents into the flood waters. There may be many dead animals. Tornadoes and earthquakes can damage containers, rip up piping, or move containers around. Large scale events such as hurricanes can cause problems with massive quantities of abandoned household hazardous waste (for example, refrigerators with Freon can't be taken to the local landfill)**(Figure 7.44, p. 372)**. If industries are affected, there may be chemical or oil spills **(Figure 7.45, p. 372)**. After the disaster, many hazardous materials containers may not be correctly placarded or labeled, resulting in identification difficulties.

Figure 7.43 Natural disasters can cause haz mat containers to be moved far from their original locations. Often, they are damaged during this time. *Courtesy of FEMA News Photos, photo by Liz Roll.*

Figure 7.44 Massive quantities of household waste can be generated after a disaster. These propane containers were collected after one such incident in the U.S. *Courtesy of FEMA News Photos, photo by Greg Henshall.*

Figure 7.45 When industries are affected, consequences can be severe. *Courtesy of FEMA News Photos, photo by Liz Roll.*

Cedar Rapids, Iowa, 2008

From June 10, 2008 until about June 18, 2008, the city of Cedar Rapids, Iowa experienced the devastating effects of a major flood **(Figure 7.46, p. 374)**. FEMA called the flooding one of the worst disasters and one of the top ten most costly disasters in United States history. Approximately 9.2 square miles (15 square kilometers) of Cedar Rapids experienced flooding. Close to the Cedar River, the water levels were more than one story high. The river depth was over 31 feet (9 m) deep on the morning of Friday, June 13, 2008. The deepest recorded depth of the river in the history of the community was 20 feet (6 m). The flooding affected over 5,000 homes, businesses, industries and government buildings in Cedar Rapids. Other communities along the Cedar River, both up and down stream, experienced the same devastating damages and harm.

Many homes, businesses, industries and government buildings had hazardous materials in them. The containers ranged in size and quantity from small cans of paint, to containers of gasoline, to compressed gas cylinders, 55 gallon (200 liter) drums, 500 gallon (1 893 liter) LPG tanks, underground gasoline storage tanks, and all kinds of containers in between. Containers of hazardous materials were swept out of buildings and traveled with the flood waters and were deposited far from their original locations.

Once the flood waters receded, property owners began to clean out flood damaged buildings and carry the debris to the side of the road for pick up by clean-up crews **(Figure 7.47, p. 374)**. Road sides were lined with different sizes of containers waiting for pick up as well as refrigerators that needed to have the refrigerant removed before they could be disposed in a land fill. Other items included lawn mowers and other gasoline fueled equipment that needed to have the fuel tanks drained before being placed in a land fill. Additionally, other containers had floated into the community from upstream and now presented problems in the Cedar Rapids area of the Cedar River **(Figure 7.48, p. 374)**.

The Cedar Rapids Fire Department Hazardous Materials Response Team and the Linn County Hazardous Materials Response Team were overwhelmed by the number of hazardous material calls and the challenges of collecting all of these hazardous material containers, identifying the products in the containers, transferring small amounts of products to larger collection containers, operating a hazard waste site collection operation, and eventually coming up with a plan to dispose of the entire mess. There were also issues of deciding who was going to pay for the clean-up. A decision was made in the Linn County Emergency Operations Center to contact and request help from the EPA Region 7 Hazardous Materials Response Team. The EPA Region 7 Team would take over most of the responsibilities for the haz mat problems with the assistance and support from the Cedar Rapids Fire Department and the Linn County hazardous Materials Response Teams and contractors that the EPA Region 7 Team brought in to help.

Upon their arrival, the EPA Team was briefed by local hazmat teams on the issues, problems, and challenges. Plans were made; responsibilities were transferred; a site was located to begin collecting, storing, and identifying containers and products in a secure fenced location; and plans were developed to dispose of the containers and products. This location became a hazardous waste site where containers were stored in diked areas, separated by DOT hazard classes. All of the rules of OSHA 1910.120 were followed at this site. This operation continued for months until the EPA completed the mission.

Another interesting hazmat challenge from the floods was determining what to do with the thousands of sand bags that were used to help control the flood waters **(Figure 7.49, p. 375)**. These sand bags were considered hazardous waste because they had been exposed to the dirty water from the Cedar River. They needed to be disposed of in a landfill.

Contributed by Rich Mahaney

These problems may overwhelm the response capabilities of the local jurisdiction. On-scene conditions may also limit defensive and offensive actions. Know who and how to call for assistance, and keep in mind that all hazardous waste rules must be followed for disposal of orphaned hazardous materials. In many cases, federal help will be needed and available; pre-incident surveys and preplanning should include consideration of potential disasters.

Figure 7.46 Flooding in Cedar Rapids, Iowa, in 2008. *Courtesy of Rich Mahaney.*

Figure 7.47 Flood-damaged debris left on the sides of the road contained all kinds of hazardous materials. *Courtesy of Rich Mahaney.*

Figure 7.48 A bridge showing some of the drums and haz mat containers washed downstream. *Courtesy of Rich Mahaney.*

Figure 7.49 Thousands of sandbags contacted contaminated flood waters. *Courtesy of Rich Mahaney.*

Summary

Unfortunately, the likelihood that emergency responders will someday be called upon to respond to a terrorist or criminal incident is higher than ever before. Natural disasters can also cause extreme conditions involving hazardous materials. Responders need to be vigilant, considering the possibility of terrorist or criminal involvement at every incident. For example, if an emergency is reported as an explosion, radiation monitoring must be conducted in case a dirty bomb is involved. An unusual number of unexplained illnesses could be caused by a biological attack. Furthermore, they need to know how to protect themselves as well as to preserve evidence for law enforcement agencies. Finally, all agencies that might be involved in a terrorist attack, illicit lab response, or natural disaster involving hazardous materials should work together to conduct a practice drill or training exercise in order to improve the response to a real emergency.

Review Questions

1. What is the definition of terrorism?

2. How is a terrorist attack different from a routine emergency?

3. Describe the different classifications of explosives.

4. What are the different types of IEDs? Describe each briefly.

5. With which types of nerve agents should first responders be familiar?

6. Describe different types of choking agents.

7. What are the categories of biological agents?

8. How can indoor biological attacks be contained? Outdoor?

9. Describe the different types of radiological devices.

10. What steps should be taken to preserve evidence and assist law enforcement?

Personal Protective Equipment

Chapter Contents

chapter 8

Key Terms

Competencies

NFPA® 472:

5.3.3(1)(a)i	5.3.3(2)(a)	6.2.3.1(1)	6.2.3.1(3)(a)ii	6.2.3.1(3)(e)
5.3.3(1)(a)ii	5.3.3(2)(b)i	6.2.3.1(2)(a)	6.2.3.1(3)(a)iii	6.2.4.1(1)
5.3.3(1)(a)iii	5.3.3(2)(b)ii	6.2.3.1(2)(b)	6.2.3.1(3)(b)	6.2.4.1(2)
5.3.3(1)(a)iv	5.3.3(2)(b)iii	6.2.3.1(2)(c)	6.2.3.1(3)(c)	6.2.4.1(3)
5.3.3(1)(a)v	5.4.4(4)	6.2.3.1(2)(d)	6.2.3.1(3)(d)i	6.2.4.1(5)
5.3.3(1)(a)vi	5.4.4(5)	6.2.3.1(2)(e)	6.2.3.1(3)(d)ii	6.2.5.1
5.3.3(1)(b)	5.4.4(6)	6.2.3.1(2)(f)	6.2.3.1(3)(d)iii	
	5.4.4(7)	6.2.3.1(3)(a)i	6.2.3.1(3)(d)iv	

Personal Protective Equipment

Learning Objectives

1. Discuss respiratory protection. [NFPA® 472, 5.3.3(1)(a)(i-vi), 5.3.3(1)(b)]

2. Discuss protective clothing and ensembles. [NFPA® 472, 5.3.3(2)(a), 5.3.3(2)(b)(i-iii), 5.4.4(4), 5.4.4(5), 6.2.3.1(1), 6.2.3.1(2)(a-f), 6.2.3.1(3)(a)(i-iii), 6.2.3.1(3)(b-c), 6.2.3.1(3)(d)(i-iv), 6.2.3.1(3)(e), 6.2.4.1(1-2), 6.2.5.1]

3. Don and doff different types of personal protective equipment (PPE). [Skill Sheet 8-1; NFPA® 472, 6.2.4.1(3)]

4. Discuss inspection, storage, testing, and maintenance of PPE. [NFPA® 472, 5.4.4(6), 5.4.4(7), 6.2.4.1(5)]

5. Given hazardous materials scenarios, determine proper PPE for each incident and report and document the decision. [Learning Activity 8-1; NFPA® 472, 6.2.3.1, 6.2.5.1]

Chapter 8
Personal Protective Equipment

Case History

On August 12, 1983 in Benicia, California a combined force, four-man team experienced equipment failure during a haz mat incident with a leaking railroad car. The railroad tank car leak was reported when AJCHEM company president walked into the Benicia Fire Department and verbally reported that a car was leaking in the park. Fire personnel initially looked for a car accident, but the information was finally relayed correctly and two engine companies responded to a railroad tank car.

Emergency response personnel in acid suits approached the tanker to conduct a size-up of the leak. Initial examinations of the rail car showed that the placards indicated that the tank car contained anhydrous methylamine, a gas. However, the consist showed it was actually dimethylamine anhydrous. After initial assessment, mutual aid was requested from the San Francisco Fire Deparment, and a combined team of fire personnel and haz mat responders worked to cap the leak and transfer the gas into cylinders for safer transport.

During the incident, measures were taken to prevent air contamination, and all haz mat team members were decontaminated between trips back to the leaking tanker car. However, after three trips into the area without incident, the four-man team experienced equipment failure. The face pieces on the acid suits of all four members failed and shattered allowing acid vapors into the suits. Each firefighter reported to decontamination immediately and removed the damaged suits promptly. While an SCBA unit was able to shield each team member's respiratory system, each individual still suffered severe dermatitis because of skin exposure. Not only did the face piece on each suit malfunction, but the chemical acted as a solvent and degraded the suit materials and adhesives at the seams to the extent that the heels of their fire boots had come unglued from the soles.

Along with the failure of the team's protective equipment, flashlights and other tools being used were damaged and became unserviceable. Further support personnel were called in from chemical specialist companies and the leak was finally stopped without further equipment failure. Once the leak was under control, the incident commander elected to put a 24-hour hold on further operations due to a large influx of visitors expected in Benicia for a fair. In the end, the tanker leak required the work of five agencies to safely contain the gas.

This incident brought to light the severe problem that a lack of standards in the chemical protective suit industry posed for emergency responders in life threatening situations. Standards for the quality and training in appropriate use of protective equipment is vital to the safety of emergency response personnel and should be regarded as such.

Personnel responding to haz mat/WMD incidents must have the appropriate personal protective equipment (PPE) to perform their mission safely and effectively. PPE may include anything from standard fire fighting protective clothing to chemical-protective clothing or body armor. An ensemble of appropriate PPE protects the skin, eyes, face, hearing, hands, feet, body, head, and respiratory system against a variety of hazardous materials, including chemical, biological, and radiological (CBR) materials used in terrorist attacks.

Unfortunately, no single set of PPE will yet protect against all hazards. It is important for first responders to understand that standard uniforms, traditional structural fire fighting clothing, and body armor offer varying degrees of limited protection against chemical hazards, whereas only body armor and bomb suits can provide limited protection against projectiles and explosives. A combination of chemical-protective clothing (CPC) and respiratory protection may offer good protection against industrial chemicals, hazardous environments, and CBR materials by protecting the routes of exposure **(Figure 8.1)**, but it will not provide adequate protection against thermal hazards or explosives. While the next generation of turnout gear will provide improved protection against CBR materials, responders must be aware of how their current equipment is likely to perform in most situations.

The correct use of PPE requires special training and instruction. When operating at the scene of an incident, operations level responders using PPE should do so in accordance with standard operating procedures, under the

Figure 8.1 An appropriate combination of respiratory protection and protective clothing will protect the routes of entry and prevent exposure to hazardous materials.

PPE Guards the Routes of Entry

**Respiratory Protection
Protects Against Inhalation
and Ingestion**

**Protective Clothing
Protects Against Skin
Contact**

guidance of a hazardous materials technician, or under the supervision of an allied professional (someone with the knowledge, skills, and competence to provide correct guidance), as appropriate.

Also, because of the potential for extreme hazards associated with CBR materials likely to be used in terrorist attacks, responders should be familiar with the various standards developed by organizations to ensure that PPE is designed to protect them at such incidents. This chapter will discuss the following topics:

- Respiratory protection
- Types of protective clothing
- PPE ensembles
- Classification, selection, inspection, testing, and maintenance of PPE

Respiratory Protection

Respiratory protection is a primary concern for first responders because inhalation is arguably the most significant route of entry for hazardous materials. When correctly worn and used, protective breathing equipment protects the body from inhaling hazardous substances. Respiratory protection is, therefore, a vital part of any PPE ensemble used at haz mat/WMD incidents **(Figure 8.2)**.

The basic types of protective breathing equipment used by responders at haz mat/WMD incidents are as follows:

- Self Contained Breathing Apparatus (SCBA)
    — Closed circuit SCBA
    — Open circuit SCBA
- Supplied-air respirators (SARs)
- Air-purifying respirators (APRs)
    — Particulate removing
    — Vapor and gas removing
    — Combination particulate and vapor-and-gas-removing
- Powered air-purifying respirators (PAPRs)

Figure 8.2 Since inhalation is one of the most dangerous routes of entry for many hazardous materials, respiratory protection is extremely important.

WARNING!
SCBA must be worn during emergency operations at terrorist/ haz mat incidents until air monitoring and sampling determines other options are acceptable.

Each type of respiratory protection equipment has limits to its capabilities. For example, self contained breathing apparatus (SCBA) offers a limited working duration based upon the quantity of air.

Responders may also need to be familiar with powered-air hoods, escape respirators, and combined respirators. The sections that follow discuss these groups of respiratory equipment (including their basic limitations) as well as U.S. and international standards for respiratory protection at CBRNE events.

Standards for Respiratory Protection at Haz Mat/WMD Incidents

National Institute for Occupational Safety and Health (NIOSH) — Part of the CDC, the federal agency responsible for conducting research and making recommendations for the prevention of work-related injury and illness.

Because of the extreme hazards associated with CBRN materials that could be used in terrorist attacks (such as military nerve agents), the U.S. Department of Homeland Security has adopted standards developed by the **National Institute for Occupational Safety and Health (NIOSH)** and the NFPA® for respiratory equipment to protect responders at haz mat/WMD incidents. NIOSH also certifies SCBA and recommends ways to select and use protective clothing and respirators at biological incidents. Depending on their location, responders may also need to be familiar with standards regarding respiratory equipment issued by the ISO (International Standards Organization), the European Union, or other authorities.

The NIOSH and NFPA® standards relating to respiratory equipment at CBRN incidents (including design, certification, and testing requirements) are as follows:

- *NIOSH Chemical, Biological, Radiological and Nuclear (CBRN) Standard for Open-Circuit Self-Contained Breathing Apparatus (SCBA)* — This standard establishes performance and design requirements to certify SCBA for use in CBRN exposures for use by first responders.

- *NIOSH Standard for Chemical, Biological, Radiological, and Nuclear (CBRN) Full Facepiece Air-Purifying Respirator (APR)* — This standard specifies minimum requirements to determine the effectiveness of full-facepiece APRs (commonly referred to as gas masks) used during entry into CBRN atmospheres that are not immediately dangerous to life and health (IDLH). Atmospheres that are above IDLH concentrations require the use of SCBA.

- *NIOSH Standard for Chemical, Biological, Radiological, and Nuclear (CBRN) Air-Purifying Escape Respirator and CBRN Self-Contained Escape Respirator* — This standard specifies minimum requirements to determine the effectiveness of escape respirators that address CBRN materials identified as inhalation hazards from possible terrorist events for use by the general working population.

- *NFPA® 1852, Standard on Selection, Care, and Maintenance of Open-Circuit Self-Contained Breathing Apparatus (SCBA)* — This standard specifies the minimum requirements for the selection, care, and maintenance of open-circuit self-contained breathing apparatus (SCBA) and combination SCBA/SAR that are used for respiratory protection during fire fighting, rescue, and other hazardous operations.

- *NFPA® 1981, Standard on Open-Circuit Self-Contained Breathing Apparatus (SCBA) for Emergency Services* — This standard specifies the minimum requirements for the design, performance, testing, and certification of open-circuit self-contained breathing apparatus (SCBA) and combination open-circuit self-contained breathing apparatus and supplied air respirators (SCBA/SAR) for the respiratory protection of fire and emergency responders where unknown, IDLH (immediately dangerous to life and health), or potentially IDLH atmospheres exist.

Responders in the U.S. should also be familiar with the following standards:

- **OSHA Regulation 29 *CFR* 1910.134, *Respiratory Protection*** — The major requirements of this OSHA Respiratory Protective Standard include: permissible practices; definitions; respiratory protection program; selection of respirators; medical evaluations; fit testing; use, maintenance, and care of respirators; identification of filters, cartridges, and canisters; training; program evaluation; and record keeping.

- **NIOSH Regulation 42 *CFR* Part 84, *Approval of Respiratory Protective Devices*** — The purpose of this NIOSH regulation is to:

  — Establish procedures and prescribe requirements that must be met in filing applications for approval by NIOSH of respirators or changes or modifications of approved respirators

  — Provide for the issuance of certificates of approval or modifications of certificates of approval for respirators that have met the applicable construction, performance, and respiratory protection requirements set forth in this part

  — Specify minimum requirements and to prescribe methods to be employed by NIOSH and by the applicant in conducting inspections, examinations, and tests to determine the effectiveness of respirators used during entry into or escape from hazardous atmospheres

Additionally, NIOSH has issued a voluntary standard for PAPRs and is working on a standard for closed-circuit SCBAs (CC-SCBA). They also provide guidance on how to retrofit previously deployed (field-deployed), NIOSH-approved full-facepiece, air-purifying respirators (APR) to chemical, biological, radiological, and nuclear approved configurations. For more detailed information about respiratory protection in general, see the IFSTA manual, **Respiratory Protection for Fire and Emergency Services**.

Self-Contained Breathing Apparatus

SCBA is an atmosphere-supplying respirator for which the user carries the breathing-air supply. The unit consists of a facepiece, pressure regulator, air hoses, compressed air cylinder, harness assembly, and end-of-service-time indicators (also known as *low-air supply* or *low-pressure alarms*). SCBA is perhaps the most important piece of PPE a responder can wear at a haz mat incident in terms of preventing dangerous exposures to harmful substances.

In the U.S., NIOSH and the Mine Safety and Health Administration (MSHA) must certify all SCBA for immediately dangerous to life or health (IDLH) atmospheres. SCBA that are not NIOSH/MSHA certified must not be used. The apparatus must also meet the design and testing criteria of NFPA® 1981 in jurisdictions that have adopted that standard by law or ordinance. In addition, American National Standards Institute (ANSI) standards for eye protection apply to the facepiece lens design and testing.

NIOSH classifies SCBA as either *closed-circuit* or *open-circuit*. Two types of SCBA are currently being manufactured in closed- or open-circuit designs: pressure-demand, or positive-pressure. SCBA may also be either a high- or low-pressure type. However, use of only positive-pressure open-circuit or closed-circuit SCBA is allowed in incidents where personnel are exposed to hazardous materials **(Figure 8.3)**.

Figure 8.3 Positive-pressure SCBA must be worn at hazmat incidents.

The advantages of using SCBA-type respiratory protection are independence, maneuverability and protection from toxic and/or asphyxiating atmospheres; however, several disadvantages are as follows:

• Weight of the units

• Limited air-supply duration

• Change in profile that may hinder mobility because of the configuration of the harness assembly and the location of the air cylinder

• Limited vision caused by facepiece fogging

• Limited communications if the facepiece is not equipped with a microphone or speaking diaphragm

NIOSH has entered into a *Memorandum of Understanding* with the National Institute of Standards and Technology (NIST), OSHA, and NFPA® to jointly develop a certification program for SCBA used in emergency response to terrorist attacks. Working with the U.S. Army Soldier and Biological Chemical Command (SBCCOM), they developed a new set of respiratory protection standards and test procedures for SCBA used in situations involving WMD. Under this voluntary program, NIOSH issues a special approval and label identifying the SCBA as appropriate for use against chemical, biological, radiological, and nuclear agents. The SCBA certified under this program must meet the following minimum requirements:

- Approval under NIOSH 42 *CFR* 84, Subpart H
- Compliance with NFPA® 1981
- Special tests under NIOSH 42 *CFR* 84.63(c):
  - Chemical Agent Permeation and Penetration Resistance Against Distilled Sulfur Mustard (HD [military designation]) and Sarin (GB [military designation])
  - Laboratory Respirator Protection Level (LRPL)

NIOSH maintains and disseminates an approval list for the SCBAs approved under this program. This list is entitled "CBRN SCBA" and contains the name of the approval holder, model, component parts, accessories, and rated duration. This list is maintained as a separate category within the NIOSH Certified Equipment List.

NIOSH authorizes the use of an additional approval label on apparatus that demonstrate compliance to the CBRN criteria. This label is placed in a visible location on the SCBA backplate (for example, on the upper corner or in the area of the cylinder neck) **(Figure 8.4)**. The addition of this label provides visible and easy identification of equipment for its appropriate use.

Figure 8.4 A NIOSH label showing compliance with CBRN criteria.

Supplied-Air Respirators

The **supplied-air respirator (SAR)** or airline respirator is an atmosphere-supplying respirator where the user does not carry the breathing air source. The apparatus usually consists of a facepiece, a belt- or facepiece-mounted regulator, a voice communications system, up to 300 feet (91 m) of air supply hose, an emergency escape pack or **emergency breathing support system (EBSS)**, and a breathing air source (either cylinders mounted on a cart or a portable breathing-air compressor) **(Figure 8.5, p. 386)**. Because of the potential for damage to the air-supply hose, the EBSS provides enough air, usually 5, 10, or 15 minutes' worth, for the user to escape a hazardous atmosphere. SAR apparatus are not certified for fire fighting operations because of the potential damage to the airline from heat, fire, or debris.

Supplied-Air Respirator (SAR) — An atmosphere-supplying respirator for which the source of breathing air is not designed to be carried by the user; not certified for fire fighting operations. Also known as an Airline Respirator.

Emergency Breathing Support System (EBSS) — Escape-only respirator that provided sufficient self-contained breathing air to permit the wearer to safely exit the hazardous area; usually integrated into an airline supplied-air respirator system.

Figure 8.5 A typical SAR with EBSS. The EBSS should provide at least 5 minutes of air in case of emergency, enough to escape the hazard area into a safe atmosphere. *Courtesy of MSA.*

NIOSH classifies SARs as Type C respirators. Type C respirators are further divided into two approved types: One type consists of a regulator and facepiece only, while the second consists of a regulator, facepiece, and EBSS. This second type may also be referred to as a SAR with escape (egress) capabilities. It is used in confined-space environments, IDLH environments, or potential IDLH environments. SARs used at haz mat or CBR incidents must provide positive pressure to the facepiece.

SAR apparatus have the advantage of reducing physical stress to the wearer by removing the weight of the SCBA. The air supply line is a limitation because of the potential for mechanical or heat damage. In addition, the length of the airline (no more than 300 feet [91 m] from the air source) restricts mobility. Problems with hose entanglement must also be addressed. Other limitations are the same as those for SCBA: restricted vision and communications.

Air-Purifying Respirators

Air-purifying respirators (APRs) contain an air-purifying filter, canister, or cartridge that removes specific contaminants found in ambient air as it passes through the air-purifying element. Based on which cartridge, canister, or filter is being used, these purifying elements are generally divided into the three following types:

- Particulate-removing APRs
- Vapor-and-gas-removing APRs

Air-Purifying Respirator (APR) — Respirator with an air-purifying filter, cartridge, or canister that removes specific air contaminants by passing ambient air through the air-purifying element; may have a full or partial facepiece.

- Combination particulate-removing and vapor-and-gas-removing APRs

APRs may be powered (PAPRs) or nonpowered. APRs do not supply oxygen or air from a separate source, and they protect only against specific contaminants at or below certain concentrations. Combination filters combine particulate-removing elements with vapor-and-gas-removing elements in the same cartridge or canister.

Respirators with air-purifying filters may have either full facepieces that provide a complete seal to the face and protect the eyes, nose, and mouth or half facepieces that provide a complete seal to the face and protect the nose and mouth **(Figure 8.6)**. Half-face respirators will NOT protect against CBR materials that can be absorbed through the skin or eyes and therefore are not recommended for use at haz mat/WMD incidents except in very specific situations (explosive attacks where the primary hazard is dust or particulates).

Figure 8.6 Half-face APRs will provide respiratory protection in accordance with the filter system used, but they are not appropriate for use in hazardous atmospheres where skin contact and absorption through the mucous membranes is a concern. *Courtesy of FEMA News Photos, photo by Jocelyn Augustino.*

Disposable filters, canisters, or cartridges are mounted on one or both sides of the facepiece. Canister or cartridge respirators pass the air through a filter, **sorbent**, **catalyst**, or combination of these items to remove specific contaminants from the air. The air can enter the system either from the external atmosphere through the filter or sorbent or when the user's exhalation combines with a catalyst to provide breathable air.

No single canister, filter, or cartridge protects against all chemical hazards. Therefore, responders must know the hazards present in the atmosphere in order to select the appropriate canister, filter, or cartridge. Responders should be able to answer the following questions before deciding to use APRs for protection at an incident:

- What is the hazard?
- What is the oxygen level?
- Is the hazard a vapor or a gas?
- Is the hazard a particle or dust?

Sorbent — A material, compound, or system that holds contaminants by adsorption or absorption. With *adsorption*, the contaminant molecule is retained on the surface of the sorbent granule by physical attraction. With *absorption*, a solid or liquid is taken up or absorbed into the sorbent material.

Catalyst — A substance that influences the rate of chemical reaction between or among other substances.

- Is there some combination of dust and vapors present?
- What concentrations are present?

Furthermore, responders should know that APRs do not protect against oxygen-deficient or oxygen-enriched atmospheres, and they must not be used in situations where the atmosphere is immediately dangerous to life and health. The three primary limitations of an APR are as follows:

- Limited life of its filters and canisters
- Need for constant monitoring of the contaminated atmosphere
- Need for a normal oxygen content of the atmosphere before use

Take the following precautions before using APRs:

- Know what chemicals/air contaminants are in the air.
- Know how much of the chemicals/air contaminants are in the air.
- Ensure that the oxygen level is between 19.5 and 23.5 percent.
- **Ensure that atmospheric hazards are below IDLH conditions.**

At haz mat/WMD incidents, APRs may be used after the hazards at the scene have been properly identified. In some circumstances, APRs may also be used in other situations (law enforcement working perimeters of the scene or EMS/medical personnel) and escape situations. APRs used for these CBRN situations should utilize a combination organic vapor/ high efficiency particulate air (OV/HEPA) cartridge (see the sections that follow).

Particulate-Removing Filters

Particulate filters protect the user from particulates (including biological hazards) in the air. These filters may be used with half facepiece masks or full facepiece masks. Eye protection must be provided when the full facepiece mask is not worn.

Particulate-removing filters are divided into nine classes, three levels of filtration (95, 99, and 99.97 percent), and three categories of filter degradation. The following three categories of filter degradation indicate the use limitations of the filter:

N — not resistant to oil

R — resistant to oil

P — used when oil or nonoil lubricants are present

Particulate-removing filters may be used to protect against toxic dusts, mists, metal fumes, asbestos, and some biological hazards **(Figure 8.7)**. High-efficiency particulate air (HEPA) filters used for medical emergencies must be 99.97 percent efficient, while 95 and 99 percent effective filters may be used depending on the health risk hazard.

Particle masks (also known as *dust masks*) are also classified as particulate-removing air-purifying filters **(Figure 8.8)**. These disposable masks protect the respiratory system from large-sized particulates. Dust masks provide very limited protection and should not be used to protect against chemical hazards or small particles such as asbestos fibers.

WARNING!

Do not wear APRs during emergency operations where unknown atmospheric conditions exist. Wear APRs only in controlled atmospheres where the hazards present are completely understood and at least 19.5 percent oxygen is present.

Figure 8.7 There may be high levels of contaminants in the air after an attack, including smoke, asbestos, and other particulates such as fiberglass. *Courtesy of FEMA News Photos, photo by Andrea Booher.*

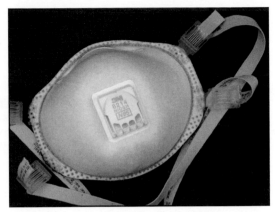

Figure 8.8 A particle mask rated N-95, meaning it is not resistant to oil and offers the lowest level of filtration. *Courtesy of the U.S. EPA.*

Vapor- and Gas-Removing Filters

As the name implies, vapor-and-gas-removing cartridges and canisters are designed to protect against specific vapors and gases. They typically use some kind of sorbent material to remove the targeted vapor or gas from the air. Individual cartridges and canisters are usually designed to protect against related groups of chemicals such as organic vapors or acid gases. Many manufacturers color-code their canisters and cartridges so it is easy to see what contaminant(s) the canister or cartridge is designed to protect against **(Figure 8.9)**. Manufacturers also provide information about contaminant concentration limitations.

Figure 8.9 Most manufacturers color-code their filters so that they are easily identified. *Courtesy of MSA.*

Powered Air-Purifying Respirator (PAPR)

The PAPR uses a blower to pass contaminated air through a canister or filter to remove the contaminants and supply the purified air to the full facepiece. Because the facepiece is supplied with air under a positive pressure, PAPRs offer a greater degree of safety than standard APRs in case of leaks or poor facial seals, and therefore may be of use at haz mat/WMD incidents for personnel conducting decontamination operations and long-term operations. Airflow also makes PAPRs more comfortable to wear for many people.

Several types of PAPR are available **(Figure 8.10, p. 390)**. Some units are supplied with a small blower and are battery operated. The small size allows users to wear one on their belts. Other units have a stationary blower (usually mounted on a vehicle) that is connected by a long, flexible tube to the respirator facepiece.

WARNING!
Do not use PAPRs in explosive or potentially explosive atmospheres.

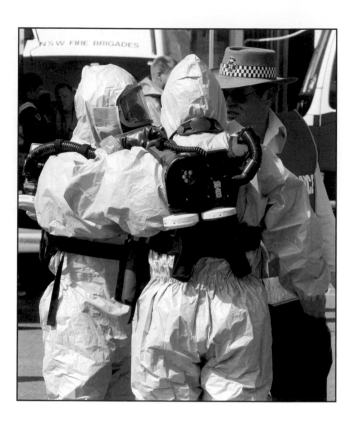

Figure 8.10 Backpack style PAPRs used in Australia. *Courtesy of New South Wales Fire Brigades.*

As with all APRs, PAPRs should only be used in situations where the atmospheric hazards are understood and at least 19.5 percent oxygen is present. PAPRs are not safe to wear in atmospheres where potential respiratory hazards are unidentified, nor should they be used during initial emergency operations before the atmospheric hazards have been confirmed. Continuous atmospheric monitoring is needed to ensure the safety of the responder.

Combined Respirators

Combination respirators include SAR/SCBA and SAR/APR **(Figure 8.11).** These respirators can provide flexibility and extend work duration times in hazardous areas. SAR/SCBAs will operate in either SAR or SCBA mode, for example, using SCBA mode for entry and exit while switching to SAR mode for extended work. SAR/APRs will also operate in either mode, but the same limitations that apply to regular APRs apply when operating in the APR mode.

Supplied-Air Hoods

Powered- and supplied-air hoods provide loose fitting, lightweight respiratory protection that can be worn with glasses, facial hair, and beards **(Figure 8.12)**. Hospitals, emergency rooms, and other organizations use these hoods as an alternative to other respirators, in part because they require no fit testing and are simple to use.

Escape Respirators

Escape respirators (sometimes called personal escape canisters or escape hoods) are designed for escaping the hot zone at a haz mat/WMD event. Escape respirators can be self-contained (usually utilizing rebreathing technology) or air-purifying (usually equipped with a combination of a HEPA filter

Figure 8.11 Combination respirators enable users to switch modes of operation between a combination of SAR, APR, PAPR, and SCBA depending on the equipment's design. *Courtesy of MSA.*

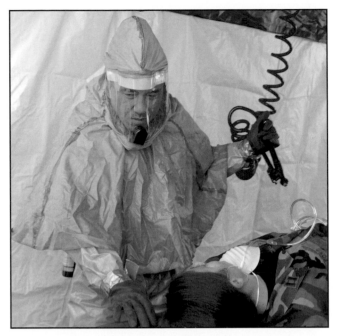

Figure 8.12 Powered- and supplied-air hoods can be worn with glasses, facial hair, and beards. *Courtesy of the U.S. Air Force, photo by Airman 1st Class Bradley A. Lail.*

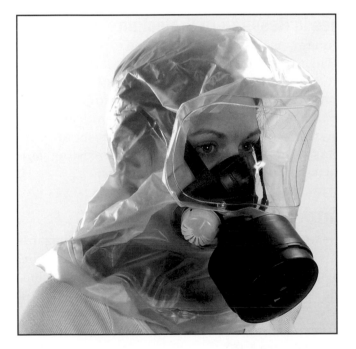

Figure 8.13 Escape respirators are designed to provide protection for a limited amount of time, such as 15 minutes. *Courtesy of MSA.*

and chemical filter such as activated carbon). Generally designed for a short duration of protection (such as 15 minutes), they are commonly designed in a hood-style. The hood-style easy to put on, has a flexible seal around the neck, and is easily donned. They can also accommodate glasses, facial hair, and beards. The filter canisters of APR-style escape respirators are usually not designed to be replaced (single use only). Some manufacturers provide escape respirator cases that can be strapped onto the body and worn as part of an emergency PPE ensemble **(Figure 8.13)**.

Respiratory Equipment Limitations

Responders must also consider the following limitations of equipment and air supply:

CAUTION

Personnel wearing respiratory euipment must have good physical conditioning, mental soundness, and emotional stability due to the physiological and psychological stresses (such as claustrophobia) of wearing PPE.

- *Limited visibility* — Facepieces reduce peripheral vision, and facepiece fogging can reduce overall vision.

- *Decreased ability to communicate* — Facepieces hinder voice communication.

- *Increased weight* — Depending on the model, the protective breathing equipment can add 25 to 35 pounds (11 kg to 16 kg) of weight to the emergency responder.

- *Decreased mobility* — The increase in weight and splinting effect of the harness straps reduce the wearer's mobility.

- *Inadequate oxygen levels* — APRs cannot be worn in IDLH or oxygen-deficient atmospheres.

- *Chemical specific* — APRs can only be used to protect against certain chemicals. The specific type of cartridge depends on the chemical to which the wearer is exposed.

Additionally, open- and closed-circuit SCBA have maximum air-supply durations that limit the amount of time a first responder has to perform the tasks at hand. Non-NIOSH certified SCBAs may offer only limited protection in environments containing chemical warfare agents.

Technological Advances in Respiratory Protection

Demands for lightweight, more comfortable SCBA have led to recent innovations in SCBA technology. Though not currently available on the market at the time of writing, next generation SCBAs may be composed of multiple thermal plastic pressure vessels that are connected **(Figure 8.14)**. These systems will have a much lower profile than existing SCBAs and weigh significantly less **(Figure 8.15)**. Additionally, they will allow increased flexibility for the user. These next generation SCBAs will meet or exceed all applicable existing standards for manufacture and testing of SCBAs at CRBN incidents.

Protective Clothing

Protective clothing must be worn whenever a wearer faces potential hazards arising from thermal hazards and chemical, biological, or radiological exposure. Skin contact with hazardous materials can cause a variety of problems, including chemical burns, allergic reactions and rashes, diseases, and absorption of toxic materials into the body. Protective clothing is designed to prevent these. Body armor and bomb suits can be worn to protect against ballistic hazards and shrapnel from explosives **(Figure 8.16)**.

No single combination or ensemble of protective equipment (even with respiratory protection), can protect against all hazards. Emergency responders must be concerned with safety when choosing and using protective clothing. For example, fumes and chemical vapors can penetrate fire fighting turnout coats and pants, so the protection they provide is not complete. Similarly, many types of chemical-protective clothing (CPC) offer little protection from fires **(Figures 6.17a and b, p. 394)**. Traditionally, body armor was not designed with chemical protection in mind.

Figure 8.14 Next generation SCBA will look significantly different than current apparatus. *Courtesy of Wayne Yoder.*

Next Generation SCBA/SCBA Profile Comparison

Figure 8.15 Next generation SCBA will have a lower profile, reduced weight, and increased flexibility.

Figure 8.16 Lance Cpl. Christopher G. West survived a vehicle bomb attack along the Iraqi-Syrian border thanks to the protection of his body armor. The protective plate insert inside the vest was strong enough to stop a sharp, foot-long piece of metal from wounding him. *Courtesy of the U.S. Marine Corps, photo by Cpl. Antonio Rosas.*

While technological advances are being made to improve the versatility of all types of PPE (for example, developing more chemical-resistant turnouts and more fire-resistant CPC), emergency responders must be aware of the limitations of the equipment they use. The sections that follow discuss the various standards that apply to protective clothing as well as the different types of clothing that will commonly be used at haz mat/WMD incidents.

Figure 8.17a and b CPC is not always flame resistant. **(a)** A manikin is dressed in CPC and subjected to a very brief flash fire. **(b)** The CPC continues to burn and melt. Some manufacturers now make CPC with increased flame resistance.

Standards for Protective Clothing and Equipment at Haz Mat/ WMD Incidents

As with respiratory protection, the U.S. Department of Homeland Security has adopted standards developed by NIOSH and NFPA® for protective clothing to be used at haz mat/WMD incidents. Primarily, these apply to clothing worn at chemical and biological incidents in regards to chemical-protective clothing (CPC). However, responders should be familiar with any standards pertaining to design, certification, and testing requirements of any type of protective clothing, including body armor, structural fire fighting gear, and bomb suits. Depending on their location, responders may also need to be familiar with standards regarding respiratory equipment issued by the ISO (International Standards Organization), the European Union, or other authorities.

The primary standards referenced by the DHS regarding PPE ensembles to be used at haz mat/WMD incidents include the following:

- *NFPA® 1991, Standard on Vapor-Protective Ensembles for Hazardous Materials Emergencies* — The purpose of this standard is to establish a minimum level of protection for emergency response personnel against adverse vapor, liquid-splash, and particulate environments during hazardous materials incidents and from specific chemical and biological terrorism agents in vapor, liquid-splash, and particulate environments during CBRN terrorism incidents. The ensemble totally encapsulates the wearer and self-contained breathing apparatus (SCBA). Class 1 ensembles, initially established under NFPA® 1994, are addressed in this standard.

- *NFPA® 1994, Standard on Protective Ensembles for First Responders to CBRN Terrorism Incidents* — NFPA® 1994 sets performance requirements for protective ensembles used in response to CBRN terrorism incidents. The standard defines three classes of ensembles (Class 2, 3, and 4) based on the protection required for different hazard types (vapors, liquids, and particulates) and airborne contaminant levels. Descriptions of the classes are:
  - **Class 2 ensembles** are intended for use at terrorism incidents involving vapor or liquid chemical or particulate hazards where the concentrations are at or above IDLH level requiring the use of CBRN compliant self-contained breathing apparatus (SCBA).
  - **Class 3 ensembles** are intended for use at terrorism incidents involving low levels of vapor or liquid chemical or particulate hazards where the concentrations are below IDLH, permitting the use of a CBRN compliant air-purifying respirator (APR) or power air-purifying respirator (PAPR).
  - **Class 4 ensembles** are intended for use at terrorism incidents involving biological or radiological particulate hazards where the concentrations are below IDLH levels permitting the use of CBRN compliant APR or PAPR. The ensembles are not tested for protection against chemical vapor or liquid permeability, gas-tightness, or liquid integrity.
- *NFPA® 1951, Standard on Protective Ensembles for Technical Rescue Incidents* — NFPA® 1951 contains performance requirements for a CBRN Technical Rescue Protective Ensemble for use during entry into CBRN atmospheres not Immediately Dangerous to Life of Health (IDLH). This CBRN protective ensemble category defines limited protection requirements for operational settings where exposure to physical, thermal, liquid, and body fluid-borne pathogen hazards and CBRN agents in vapor, liquid-splash, and particulate forms could be encountered.
- *NFPA® 1971, Standard on Protective Ensembles for Structural Fire Fighting and Proximity Fire Fighting* — NFPA® 1971 includes optional protection from CBRN hazards. Only complete ensembles certified as compliant with these additional optional requirements provide this specified level of CBRN protection. The protection levels set in the NFPA® 1971 CBRN option are based on the Class 2 requirements contained in NFPA® 1994.
- *NFPA® 1851, Standard on Selection, Care, and Maintenance of Protective Ensembles for Structural Fire Fighting and Proximity Fire Fighting* — This standard specifies the minimum selection, care, and maintenance requirements for structural fire fighting protective ensembles.
- *NFPA® 1992, Standard on Liquid Splash-Protective Ensembles and Clothing for Hazardous Materials Emergencies* — This standard specifies minimum design, performance, certification, and documentation requirements; test methods for liquid splash-protective ensembles and liquid splash-protective clothing; and additional optional criteria for chemical flash fire protection.
- *NFPA® 1999, Standard on Protective Clothing for Emergency Medical Operations* — This standard establishes minimum performance requirements for ensembles and ensemble elements to protect first responders from contact with blood- and body-fluid-borne pathogens when providing victim or patient care during emergency medical operations.

- *NFPA® 1975, Standard on Station/Work Uniforms for Fire and Emergency Services* — This standard specifies requirements for the design, performance, testing, and certification of nonprimary protective station/work uniforms and the individual garments comprising station/work uniforms.

- *NFPA® 1982, Standard on Personal Alert Safety Systems (PASS)* — Defines minimum performance criteria, functioning, and test methods for Personal Alert Safety Systems to be used by firefighters engaged in rescue, fire fighting, and other hazardous duties.

- *NFPA® 2112, Standard on Flame-Resistant Garments for Protection of Industrial Personnel Against Flash Fire* — Specifies the minimum design, performance, and certification requirements, and test methods for new flash fire protective garments.

- *NFPA® 2113, Standard on Selection, Care, Use, and Maintenance of Flame-Resistant Garments for Protection of Industrial Personnel Against Flash Fire* — Specifies the minimum requirements for the selection, care, use, and maintenance of flash fire protective garments meeting the requirements of NFPA® 2112, Standard on Flash Fire Protective Garments or Industrial Personnel.

- *ANSI/ISEA 105-2005, American National Standard for Hand Protection Selection Criteria* — This standard provides guidance for selecting the correct gloves that will protect workers and assist employers in compliance with OSHA regulations.

- *ANSI/ISEA 107-2004, American National Standard for High-Visibility Safety Apparel and Headwear* — This standard provides a uniform, authoritative guide for the design, performance specifications, and use of high-visibility and reflective apparel including vests, jackets, bib/jumpsuit coveralls, trousers and harnesses.

- *ANSI Z87.1-2003, American National Standard for Occupational and Educational Personal Eye and Face Protection Devices* — This standard establishes performance criteria and testing requirements for devices used to protect the eyes and face from injuries from impact, non-ionizing radiation and chemical exposure in workplaces and schools.

- *ANSI Z89.1-2003, American National Standard for Industrial Head Protection* — This standard provides performance and testing requirements for industrial helmets, commonly known as hard hats.

In the U.S. the following guidance documents and related consensus standards also apply:

- **OSHA Regulation 29** *CFR 1910.120, Hazardous Waste Operations and Emergency Response (HAZWOPER) Standard* — This Federal regulation applies to five distinct groups of employers and their employees. This includes any employees who are exposed, or potentially exposed to hazardous waste including emergency response operations for releases of, or substantial threat of the release of, hazardous substances regardless of the location.

- **OSHA Regulation 29** *CFR 1910.132, Personal Protective Equipment* — This standard applies to personal protective equipment for eyes, face, head, and extremities and protective clothing, respiratory devices, and protective shields and barriers. The major requirements include: permissible practices;

definitions; hazard assessment and equipment selection; training; and the proper care, maintenance, useful life, and disposal; program evaluation; and record keeping.

- **EPA Regulation 40 *CFR* Part 311,** ***Worker Protection*** — The EPA promulgated a standard identical to 29 *CFR* 1910.120 (OSHA's HAZWOPER Standard) to protect employees of State and local governments engaged in hazardous waste operations in States that do not have an OSHA-approved State plan.

OSHA Regulation 29 *CFR* 1910.156, Fire Brigades, identifies PPE requirements for industrial fire brigades. In many states, this also applies to fire departments.

At the time of this writing, the U.S. National Institute of Justice (NIJ) is developing a new CBRN ensemble standard aimed especially at garments for use by law enforcement. This new standard will be based on NFPA® 1994. It will address specific law enforcement concerns such as stealth issues (how much noise the CPC creates), durability (reinforced knees and elbows so kneeling and crawling will be possible), and dexterity (making it possible to reload a weapon). It will specify design and performance issues, test methods, labeling and information requirements, and others.

Structural Firefighters' Protective Clothing

Structural fire fighting clothing is not a substitute for chemical-protective clothing; however, it does provide some protection against many hazardous materials. The atmospheres in burning buildings, after all, are filled with toxic gases, and modern structural firefighters' protective clothing with SCBA provides adequate protection against some of those hazards **(Figure 8.18)**. The multiple layers of the coat and pants may provide short-term exposure protection from such materials as liquid chemicals; however, to avoid harmful exposures, responders must recognize the limitations of this level of protection.

Figure 8.18 Structural fire-fighting protective clothing will provide limited protection against many hazardous materials.

For example, structural fire fighting clothing is neither corrosive-resistant nor vapor-tight. Liquids can soak through, acids and bases can dissolve or deteriorate the outer layers, and gases and vapors can penetrate the garment **(Figure 8.19)**. Gaps in structural fire fighting clothing occur at the neck, wrists, waist, and the point where the pants and boots overlap.

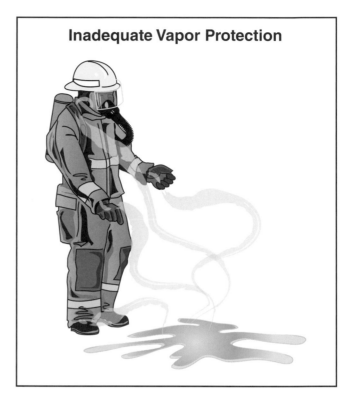

Inadequate Vapor Protection

Figure 8.19 While wearing structural fire-fighting protective clothing, liquids can soak through to come in contact with skin; acids and bases can dissolve or deteriorate the outer layers; and vapors, fumes, and gases can penetrate through gaps in the material and ensemble.

Besides knowing what can deteriorate or destroy protective clothing, responders should be aware that some hazardous materials can permeate (pass through at the molecular level) and remain in the protective equipment. Chemicals absorbed into the equipment can subject the wearer to repeated exposure or to a later reaction with another chemical. In addition, chemicals can permeate the rubber or neoprene in boots, gloves, kneepads, and SCBA facepieces making any of them unsafe for use. It may be necessary to discard any equipment exposed to permeating types of chemicals.

While there is much debate among experts as to the degree of protection provided by structural fire fighting protective clothing (and SCBA) at incidents involving chemical weapons, there may be circumstances under which it will provide adequate protection for short-term duration (less than five minutes) operations such as an immediate rescue **(Figure 8.20)**. Agency emergency response plans and SOPs should specify the conditions and circumstances under which it is appropriate for emergency responders to rely on firefighter structural protective clothing and SCBA during operations at haz mat/WMD incidents. In general, structural fire fighting protective clothing may be appropriate for use at haz mat/WMD incidents involving chemical weapons when the following conditions are met:

- Contact with splashes of extremely hazardous materials (such as chemical nerve agents) is unlikely.

Figure 8.20 Some jurisdictions allow the use of fire-fighting protective clothing and SCBA to perform rescues (involving short-term exposures, i.e. less than 5 minutes) at haz mat/WMD incidents.

- The CBR hazards have been identified and they will not rapidly damage or permeate structural fire fighting protective clothing.

- Total atmospheric concentrations do not contain high levels of chemicals that are toxic to the skin, and there are no adverse effects from chemical exposure to small areas of unprotected skin.

- There is a chance of fire or there is a fire (for example, a flammable liquid fire), and this type of protection is appropriate.

- When structural fire fighting protective clothing is the only PPE available; chemical-protective clothing is not immediately available; and the incident commander decides it is appropriate after conducting a risk assessment.

Structural fire fighting protective clothing will provide protection against thermal damage in an explosive attack, but limited or no protection against projectiles, shrapnel, and other mechanical effects from a blast. It will provide adequate protection against some types of radiological materials, but not others. In cases where biological agents are strictly respiratory hazards, structural fire fighting protective clothing with SCBA may provide adequate protection. However, in any case where skin contact is potentially hazardous, it is not sufficient. Materials must be properly identified in order to make this determination, and any time a CBR attack is suspected but not positively identified, it should be assumed that responders wearing only structural fire fighting protective clothing with SCBA are at some level of increased exposure risk.

Next Generation Firefighters' Protective Clothing

In an effort known as Project HEROES® (Homeland Emergency Response Operational and Equipment Systems) launched by the IAFF (International Association of Fire Fighters), next generation turnout gear is being designed with greater CBRN protection in mind **(Figure 8.21)**. Better closures and interfaces will prevent exposure to hazardous materials penetrating through gaps **(Figure 8.22** and **Figure 8.23, p. 400)**. A layer of chemical-resistant material added to the fabric will provide a barrier to CBRN materials and will improve problems with permeation, penetration, and degradation (see Penetration, Permeation, and Degradation section). New ensembles will meet the requirements for CBRN requirements in NFPA® 1971. Advancements in design and technology may be monitored through manufacturers' information networks and resources such as the Responders' Knowledge Database website.

Figure 8.21 Project HEROES® ensemble. SCBA exhalation air is captured to provide positive pressure to prevent inward leakage. *Courtesy of IAFF and Morning Pride Manufacturing.*

Figure 8.22 The SCBA facepiece seals with a flexible gasket. The coat closure zipper extends upward to seal the hood around the facepiece to prevent any gaps. *Courtesy of IAFF and Morning Pride Manufacturing.*

Figure 8.23 Specially designed glove/sleeve rings use strong, heat-resistant magnets to create a seal. *Courtesy of IAFF and Morning Pride Manufacturing.*

High-Temperature Protective Clothing

High-temperature protective clothing is designed to protect the wearer from short-term exposures to high-temperature in situations where heat levels exceed the capabilities of standard fire-fighting protective clothing. This type of clothing is usually of limited use in dealing with chemical hazards. Two basic types of high-temperature clothing that are available are as follows:

1. ***Proximity suits*** — Permit close approach to fires for rescue, fire-suppression, and property-conservation activities such as in aircraft rescue and fire fighting or other firefighting operations involving flammable liquids **(Figure 8.24)**. Such suits provide greater heat protection than standard structural fire-fighting protective clothing.

Figure 8.24 Proximity suits are frequently used in aircraft rescue and fire fighting or other firefighting operations involving flammable liquids. *Courtesy of William D. Stewart.*

2. *Fire-entry suits* — Allow a person to work in total flame environments for short periods of time; provide short-duration and close-proximity protection at radiant heat temperatures as high as 2,000°F (1 093°C). Each suit has a specific use and is not interchangeable. Fire-entry suits are not designed to protect the wearer against chemical hazards.

Several limitations to high-temperature protective clothing are as follows:

- Contributes to heat stress by not allowing the body to release excess heat
- Is bulky
- Limits wearer's vision
- Limits wearer's mobility
- Limits communication
- Requires frequent and extensive training for efficient and safe use
- Is expensive to purchase
- Integrity of suit is designed for limited exposure time.

Chemical-Protective Clothing

The purpose of chemical-protective clothing and equipment is to shield or isolate individuals from the chemical, physical, and biological hazards that may be encountered during hazardous materials operations. CPC is made from a variety of different materials, *none of which protects against all types of chemicals.* Each material provides protection against certain chemicals or products, but only limited or no protection against others. The manufacturer of a particular suit must provide a list of chemicals for which the suit is effective. Selection of appropriate CPC depends on the specific chemical and on the specific tasks to be performed by the wearer.

CPC is designed to afford the wearer a known degree of protection from a known type, concentration, and length of exposure to a hazardous material, but only if it is fitted properly and worn correctly. Improperly worn equipment can expose and endanger the wearer.

Most protective clothing is designed to be impermeable to moisture, thus limiting the transfer of heat from the body through natural evaporation. This can contribute to heat disorders in hot environments. Other factors include the garment's degradation, permeation, and penetration abilities and its service life. A written management program regarding selection and use of CPC is recommended. Regardless of the type of CPC worn at an incident, it must be decontaminated before storage or disposal. Responders who may be called upon to wear CPC must be familiar with (and comfortable going through) their local procedures for technical decontamination (see Chapter 9, Decontamination).

Design and testing standards generally recognize two types of CPC: liquid-splash protective clothing and vapor-protective clothing. The sections that follow describe these two types and, in addition, discuss operations where CPC is required, written management programs that specify CPC use, the ways in which CPC can be damaged, and considerations for the service life of CPC.

WARNING!
No single type of CPC protects against all chemical hazards.

WARNING!
Responders must have sufficient training to operate in conditions requiring the use of chemical-protective clothing.

Liquid-Splash Protective Clothing — Chemical-protective clothing designed to protect against liquid splashes per the requirements of NFPA® 1992, *Standard on Liquid Splash-Protective Suits for Hazardous Chemical Emergencies;* part of an EPA Level B ensemble.

Liquid-Splash Protective Clothing

Primarily, **liquid-splash protective clothing** is designed to protect users from chemical liquid splashes but not against chemical vapors or gases **(Figure 8.25)**. NFPA® 1992 sets the minimum design criteria for this type of clothing. Liquid-splash protective clothing can be encapsulating or nonencapsulating.

Figure 8.25 While liquid-splash protective clothing is appropriate for use at many haz mat incidents, it is not designed to be gas- and vapor-tight. *Courtesy of the U.S. Air Force, photo by Airman 1st Class Jason Epley.*

Encapsulating — Completely enclosed or surrounded as in a capsule.

An **encapsulating** suit is a single, one-piece garment that protects against splashes or, in the case of vapor-protective encapsulating suits, also against vapors and gases. Boots and gloves are sometimes separate, or attached and replaceable. Two primary limitations to fully encapsulating suits are as follows:

1. Impairs worker mobility, vision, and communication

2. Traps body heat necessitating a cooling vest, particularly when SCBA is also worn

A *nonencapsulating* suit commonly consists of a one-piece coverall, but sometimes is composed of individual pieces such as a jacket, hood, pants, or bib overalls. Gaps between pant cuffs and boots and between gloves and sleeves are usually taped closed. Limitations to nonencapsulating suits include the following:

● Protects against splashes and dusts but not against gases and vapors

● Does not provide full body coverage: parts of head and neck are often exposed

● Traps body heat and contributes to heat stress

Neither encapsulating and nonencapsulating liquid-splash protective clothing are resistant to heat or flame exposure, nor do they protect against projectiles or shrapnel. The material of liquid-splash protective clothing is made from the same types of material used for vapor-protective suits (see following section). Liquid splash-protective clothing must be tested for penetration resistance to the following chemicals listed in NFPA® 1992 Section 7.1.3:

● Acetone

● Ethyl acetate

● 50 percent w/w sodium hydroxide

● 93.1 percent w/w sulfuric acid

● Tetrahydrofuran

● Dimethylformanide

● Nitrobenzene

NOTE: Most manufacturers test their materials against far more chemicals than the minimum required.

When used as part of a protective ensemble, liquid-splash protective ensembles may use an SCBA, an airline (supplied-air respirator [SAR]), or a full-face, air-purifying, canister-equipped respirator. Class 3 ensembles described in NFPA® 1994 use liquid-splash protective clothing. This type of protective clothing is also a component of EPA Level B chemical protection ensembles.

Vapor-Protective Clothing

Vapor-protective clothing is designed to protect the wearer against chemical vapors or gases and offers a greater level of protection than liquid-splash protective clothing, **(Figure 8.26)**. NFPA® 1991 specifies requirements for a minimum level of protection for response personnel facing exposure to specified chemicals. This standard sets performance requirements for vapor-tight, totally encapsulating chemical-protective (TECP) suits and includes rigid chemical-resistance and flame-resistance tests and a permeation test against 21 challenge chemicals. NFPA® 1991 also includes standards for performance tests in simulated conditions.

Vapor-Protective Clothing — Gas-tight chemical-protective clothing designed to meet NFPA® 1991, *Standard on Vapor-Protective Suits for Hazardous Chemical Emergencies*; part of an EPA Level A ensemble.

Figure 8.26 Vapor-protective clothing provides the highest degree of protection against hazardous gases and vapors (for example, toxic and corrosive gases). *Courtesy of the U.S. Air Force, photo by Senior Airman Taylor Marr.*

Figure 8.27 Vapor-protective clothing can significantly impair vision, mobility, and communication.

Vapor-protective ensembles must be worn with positive-pressure SCBA or combination SCBA/SAR. Vapor-protective ensembles are components of Class 1 and 2 ensembles to be used at chemical and biological haz mat/WMD incidents as specified in NFPA® 1994. These suits are also primarily used as part of an EPA Level A protective ensemble, providing the greatest degree of protection against respiratory, eye, or skin damage from hazardous vapors, gases, particulates, sudden splash, immersion, or contact with hazardous materials. Limitations to vapor-protective suits are as follows:

- Does not protect the user against all chemical hazards
- Impairs mobility, vision, and communication **(Figure 8.27)**
- Does not allow body heat to escape, so can contribute to heat stress, which may require the use of a cooling vest

Vapor-protective ensembles are made from a variety of special materials. No single combination of protective equipment and clothing is capable of protecting a person against all hazards. NFPA® 1991 requires, as a minimum, that the suit be certified to provide minimum protection from the chemicals listed in Section 7.2.1.

Vapor-protective suit materials are tested for permeation resistance against the following chemicals:

- Acetone
- Acetonitrile
- Anhydrous ammonia gas
- 1,3-Butadiene gas
- Carbon disulfide
- Chlorine gas
- Dichloromethane
- Diethyl amine
- Dimethyl formamide
- Ethyl acetate
- Ethylene oxide gas
- Hexane
- Hydrogen chloride gas
- Methanol
- Methyl chloride gas
- Nitrobenzene
- Sodium hydroxide
- Sulfuric acid
- Tetrachloroethylene
- Tetrahydrofuran
- Toluene

Operations Requiring Use of Chemical-Protective Clothing

Responders should also be familiar with the circumstances in which chemical-protective clothing must be worn. Without regard to the level of training required to perform them, U.S. OSHA identifies the following emergency response operations that may require the use of CPC:

- **Site survey** — Initial investigation of a hazardous materials incident (including terrorist attacks), usually characterized by a large degree of uncertainty about the materials involved that mandates the highest levels of protection **(Figure 8.28)**.

Figure 8.28 CPC may be required when conducting the initial site survey, particularly in situations when there is a great deal of uncertainty about the hazardous materials involved.

- **Rescue** — Entering a hazardous environment for the purpose of removing an exposure victim. Special considerations must be given to how the selected protective clothing may affect the ability of the wearer to perform rescue and handle the contamination and decontamination of the victim.

- *Spill mitigation* — Entering a hazardous environment to prevent a potential spill or reduce the hazards from an existing spill (for example, applying a chlorine kit on a sabotaged railroad tank car). Protective clothing must accommodate required tasks without sacrificing adequate protection.

- *Emergency monitoring* — Outfitting personnel in protective clothing for the primary purpose of observing an incident without entry into the hot zone.

- *Decontamination* — Applying decontamination procedures to personnel or equipment leaving the site; in general, requires a lower level of protective clothing than required for those working in the hot zone.

- *Evacuation* — Evacuating people downwind of a scene when potential for CBR exposure exists.

If responders are involved in any of these activities, consideration must be given to what type of protective equipment is necessary given the known and/or unknown hazards present at the scene.

Written Management Program

All emergency responders and organizations who routinely select and use CPC should establish a written Chemical-Protective Clothing/Respiratory Protection Management Program. A written management program includes policy statements, procedures, and guidelines. Copies must be made available to all personnel who may use CPC in the course of their duties or job.

The two basic objectives of any management program are protecting the wearer from safety and health hazards and preventing injury to the wearer from incorrect use or malfunction of the CPC. To accomplish these goals, a comprehensive CPC management program includes the following elements:

- Hazard identification
- Medical monitoring
- Environmental surveillance
- Selection, care, testing, and maintenance
- Training

Permeation, Degradation, and Penetration

The effectiveness of CPC can be reduced by three actions: permeation, degradation, and penetration. These are also characteristics that must be considered when choosing and using protective ensembles.

Permeation. Permeation is a process that occurs when a chemical passes through a fabric on a molecular level. In most cases, there is no visible evidence of chemicals permeating a material. The rate at which a compound permeates CPC depends on factors such as the chemical properties of the compound, nature of the protective barrier in the CPC, and concentration of the chemical on the surface of the CPC **(Figure 8.29, p. 406)**. Most CPC manufacturers provide charts on breakthrough time (time it takes for a chemical to permeate the material of a protective suit) for a wide range of chemical compounds. Permeation data also includes information about the permeation rate or the speed (or rate) at which the chemical moves through the CPC material after it breaks through.

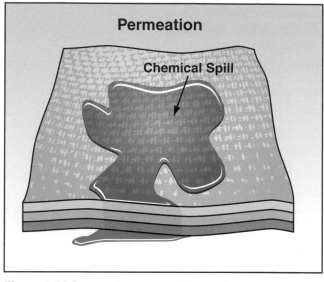

Figure 8.29 Permeation occurs when a chemical passes through a fabric or material on a molecular level. This process often goes unnoticed.

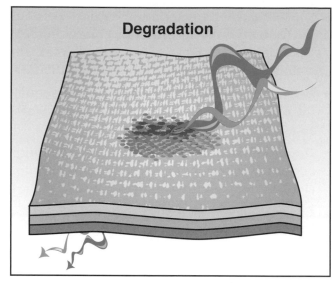

Figure 8.30 An acid eating away the outer layers of structural fire-fighting protective clothing is an example of chemical degradation.

Chemical degradation. This process occurs when the characteristics of a material are altered through contact with chemical substances. Examples include cracking, brittleness, and other changes in the structural characteristics of the garment **(Figure 8.30)**. The most common observations of material degradation are discoloration, swelling, loss of physical strength, or deterioration.

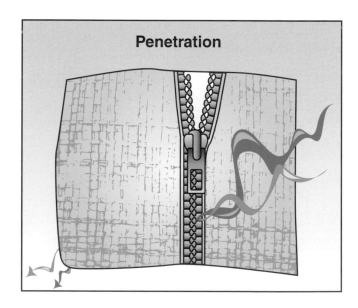

Figure 8.31 Hazardous materials can penetrate PPE through gaps, tears, punctures, or other openings.

Penetration. Penetration is a process that occurs when a hazardous material enters an opening or a puncture in a protective material. Rips, tears, and cuts in protective materials — as well as unsealed seams, buttonholes, and zippers — are considered penetration failures. Often such openings are the result of faulty manufacture or problems with the inherent design of the suit **(Figure 8.31)**.

Service Life

Each piece of CPC has a specific service life over which the clothing is able to adequately protect the wearer. Protective clothing may be labeled as *reusable* (multiuse) for repeated use, *limited use* (not disposable), or *disposable* for one-time use. For example, a Saranex/Tyvek® garment may be designed to be a coverall (covering the wearer's torso, arms, and legs) intended for liquid-splash protection and single use. Testing and inspection per manufacturer's guidelines are necessary to ensure CPC is still useable. Other suits are designed to be reused multiple times. Nevertheless, extensive contamination of any garment may require its removal from service.

All types require decontamination when the wearer leaves a contaminated area. However, reusable suits must receive extra decontamination measures to ensure they are clean for the next user. Chapter 9, Decontamination, provides more information about contamination and decontamination of CPC.

Body Armor

Body armor is designed to protect against ballistic threats. It is commonly used by law enforcement personnel, but some fire service and EMS agencies have also started wearing it, particularly when operating in dangerous situations or areas where attacks might be likely. Body armor prevents a bullet from penetrating the body because the fibers from which the armor is constructed essentially work as a very tightly woven net. Body armor should always be replaced if it has been impacted or damaged.

In the U.S., body armor has been divided by the National Institute of Justice into six different categories (spanning 4 levels) based on the level of protection provided. Type I (or Level I) body armor provides the least amount of protection, while Type IV provides the greatest. Type III is the first level to utilize plates.

Bomb Disposal Suits

Bomb disposal suits must provide full body protection against fragmentation, overpressure, impact, and heat. Normally designed to meet appropriate military specifications, they incorporate high-tech materials and ballistic plates in a head-to-toe ensemble **(Figure 8.32)**. Helmets are usually designed with built-in communications capabilities and forced-air ventilation systems, some of which also provide protection/filtration against CBR materials. Bomb suits are very heavy and significantly impair dexterity and range of motion. New technology is being incorporated into next generation bomb suits to improve protection against CBR materials.

PPE Ensembles, Classification, and Selection

The approach in selecting PPE must encompass an ensemble of clothing and equipment items that are easily integrated to provide both an appropriate level of protection and still allow one to perform activities involving hazardous materials/WMD. For example, simple protective clothing such as gloves and a work uniform in combination with a faceshield (or safety goggles) may be sufficient to prevent exposure to certain etiological agents (such as bloodborne pathogens). At the other end of the spectrum, the use of vapor-protective, totally-encapsulating suits combined with positive-pressure SCBA is considered

Figure 8.32 Bomb suits incorporate high-tech materials and ballistic plates in a head-to-toe ensemble. *Courtesy of the U.S. Marine Corps, photo by Cpl. Brian A. Tuthill.*

Level A Protection —
Highest level of skin, respiratory, and eye protection that can be afforded by personal protective equipment (PPE) as specified by the U.S. Environmental Protection Agency (EPA); consists of positive-pressure self-contained breathing apparatus, totally encapsulating chemical-protective suit, inner and outer gloves, and chemical-resistant boots.

Level B Protection —
Personal protective equipment that affords the highest level of respiratory protection, but a lesser level of skin protection. Consists of positive-pressure self-contained breathing apparatus, liquid-splash protective chemical-protective suit, inner and outer gloves, and chemical-resistant boots.

the minimum level of protection necessary when dealing with extremely hazardous vapors, gases, or particulates of material that are harmful to skin or capable of being absorbed through the skin.

While the EPA has established a set of chemical-protective PPE ensembles providing certain levels of protection that are commonly used by fire and emergency service organizations, other organizations such as law enforcement, industrial responders, and the military may have their own standard operating procedures or equivalent procedures guiding the choice and use of appropriate combinations of PPE. A special weapons attack team (SWAT) may be equipped with a far different PPE ensemble than a firefighter, haz mat technician, or environmental cleanup person responding to the same terrorist incident.

The sections that follow describe a variety of factors that concern PPE ensembles worn at haz mat incidents including the following:

- EPA standards
- Specifics found at the scene
- Typical ensembles worn by various responders
- Health, safety, and medical considerations

U.S. EPA Levels of Protection

The U.S. EPA has established the following levels of protective equipment to be used at incidents involving hazardous materials/WMD: **Level A**, **Level B**, **Level C**, and **Level D (Figures 6.33 a-d)**. NIOSH, OSHA, and the U.S. Coast Guard (USCG) also recognize these levels. They can be used as the starting point for ensemble creation; however, each ensemble must be tailored to the specific situation in order to provide the most appropriate level of protection.

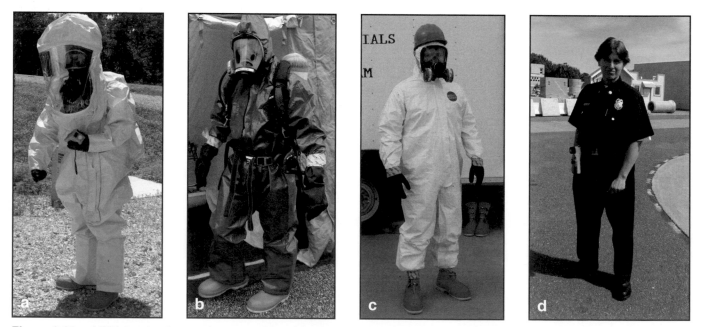

Figure 8.33a-d EPA levels of protective ensembles. **(a)** Level A, **(b)** Level B, **(c)** Level C, **(d)** Level D.

Selecting protective clothing and equipment by how they are designed or configured alone is not sufficient to ensure adequate protection at haz mat incidents. Just having the right components to form an ensemble is not enough. Responders must understand that the EPA levels of protection do not define or specify what performance (for example, vapor protection or liquid-splash protection) the selected clothing or equipment must offer, and they do not identically mirror the performance requirements of NFPA® performance standards. **Table 8.1** matches EPA levels of protection with the NFPA®'s applicable performance standards. Refer to **Skill Sheet 8-1** for steps on how to don and doff examples of PPE from each EPA level.

Level C Protection — Personal protective equipment that affords a lesser level of respiratory and skin protection than levels A or B. Consists of full-face or half-mask APR, hooded chemical-resistant suit, inner and outer gloves, and chemical-resistant boots.

Level D Protection — Personal protective equipment that affords the lowest level of respiratory and skin protection. Consists of coveralls, gloves, and chemical-resistant boots or shoes.

Table 8.1
NFPA®/EPA Level Match

Performance-Based Standard	EPA Level
NFPA® 1991 (2005 Edition) worn with NIOSH CBRN SCBA	A
NFPA® 1994 (2007 Edition) Class 2 worn with NIOSH CBRN SCBA	B
NFPA® 1971 (2007 Edition) with CBRN option worn with NIOSH CBRN SCBA	B
NFPA® 1994 (2007 Edition) CLASS 3 worn with NIOSH CBRN APR/PAPR	C
NFPA® 1994 (2007 Edition) Class 4 worn with NIOSH CBRN APR/PAPR	C
NFPA® 1951 (2007 Edition) CBRN technical rescue ensemble worn with NIOSH CBRN APR/PAPR	C

Level A

The Level A ensemble provides the highest level of protection against vapors, gases, mists, and particles for the respiratory tract, eyes, and skin. Operations-Level responders are generally not allowed to operate in situations requiring Level A protection. However, Operations-Level personnel must be appropriately trained to wear Level A PPE if they are required to wear it. The elements of Level A ensembles are as follows:

- *Components* — Ensemble requirements are as follows:

  — Positive-pressure, full facepiece, SCBA, or positive-pressure airline respirator with escape SCBA, approved by NIOSH

  — Vapor-protective suits: TECP suits constructed of protective-clothing materials that meet the following criteria:

    ⊚ Cover the wearer's torso, head, arms, and legs

    ⊚ Include boots and gloves that may either be an integral part of the suit or separate and tightly attached

    ⊚ Enclose the wearer completely by itself or in combination with the wearer's respiratory equipment, gloves, and boots

    ⊚ Provide equivalent chemical-resistance protection for all components of a TECP suit (such as relief valves, seams, and closure assemblies)

    ⊚ Meet the requirements in NFPA® 1991

- Coveralls (optional)
- Long underwear (optional)
- Chemical-resistant outer gloves
- Chemical-resistant inner gloves
- Chemical-resistant boots with steel toe and shank
- Hardhat (under suit) (optional)
- Disposable protective suit, gloves, and boots (can be worn over totally encapsulating suit, depending on suit construction)
- Two-way radios (worn inside encapsulating suit)

- *Protection Provided* — Highest available level of respiratory, skin, and eye protection from solid, liquid, and gaseous chemicals.

- *Use* — Ensembles are used in the following situations:
  - Chemical hazards are unknown or unidentified.
  - Chemical(s) have been identified and have high level of hazards to respiratory system, skin and eyes.
  - Site operations and work functions involve a high potential for splash, immersion, or exposure to unexpected vapors, gases, or particulates of material that are harmful to skin or capable of being absorbed through the intact skin.
  - Substances are present with known or suspected skin toxicity or carcinogenicity.
  - Operations that are conducted in confined or poorly ventilated areas.

Level B

Level B protection requires a garment that includes an SCBA or a supplied-air respirator and provides protection against splashes from a hazardous chemical. This ensemble is worn when the highest level of respiratory protection is necessary but a lesser level of skin protection is needed. A Level B ensemble provides liquid-splash protection, but little or no protection against chemical vapors or gases to the skin. Level B CPC may be encapsulating or nonencapsulating. The elements of Level B ensembles are as follows:

- *Components* — Ensemble requirements are as follows:
  - Positive-pressure, full facepiece, SCBA, or positive-pressure airline respirator with escape SCBA approved by NIOSH
  - Hooded chemical-resistant clothing that meets the requirements of NFPA® 1992 (overalls and long-sleeved jacket, coveralls, one- or two-piece chemical-splash suit, and disposable chemical-resistant overalls)
  - Coveralls (optional)
  - Chemical-resistant outer gloves
  - Chemical-resistant inner gloves
  - Chemical-resistant boots with steel toe and shank
  - Disposable, chemical-resistant outer boot covers (optional)
  - Hardhat (outside or on top of nonencapsulating suits or under encapsulating suits)

— Two-way radios (worn inside encapsulating suit or outside nonencapsulating suit)

— Faceshield (optional)

- **Protection Provided** — Ensembles provide the same level of respiratory protection as Level A but have less skin protection. Ensembles provide liquid-splash protection, but no protection against chemical vapors or gases.

- **Use** — Ensembles are used in the following situations:

  — Type and atmospheric concentration of substances have been identified and require a high level of respiratory protection but less skin protection.

  — Atmosphere contains less than 19.5 percent oxygen or more than 23.5 percent oxygen.

  — Presence of incompletely identified vapors or gases is indicated by a direct-reading organic vapor detection instrument, but the vapors and gases are known not to contain high levels of chemicals harmful to skin or capable of being absorbed through intact skin.

  — Presence of liquids or particulates is indicated, but they are known not to contain high levels of chemicals harmful to skin or capable of being absorbed through intact skin.

Level C

Level C protection differs from Level B in the area of equipment needed for respiratory protection. Level C is composed of a splash-protecting garment and an air-purifying device (APR or PAPR). Level C protection includes any of the various types of APRs. Emergency response personnel would not use this level of protection unless the specific material is known, it has been measured, and this protection level is approved by the IC after all qualifying conditions for APRs and PAPRs have been met (that is, the product is known, an appropriate filter is available, the atmospheric oxygen concentration is between 19.5 to 23.5 percent, and the atmosphere is not IDLH). Periodic air monitoring is required when using this level of PPE. The elements of Level C ensembles are as follows:

- **Components** — Ensemble requirements are as follows:

  — Full-face or half-mask APRs, NIOSH approved

  — Hooded chemical-resistant clothing (overalls, two-piece chemical-splash suit, and disposable chemical-resistant overalls)

  — Coveralls (optional)

  — Chemical-resistant outer gloves

  — Chemical-resistant inner gloves

  — Chemical-resistant boots with steel toe and shank

  — Disposable, chemical-resistant outer boot covers (optional)

  — Hardhat

  — Escape mask (optional)

  — Two-way radios (worn under outside protective clothing)

  — Face shield (optional)

- ***Protection Provided*** — Ensembles provide the same level of skin protection as Level B but have a lower level of respiratory protection. Ensembles provide liquid-splash protection but no protection from chemical vapors or gases on the skin.

- ***Use*** — Ensembles are used in the following situations:

  — Atmospheric contaminants, liquid splashes, or other direct contact will not adversely affect exposed skin or be absorbed through any exposed skin.

  — Types of air contaminants have been identified, concentrations have been measured, and an APR is available that can remove the contaminants.

  — All criteria for the use of APRs are met.

  — Atmospheric concentration of chemicals does not exceed IDLH levels. The atmosphere must contain between 19.5 and 23.5 percent oxygen.

Level D

Level D ensembles consist of typical work uniforms, street clothing, or coveralls. Firefighter structural protective clothing is also considered Level D because, per the EPA definition, it is not considered chemical-protective clothing. Level D protection can be worn only when no atmospheric hazards exist. The elements of Level D ensembles are as follows:

- ***Components*** — Ensemble requirements are as follows:

  — Coveralls

  — Gloves (optional)

  — Chemical-resistant boots/shoes with steel toe and shank

  — Disposable, chemical-resistant outer boot covers (optional)

  — Safety glasses or chemical-splash goggles

  — Hardhat

  — Escape device in case of accidental release and the need to immediately escape the area (optional)

  — Faceshield (optional)

- ***Protection Provided*** — Ensembles provide no respiratory protection and minimal skin protection.

- ***Use*** — Ensembles may not be worn in the hot zone and are not acceptable for haz mat emergency response above the Awareness Level. Level D ensembles are used when both of the following conditions exist:

  — Atmosphere contains no hazard.

  — Work functions preclude splashes, immersion, or the potential for unexpected inhalation of or contact with hazardous levels of any chemicals.

PPE Selection Factors

The risks and potential hazards present at an incident will determine the PPE needed. Many available sources can be consulted to determine which type and what level of PPE to use at haz mat incidents/terrorist attacks depending on the circumstances and hazards at the scene. First-arriving responders often rely upon information in the *Emergency Response Guidebook (ERG)* to determine the minimum type of protection required for defensive operations. SOPs may also provide guidance for situations involving rescue and initial responses.

NOTE: The *ERG* will only reference EPA ensemble Levels A and B, since SCBA is always recommended.

In general, the higher the level of PPE, the greater the associated risks. For any given situation, select equipment and clothing that provide an adequate level of protection. **Overprotection as well as underprotection can be hazardous and should be avoided.**

Most emergency responders will not make the decisions about which PPE, particularly CPC, is worn by personnel entering the hot zone, but it is nevertheless important for them to understand the selection process. General selection factors that need to be considered are as follows:

- *CBRN hazards* — Chemicals, biological agents, and radioactive materials present a variety of dangers including chemical hazards such as toxicity, corrosiveness, flammability, reactivity, and oxygen deficiency. Depending on what materials are present, any combination of hazards may need to be protected against.

- *Physical environment* —The choice of ensemble components must account for varied conditions:

  — Incidents may occur at industrial settings, on the highways, or in residential areas.

  — Exposure may occur either indoors or outdoors.

  — Environments may be extremely hot, cold, or moderate.

  — Exposure sites may be relatively uncluttered or rugged (which presents a number of physical hazards).

  — Incident resolution activities may involve entering confined spaces, lifting heavy items, climbing ladders, or crawling on the ground.

- *Exposure duration* —The protective qualities of ensemble components may be limited by a variety of factors, including exposure levels, material chemical resistance, and air supply. The decision for determining how long to use an ensemble must be made by assuming the worst-case exposure so that safety margins can be added to increase the protection available to personnel.

- *Available protective clothing or equipment* — An array of different clothing or equipment should be available to personnel to meet all intended applications. Reliance on one particular clothing type or equipment item may severely limit the ability to handle a broad range of hazardous materials or chemical exposures. In its acquisition of equipment and protective clothing, the responsible authority should attempt to provide a high degree of flexibility while choosing protective clothing and equipment that is easily integrated and provides protection against each conceivable hazard.

- *Compliance with regulations* — Agencies responsible for responding to CBR incidents should select equipment in accordance with regulatory standards for response to such incidents (such as NIOSH standards and NFPA® 1994).

Protective clothing selection factors include the following:

- *Clothing design* — Manufacturers sell clothing in a variety of styles and configurations. Design considerations include the following:
  - Clothing configuration
  - Seam and closure construction
  - Components and options
  - Sizes
  - Ease of donning and doffing
  - Clothing construction
  - Accommodation of other selected ensemble equipment
  - Comfort
  - Restriction of mobility

- *Material chemical resistance* — The chosen material(s) must resist permeation, degradation, and penetration by the respective chemicals. Mixtures of chemicals can be significantly more aggressive towards protective clothing materials than any single chemical alone. One permeating chemical may pull another with it through the material. Other situations may involve unidentified substances. *Details:*
  - Very little test data are available for chemical mixtures. If clothing must be used without test data, clothing that demonstrates the best chemical resistance against the widest range of chemicals should be chosen.
  - In cases of chemical mixtures and unknowns, serious consideration must be given to selecting protective clothing.

- *Physical properties* — Clothing materials may offer wide ranges of physical qualities in terms of strength, resistance to physical hazards, and operation in extreme environmental conditions. Comprehensive performance standards (such as those from NFPA®) set specific limits on these material properties, but only for limited applications such as emergency response. Users may also need to ask manufacturers the following questions:
  - Does the material have sufficient strength to withstand the physical strength of the tasks at hand?
  - Will the material resist tears, punctures, cuts, and abrasions?
  - Will the material withstand repeated use after contamination and decontamination?
  - Is the material flexible or pliable enough to allow users to perform needed tasks?
  - Will the material maintain its protective integrity and flexibility under hot and cold extremes?
  - Is the material flame-resistant or self-extinguishing (if these hazards are present)?
  - Are garment seams in the clothing constructed so they provide the same physical integrity as the garment material?

- *Ease of decontamination* — The degree of difficulty in decontaminating protective clothing may dictate whether disposable clothing, reusable clothing, or a combination of both is used.

- *Ease of maintenance and service* — The difficulty and expense of maintaining equipment should be considered before purchase.

- *Interoperability with other types of equipment* — Interoperability issues should be considered, for example, whether or not communications equipment can be integrated into the ensemble.

- *Cost* — Protective clothing end users must endeavor to obtain the broadest array of protective equipment they can buy with available resources to meet their specific applications.

Typical Ensembles of Response Personnel

The ensemble worn at an incident will vary depending on the mission of the responder. PPE for US&R personnel will differ from that of haz mat response teams, and so forth. However, responders of any discipline must be aware of what hazards are present at the incident, and what PPE is necessary to protect against the hazards to which they may be exposed. For example, if respiratory hazards exist at the incident, all personnel who might be exposed to these hazards must wear respiratory protection regardless of their mission. It is important that personnel who may need to use such PPE be trained to do so. The sections that follow will outline some of the ensembles used by emergency responders at haz mat/WMD incidents, keeping in mind that the nature of the incident will dictate the PPE requirements.

Fire Service Ensembles

Fire service personnel will wear ensembles appropriate for their mission at the incident, including typical fire fighting operations (such as fire extinguishment), hazardous materials response, and urban search and rescue (US&R). **Table 8.2, p. 416** shows a conservative estimate of the effectiveness of typical fire service PPE ensembles in the hot zone of haz mat/WMD incidents. EMS ensembles are discussed in a later, separate section.

The majority of responders will initially be wearing structural fire fighting protective clothing ensembles (turnout gear) that may offer only limited protection against haz mat/WMD hazards. These ensembles may be appropriate for conducting some operations (such as rescue) at haz mat/WMD incidents given appropriate protective measures such as limited exposure times.

Responders trained to use CPC at haz mat events may don EPA Level A or B ensembles as described in previous sections. Chemical-protective ensembles must be designed to protect the wearer's upper and lower torso, head, hands, and feet. Ensemble elements must include protective garments, protective gloves, and protective footwear. Ensembles must accommodate appropriate respiratory protection.

Law Enforcement Ensembles

A 1999 assessment conducted by the U.S. Army SBCCOM concluded that law enforcement and EMS personnel can be equipped with an effective low-cost clothing ensemble when responding to incidents of chemical-warfare terrorism. An ensemble consisting of a high-quality respirator, butyl rubber gloves, and a commercial chemical overgarment (elastic wrists and hood closures with built-in boots) will provide some liquid-droplet and vapor protection to

Table 8.2
Effectiveness of Typical Fire Service PPE Ensembles
in the Hot Zone of CBRNE Incidents

Fire Service Ensembles	Chemical Warfare Agents	TIMs	Biological Agents	Radiological Hazards	Explosives /Ballistics	Incendiaries /Fires
Standard structural-firefighting ensemble including SCBA*	Inadequate for extended hot zone use**	Inadequate for extended hot zone use**	Varies Inadequate for incidents in which agents or dissemination methods are unidentified or may still be occuring May be adequate in circumstances where agent and dissemination methods are known	Adequate for Alpha and Beta radiation Inadequate for Gamma radiation	Inadequate for protection against explosives and ballistics Adequate for operations after an explosion not involving other CBR hazards	Adequate
Haz-Mat/ Chemical Protective Ensembles	EPA Level A and B (NFPA® 1994 Class 1, 2 and 3) as appropriate	EPA Level A, B, or C Class 1, 2 and 3 as appropriate	EPA Level A, B, or C (NFPA® 1994 Class 1, 2 and 3) as appropriate	Adequate for Alpha and Beta radiation Inadequate for Gamma radiation	Inadequate for protection against explosives and ballistics Adequate for operations after an explosion involving other CBR hazards as applicable	Inadequate
US&R Ensembles (without turnout gear)	Inadequate	Inadequate	Inadequate	Adequate for Alpha radiation with appropriate respiratory protection Inadequate for Beta and Gamma radiation	Inadequate for protection against explosives and ballistics Adequate for rescue and mitigation operations after an explosion not involving other CBR hazards	Inadequate

*Not including turnout gear designed with improved CBR protection

**May be adequate for short duration exposures in certain situations (for example, during rescue operations, as determined by the incident commander, SOPs, or emergency response plan, etc.), and depending on the agent.

the responder. However, it must be emphasized that this clothing ensemble is *NOT* adequate protection for law enforcement personnel in areas where significant levels of chemical agent vapor concentration might be present such as in the hot zone or downwind areas.

SWAT teams, bomb squads, evidence recovery teams, and other specialty units operating in the hot zone would require higher levels of PPE **(Figure 8.34)**. Such PPE may be worn with body armor. **Table 8.3** shows a conservative estimate of the effectiveness of typical law enforcement PPE ensembles in the hot zone of haz mat/WMD incidents.

Figure 8.34 Law enforcement personnel operating in hazardous areas will need increased PPE, often including respiratory protection. *Courtesy of MSA.*

Table 8.3
Effectiveness of Typical Law Enforcement PPE in the Hot Zone of CBRNE Incidents

Law Enforcement PPE	Chemical Warfare Agents	TIMs	Biological Agents	Radiological Agents	Explosives/ Ballistics	Incenniaries/ Fires
Body Armor (w/ duty uniform)	Inadequate	Inadequate	Inadequate	Inadequate	Protection provided per type of armor	Inadequate
Haz Mat / Chemical Protective Ensembles	EPA Level A and B (NFPA® 1994 Class 1, 2 and 3) as appropriate	EPA Level A, B, or C (NFPA® 1994 Class 1, 2 and 3) as appropriate	EPA Level A, B, or C (NFPA® 1994 Class 1, 2 and 3) as appropriate	Adequate for Alpha and Beta radiation Inadequate for Gamma radiation	Inadequate for protection against explosives or ballistics Adequate for operations after an explosion involving other CBR hazards as applicable	Inadequate
Bomb Suits*	Inadequate	Inadequate	Varies Inadequate for incidents in which agents or dissemination methods are unidentified or may still be occurring	Adequate for Alpha and Beta radiation with appropriate respiratory protection Inadequate for Gamma radiation	Inadequate	Adequate for flash fires

*Not including bomb suits designed with improved CBR protection

EMS Ensembles

EMS PPE must provide blood- and body-fluid pathogen barrier protection to whatever parts of the body they cover. PPE ensembles should include outer protective garments, gloves, footwear, and face protection. While no partial protection is allowed for the EMS PPE, the items might be configured to cover only part of the upper or lower torso such as arms with sleeve protectors, torso front with apron-styled garments, and face with faceshields.

As with law enforcement, EMS personnel not working in the hot zone may have adequate protection using a low-cost ensemble consisting of a high-quality respirator, butyl rubber gloves, and a commercial chemical overgarment (elastic wrists and hood closures with built-in boots) that provides some liquid-droplet and vapor protection **(Figure 8.35)**. This level of protection may or may not be adequate for personnel conducting triage and decontamination operations in the warm zone, depending on circumstances.

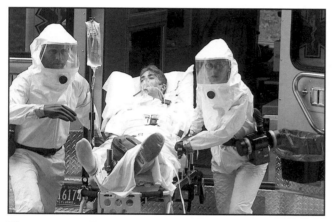

Figure 8.35 EMS ensembles consisting of high-quality respirators, butyl rubber gloves, and commercial chemical overgarments. *Courtesy of MSA.*

Figure 8.36 MOPP ensembles consist of an overgarment, mask, hood, overboots, and protective gloves. *Courtesy of the U.S. Army, photo by Sgt. Scott Kim.*

Mission-Oriented Protective Posture (MOPP)

The U.S. military uses MOPP ensembles to protect against chemical, biological, and radiological hazards. MOPP ensembles consist of an overgarment, mask, hood, overboots, and protective gloves **(Figure 8.36)**. MOPPs provide 6 flexible levels of protection (0-4, plus Alpha) based on threat level, work rate for the mission, temperature and humidity **(Table 8.4)**. The higher the MOPP level, the greater the protection (and the lower the rate of work productivity and efficiency). MOPP Level Alpha is designed for use in situations upwind from any threat with little danger of exposure to hazardous vapors.

Permeable garments such as the **joint service lightweight integrated suit technology (JSLIST)** protective overgarment provide protection against liquid, solid, and/or vapor CB agents and radioactive alpha and beta particles. JSLIST overgarments are manufactured of lightweight 50% Nylon and 50% cotton ripstop water repellant permeable materials. They are equipped with a charcoal/carbon lining designed to absorb harmful materials, much like the carbon filter/canister on an APR. The JSLIST protective garment will be degraded when in contact with certain solvents such as sweat and petroleum products. JSLIST can be laundered up to six times for personal hygiene. The JSLIST ensemble may be worn 45 consecutive days with a total out of the bag

Joint Service Lightweight Integrated Suit Technology (JSLIST) — A chemical-protective universal, lightweight, two-piece, front-opening suit that can be worn as an overgarment or as a primary uniform over underwear. The JSLIST liner consists of a non-woven front laminated to activate carbon spheres and bonded to a knitted back that absorbs chemical agents.

Table 8.4
MOPP Levels

Non-firefighter MOPP Levels

	MOPP 0	MOPP 1	MOPP 2	MOPP 3	MOPP 4
JSLIST	Carried	Worn	Worn	Worn	Worn
Protective Mask	Carried	Carried	Carried	Worn	Worn
Cotton Insert Gloves	Carried	Carried	Carried	Carried	Worn
Butyl Rubber Gloves	Carried	Carried	Carried	Carried	Worn
Protective Overgarment Boots	Carried	Carried	Worn	Worn	Worn

Firefighter MOPP Levels

	MOPP 0	MOPP 1	MOPP 2	MOPP 4 Firefighting Mode
SLIST	Carried	Worn	Worn	Worn
Nomex Hood	Carried	Worn	Worn	Worn
Firefighting Protective Trousers	Carried	Carried	Worn	Worn
Firefighting Protective Jacket	Carried	Carried	Carried	Worn
Firefighting Protective Footwear	Carried	Carried	Worn	Worn
CW Mask	Carried	Carried	Worn	Worn
Fire & Chemical Protective Gloves	Carried	Carried	Worn	Worn
Cotton Insert Gloves *	Carried	Carried	Worn	Worn
Butyl Rubber Gloves *	Carried	Carried	Worn	Worn
Firefighting Gloves *	Carried	Carried	Carried	Worn
Structural ARFF Helmet	Carried	Carried	Carried	Worn
SCBA	Carried	Carried	Carried	Worn

* Per current military Technical Orders

available usage of 120 days. The protective mask is a hooded APR, and the boots and gloves are made of Butyl rubber.

Firefighters who wear the JSLIST will find that the MOPP levels are modified to suit their capabilities. With the additional wear and tear of fire fighting protective equipment commonly found by firefighters wearing the JSLIST, firefighters will use only 3 MOPP levels (0,1,4). When the typical wearer is in MOPP 2, firefighters will remain in MOPP 1 until directed to proceed to a higher level. When MOPP 3 is directed, firefighters will immediately proceed from MOPP 1 and don the appropriate MOPP 4 attire. When a need to conduct fire fighting operations arises, firefighters will proceed to MOPP 4 firefighter mode where they will don their fire fighting protective jacket and SCBA.

Safety, Climate Concerns, and Health Issues
Most types of PPE inhibit the body's ability to disperse heat, which is magnified because an emergency responder is usually performing strenuous work while wearing the equipment. Thus, wearing PPE usually increases responders' risks of developing heat-related disorders. However, when working in cold climates,

Heat Stroke — Heat illness caused by heat exposure, resulting in failure of the body's heat regulating mechanism; symptoms include (a) high fever of 105 to 106°F (40.5°C to 41.1°C), (b) dry, red, and hot skin, (c) rapid, strong pulse, and (d) deep breaths or convulsions; may result in coma or possibly death. Also called Sunstroke.

Heat Exhaustion — Heat illness caused by exposure to excessive heat; symptoms include weakness, cold and clammy skin, heavy perspiration, rapid and shallow breathing, weak pulse, dizziness, and sometimes unconsciousness.

Heat Cramps — Heat illness resulting from prolonged exposure to high temperatures; characterized by excessive sweating, muscle cramps in the abdomen and legs, faintness, dizziness, and exhaustion.

Heat Rash — Condition that develops from continuous exposure to heat and humid air; aggravated by clothing that rubs the skin; reduces the individual's tolerance to heat.

considerations must be taken to protect responders from cold-related disorders, as well. For example, CPC is not designed to provide insulation against the cold. It is important to take preventive measures to reduce the effects of any temperature extreme. Medical monitoring of responders is required when they are at risk because of environmental hazards.

Heat Disorders

Wearing PPE or other special full-body protective clothing puts the wearer at considerable risk of developing heat stress. This stress can result in health effects ranging from transient heat fatigue to serious illness (heat stroke) or death.

First responders need to be aware of several heat disorders, including **heat stroke** (the most serious, see safety box), **heat exhaustion**, **heat cramps**, and **heat rashes**. In addition, they should know how to prevent the effects of heat exposure.

Heat Stroke

Heat stroke occurs when the body's system of temperature regulation fails and body temperature rises to critical levels. This condition is caused by a combination of highly variable factors, and its occurrence is difficult to predict. **Heat stroke is a serious medical emergency and requires immediate medical treatment and transport to a medical care facility.** The primary signs and symptoms of heat stroke are as follows:

- Confusion
- Irrational behavior
- Loss of consciousness
- Convulsions
- Lack of sweating (usually)
- Hot, dry skin
- Abnormally high body temperature (for example, a rectal temperature of 105.8°F [41°C])

When the body's temperature becomes too high, it causes death. The elevated metabolic temperatures caused by a combination of workload and environmental heat load, both of which contribute to heat stroke, are also highly variable and difficult to predict. If a first responder shows signs of possible heat stroke, obtain professional medical treatment immediately.

Heat-Exposure Prevention

Responders wearing protective clothing need to be monitored for the effects of heat exposure. Methods to prevent and/or reduce the effects of heat exposure include the following:

- *Fluid consumption* — Use water or commercial body-fluid-replenishment drink mixes to prevent dehydration. First responders should drink generous amounts of fluids both before and during operations. Drinking 7 ounces (200 ml) of fluid every 15 to 20 minutes is better than drinking large quantities once an hour. Balanced diets normally provide enough salts to avoid cramping problems. *Details:*

  — Before working, drinking chilled water is good.

— After a work period in protective clothing and an increase in core temperature, drinking room-temperature water is better. It is not as severe a shock to the body.

- *Air cooling* — Wear long cotton undergarments or similar types of clothing to provide natural body ventilation. Once PPE has been removed, blowing air can help to evaporate sweat, thereby cooling the skin. Wind, fans, blowers, and misters can provide air movement. However, when ambient air temperatures and humidity are high, air movement may provide only limited benefit.

- *Ice cooling* — Use ice to cool the body; however care must be taken not to damage skin with direct contact with ice, as well as to not cool off an individual too quickly. Ice will also melt relatively quickly. Ice cooling vests are available.

- *Water cooling* — Use water to cool the body. When water (even sweat) evaporates from skin, it cools. Provide mobile showers and misting facilities or evaporative cooling vests. Water cooling becomes less effective as air humidity increases and water temperatures rise.

- *Cooling vests* —Wear cooling vests beneath PPE. Cooling vest technologies may use ice, fluids, evaporation, gels, or phase change cooling technology. Unlike the lower temperatures provided by ice or gel vests, phase change cooling technology vests interact with body heat to maintain the garment at a consistent temperature of 59°F (15°C). These vests may be bulky, cumbersome, and they may impair movement. Forced air cooling vest systems blow air through tubes close to the skin to cool the body. **NOTE:** Use of cooling vests is being reviewed in Canada and the U.S. due to various health concerns, and several haz mat teams have disallowed them.

- *Rest/rehab areas* — Provide shade, humidity changes (misters), and air-conditioned areas for resting.

- *Work rotation* — Rotate responders exposed to extreme temperatures or those performing difficult tasks frequently.

- *Proper liquids* — Avoid liquids such as alcohol, coffee, and caffeinated drinks (or minimize their intake) before working. These beverages can contribute to dehydration and heat stress.

- *Physical fitness* — Encourage responders to maintain good physical fitness.

Cold Disorders

Cold temperatures caused by weather and/or other conditions such as exposure to cryogenic liquids are also environmental factors that must be considered when selecting PPE. Prolonged exposure to freezing temperatures can result in health problems as serious as **trench foot**, **frostbite**, and **hypothermia**. Protection from the cold must be a priority when conditions warrant.

The four primary environmental conditions that cause cold-related stress are low temperatures, high/cool winds, dampness, and cold water. Wind chill, a combination of temperature and velocity, is a crucial factor to evaluate when working outside. For example, when the actual air temperature of the wind is 40°F (4.4°C) and its velocity is 35 mph (56 kmph), the exposed skin experiences conditions equivalent to the still-air temperature of 11°F (-12°C). A dangerous situation of rapid heat loss may arise for any individual exposed to high winds and cold temperatures.

Trench Foot — Foot condition resulting from prolonged exposure to damp conditions or immersion in water; symptoms include tingling and/or itching sensation, pain, swelling, cold and blotchy skin, numbness, and a prickly or heavy feeling in the foot. In severe cases, blisters can form followed by skin and tissue dying and falling off.

Frostbite — Local freezing and tissue damage due to prolonged exposure to extreme cold.

Hypothermia — Abnormally low or decreased body temperature.

Medical Monitoring

It is very important to provide ongoing medical monitoring of responders who may be at risk because of environmental hazards (heat/cold stresses) as well as potential exposure to hazardous materials. Medical monitoring should be conducted before responders wearing chemical liquid-splash or vapor-protective clothing enter the warm and hot zones (pre-entry monitoring) as well as after leaving these zones (post-entry monitoring) as directed by the authority having jurisdiction. The evaluation will check such things as vital signs, hydration, skin, mental status, and medical history. Specifically, this check will measure:

- Blood pressure
- Weight
- Respirations
- Pulse
- Core body temperature
- ECG (electrocardiogram, a test that measures heart rhythm)

Each organization needs to establish written medical monitoring guidelines that establish minimum and maximum values for these evaluations. A postmedical monitoring follow-up is also recommended.

Safety and Emergency Procedures

In addition to issues such as cooling, preventing dehydration, and medical monitoring, there are other safety and emergency issues involved with wearing PPE. The safety briefing will cover relevant information such as the status of the incident (based on the preliminary evaluation and subsequent updates), the hazards identified, a description of the site, the tasks to be performed, and the expected duration of the tasks. It must also cover the PPE requirements, monitoring requirements, notification of identified risks, and any other pertinent information. After using PPE at an incident, it is important to fill out any associated reports or documentation as required by the AHJ.

Anytime a limited air supply such as SCBA is worn, air management is an important consideration. Time taken to walk to the incident, time taken to return from the incident, decon time, safety time (extra time allocated for emergency use), and work time must be calculated **(Figure 8.37)**. Air must be allocated for these estimated times. Many organizations have SOPs that explain calculations for doing this and/or designate maximum entry times (such as 20 minutes) based on the air supply available.

Communications are also a concern. Have pre-designated hand signals, motions, and gestures to communicate in case of problems, for example, a way to ask if everything is okay followed by a thumbs up or thumbs down. Signals for emergencies should also be designated (such as loss of air supply, medical emergency, or suit failure). If possible, entry teams, backup personnel, and appropriate safety personnel at the scene should have their own designated radio channel. Responders should always operate with buddy systems and with backups dressed in appropriate PPE (same level as the entry team) standing by. Responders should also be familiar with established evacuation signals.

Figure 8.37 Air management must consider the time taken to travel to and from the work zone. *Courtesy of the U.S. Navy, photo by JO2 Mark Schultz.*

In case of an emergency such as loss of air supply, suit integrity, or an injury and illness, all responders should follow local protocols for such situations. Typically, these protocols will involve notifying the appropriate personnel (such as the Entry Team Leader and/or Haz Mat Safety Officer) and exiting the hot zone as quickly as possible. If air supply is lost while wearing a vapor-protective suit, there is a limited amount of air in the suit itself that can be breathed if the SCBA facepiece or regulator is removed.

Responders using PPE at haz mat incidents must be familiar with their local procedures for going through the technical decontamination process (see Chapter 9, Decontamination). For more information on emergency procedures, see **Hazardous Materials Managing the Incident**, 3rd edition, by Noll, Hildebrand, and Yvorra.

PPE Inspection, Storage, Testing, and Maintenance

When wearing protective clothing and equipment, the end user must take all necessary steps to ensure that the protective ensemble performs as expected. During emergencies is not the right time to discover discrepancies in the protective clothing or respiratory protection. Following a standard program for inspection, proper storage, maintenance, and cleaning along with realizing PPE limitations is the best way to avoid exposure to dangerous materials during an emergency response. All inspections, testing, and maintenance of PPE must be conducted in accordance with manufacturer's recommendations.

Records must be kept of all inspection procedures. Periodic review of these records can provide an indication of protective clothing or equipment that requires excessive maintenance and can also serve to identify clothing or equipment that is susceptible to failure.

PPE must be stored properly to prevent damage or malfunction from exposure to dust, moisture, sunlight, damaging chemicals, extreme temperatures (hot and cold), and impact. Procedures are needed for both initial receipt of equipment and after use or exposure of that equipment. Many manufacturers specify recommended procedures for storing their products. Follow these procedures to avoid equipment failure resulting from improper storage.

Respiratory equipment is initially inspected when it is purchased. Once the equipment is placed into service, the organization's personnel perform periodic inspections. Operational inspections of respiratory protection equipment occur after each use, daily or weekly, monthly, and annually. The organization must define the frequency and type of inspection in the respiratory protection policy, and they should follow manufacturer's recommendations. The care, cleaning, and maintenance schedules of respiratory protection equipment should be based on the manufacturer's recommendation, NFPA® standards, or OSHA requirements. For more information on maintenance, testing, inspection, storage, and documentation of respiratory protection, see IFSTA's **Respiratory Protection for Fire and Emergency Services.**

Summary

Personal protective equipment is needed to protect emergency responders from the hazards present at haz mat and WMD incidents. However, no one type of PPE can protect against all hazards, so responders must select and use ensembles appropriate to their mission at the incident, based on the hazards and risks present.

PPE ensembles usually combine respiratory protective equipment such as SCBA or respirators with protective clothing such as firefighter protective gear, chemical-protective clothing and/or body armor. All protective clothing used at haz mat/WMD incidents should meet recognized standards such as those published by NIOSH or the NFPA® for use at such incidents.

Review Questions

1. What types of respiratory protection are used by responders at haz mat/WMD incidents? Describe each.

2. What are the advantages and disadvantages of SCBA?

3. What types of protective clothing may be used by responders at haz mat/WMD incidents? Describe each.

4. What are the limitations of using high-temperature protective clothing?

5. What kinds of operations require the use of chemical-protective clothing?

6. Describe the U.S. EPA levels of protection.

7. What factors determine the selection of PPE?

8. What types of ensembles may be used at haz mat/WMD incidents?

9. How can heat exposure be prevented when working in PPE?

10. How should PPE be stored?

Level A Suit

Step 1: Perform a visual inspection of PPE and SCBA for damage or defects.

Step 2: Don SCBA. Ensure that the cylinder valve is fully open and that all straps are secured.

Step 3: Don SCBA facepiece and ensure a proper fit and seal.

Step 4: Don Level A suit, placing legs into suit and pulling suit up to waist. Secure inner belt.

Step 5: Don outer boots.

Step 6: Don inner gloves.

Step 7: Place arms in suit and hands into outer gloves.

Step 8: Attach SCBA regulator to facepiece and make sure SCBA is functioning properly.

Step 9: Don hard hat and pull protective hood over head.

Note: If your AHJ requires a hard hat, don before pulling hood over head.

Step 10: Zip suit enclosure and secure zipper flap.

Step 11: Perform work assignment.

Step 12: After assignment has been performed, proceed to decontamination line.

Step 13: Undergo technical decontamination per AHJ's SOPs.

Step 14: Doff suit and SCBA according to AHJ SOPs, avoiding contact with outer suit or surfaces that may be contaminated.

Level B Suit

Step 1: Perform a visual inspection of PPE for damage or defects.

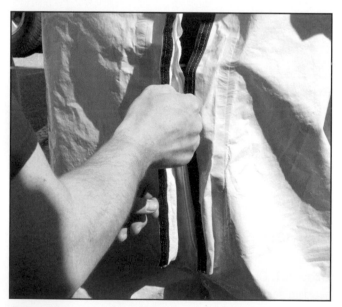

Step 2: Don non-encapsulating Level B PPE and secure closures.

Step 3: Don work boots.

Step 4: Pull suit leg opening over the top of the work boots.

Step 5: Don SCBA. Ensure that the cylinder valve is fully open and that all straps are secured.

Step 6: Don SCBA facepiece and ensure a proper fit and seal.

Step 7: Pull suit hood up completely so that facepiece straps and skin are not exposed.

Note: If your AHJ requires a hard hat, don after pulling hood over head.

Step 8: Don inner protective gloves.

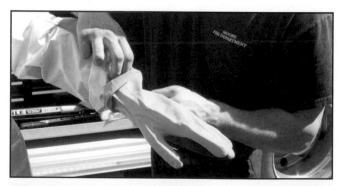

Step 9: Don outer protective gloves.

Step 10: Pull suit sleeves over the outside of the gloves.

Step 11: Attach SCBA regulator to facepiece and make sure SCBA is functioning properly.

Step 12: Perform work assignment.

Step 13: After assignment has been performed, proceed to decontamination line.

Step 14: Undergo technical decontamination per AHJ's SOPs.

Step 15: Doff suit and SCBA according to AHJ SOPs, avoiding contact with outer suit or surfaces that may be contaminated.

8-1

Don and doff different types of personal protective equipment (PPE).

Level C Suit

Step 1: Perform a visual inspection of PPE for damage or defects.

Step 2: Don non-encapsulating Level C PPE and secure closures.

Step 3: Don work boots.

Step 4: Pull suit leg opening over the top of the work boots.

Step 5: Ensure that the canister is compatible to the hazardous material being encountered.

Step 6: Attach canister to facepiece.

Step 7: Don facepiece and ensure a proper fit and seal.

Step 8: Pull suit hood up completely so that facepiece straps and skin are not exposed.

Step 9: Don inner protective gloves.

Step 10: Don outer protective gloves.

Step 11: Pull suit sleeves over the outside of the gloves.

Step 12: Attach SCBA regulator to facepiece and make sure SCBA is functioning properly.

Step 13: Perform work assignment.

Step 14: After assignment has been performed, proceed to decontamination line.

Step 15: Undergo technical decontamination per AHJ's SOPs.

Step 16: Doff suit and SCBA according to AHJ SOPs, avoiding contact with outer suit or surfaces that may be contaminated.

Structural Fire-Fighting PPE

Step 1: Perform a visual inspection of PPE for damage or defects.

Step 2: Don protective trousers and boots.

Step 3: Don protective hood, pulling hood down around neck and exposing head.

Step 4: Don protective coat.

Step 5: Don SCBA. Ensure that the cylinder valve is fully open and that all straps are secured.

Step 6: Don SCBA facepiece and ensure a proper fit and seal.

Step 7: Attach SCBA regulator to facepiece and make sure SCBA is functioning properly.

Step 8: Pull hood up completely so that facepiece straps and skin are not exposed.

Step 9: Don helmet and secure.

Step 10: Don gloves.

Step 11: Ensure that all fasteners, straps, buckles, etc. are fastened.

Step 12: Ensure that no skin is exposed.

Step 13: Perform work assignment.

Step 14: After assignment has been performed, proceed to decontamination line.

Step 15: Undergo technical decontamination per AHJ's SOPs.

Step 16: Doff PPE and SCBA according to AHJ SOPs, avoiding contact with outer suit or surfaces that may be contaminated.

Decontamination

Chapter Contents

chapter **9**

Key Terms

Competencies

NFPA® 472:	6.2.4.1(4)	6.3.3.2(5)	6.4.3.2(1)	6.4.3.2(2)(i)	6.4.4.1(2)
5.3.4(1)	6.3.3.1	6.3.4.1	6.4.3.2(2)(a)	6.4.3.2(2)(j)	6.4.4.2(1)
5.3.4(2)	6.3.3.2(1)	6.3.4.2	6.4.3.2(2)(b)	6.4.3.2(2)(k)	6.4.4.2(2)
5.3.4(3)	6.3.3.2(2)	6.3.5.1	6.4.3.2(2)(c)	6.4.3.2(2)(l)	6.4.5.1
5.3.4(4)	6.3.3.2(2)(a)	6.3.6.1(1)	6.4.3.2(2)(d)	6.4.3.2(3)	6.4.6.1(1)
5.3.4(5)	6.3.3.2(2)(b)	6.3.6.1(2)	6.4.3.2(2)(e)	6.4.3.2(4)	6.4.6.1(2)
5.3.4(6)	6.3.3.2(2)(c)	6.3.6.1(3)	6.4.3.2(2)(f)	6.4.3.2(5)	6.4.6.1(3)
5.4.1(4)	6.3.3.2(3)	6.3.6.1(4)	6.4.3.2(2)(g)	6.4.3.2(6)	6.4.6.1(4)
6.2.3.1(3)(f)	6.3.3.2(4)	6.4.3.1	6.4.3.2(2)(h)	6.4.4.1(1)	6.9.4.1.1(2)

Decontamination

Learning Objectives

1. Define decontamination. [NFPA® 472, 5.3.4(1), 5.3.4(2)]

2. Identify various decontamination methods. [NFPA® 472, 5.3.4(3), 5.3.4(4)]

3. Discuss general guidelines for decon operations.

4. Describe the different types of victims that may receive decontamination.

5. Describe emergency decontamination. [NFPA® 472, 5.3.4(5), 5.3.4(6)]

6. Perform emergency decontamination. [Skill Sheet 9-1; NFPA® 472, 5.4.1(4)]

7. Describe technical decontamination. [NFPA® 472, 6.2.3.1(3)(f), 6.4.3.2(1-6), 6.4.4.1(1-2)]

8. Set up and implement a technical decontamination corridor and undergo decontamination. [Skill Sheet 9-2; NFPA® 472, 6.4.3.1, 6.4.4.2(1-2), 6.9.4.1.1(2)]

9. Perform technical decontamination on a non-ambulatory victim. [Skill Sheet 9-3; NFPA® 472, 6.2.4.1(4)]

10. Discuss mass decontamination. [NFPA® 472, 6.3.3.2(1), 6.3.3.2(2)(a-c), 6.3.3.2(3-5)]

11. Perform mass decontamination. [Skill Sheet 9-4; NFPA® 472, 6.3.4.2]

12. Determine the effectiveness of decontamination operations. [NFPA® 472, 6.3.5.1, 6.4.5.1]

13. Explain how to implement decontamination. [NFPA® 472, 6.3.6.1(1-4), 6.4.6.1(1-4), 6.9.4.1.1(2)]

Chapter 9
Decontamination

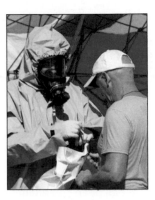

Case History

On October 12, 2001, during the height of the U.S. anthrax attacks, a Michigan State University (MSU) police dispatcher received a call from Debbie Conlin, an employee in Linton Hall, who reported that opening a letter had caused her to have a burning sensation in her throat. Unfortunately, the dispatcher confused this information with a call received 16 minutes earlier from an employee in the University Club (a different location on campus) who reported she had received an envelope with a suspicious white powder in it a month before. The dispatcher proceeded to notify the FBI; the MSU Office of Radiation, Chemical, and Biological Safety; and the East Lansing Fire Department of a white powder incident at Linton Hall.

Because the two calls were confused, firefighters from East Lansing arrived at Linton Hall expecting a white powder incident, following procedures for a biological or chemical threat received in the mail. Even though Conlin and her coworkers tried to explain that there had been no powder in the letter she reported, emergency responders decided to treat the situation as though there had been. Eight employees in the Linton Hall office and seven other employees on adjoining floors were forced to strip and be decontaminated in a makeshift cleansing station, being scrubbed down by male police officers and firefighters.

In the ensuing lawsuit brought by several of the women against the East Lansing FD, the women testified that they felt demeaned and traumatized by the experience. Those chosen to be decontaminated appeared to be at random, and the fact that police officers and firefighters not actively assisting in the decontamination process were standing adjacent to the naked women was highlighted as particularly disturbing. While precautions were taken to block doorways leading to the outside, none were taken to shield the windows in the hall from the stairwells leading to other levels of the building. This resulted in other employees, who were not being decontaminated, seeing the women naked during the cleaning process.

The women also were given conflicting information regarding the best way to prevent continued contamination. The emergency responders on the way to the hospital said to only take cold showers, while those at the hospital forced them to take hot showers. Upon returning to Linton Hall after this experience the women were told to wash their clothes as normal and no further follow up was offered initially.

The confusion in the dispatcher's information, the conflicting advice from emergency personnel, and the disregard for the privacy of the women being decontaminated all contributed to a haz mat incident that ended badly. The lessons learned from this incident have lead to changes in MSU's policies and procedures when dealing with situations like this. However, better communication and sensitivity to the personal rights of the women in the situation could have prevented the problems that did occur.

Figure 9.1 Decontamination is performed to remove hazardous materials from victims, responders, and anything else that has been contaminated or potentially contaminated. *Courtesy of Boca Raton Fire Rescue.*

Contamination is the transfer of a hazardous material to persons, equipment, and the environment in greater than acceptable quantities. Decontamination (commonly referred to as *decon*) or contamination reduction is the process of removing hazardous materials to prevent the spread of contaminants beyond a specific area and reduce contamination to levels that are no longer harmful. Decontamination (decon) is an essential activity that must be considered at any hazardous materials or terrorism incident to ensure the safety of emergency responders and the public. Decon operations minimize harmful exposures and reduce or eliminate the spread of contaminants. Decontamination is performed at haz mat/WMD incidents to remove hazardous materials from responders, victims, PPE, tools, equipment, and anything else that has been contaminated **(Figure 9.1)**. Everyone and everything in the hot zone is subject to contact with the hazardous material and can become contaminated. Because of this potential, anything that goes into the hot zone passes through a decon area when leaving the zone.

Decon also provides victims with psychological reassurance. Some individuals who have been potentially exposed to hazardous materials may develop psychologically-based symptoms (i.e., shortness of breath, anxiety) even if they have not actually been exposed to harmful levels of contamination. Conducting decon can reduce or prevent these types of problems.

Although emergency responders may have considerable experience with decon at haz mat incidents, performing decon at a terrorist incident may require some changes to the procedures used. Haz mat/WMD incidents may involve large numbers of people that have to be quickly assessed for injury or exposure and then passed through a decon corridor for treatment or safe sheltering away from the incident area (mass decon). Also, since a terrorist incident must be treated as a crime scene, any clothing, equipment, or contaminated materials have to be protected as evidence and handled in accordance with locally adopted procedures.

The type of decon operations conducted at an incident will be determined by the number of persons requiring decon, the type of hazardous materials involved, weather (washing off contaminants with a hose stream may not be a viable option in subzero temperatures), personnel available, and a variety of other factors. However, regardless of the many variables that may be encountered at the incident, the basic principles of any decontamination operation are easy to summarize:

1. Get it off.

2. Keep it off.

3. Contain it (prevent cross-contamination).

It is important to continually assess the effectiveness of any decontamination operation. If monitoring determines that the selected method is not working, a different technique must be tried. Before initiating decontamination, the answers to the following questions should be considered:

- Do victims need to be decontaminated immediately or can they wait?

- Is it safe to conduct decon?

- What alternative decon methods are available?

- Are there adequate resources to conduct the operation? If not, can additional resources be obtained in a timely fashion?

- What is the time limit available to conclude decon before the victims further deteriorate?

- Is the equipment you are attempting to decontaminate going to be useable again and/or is it more cost effective to simply dispose of? Does decon save money or add value?

Because decon procedures, terminology, and other details may differ greatly from organization to organization, responders must be familiar with their organization's decon policies and procedures and how decon operations are implemented within the AHJ's incident command system. This chapter discusses decontamination methods and types, and how to implement decontamination procedures. It will discuss **emergency decon**, **mass decon**, and **technical decon** separately.

Decontamination Methods

Decontamination methods can be divided into the four broad categories: Wet or dry methods and physical or chemical methods. Decontamination methods vary in their effectiveness for removing different substances, and many factors may play a part in the selection decision.

The most effective means of decontamination is often as simple as the removal of the outer clothing or PPE that has been contaminated by the hazardous material. It is estimated that simply taking clothes off can remove a high percentage of the contaminant.

Additionally, flushing the contaminated surface with water is effective at removing the harmful substance or sufficiently diluting it to a safe level. For this reason, removal of contaminated clothing/PPE and flushing with water is usually sufficient for emergency and mass decon. Technical decon requires additional effort to meet the objective of thoroughly removing all contaminants

Emergency Decon — The physical process of immediately reducing contamination of individuals in potentially life-threatening situations with or without the formal establishment of a decontamination corridor.

Mass Decon — Process of decontaminating large numbers of people in the fastest possible time to reduce surface contamination to a safe level. It is typically a gross decon process utilizing water or soap and water solutions to reduce the level of contamination.

Technical Decon — A planned and systematic process of reducing contamination to a level that is As Low As Reasonably Achievable (ALARA). Technical decon operations are normally conducted in support of emergency responder recon and entry operations at a hazardous materials incident, as well as for handling contaminated patients at medical facilities.

and involves washing with water and some sort of soap, detergent, or chemical solution. The decision whether to perform emergency or technical decon is determined based on the hazardous material involved and the urgency in removing the victim from the contaminated environment.

Wet and Dry

As their names imply, wet and dry methods are categorized by whether they use water or other solutions as part of the decon process. Wet methods usually involve washing the contaminated surface with solutions or flushing with a hose stream or safety shower, whereas dry methods such as scraping, brushing, and absorption do not.

Wet methods may necessitate the collection of runoff water in wading pools or other liquid-retaining (containment) devices **(Figure 9.2)**. Collected or containerized water may need to be analyzed for treatment and disposal. Disposal of runoff water and residue from decon operations must be properly disposed of in accordance with applicable laws and regulations. Proper authorities must be notified and consulted during this process. In some cases, wet methods may be difficult or impractical to use due to environmental or weather conditions. Life safety must take precedence over environmental considerations (for example, in mass decon situations).

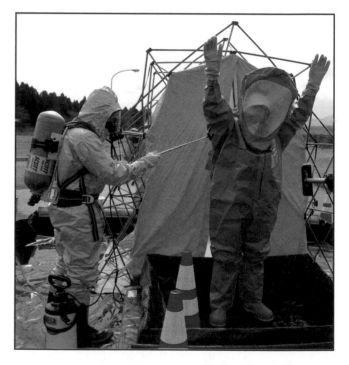

Figure 9.2 Potentially contaminated runoff water from wet decon methods may need to be contained in wading pools or other catch basins in order to protect the environment. *Courtesy of the U.S. Marines, photo by Warren Peace.*

Dry methods may be as simple as removing contaminated clothing and putting it into a suitable plastic bag (or recovery drum) or allowing the contaminant to evaporate. Other dry methods include vacuuming or brushing a powder or dust from a contaminated surface, scraping a material off, or using sticky tape (or a sticky pad) to clean or wipe off contamination. Dry methods have the advantage of not creating large amounts of contaminated liquid runoff (although absorption can result in a greater amount of contaminated material in general), and they may be accomplished through the systematic removal of disposable PPE while avoiding contact with any contaminants. Dry methods may be used during cold weather operations when wet methods are difficult to implement.

Dry materials can also be used to remove liquid chemicals by absorption. Such materials include clay, sawdust, flour, dirt, Fuller's earth, tissue paper, charcoal, silica gel, paper towels and sponges **(Figure 9.3)**. Once used, these materials must be treated as contaminated waste and disposed of accordingly.

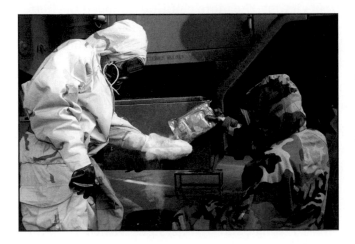

Figure 9.3 Dry decon can be conducted using materials such as M100 reactive sorbent powder. *Courtesy of the U.S. Army, photo by Staff Sgt. Fredrick P. Varney, 133rd Mobile Public Affairs Detachment.*

Physical and Chemical

Physical methods of decontamination remove the contaminant from a contaminated person without changing the material chemically (although with wet methods it may be diluted). The contaminant is then contained (when practical) for disposal. Examples of physical decontamination methods include absorption, adsorption, brushing and scraping, dilution, evaporation, isolation and disposal, washing, and vacuuming.

Chemical methods are used to make the contaminant less harmful by changing it through some kind of chemical process. For example, using bleach to sanitize tools and equipment that have been exposed to potentially harmful etiological agents is a form of chemical decontamination because the organisms are actually killed by the bleach. Examples of chemical methods include chemical degradation, sanitization, disinfection, sterilization, neutralization, and solidification.

General Guidelines for Decon Operations

General guidelines for decon operations include the following:

- Ensure technical decon setup is operational before entry personnel enter the hot zone.
- Begin emergency/mass decon operations quickly; the speed necessary will be determined by the material and type of incident involved (for example, chemical agents may need immediate removal whereas biological agents may not).
- Always wear appropriate PPE.
- Avoid contacting hazardous materials, including contaminated victims.

- Decon operations may be coupled with an initial separation of victims into ambulatory/non-ambulatory and male/female.

- Assess all victims believed to have been in the hot zone to determine the need for decontamination before moving them to the cold zone. Decontaminate as necessary.

- Establish clearly designated decon entry points so that victims and responders both know where to go.

- When conducting decon of victims, the more clothing removed the better (disrobing is effective decon by itself). Unless a victim is soaked in something that would have penetrated outer clothing and into their underwear, there is no real need to have people disrobe completely.

- Decontaminate all emergency response personnel who have been in the hot zone before moving to the cold zone.

- Decon emergency responders separately from victims (establish separate decon lines when possible and practical).

- Establish a medical triage and treatment area just outside the decon zone so that persons exiting the decon area can be evaluated for injuries and exposure-related medical symptoms.

- Communicate with victims by using hand signals, signs with pictures, apparatus public address systems, megaphones or other methods to direct them to decon gathering areas as well as through the decon process itself. It is very important to provide clear and easily understood directions since people may be traumatized and/or suffering from exposures **(Figure 9.4)**.

Figure 9.4 It is very important to communicate clearly with potentially traumatized victims. Use hand signals, signs with pictures, apparatus public address systems, megaphones or other methods to provide simple instructions on where to go and what to do. *Courtesy of New South Wales Fire Brigades.*

- Provide privacy whenever possible (including from overhead vantage points, for example, from circling news helicopters or upper stories of nearby buildings).

- Provide warm water for washing, if possible. If water is cold, allow victims to gradually get wet in order to acclimate to the temperature and avoid cold shock.

What This Means To You

In mass decon situations, you will wear whatever PPE you have available when you arrive on the scene. In most situations, fire service personnel will wear structural firefighter protective clothing and SCBA **(Figure 9.5)**. As more information about the hazardous material involved is gathered, adjustments can be made accordingly.

In technical decon situations where the hazardous material has been identified, use NIOSH guides and manufacturer's recommendations to determine appropriate chemical protective clothing and respiratory protection. Often, those conducting decon are dressed in an ensemble classified one level below that of the entry team **(Figure 9.6)**. Thus, if the entry team is dressed in an Environmental Protection Agency (EPA) Level A ensemble, the decon team is dressed in Level B. In some cases, if you are the first person working in the decon line, you may need to be dressed at the same level as the entry team. In either case, chemical gloves are necessary; fire-fighting gloves should not be used in decontamination procedures **(Figure 9.7)**. Because there is a possibility that you may become contaminated while working decontamination operations, you will need to pass through decontamination before leaving the corridor.

Decon supervisors should always have a plan for responders who may run low on air involving SCBA during decon operations. Consult your SOPs and emergency response plans for additional information.

Figure 9.5 In most mass decon situations, fire service responders will start off wearing firefighter protective clothing and SCBA.

Figure 9.6 In technical decon scenarios, those conducting decon are often dressed in an ensemble classified one level below that of the entry team. *Courtesy of FEMA News Photos, photo by Win Henderson.*

Figure 9.7 Leather gloves can absorb contaminants and should not be worn by individuals conducting decon operations.

- Preserve and record belongings of victims decontaminated for future identification of victims and forensic examination.

- Provide the victims and responders with clean/alternative clothing to maintain their privacy and protect them from the weather.

Recipients of Decontamination Operations

Decon supervisors must plan operations according to the needs of the individuals needing to undergo decon. These may include responders, victims (both ambulatory and nonambulatory), pets, and other animals **(Figure 9.8)**.

Figure 9.8 Because past disasters have shown that many individuals are unwilling to abandon pets even in life-threatening situations, agencies should have a plan in place for managing a variety of animals in evacuation and decon situations. *Courtesy of FEMA News Photos, photo by Michael Rieger.*

Victims may be of all ages and capabilities, with any number of disabilities, from a variety of cultures (who may or may not understand the local language). They may not be able to hear, see, or understand directions from responders. SOPs should provide guidance on how to conduct triage operations and manage decon operations for ambulatory and non-ambulatory persons.

Triage

Triage — System used for sorting and classifying accident casualties to determine the priority for medical treatment and transportation.

Procedures for conducting **triage** of patients should be predetermined within the local emergency response plan. In most instances, triage will be conducted in the cold zone after decontamination has been performed. However, in some cases (such as at explosive incidents) it may be conducted in the hot zone prior to decon. Prioritization of nonambulatory patients for decontamination can be done using medical triage systems such as START (Simple Triage and Rapid Treatment/Transport). **Figure 9.9** provides four START Categories as provided in the SBCCOM document, *Guidelines for Mass Casualty Decontamination During a Terrorist Chemical Agent Incident.*

START Medical Triage System

START Category	Decon Priority	Classic Observations	Chemical Agent Observations
IMMEDIATE Red Tag	1	Respiration is present only after repositioning the airway. Applies to victims with respiratory rate >30. Capillary refill delayed more than 2 seconds. Significantly altered level of consciousness.	• Serious signs/symptoms • Known liquid agent contamination
DELAYED Yellow Tag	2	Victim displaying injuries that can be controlled/treated for a limited time in the field.	• Moderate to minimal signs/symptoms • Known or suspected liquid agent contamination • Known aerosol contamination • Close to point of release
MINOR Green Tag	3	Ambulatory, with or without minor traumatic injuries that do not require immediate or significant treatment.	• Minimal signs/symptoms • No known or suspected exposure to liquid, aerosol, or vapor
DECEASED/ EXPECTANT Black Tag	4	No spontaneous effective respiration present after an attempt to reposition the airway.	• Very serious signs/symptoms • Grossly contaminated with liquid nerve agent • Unresponsive to autoinjections

Figure 9.9 These four START categories can be used to triage patients at chemical agent incidents. *Courtesy of the U.S. Army Soldier and Biological Chemical Command (SBCCOM).*

Ambulatory

Victims who are able to understand directions, talk, and walk unassisted are considered ambulatory, and they should be directed to an area of safe refuge within the isolation perimeter to await prioritization for decontamination. **Figure 9.10** provides a sample layout for ambulatory decon.

Several of the following factors may influence the priority for ambulatory patients:

- Victims with serious medical symptoms (such as shortness of breath or chest tightness)
- Victims closest to the point of release
- Victims reporting exposure to the hazardous material
- Victims with evidence of contamination on their clothing or skin
- Victims with conventional injuries (broken bones, open wounds, and the like)

Figure 9.10 A sample layout for ambulatory decon. Responders should be familiar with the system used by their agency. *Courtesy of Doug Weeks.*

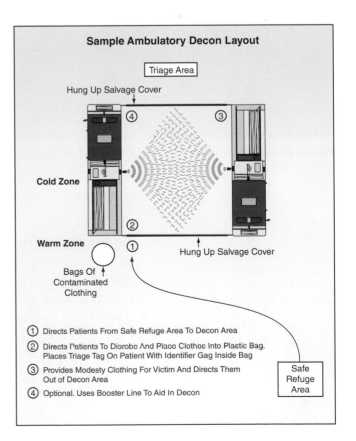

Sample Ambulatory Decon Layout

Triage Area

Hung Up Salvage Cover

Cold Zone

Warm Zone

Bags Of Contaminated Clothing

Hung Up Salvage Cover

Safe Refuge Area

① Directs Patients From Safe Refuge Area To Decon Area

② Directs Patients To Disrobe And Place Clothes Into Plastic Bag. Places Triage Tag On Patient With Identifier Gag Inside Bag

③ Provides Modesty Clothing For Victim And Directs Them Out of Decon Area

④ Optional. Uses Booster Line To Aid In Decon

Emergency Decon of Ambulatory Patients at Incidents Involving Chemical Agents

At incidents involving chemical agents, very little time may be available to successfully conduct decon. For example, after skin contact with nerve agents, industrial chemical agents, and vesicants, emergency decon should be conducted immediately (within minutes). It is therefore vital that rescuers conduct rapid extraction and prioritization of ambulatory victims as quickly and efficiently as possible.

Ambulatory victims showing visible signs of contamination or symptoms of exposure to chemical agents should be directed to undergo emergency decon **(Figure 9.11)**. Emergency decon may precede additional mass decon procedures (for example, if portable decon tents or trailers are not yet set-up or available for use) and generally involves disrobing and flushing in high volume, low pressure showers (hand-held hose lines, or side-by-side apparatus, **Figure 9.12**). Clothing and personal items may be bagged, but when victims require emergency decon, time should not be taken to tag items or conduct medical evaluations. The goal of emergency decon is to remove contaminants quickly. After undergoing emergency decon, victims may be directed to undergo additional mass decon, or they may be allowed to dry and redress (in provided clean clothing) as appropriate before proceeding for medical evaluation and treatment. In most cases where emergency decon and mass decon are needed, dilution is the solution.

Figure 9.11 During emergency decon, life safety is the first priority. *Courtesy of New South Wales Fire Brigades.*

Figure 9.12 Use large volumes of water at low pressures to dilute contaminants and wash them away. This configuration is used by the Los Angeles Fire Department.

Nonambulatory

Nonambulatory patients are victims or responders who are unconscious, unresponsive, or unable to move unassisted. These patients may be more seriously injured than ambulatory patients. They may have to remain in place if sufficient personnel are not available to remove them from the hot zone. **Figure 9.13** provides an example of a decon corridor layout for nonambulatory patient decon.

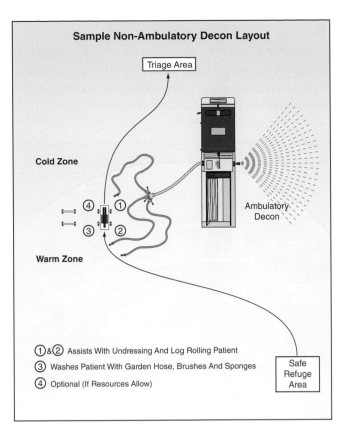

Sample Non-Ambulatory Decon Layout

Triage Area

Cold Zone

Warm Zone

Ambulatory Decon

① & ② Assists With Undressing And Log Rolling Patient

③ Washes Patient With Garden Hose, Brushes And Sponges

④ Optional (If Resources Allow)

Safe Refuge Area

Figure 9.13 A sample layout for non-ambulatory decon. *Courtesy of Doug Weeks.*

Deceased

Responders must be prepared for the reality of handling deceased victims. Normally, removal of deceased victims from the hot zone will be delayed until all viable victims have been removed. Responders must consider ethical issues when removing the deceased and handle them with the utmost level of respect and dignity. Once deceased victims have been removed from the hot zone, it is important that decon operations be completed prior to the transfer of the deceased to the medical examiner.

As a general rule, deceased victims at the scene of an event should remain untouched. Emergency services personnel should at all times be mindful of the need to preserve the incident scene and conduct operations with minimal disturbance and in consultation with those tasked with forensic evidence collection. At the appropriate time, the law enforcement agency (or designated authority) having jurisdiction will make a determination as to how victim remains will be managed.

Handling large numbers of deceased victims may be beyond the capabilities of local emergency response personnel. Specialty response teams (i.e. disaster mortuary teams [DMORT]) may be requested to assist in these types of incidents. In the U.S. and Canada, these teams must be requested through the appropriate emergency management office in order to activate the assistance request. An on-scene morgue facility may have to be established if the incident involves large numbers of deceased victims.

Emergency Decon

The goal of emergency decontamination is to remove the threatening contaminant from the victim as quickly as possible — there is no regard for the environment or property protection. Emergency decon may be necessary for both victims and rescuers **(Figure 9.14)**. If either is contaminated, individuals

Figure 9.14 Both victims and responders may need emergency decon.

are stripped of their clothing and washed quickly. Victims may need immediate medical treatment, so they cannot wait for a formal decontamination corridor to be established. The following situations are examples of instances where emergency decontamination is needed:

- Failure of protective clothing
- Accidental contamination of emergency responders
- Heat illness or other injury suffered by emergency workers in the hot zone
- Immediate medical attention is required

Emergency decontamination has definite limitations. Removal of all contaminants may not occur, and a more thorough decontamination must follow. Emergency decontamination can harm the environment. If possible, measures must be taken to protect the environment, but such measures should not delay life-saving actions. The advantage of eradicating a life-threatening situation far outweighs any negative effects that may result.

There are times when what appears to be a normal incident really involves hazardous materials. Emergency responders may become contaminated before they realize what the situation really is or because they have been targeted by terrorists. When these situations occur, emergency responders need to withdraw immediately. They need to follow local procedures for emergency decon. They should remain isolated until someone with the proper expertise and monitoring equipment can ensure that they have been adequately decontaminated. Emergency decontamination procedures may differ depending on the circumstances and hazards present at the scene. Refer to **Skill Sheet 9-1** for steps in conducting emergency decontamination.

ℹ️

Emergency Decontamination: Advantages and Limitations

Advantages:

- Fast to implement
- Requires minimal equipment (usually just a water source such as a hoseline)
- Reduces contamination quickly
- Does not require a formal contamination reduction corridor or decon process

Limitations:

- Does not always totally decontaminate the victim
- Can create contaminated runoff that can harm the environment and other exposures

Technical Decon

Technical decontamination uses chemical or physical methods to thoroughly remove or neutralize contaminants from responders' PPE (primarily entry team personnel) and equipment **(Figure 9.15)**. It may also be used on incident victims in non-life-threatening situations. Operations-Level responders involved in technical decon operations must do so under the guidance of a haz mat technician, SOPs, or an **allied professional**. Responders must be familiar with the AHJ's procedures for implementing technical decon within the incident command system. During technical decon operations, Operations-Level responders will typically act as follows:

- Protect themselves by dressing in appropriate PPE
- Establish a water supply
- Set up the decon corridor
- Establish perimeters

Allied Professional — Individual with the training and expertise to provide competent assistance and direction at haz mat and WMD incidents.

Figure 9.15 Technical decon uses chemical or physical methods to thoroughly remove or neutralize contaminants from entry team members. It may also be used on victims in non-life-threatening situations. *Courtesy of the U.S. Air Force, photo by Chiaki Iramina.*

- Perform physical decontamination activities (scrubing, washing, spraying, etc.)
- Assist in the undressing/removal of PPE or clothing of individuals going through the decon line **(Figure 9.16)**

Figure 9.16 Ops-Level responders may perform a variety of functions in a tech decon line, including assisting in the removal of PPE or clothing of individuals undergoing decon.

- Assist individuals going through the decon process
- Perform other duties per SOPs and training

Technical decon is usually conducted within a formal decon line or corridor. The contaminants involved at the incident determine the type and scope of technical decon. Resources for determining the correct technical decon procedures include the following:

- NIOSH Pocket Guide
- Safety data sheets (SDSs)
- Emergency response centers (CHEMTREC, CANUTEC, SETIQ, etc.)
- Pre-incident plans
- Technical experts
- *ERG* (may tell responders if material is water soluble or water reactive)
- Other books, reference sources, computer programs, data bases
- Poison Control Centers

Technical decon may use the following techniques:

- *Absorption* — Process of picking up liquid contaminants with absorbents. Some examples of absorbents used in decon are diatomaceous earth, baking powder, ashes, activated carbon, vermiculite, soil (not recommended by many experts), or other commercially available materials. Many absorbents are inexpensive and readily available, but expensive to dispose of. They work extremely well on flat surfaces.

- *Adsorption* — Process in which a hazardous liquid interacts with (or is bound to) the surface of a sorbent material (such as activated charcoal). Adsorbents tend to not swell like absorbents, and it is important to make sure that the adsorbent used is compatible with the hazardous material in order to avoid potentially dangerous reactions.

- *Brushing and scraping* — Process of removing large particles of contaminant or contaminated materials such as mud from boots or other PPE. Generally, brushing and scraping alone is not sufficient decontamination. This technique is used before other types of decon.

- *Chemical degradation* — Process of using another material to change the chemical structure of a hazardous material. For example, household liquid bleach is commonly used to neutralize spills of etiological agents. The interaction of the bleach with the agent almost instantaneously kills the dangerous germs and makes the material safer to handle. Several of the following materials are commonly used to chemically degrade a hazardous material, typically not used on living tissue:

  — Household bleach (sodium hypochlorite)

  — Isopropyl alcohol

  — Hydrated lime (calcium oxide)

  — Household drain cleaner (sodium hydroxide)

  — Baking soda (sodium bicarbonate)

  — Liquid detergents

- *Dilution* — Process of using water to flush contaminants from contaminated victims or objects and diluting water-soluble hazardous materials to safe levels. Dilution is advantageous because of the accessibility, speed, and economy of using water. However, there are disadvantages as well. Depending on the material, water may cause a reaction and create even more serious problems. Additionally, runoff water from the process is still contaminated and may have to be confined and disposed of properly. The amount of water needed for dilution may be impractical in some circumstances.

- *Evaporation* — Some hazardous materials evaporate quickly and completely. In some instances, effective decontamination can be accomplished by simply waiting long enough for the materials to evaporate. Evaporation is generally not a technique used during emergency operations. However, it can be used on tools and equipment when extending exposure time is not a safety issue.

- *Isolation and disposal* — This process isolates the contaminated items (such as clothing, tools, or equipment) by collecting them in some fashion and then disposing of them in accordance with applicable regulations and laws. All equipment that cannot be decontaminated properly must be disposed of correctly. All spent solutions and wash water must be collected and disposed of properly. Disposing of equipment may be easier than decontaminating it; however, disposal can be very costly in circumstances where large quantities of equipment have been exposed to a material.

- *Neutralization* — Process of changing the pH of a corrosive, raising or lowering it towards 7 (neutral) on the pH scale. Neutralization should not be performed on living tissue.

- *Sanitization, disinfection, or sterilization* — Processes that render etiological contaminates harmless:

  — *Sanitization:* Reduces the number of microorganisms to a safe level (such as by washing hands with soap and water).

  — *Disinfection:* Kills most of the microorganisms present. In a decon setting, a variety of chemical or antiseptic products may be used to accomplish disinfection. Most first responders are familiar with the disinfection procedures used to kill bloodborne pathogens such as wiping contaminated surfaces with a bleach solution.

  — *Sterilization:* Kills all microorganisms present. Sterilization is normally accomplished with chemicals, steam, heat, or radiation. While sterilization of tools and equipment may be necessary before they are returned to service, this process is usually impossible or impractical to do in most onsite decon situations. Such equipment will normally be disinfected on the scene and then sterilized later.

- *Solidification* — Process that takes a hazardous liquid and treats it chemically so that it turns into a solid.

- *Vacuuming* — Process using high efficiency particulate air (HEPA) filter vacuum cleaners to vacuum solid materials such as fibers, dusts, powders, and particulates from surfaces. Regular vacuums are not used for this purpose because their filters are not fine enough to catch all of the material.

- *Washing* — Process similar to dilution in that they are both wet methods of decontamination. However, washing also involves using prepared solutions such as solvents, soap, and/or detergents mixed with water in order to make the contaminant more water-soluble before rinsing with plain water **(Figure 9.17)**. The difference is similar to simply rinsing a dirty dinner dish in the sink versus washing it with a dishwashing liquid. In some cases the former may be sufficient, in others the latter may be necessary. Washing is an advantageous method of decontamination because of the accessibility, speed, and economy of using water and soap. As with the dilution process, runoff water from washing may need to be contained and disposed of properly.

NOTE: Table 9.1 provides a list of advantages and disadvantages for each of these decon techniques.

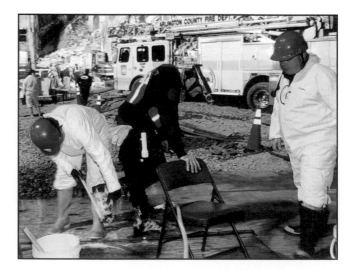

Figure 9.17 Washing uses prepared solutions such as solvents, soap, or detergents to make contaminants more water-soluble. *Courtesy of FEMA News Photos, photo by Jocelyn Augustino.*

Table 9.1
Advantages and Disadvantages of Technical Decon Methods

Method	Advantages	Disadvantages
Absorption	• Many absorbent materials are inexpensive and readily available • Can be used as part of dry decon operations • Effective on flat surfaces	• Do not alter the hazardous material • Ineffective for decontaminating protective clothing and vertical surfaces • Disposal of contaminated absorbent materials may be problematic and expensive • Absorbent materials may increase in weight and/or volume as they absorb the haz mat • Absorbent materials must be compatible with the hazardous material
Adsorption	• Contains the hazardous material better than absorbent materials • Transportation of materials to disposal is simplified • Off-gasing (release of vapors/gases) is effectively reduced • Adsorptive materials do not swell	• Process can generate heat • Application typically limited to remediation of shallow liquid spills • Adsorptive materials are expensive • Adsorptive material must be compatible with the hazardous material (they are product specific)
Chemical Degradation	• Can reduce cleanup costs • Reduces risk posed to the first responder when dealing with biological agents • Often utilizes commonly available, inexpensive materials such as bleach, isopropyl alcohol, or baking soda • Utilizes products that are readily available	• Takes time to determine the right chemical to use (which should be approved by a chemist) and set up the decon process • Can cause violent reactions if done incorrectly and may create heat and toxic vapors • Rarely used to decontaminate people
Dilution	• Lessens the degree of hazard present by reducing the concentration of the hazardous material • Easy to implement (water is usually readily available) • Is very effective in many circumstances requiring decon • Can be used to decon large pieces of equipment/apparatus	• Can't be used on materials that react adversely to water • May be problematic in cold weather • May create large amounts of contaminated run-off • May be impractical because of the amount of water required for effective dilution
Disinfection	• Kills most of the biological organisms present • Can be used on site • Can be accomplished using a variety of chemical or antiseptic products • Disinfecting agent may be as simple as antibacterial soap or detergent	• Limited to biological decon only • May be difficult to decon large pieces of equipment/apparatus • Disinfecting agent may be toxic or harmful

Continued

Table 9.1 (concluded)

Method	Advantages	Disadvantages
Evaporation	• No additional materials necessary • No runoff collection necessary • No (or very limited) expense incurred	• Applicable for a very limited number of chemicals • Generally limited to decon of tools and equipment, not people • May be dramatically affected by weather conditions (including wind, temperature, humidity, and rain) • Hazardous vapors may travel and cause problems • May require a long time to complete • May not be acceptable method to use depending on applicable laws and regulations
Isolation and Disposal	• Isolation can be quick and effective • Easily achieved with containers such as isolation drums, heavy plastic bags, and other means of containment	• Disposal and transport costs may be extremely high • May require replacement of equipment and PPE that cannot be decontaminated and placed back in service
Neutralization	• Chemically alters the hazardous material to reduce the degree of hazard present • Effective on most corrosives and some poisons • Neutralizing agents are readily available (soda ash, vinegar)	• May be very difficult to successfully implement • Rarely done on living tissue • May require large quantities of neutralizing agents • May create violent chemical reaction including the release of heat and hazardous vapors • Preplanning is usually necessary
Solidification	• Solids are easier to contain than liquids and gases • Reduces the amount of vapor production and off gasing • Easier to clean up	• Require specialized materials to implement
Sterilization	• Kills all microorganisms present	• Difficult or impossible to do onsite
Vacuuming	• Effective at removing dust and particulates • Effective indoors • Dry method, useful for cold weather operations in some situations	• Requires specialized vacuums equipped with hepa filters • May require high risk, negative air containment for decon area • Removing liquid chemical contamination requires special equipment • May require additional decon procedures to ensure complete decontamination (for example, washing) • Can't be used to decontaminate materials that react adversely to contact with water • May be problematic in cold weather • May create large amounts of contaminated run-off
Washing	• Quick and easy to implement (water is usually readily available) • Soap is readily available and inexpensive • Typically more effective than dilution alone • Is very effective in many circumstances requiring decon • Can be used to decon large pieces of equipment/apparatus	• Can't be used to decontaminate materials that react adversely to contact with water • May be problematic in cold weather • May create large amounts of contaminated run-off

Emergency responders must know what to do when assigned to a decontamination corridor or line. They should be briefed before being assigned. Technical decon corridors vary in the number of stations, depending on the needs of the situation **(Figure 9.18)**. Corridors may be set up for wet methods or dry methods. **Table 9.2, p. 452** is a sample technical decon checklist. Procedures for conducting decon operations in support of entry operations as well as procedures for technical decon of ambulatory and nonambulatory victims are provided in **Skill Sheets 9-2** and **9-3**.

As with mass decon, monitoring should be conducted to determine if decon operations are effective. In some cases, equipment may need to be disposed of due to permeation of contaminants or other factors. Such equipment must be removed from service and properly contained before disposal according to established policies and procedures.

Figure 9.18 The number of stations in a technical decon corridor will vary depending on the needs of the situation and SOPs.

Figure 9.19 Mass decon is initiated to expedite decon of a large number of people. *Courtesy of David Lewis.*

Mass Decon

Mass decontamination is the physical process of rapidly reducing or removing contaminants from multiple persons (victims and responders) in potentially life-threatening situations. Mass decon is initiated when the number of victims and time constraints do not allow the establishment of an in-depth decontamination process (such as technical decon) **(Figure 9.19)**. All agencies should have a mass decon plan as part of their overall emergency response plan. Operations-Level responders involved in mass decon operations must do so under the guidance of a haz mat technician, SOPs, or an allied professional. To determine the correct mass decon procedure, first responders must be familiar with established SOPs, emergency response plans, training, and skills learned during drills/exercises, and preplans.

The scene of an incident requiring the use of mass decon may be quite chaotic and difficult to control. To combat the chaos of the incident, responders should take the following actions:

● Communicate with victims by using hand signals, signs with pictures, apparatus public address systems, megaphones or other methods to direct them to decon gathering areas as well as through the decon process itself.

Table 9.2
Sample Decon Checklist

Date: **Location:**

- ☐ Initial briefing from the team leader
- ☐ Incident profile
- ☐ Decon solution and method
- ☐ PPE

Personnel Assignments

Decon Officer

[ ] Identified by vest

- ☐ All personnel monitored by Medical Branch

Decon Site Selection Criteria

- ☐ Decon is located in Warm Zone at exit from Hot Zone
- ☐ Decon area located uphill/upwind from Hot Zone
- ☐ Decon area level or sloped toward Hot Zone
- ☐ Water supply available

Decon Site Setup

- ☐ Area clearly marked with traffic cones and barrier tape to be secure against unauthorized entry
- ☐ Entry and exit points marked
- ☐ Emergency corridor established and clearly marked
- ☐ Runoff contained (tarp, plastic sheeting, dikes)
- ☐ Gross decon shower(s) setup
- ☐ Water supply established
- ☐ Containment basins and pools arranged in proper order
- ☐ Disposal containers in place for PPE and equipment drop
- ☐ Decon solutions mixed
- ☐ Brushes, hand sprayers, hoses and equipment in place
- ☐ Tool drop set up
- ☐ Spare SCBA cylinders available
- ☐ Relief personnel available

Branch Officers Briefing

- ☐ Preparation of branch status report
- ☐ Evaluation of branch readiness for mitigation plan

Entry/Decon Operations

- ☐ Decon and entry personnel briefed on hazards
- ☐ Emergency procedures and hand signals reviewed and understood
- ☐ Decon and entry personnel briefed on decon procedures
- ☐ Decon corridor complete
- ☐ Decon personnel on air
- ☐ Monitored for adequate relief personnel

Termination

- ☐ Disposable/contaminated materials isolated, bagged, and containerized
- ☐ All containers sealed, marked, and isolated
- ☐ All team equipment cleaned and accounted for

Source: Department of Fire Services, Office of Public Safety, Commonwealth of Massachusetts.

- Provide clear, easily understood, short, specific directions since people may be traumatized and/or suffering from exposures.
- Use barrier tape, traffic cones, or other highly visible means to mark decon corridors (see Decon Corridor Layout section). Crowd control methods were discussed in Chapter 6, Isolation section.

Goal of Mass Decon
Do the greatest good for the greatest number!

Mass decon methods include dilution, isolation, and washing. As discussed in the technical decon section, each of these methods have their own advantages and limitations. Washing with a soap-and-water solution or universal decontamination solution will remove many hazardous chemicals and WMD agents; however, availability of such solutions in sufficient quantities cannot always be ensured. Therefore, mass decon can be most readily and effectively accomplished with a simple water shower system that merely dilutes the hazardous product and physically washes it away. Mass decon uses large volumes of low-pressure water to quickly reduce the level of contamination.

Mass decon showers should use a high volume, low pressure of water delivered in a fog pattern to ensure the showering process physically removes the hazardous material. The actual showering time is an incident-specific decision but may be as long as 2 to 3 minutes per individual under ideal situations. When large numbers of potential victims are involved and queued for decon, showering time may be significantly shortened, however, this should be determined by post decon monitoring (see Evaluating Effectiveness of Decon Operations section). This time may also depend upon the volume of water available in the showering facilities.

Emergency responders should not overlook existing facilities when identifying means for rapid decontamination methods. For example, although water damage to a facility might result, the necessity of saving victims' lives would justify the activation of overhead fire sprinklers for use as showers. Similarly, having victims wade and wash in water sources such as public fountains, chlorinated swimming pools, or swimming areas, provides an effective, high-volume decontamination technique, although consideration must be given to the persistence of chemical agents in contained and contaminated water.

It is recommended that all victims undergoing mass decon remove clothing at least down to their undergarments before showering **(Figure 9.20, p. 454)**. Removal of clothing can remove significant amounts of the contaminant materials. Victims should be encouraged to remove as much clothing as possible, proceeding from head to toe. Contaminated clothing should be isolated in drums, appropriate bags, or other containers for later disposal **(Figure 9.21, p. 454)**.

WARNING!
Never delay decon while waiting for additional resources to arrive unless an assessment has been made that further injury or exposure will not occur.

Removal of Clothing
There may be circumstances (such as with particulate contaminants) in which removal of clothing before showering actually increases the potential risk of exposure. These risks must be evaluated before decon methods are implemented. For example, with some radiological materials and biological agents, it is best to dampen clothing before removal before showering. This limits the potential of aerosolizing the agents.

Figure 9.20 In most cases, disrobing and showering will provide effective removal of contaminants. Victims should be encouraged to remove as much clothing as possible, while remaining sensitive to privacy and modesty issues. *Courtesy of Glen Rudner.*

Figure 9.21 Victims should be instructed to remove personal belongings such as jewelry, cell phones, wallets, and other valuable personal belongings as well as their outer clothing.

Many innovations and products have been developed to assist in mass decon operations, from decon trailers and portable tents (that help alleviate privacy concerns) to portable water heaters, disposable coveralls, and tagging and bagging systems **(Figures 9.22a-d)**. The use of trailers and portable tents are best suited to long-duration incidents, incidents where weather may require them, and incidents where immediate decon is not as vital (such incidents involving biological hazards). Emergency responders should be familiar with the equipment and resources available as well as the mass decon procedures that their agency uses. **Figures 9.23a and b, p. 456** provide examples of apparatus placement for generic mass decon schematics, however, many agencies have tents or trailers available for use.

To determine victim priority during decon, responders must consider factors related both to medical needs and decontamination. For maximum effectiveness, it is recommended that patients be divided into two groups: ambulatory and nonambulatory.

This division may require establishing separate decon areas for each group to avoid slowing down the progression of ambulatory patients through the decon area. Incidents involving a large number of incapacitated victims may require additional resources for separate decon corridors since nonambulatory victims will not be able to walk through the decon line **(Figure 9.24, p. 457)**. A separate decon line for emergency response personnel should also be provided.

If there are adequate resources available and the situation allows for additional time, it may be beneficial to separate victims by gender for privacy reasons; however, families should be allowed to stay together. Children, the elderly, and/or the disabled should not be separated from their parents or caretakers. Procedures for conducting mass decon are provided in **Skill Sheet 9-4**.

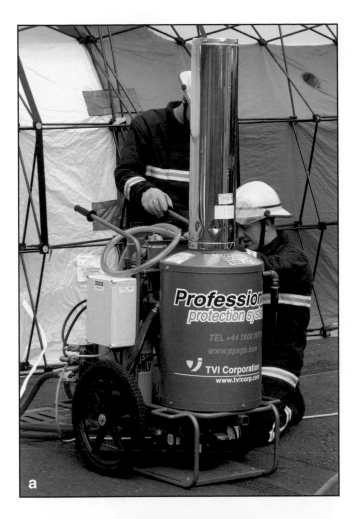

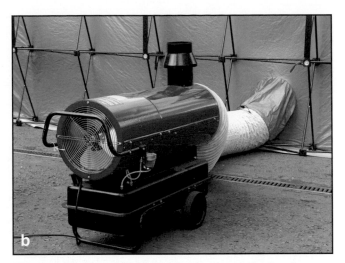

Figure 9.22a Portable water heaters can make the decon process much more comfortable, particularly in cool weather.

Figure 9.22b Forced air heaters can warm the interior of a decon tent.

Figure 9.22c Pre-assembled decon kits may include everything from individual bags and ID tags to disposable garments, towels, and shoes. Bag and tag personal items so they may be returned to their owners. *Courtesy of New South Wales Fire Brigades.*

Figure 9.22d After showering, provide victims with clean, dry replacement clothing such as disposable coveralls. *Courtesy of New South Wales Fire Brigades.*

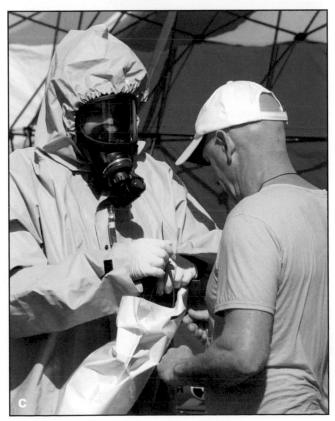

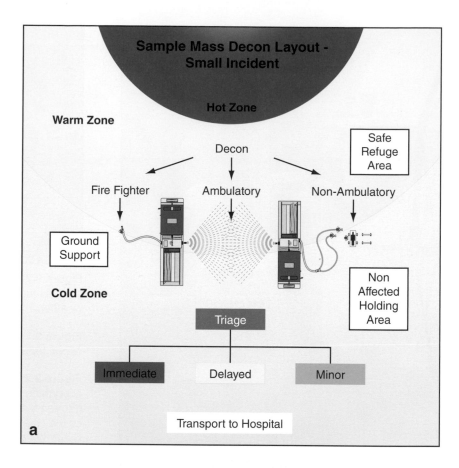

Figure 9.23a and b Sample mass decon schematics for small and large incidents. *Courtesy of Doug Weeks.*

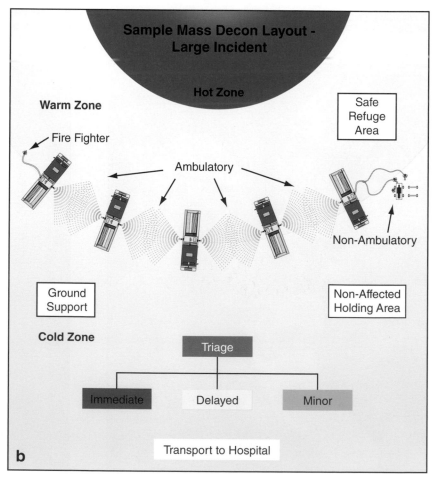

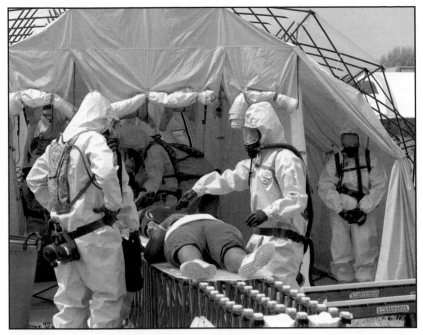

Figure 9.24 A system of rollers can be used to move non-ambulatory victims through the decon corridor. *Courtesy of the U.S. Air Force, photo by Tech. Sgt. Todd Pendleton.*

Figure 9.25 The effectiveness of decon operations should be checked using appropriate monitoring and detection equipment. *Courtesy of New South Wales Fire Brigades.*

Advantages and Limitations of Mass Decon

Advantages —

- Accommodates large numbers of people
- Can be implemented quickly using limited amount of personnel and equipment
- Reduces contamination quickly

Limitations —

- Does not always totally decontaminate the victim
- Relies on the cooperation of the victim
- Can create contaminated runoff that can harm the environment and other exposures

Evaluating Effectiveness of Decon Operations

Effectiveness of decon operations must be evaluated. This may be done through use of monitoring and detection devices or other equipment as well as visually **(Figure 9.25)**. Generally, the hazardous materials involved will determine the technology or device needed to perform the operations. See Chapter 11, Air Monitoring and Sampling, for more information on technology used to test for hazardous materials.

If large numbers of people are involved, individuals should be briefly checked after they have gone through the decon process, otherwise check each individual more carefully. These checks should be done as they exit the decon corridor. If contamination is detected, individuals must be redirected through the decon process.

Victims still complaining of symptoms or effects should be checked (or rechecked) for contaminants. If the effectiveness of decon is called into question, victims should go through decon again prior to transport.

Tools and equipment will normally need to be stored in the decon area until the emergency phase of the operation is completed. After being decontaminated they will need to be checked to ensure all contamination has been removed before being placed back in service. Apparatus will also need to undergo decon if they have been exposed or potentially exposed to hazardous materials. The same monitoring and detection equipment used to determine effectiveness of decon on victims and responders may be used on equipment, tools, and apparatus.

Decon Implementation

Many things must be considered when implementing decontamination. An appropriate site must be selected, the number of stations and setup of the decon corridor or line during technical decon must be decided, methods for collecting evidence must be determined, and termination procedures must be followed.

Site Selection

The following factors are considered when choosing a decontamination site:

- *Wind Direction* — The decontamination site needs to be upwind of the hot zone to help prevent the spread of airborne contaminants into clean areas. If the decontamination site is improperly located downwind, wind currents will blow mists, vapors, powders, and dusts toward responders and victims. During long term operations, the local weather service can provide assistance in predicting changes in the wind direction and weather.

- *Weather* — Ideally, during cold weather, the site should be protected from blowing winds, especially near the end of the corridor. Victims should be shielded from cold winds when they are removing protective clothing.

- *Accessibility* — The site must be away from the hazards, but adjacent to the hot zone so that persons exiting the hot zone can step directly into the decontamination corridor. An adjacent site eliminates the chance of contaminating clean areas. It also puts the decontamination site as close as possible to the actual incident. Time is a major consideration in the selection of a site. The less time it takes personnel to get to and from the hot zone, the longer personnel can work. Four crucial time periods are as follows:
  - Travel time in the hot zone
  - Time allotted to work in the hot zone
  - Travel time back to the decontamination site
  - Decontamination time

- *Terrain and surface material* — The decontamination site ideally is flat or slopes toward the hot zone; thus, anything that may accidentally get released in the decontamination corridor would drain toward or into the contaminated hot zone and persons leaving the decon corridor would enter into a clean area. If the site slopes away from the hot zone, contaminants could flow into a clean area and spread contamination. Finding the perfect topography is not always possible, and first responders may have to place some type of barrier to ensure confinement of an unintentional release. *Details:*

— Diking around the site prevents accidental contamination escaping.

— It is best if the site has a hard, nonporous surface to prevent ground contamination.

— When a hard-surfaced driveway, parking lot, or street is not accessible, some type of impervious covering may be used to cover the ground. Salvage covers or plastic sheeting will prevent contaminated water from soaking into the earth.

— Covers or sheeting should be used to form the technical decontamination corridor regardless of whether the surface is porous **(Figure 9.26)**.

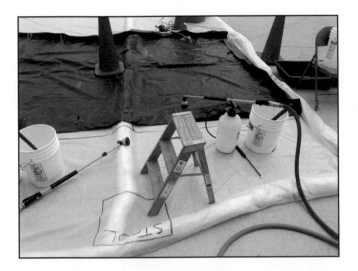

Figure 9.26 Tarps, plastic sheeting, or salvage covers should be used for flooring, even when the decon corridor is set up on hard surfaces such as concrete or asphalt.

- *Lighting (and electrical supply)* — The decontamination corridor should have adequate lighting to help reduce the potential for injury to personnel in the area. Selecting a decontamination site illuminated by streetlights, floodlights, or other type of permanent lighting reduces the need for portable lighting. If permanent lighting is unavailable or inadequate, portable lighting will be required. Ideally, the decontamination site will have a ready source of electricity for portable lighting (as well as heaters, water heaters, and other needs). However, if such a source is not available, portable generators will be needed.

- *Drains and waterways* — Locating a decontamination site near storm and sewer drains, creeks, ponds, ditches, and other waterways should be avoided (unless the sewer system is approved for use as a contained system that can be managed and neutralized). If this situation is not possible, a dike can be constructed to protect the storm drain opening, or a dike may be constructed between the site and a nearby waterway. Protect all environmentally sensitive areas if possible but never delay decon to protect the environment if the delay will increase injury to those affected by the event.

- *Water supply* — Water must be available at the decontamination site if wet decon is used.

Preplans should include pre-designated areas for mass decon at locations likely to be target by terrorists such as government buildings and stadiums. Hospitals must also have plans to decon potentially large numbers of victims who self-present at emergency rooms.

Decon Corridor Layout

Establish the decontamination corridor before performing any work in the hot zone. First responders are often involved with setting up and working in the decontamination corridor. The types of decontamination corridors vary as to the numbers of sections or steps used in the decontamination process. Corridors can be straightforward and require only a few steps, or they can be more complex and require a handful of sections and a dozen or more steps. Emergency responders must understand the process and be trained in setting up the type of decontamination required by different materials. Some factors to consider are as follows:

- *Ensure privacy* — Decon tents or decon trailers allow more privacy for individuals going through the decon corridor. Decon officers and ICs need to be particularly sensitive to the needs of women being asked to remove their clothing in front of men (regardless of whether they are victims or other emergency responders). Lawsuits have resulted from situations in which women have felt uncomfortable or even humiliated while going through decon. Providing a private, restricted area such as a tent or trailer in which to conduct decon may prevent similar litigation **(Figure 9.27)**. Use female responders to assist whenever possible when decontaminating women.

Figure 9.27 It is important to allow men and women to shower separately if possible. However, family units and others with personal ties (such as children with a babysitter, or an elderly person with a caregiver) should not be separated if they wish to stay together. *Courtesy of New South Wales Fire Brigades.*

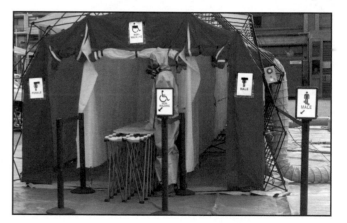

- *Bag and tag contaminated clothing/effects* — Various methods can be used. Place clothing and/or personal effects in bags and label the bags with the person's name or other identifier whenever possible. Separate personal effects when possible (wallets, rings, watches, identification cards, and the like) into clear plastic bags clearly marked with the person's name or a unique identifying number (triage tag, ticket, and the like). These items may need to be decontaminated before being returned. Have some sort of system in place to label or mark all personal effects so that they can be returned to their proper owners after the incident without confusion. All bags that contain contaminated clothing should remain in the warm zone on the dirty side of the decon line. Commercial tagging systems may be used for this purpose or multiple-part plastic hospital identification bracelets for example.

Personal Belongings

Be aware that many individuals will find it very stressful to be separated from their personal belongings. For this reason, responders assisting with disrobing should be sympathetic and empathic about the process to help alleviate anxiety.

To the degree possible, keep track of the status of people and personal belongings. This is important for public information (victim relatives will want to know who went where and why) as well as for crime scene investigation at incidents involving terrorism.

The decontamination corridor may be identified with barrier tape, safety cones, or other items that are visually recognizable **(Figure 9.28)**. Coverings such as a salvage covers or plastic sheeting may also be used to form the corridor. Aside from delineating the corridor and providing privacy, protective covering ensures against environmental harm if contaminated rinse water splashes from a containment basin. Containment basins can be constructed of salvage covers and fire hose or ladders. Some organizations use wading pools or portable drafting tanks as containment basins. Also needed at the site are recovery drums or other types of containers and plastic bags for stowing contaminated tools and PPE.

Figure 9.28 The decon corridor should be clearly marked by barrier tape, safety cones, or other means.

Figure 9.29 Everyone leaving the hot zone, including law enforcement and military personnel, must undergo decon. *Courtesy of the U.S. Air Force, photo by Staff Sgt. C. Todd Lopez.*

Other Implementation Considerations

Performing decon operations on law enforcement and military personnel offers a unique challenge to incident operations. Law enforcement and military personnel leaving the hot zone must undergo decon **(Figure 9.29)**. These personnel are often carrying weapons and will not release them to civilian personnel during decon operations. It may be necessary to include a hazmat trained law enforcement officer involved in the decon operations with the sole responsibility of decontaminating weapons and ensuring their security as the operation proceeds. Special consideration must be given to the decon of weapons, ammunition, and other equipment that could be damaged by exposure to liquid decon solutions or water (**Figure 9.30**). Decon plans must take this equipment into consideration in accordance with local policies and procedures.

Figure 9.30 Local policies must establish procedures for managing potentially contaminated weapons, ammunition, and other equipment that could be damaged by exposure to liquid decon solutions or water.

Another decontamination corridor for armed emergency services personnel who are leaving the hot zone may be established. Weapons are placed in a hazmat recovery bin under the supervision of a suitably protected (the correct level of PPE) law enforcement officer as officers disarm and go through decon.

Emergency response personnel should take precautions (such as putting protective booties on dogs' feet) before taking canines into the hot zone of haz mat/WMD incidents and should have their own procedures to decontaminate their animals. However, these animals will be processed through the decontamination corridor, and fire department personnel may have to assist in the decontamination of animals (**Figure 9.31**).

Criminal suspects may need to be decontaminated. The suspect must be supervised by law enforcement throughout this process. If conducting technical decon, the suspect will go through the same decon steps established for responders and other victims (**Figure 9.32**). Consideration must be given to whether or not handcuffs must be removed and decontaminated. Follow departmental procedures for decontaminating criminal suspects.

NOTE: At haz mat/WMD incidents, there may be requests to decontaminate animals and pets. Contingency plans should include guidelines for decontamination of animals and pets.

Cold Weather Decon

Conducting wet decon operations in freezing weather can be difficult to execute safely. Even if showers utilize warmed water, run-off water can quickly turn to ice creating a serious slipping hazard for both victims and responders. If warm water is not available, susceptible individuals (elderly, very young, individuals with chemical injuries or pre-existing health conditions such as diabetes) can suffer cold shock or hypothermia.

Figure 9.31 Service dogs leaving the hot zone will need to undergo decon. *Courtesy of FEMA News Photos, photo by Jocelyn Augustino.*

Figure 9.32 Criminal suspects will go through the same decon steps established for responders and other victims. Provisions must be made to ensure careful supervision during this process.

If temperatures are 64°F (18°C) degrees or lower, consideration should be given to protecting victims from the cold. Answering the following questions will provide information on how best to protect victims:

- Are wet methods necessary, or can disrobing and dry methods accomplish effective decon?

- Is wind chill a factor?

- Is shelter available for victims during and after decon?

- Is it possible to conduct decon indoors (sprinkler systems, indoor swimming pools, locker room showers, etc.)?

- If decon will be conducted indoors (at preplanned facilities, for example), how will victims be transported?

- If decon must be conducted outside in freezing temperatures, how will icy conditions be managed (ie. sand, sawdust, kitty litter, salt)?

Individuals who have been exposed to deadly levels of chemical agents should undergo emergency decon immediately, regardless of ambient temperatures. They should disrobe and thoroughly shower. Dry clothing and warm shelter should be provided as soon as possible after showering.

WARNING!
Individuals who have been exposed to deadly levels of chemical agents should undergo emergency decon immediately, regardless of ambient temperatures.

Figure 9.33 Decontamination of evidence will need to be conducted under the direction and supervision of qualified law enforcement officials.

Evidence Collection and Decontamination

Collection, preservation, and sampling of evidence will be performed under the direction of law enforcement, per established procedures. Decontamination issues associated with these activities will also be determined in conjunction with law enforcement.

Evidence collected on the scene by law enforcement personnel must be appropriately packaged (for example, in approved bags or other evidence containers). Only the exterior of the packaging will be decontaminated as it passes from the hot zone to the cold zone **(Figure 9.33)**. When evidence passes through the decon corridor, chain of custody must be documented in writing.

Termination

After concluding decon activities, a debriefing needs to be held for those involved in the incident as soon as is practical. In some cases, return of personal items may be a law enforcement function because of the evidentiary issues involved. There may be circumstances in which personal effects are immediately returned to the persons undergoing decon. Provide exposed persons with as much information as possible about the delayed health effects of the hazardous materials involved in the incident.

Additional reports and supporting technical documentation such as incident reports, after action reports, and regulatory citations may be required by emergency response plans and/or SOPs. Exposure records may also need to be filled out and filed.

Exposure records are required for all first responders who have been exposed or potentially exposed to hazardous materials. Follow agency SOPs for filling out exposure records. Information included on the exposure report might include activities performed, product involved, reason for being there, equipment failures or malfunction of PPE, hazards associated with the product, symptoms experienced, monitoring levels in use, circumstances of exposure, and other pertinent information. Follow-up examinations should be scheduled with medical personnel if necessary. The individual, the individual's personal physician, and the individual's employer need to keep copies of these exposure records for future reference.

An activity log must be maintained during the incident or put together afterwards as appropriate. At a minimum, information for the activity log should be captured during the incident debrief. The activity log may be preformatted, and at a minimum must document the chronology of the events and activities that occurred during the incident and decon procedure.

In the U.S., OSHA standard 29 CFR 1910.1020 (Access to Employee Exposure and Medical Records) should be followed as a guide for requirements involving medical records and maintaining exposure reports. SOPs should spell out additional requirements for local recordkeeping and reports.

Summary

Any time hazardous materials are involved in an incident, contamination becomes a concern. Contamination with hazardous materials can lead to exposure, which can in turn cause harm, depending on the hazards of the material and the nature of the exposure. Decontamination is conducted to prevent the spread of contaminants from the hot zone to other areas, and it is done to reduce the level of contamination to levels that are not harmful.

Emergency decon is conducted in life-threatening situations. Mass decon will normally be conducted when large number of victims are involved. Technical decon is typically conducted on emergency responders and at incidents when very few victims are involved.

Review Questions

1. What is decontamination?

2. What questions should be answered before initiating decontamination?

3. What are the different methods of decontamination?

4. What are the general guidelines for decontamination operations?

5. Describe the different types of recipients of decontamination.

6. What is emergency decontamination? When might it be used?

7. What techniques might be used for technical decontamination?

8. What are the advantages and disadvantages of mass decontamination?

9. What factors should be considered when choosing a decontamination site?

10. How does cold weather decontamination differ from normal decontamination?

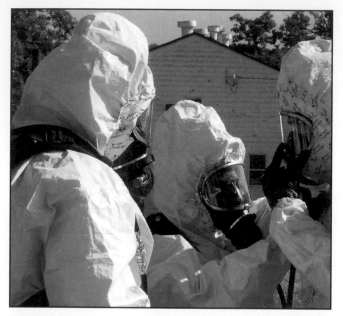

Step 1: Ensure that all responders involved in control functions are wearing appropriate PPE for performing emergency decontamination operations.

Step 2: Remove the victim from the contaminated area.

Step 4: Remove victim's clothing and/or PPE rapidly – if necessary, cutting from the top down in a manner that minimizes the spread of contaminants.

Step 5: Perform a quick cycle of head-to-toe rinse, wash, and rinse.

Step 6: Transfer the victim to treatment personnel for assessment, first aid, and medical treatment.

Step 7: Ensure that ambulance and hospital personnel are informed about the contaminant involved.

Step 8: Decontaminate tools.

Step 9: Proceed to decontamination line for decontamination.

Step 3: Wash immediately any contaminated clothing or exposed body parts with flooding quantities of water.

Set Up a Technical Decontamination Corridor

Step 1: Ensure that all responders are wearing appropriate PPE for establishing the decon corridor and performing technical decontamination operations.

Step 2: Set up the decontamination corridor uphill and upwind from the hot zone, away from remote drains or waterways.

Step 3: Mark entry and exit of decontamination corridor so that they are clearly identified.

Step 4: Set up ground cover (tarp or salvage cover) for secondary contamination and runoff containment.

Step 5: Establish a tool-drop station in the hot zone at the entry to the decontamination corridor.

Note: The number of stations in the decon corridor will vary depending on the needs of the incident and local SOPs. Law enforcement personnel may need a separate decontamination station for tactical equipment.

Step 6: Establish a gross decontamination station after the tool-drop station in the decontamination corridor.

Step 7: Establish a secondary decontamination station including appropriate cleaning solution(s) as set forth by the haz mat technician, SOPs, or allied professional.

Step 8: Establish a PPE removal station with waste disposal containers for contaminated PPE.

Step 9: Establish a respiratory protection removal station.

Note: Steps 8 and 9 may need to be reversed, depending on the PPE worn.

Step 10: Establish an undergarment removal station with waste disposal containers for contaminated clothing.

Step 11: Establish shower and clothing change station.

Step 12: Establish a medical evaluation area.

Step 13: Perform technical decontamination operations for persons according to assigned tasks and the AHJ's SOPs.

SHEETS

9-2

Set up and implement a technical decontamination corridor and undergo decontamination.

Undergo Technical Decontamination, Ambulatory

Step 1: Proceed to the first station in the decontamination corridor.

Step 2: Drop tools in collection container.

Step 3: Undergo gross decontamination.

Step 4: Undergo secondary decontamination wash.

Step 5: Remove outer PPE. Place PPE in waste container.

Step 6: Remove respiratory protection, removing face-piece last.

Note: Steps 5 and 6 may need to be reversed, depending on the PPE worn.

Step 7: Remove undergarments.

Step 8: Shower and wash thoroughly from the top down.

Step 9: Monitor for additional contamination using the appropriate detection device.

Note: If contamination is detected, repeat decontamination wash and/or change decontamination method, as appropriate.

Step 10: Proceed to medical evaluation station.

Step 1: Ensure that all responders are wearing appropriate PPE for performing technical decontamination operations.

Step 2: Establish technical decontamination corridor for non-ambulatory decontamination according to the AHJ's SOPs.

Step 3: Establish an initial triage point to evaluate and direct persons.

Step 4: Perform lifesaving intervention.

Step 5: Transfer the victim to the non-ambulatory wash area of the decontamination station on an appropriate backboard/litter device.

Step 6: Remove all clothing, jewelry, and personal belongings, and place in appropriate containers. Decontaminate as required, and safe-guard. Use plastic bags with labels for identification.

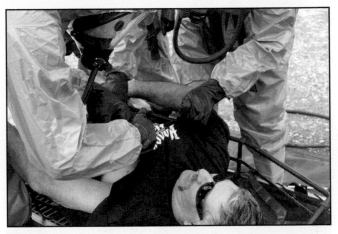

Step 7: Carefully undress non-ambulatory persons, and avoid spreading the contamination when undressing. Do not touch the outside of the clothing to the skin. If biological agents are suspected, a fine water mist can be applied to trap the agent in the clothing and prevent the spread of contamination.

Step 8: Completely wash the victim's entire body using handheld hoses, sponges, and/or brushes and then rinse.

Step 9: Clean the victim's genital area, armpits, folds in the skin, and nails with special attention. If conscious, instruct the victim to close his/her mouth and eyes during wash and rinse procedures.

Step 10: Transfer the victim from the wash and rinse stations to a drying station after completing the decontamination process. Ensure that the victim is completely dry.

Step 11: Monitor for additional contamination using the appropriate detection device.

Note: If contamination is detected, repeat decontamination wash and/or change decontamination method, as appropriate.

Step 12: Have on-scene medical personnel reevaluate the victim's injuries.

Step 1: Ensure that all responders are wearing appropriate PPE for performing mass decontamination operations.

Step 2: Prepare fire apparatus for use during mass decontamination.

Step 3: Set fire nozzle to fog pattern.

Step 4: Instruct all victims to go through mass decontamination.

Note: Non-ambulatory victims should be carried or moved through the decontamination process according to the AHJ's SOPs. Refer to Skill Sheet 9-3 for specific instructions on performing decontamination of non-ambulatory victims.

Step 5: Instruct victims to remove contaminated clothing, ensuring that victims to not come into further contact with any contaminants.

Step 6: Instruct victims to keep arms raised as they proceed through the wash area.

Step 7: Monitor for additional contamination using the appropriate detection device.

Note: If contamination is detected, repeat decontamination wash and/or change decontamination method, as appropriate.

Step 8: Instruct victims to move to a clean area to dry off.

Step 9: Send victims for medical treatment.

Step 10: Inform EMS personnel of contaminant involved and its hazards, if known.

Step 11: Document activity log.

Product Control

Chapter Contents

chapter 10

Key Terms

Competencies

NFPA® 472:	6.6.3.1(2)(e)	6.6.3.2	6.6.4.1(3)(a)	6.6.4.1(3)(g)	6.6.4.2
6.6.3.1(1)	6.6.3.1(2)(f)	6.6.4.1(1)	6.6.4.1(3)(b)	6.6.4.1(3)(h)	
6.6.3.1(2)(a)	6.6.3.1(2)(g)	6.6.4.1(2)(a)	6.6.4.1(3)(c)	6.6.4.1(3)(i)	
6.6.3.1(2)(b)	6.6.3.1(2)(h)	6.6.4.1(2)(b)	6.6.4.1(3)(d)	6.6.4.1(3)(j)	
6.6.3.1(2)(c)	6.6.3.1(2)(i)	6.6.4.1(2)(c)	6.6.4.1(3)(e)	6.6.4.1(4)	
6.6.3.1(2)(d)	6.6.3.1(2)(j)	6.6.4.1(2)(d)	6.6.4.1(3)(f)	6.6.4.1(5)	

Product Control

Learning Objectives

1. Describe each of the various spill control tactics. [NFPA® 472, 6.6.3.1(1), 6.6.3.1(2)(a-j), 6.6.3.2, 6.6.4.1(2)(a-d)]

2. Perform absorption/adsorption. [Skill Sheet 10-1; NFPA® 472, 6.6.4.1(3)(a-b)]

3. Perform damming operations. [Skill Sheet 10-2; NFPA® 472, 6.6.4.1(3)(c)]

4. Perform diking operations. [Skill Sheet 10-3; NFPA® 472, 6.6.4.1(3)(d)]

5. Perform diversion operations. [Skill Sheet 10-4; NFPA® 472, 6.6.4.1(3)(f)]

6. Perform retention operations. [Skill Sheet 10-5; 6.6.4.1(3)(g)]

7. Perform vapor suppression. [Skill Sheet 10-6; 6.6.4.1(1), 6.6.4.1(3)(j)]

8. Perform vapor dispersion. [Skill Sheet 10-7; NFPA® 472, 6.6.4.1(3)(i)]

9. Perform dilution operations. [Skill Sheet 10-8; NFPA® 472, 6.6.4.1(3)(e)]

10. Discuss leak control. [NFPA® 472, 6.6.4.1(4-5), 6.6.4.2]

11. Perform remote valve shutoff. [Skill sheet 10-9; NFPA® 472, 6.6.4.1(3)(h)]

12. Explain fire control.

Chapter 10
Product Control

Case History

In Iowa, on April 9, 1998, a call came into the fire department late at night regarding a fire at a turkey farm. Two people riding an off-road vehicle had struck and broken one of two fixed, metal pipes between a propane tank and two vaporizers (devices that receive lique-fied petroleum-gas in liquid form and add sufficient heat to convert the liquid to a gaseous state). As the liquid propane spewed from the pipe, the operator of the off-road vehicle drove away to call 911. The propane vapors, which have a vapor density of 1.53 and are thus heavier than air, spread along the ground and were eventually ignited by the pilot flame at the vaporizers. Burning propane vapors spread throughout the area and began to impinge on the tank, causing the pressure relief valve to activate and send burning propane flames high into the air.

Upon arrival at the fire scene, the Incident Commander made an assessment of the burning tank. Fire had engulfed the propane tank, and it was venting burning propane vapors via two pressure relief vent pipes located on top of the tank. Also, the tank's pressure relief valve was emitting an extremely loud noise similar to a jet engine. The tank was manufactured in 1964 and the tank's shell was constructed of 3/4-inch (20 mm) carbon steel. The tank was 42 feet, 2 inches (13 m) long, with an inside diameter of 106 inches (3 m) and an 18,000 gallon (68 137 L) capacity, and it rested on two concrete supports **(Figure 10.1, p. 476)**. The tank was cylindrical and contained about 10,000 gallons (37 854 L) of liquid propane. The tank had been fitted with two internal spring-type pressure relief valves which would vent to the atmosphere via two 2-inch (51 mm) diameter pressure relief vent pipes when the internal tank pressure reached a set limit. The tank also had a fixed metal piping system between the tank and two vaporizers located on the ground about 35 feet (11 m) away. The piping was 3/4-inches (20 mm) in diameter, positioned about 36 inches (1 m) above the ground, was unprotected and had not been fitted with an excess flow valve.

After seeing the flames and hearing the high-pitched sound being emitted by pressure relief vent pipes located on the top west section of the tank, a decision was made to allow the tank to burn itself out and to try to save the adjacent buildings. The fire fighters positioned themselves in various areas in a semicircle north, northeast and northwest of the tank. The two victims were about 105 feet (32 m) away from the tank spraying one of the buildings with water when a BLEVE occurred. The BLEVE ripped the tank into four parts, each flying in a different direction. One part of the tank traveled in a northwest direction toward the two victims, striking them and killing them instantly. Six other fire fighters and a deputy sheriff received varying degrees of burns and assorted injuries.

Source: NIOSH

Figure 10.1 If not cooled by water streams, propane tanks can BLEVE when involved in fires. *Courtesy of Rich Mahaney.*

In order to understand the principles behind product control, it is useful to return to the General Hazardous Materials Behavior Model. Container stress can result in a breach or rupture. When the container is breached, it may release its contents. The released material then disperses according to its chemical and physical properties, topography, prevailing weather conditions and the amount and duration of the release. Depending on its hazardous properties, the material can then engulf and harm whatever it contacts. If the material is flammable, it may contact an ignition source and ignite or explode.

Spill-control tactics are used to confine a hazardous material that has already been released from its container. These tactics reduce the amount of contact the product makes with people, property, and the environment thereby limiting the amount of potential harm. Spill-control actions are generally defensive in nature **(Figure 10.2)**.

Figure 10.2 Defensive spill-control tactics are used to contain hazardous materials that have already escaped their containers. *Courtesy of Water-gate.com.*

Figure 10.4 Fire-control tactics are used to extinguish fires and prevent ignition of hazardous materials. *Courtesy of Rich Mahaney.*

Figure 10.3 Offensive leak-control tactics are used to contain the material by preventing it from escaping its original container or moving it to another undamaged container like an overpack drum. Leak-control often involves a higher degree of risk than spill-control tactics. *Courtesy of the CDC, photo by Jim Gathany.*

Leak-control tactics are used to contain the product in its original (or another) container, thereby preventing it from escaping. Most leak-control tactics are offensive, performed by hazardous materials technicians and specialists **(Figure 10.3)**.

Fire-control tactics are aimed at extinguishing fires and preventing ignition of flammable materials. Fire-control tactics may be offensive or defensive, depending on the situation **(Figure 10.4)**.

Responders should be familiar with their AHJ's procedures for product control as specified in SOPs and emergency response plans. As with all haz mat incidents, the risks and hazards present at the incident will determine the type of PPE needed and the equipment necessary (see Chapter 8, Personal Protective Equipment). In many cases, after performing product control activities, responders will need to undergo technical decon (see Chapter 9, Decontamination). This chapter will discuss spill-control tactics, leak-control actions, and basic fire-control tactics that may be performed by Operations-Level responders.

Spill Control

The main priority of spill control is confinement and the prevention of further contamination or contact with the hazardous material. For this reason, it is often just called **confinement**. Some spill-control tactics such as neutralization and dispersion minimize the amount of harm caused by contact with the material. Spill control is primarily a defensive operation with the most important issue being the safety of the responders performing these actions.

Confinement — (1) Process of controlling the flow of a spill and capturing it at some specified location. (2) Operations required to prevent fire from extending from the area of origin to uninvolved areas or structures.

Spills may involve gases, liquids, or solids, and the product involved may be released into the air (as a vapor or gas), into water, and/or onto a surface. The type of dispersion determines the defensive measure(s) needed to control it. For example, in the event of a flammable liquid spill, first responders must address not only the liquid spreading on the ground, but the vapors being released into the air as well.

Building dams or dikes near the source, catching the material in another container, or directing (diverting) the flow to a remote location for collection are all methods for confining some hazardous materials. Before using equipment to confine spilled materials, ICs need to seek advice from technical sources to determine if the spilled materials will adversely affect the equipment. Large or rapidly spreading spills may require the use of heavy construction-type equipment, floating confinement booms, or special sewer and storm drain plugs.

Spill control is not restricted to controlling liquids. Dusts, vapors, and gases can also be confined. A protective covering consisting of a fine spray of water, a layer of earth, plastic sheets, or a salvage cover can keep dusts from blowing about at incidents. Foam blankets can be used on liquids to reduce the release of vapors. Strategically placed water streams can direct gases, or allow the water to absorb them. Reference sources provide the proper procedures for confining gases. The material type, rate of release, speed of spread, number of personnel available, tools and equipment needed, weather, and topography dictate confinement efforts.

Operations-Level responders are expected to take protective actions but not to physically stop the release unless it can be done from a safe location by closing a remote shutoff valve (see Leak Control section). Defensive spill-control tactics that confine hazardous materials include the following:

- Absorption
- Adsorption
- Blanketing/covering
- Dam, dike, diversion, and retention
- Vapor suppression

Rather than attempting to confine the dispersion, some defensive spill control tactics are aimed at reducing the amount of harm the material causes by diluting the concentration or changing its physical and/or chemical properties. These tactics include the following:

- Vapor dispersion
- Ventilation
- Dispersion
- Dilution
- Dissolution
- Neutralization

NOTE: Neutralization is considered a Tech-Level offensive containment tactic by many experts. However, neutralization is aimed at reducing or eliminating the chemical hazard of the material rather than physically containing it. The tactics of dilution, neutralization, and dissolution are used only infrequently at haz mat incidents and under very specific circumstances.

CAUTION

Spill-control actions are undertaken only if first responders are reasonably ensured that they will not come in contact with or be exposed to the hazardous material.

Table 10.1 provides a summary of the spill control tactics that can be used on different types of releases and their resulting dispersions. It also provides an example of a task related to one of the appropriate tactics.

	Table 10.1 Spill Control Tactics Used According to Type of Release		
Type of Release	**Type of Dispersion**	**Spill Control Tactics**	**Task Example**
Liquid: Airborne Vapor	Hemispheric, cloud, plume, or cone	• Vapor Suppression • Ventilation • Vapor Dispersion • Dissolution	Cover spill with vapor suppressing foam (Vapor Suppression)
Liquid: Surface	Stream	• Diking • Diversion • Retention • Adsorption • Absorption	Dig a ditch to divert a spill away from a stream (Diversion)
Liquid: Surface	Pool	• Absorption • Adsorption (for shallow spills) • Neutralization	Cover spill with an absorbent pillow (Absorption)
Liquid: Surface	Irregular	• Dilution • Absorption • Neutralization	Spray slightly contaminated surfaces with water (Dilution)
Liquid: Waterborne Contamination	Stream or pool	• Damming • Diversion • Retention • Absorption • Dispersion	Place absorbent booms across a river (Absorption)
Solid: Airborne Particles	Hemispheric, cloud, plume, or cone	• Particle Dispersion/ Ventilation • Particle Suppression (wetting material) • Blanketing/Covering	Set up ventilation fans (Particle Dispersion/Ventilation)
Solid: Surface	Pile	• Blanketing/Covering • Vacuuming	Cover spilled material with a tarp or salvage cover (Blanketing/Covering)
Solid: Surface	Irregular	• Blanketing/Covering • Dilution • Dissolution	Spray scattered sprinkles of corrosive powders or dusts with water (Dilution)
Gas: Airborne Gas	Hemispheric, cloud, plume, or cone	• Ventilation • Vapor Dispersion • Dissolution	Spray leaking cylinder with fog stream (Dissolution)

Absorption

Absorption is a physical and/or chemical event occurring during contact between materials that have an attraction for each other. This event results in one material being retained within the other. The bulk of the material being absorbed enters the cell structure of the absorbing medium. An example of absorption is soaking an axe head in water to make the handle swell. Some of the materials typically used as absorbents are sawdust, clays, charcoal, and polyolefin-type fibers **(Figure 10.5)**. The absorbent is spread directly onto the hazardous material or in a location where the material is expected to flow. After use, absorbents must be treated and disposed of as hazardous materials themselves because they retain the properties of the materials they absorb. For more information about performing absorption, see **Skill Sheet 10-1**.

Figure 10.5 Absorbents are used to soak up hazardous materials.

Adsorption

Adsorption is different from absorption in that the molecules of the hazardous material physically adhere to the adsorbent material rather than being absorbed into the inner spaces of an absorbent material. Adsorbents tend to not swell like absorbents, and they are often organic-based materials such as activated charcoal or carbon. Adsorbents are primarily used to control shallow liquid spills. It is important to make sure that the adsorbent used is compatible with the spilled material in order to avoid potentially dangerous reactions. For more information about performing adsorption, see Skill Sheet 10-1.

Blanketing/Covering

This spill-control measure involves blanketing or covering the surface of the spill to prevent dispersion of materials such as powders or dusts. Blanketing or covering of solids can be done with tarps, plastic sheeting, salvage covers, or other materials (including foam), but consideration must be given to compatibility between the material being covered and the material covering it **(Figure 10.6)**. Covering may also be done as a form of temporary mitigation for radioactive and biological substances.

Figure 10.6 Blanketing covers the hazardous material(s) with a tarp, sheeting, or salvage cover. The material used must be compatible with the hazardous material(s) being blanketed. *Courtesy of the U.S. Department of Defense, photo by Senior Amn Christopher J. Wiant.*

Blanketing of liquids is essentially the same as vapor suppression (see Vapor Suppression section), because it primarily uses an appropriate aqueous (water) foam agent to cover the surface of the spill. Operations-Level responders may or may not be allowed to perform blanketing/covering actions, depending on the hazards of the material, the nature of the incident, and the distance from which they must operate to ensure their safety.

Dam, Dike, Diversion, and Retention

Diking, damming, diverting, and retaining are ways to confine a hazardous material. These actions are taken to control the flow of liquid hazardous materials away from the point of discharge. Responders can use available earthen materials or materials carried on their response vehicles to construct curbs that direct or divert the flow away from gutters, drains, storm sewers, flood-control channels, and outfalls **(Figure 10.7)**. In some cases, it may be desirable to direct the flow into certain locations in order to capture and retain the material for later pickup and disposal. Dams may be built that

Figure 10.7 Diverting hazardous materials from storm drains may be necessary to protect the environment.

permit surface water or runoff to pass over (or under) the dam while holding back the hazardous material **(Figure 10.8)**. Any construction materials that contact the spilled material must be properly disposed of. See **Skill Sheets 10-2, 10-3, 10-4,** and **10-5** for instructions on how to perform damming, diking, diversion, and retention.

Figure 10.8 Dams may be constructed to trap materials that are heavier or lighter than water, depending on their specific gravity.

Vapor Suppression

Vapor suppression is the action taken to reduce the emission of vapors at a haz mat spill. Fire-fighting foams are effective on spills of flammable and combustible liquids if the foam concentrate is compatible with the material **(Figure 10.9)**. All foam concentrates in use today must be proportioned (mixed with water) and aerated (mixed with air) before they can be used. These mechanical foam concentrates are divided into two general categories: those intended for use on Class A fuels (ordinary combustibles) and those for use on Class B fuels (flammable and combustible liquids). Foam concentrates must match the fuel to which they are applied to be effective. For that reason, this section will focus on **Class B foam concentrates** which can be used for vapor suppression.

Class B Foam Concentrate — Foam fire-suppression agent that is designed for use on un-ignited or ignited Class B flammable or combustible liquids.

Figure 10.9 Vapor suppression is used to prevent vapors from escaping the material. In this case, foam is used on a flammable liquid.

There are significant differences in Class B foams. For example, concentrates designed solely for hydrocarbon fires (such as regular fluoroprotein and regular **Aqueous Film Forming Foam [AFFF]**) will not extinguish **polar solvent** (alcohol-type fuel) fires regardless of the concentration at which they are used. Water-miscible materials such as alcohols, esters, and ketones destroy regular fire-fighting foams and require an alcohol-resistant foam agent. However, foam concentrates that are intended for polar solvents may be used on hydrocarbon fires. The *ERG* provides guidance on when alcohol-resistant foam should be used for a particular material **(Figure 10.10)**.

Aqueous Film Forming Foam (AFFF) — Synthetic foam concentrate that, when combined with water, can form a complete vapor barrier over fuel spills and fires and is a highly effective extinguishing and blanketing agent on hydrocarbon fuels.

ERG2008	FLAMMABLE LIQUIDS - CORROSIVE	GUIDE 132

EMERGENCY RESPONSE

FIRE
• **Some of these materials may react violently with water.**
Small Fire • Dry chemical, CO_2, water spray or alcohol-resistant foam.
Large Fire • Water spray, fog or alcohol-resistant foam.
• Move containers from fire area if you can do it without risk.
• Dike fire-control water for later disposal; do not scatter the material.
• Do not get water inside containers.
Fire involving Tanks or Car/Trailer Loads
• Fight fire from maximum distance or use unmanned hose holders or monitor nozzles.
• Cool containers with flooding quantities of water until well after fire is out.

Figure 10.10 The *ERG* can be consulted to determine the appropriate type of foam to use.

Producing High-Quality Foam

Four elements are necessary to produce high-quality foam: foam concentrate, water, air, and mechanical agitation. These elements must be blended in the correct ratios. Removing any element results in either no foam or poor-quality foam. Finished foam is produced in two stages. First, water is mixed with foam liquid concentrate to form a foam solution (proportioning stage). Second, the foam solution passes through the piping or hose to a foam nozzle or sprinkler that aerates the foam solution to form finished foam (aeration stage). Aeration produces an adequate amount of foam bubbles to form an effective foam blanket. Proper aeration also produces uniform-sized bubbles that form a long lasting blanket. A good foam blanket maintains an effective cover over a fuel.

WARNING!
First responders performing vapor-suppression tactics must constantly be aware of the threat of ignition of the spilled material. First responders must avoid the vapors at all times while applying foam.

What This Means To You

Once you've identified a hazardous material, you can use the orange guide in the *ERG* to tell you what type of foam you should use. For polar/water-miscible liquids, it recommends alcohol-resistant foam. For nonpolar/water-immiscible liquids, it recommends regular foam.

Various manufacturers use different terms to identify their foam concentrates that are effective on polar-solvents. Look for these letters before the name on the foam concentrate container:

• ARC (alcohol-resistant concentrate)

• PSL (polar-solvent liquid)

• ATC (alcohol-type concentrate) to use on polar-solvent liquids.

First responders must be familiar with the use and application of the following types of foam:

- *Fluoroprotein foam* — Used in storage tank, subsurface firefighting. Is effective on hydrocarbon fuels. Not effective on polar solvents. Derived from protein foam concentrates to which fluorochemical surfactants are added producing finished foam that flows across fuel surfaces rapidly. Fluoroprotein foam concentrates can be formulated to be alcohol-resistant. The foam bubble is protected from fuel saturation. Fluoroprotein foam concentrates have the following characteristics:

  — Has a very high degree of heat resistance and water retention

  — Maintains rather low viscosity at low temperatures

  — Compatible with simultaneous application of dry-chemical extinguishing agents

- *Aqueous film forming foam (AFFF)* — The most commonly used foam concentrate today **(Figure 10.11)**. AFFF foams have the following characteristics:

  — Can be used with either freshwater or saltwater

  — Premixable in portable fire extinguishers and apparatus water tanks

  — Suitable for subsurface injection

  — Fair penetrating capabilities in baled storage fuels or high surface-tension fuels such as treated wood

  — Compatible with dry-chemical extinguishing agents

  — Rather fast draining (Reapply AFFF finished foam often to maintain hot-spill security.)

  — As the quantity of ethanol and other gasoline additives goes up, nonalcohol-resistant AFFF becomes less effective

Figure 10.11 Aqueous film forming foam (AFFF) is one of the most commonly used foams.

Alcohol-Resistant AFFF Concentrate (AR-AFFF) — Aqueous film forming foam that is designed for use with polar solvent fuels.

- *Alcohol-resistant AFFF (AR-AFFF)* — Another class of AFFF concentrates is composed of alcohol-resistant concentrates. When **alcohol-resistant AFFF concentrates** are applied to polar solvent fuels, they create a membrane between the fuel and the finished foam. This membrane separates the water in the foam blanket from the attack of the solvent. Other characteristics are similar to AFFF listed previously.

- *High-expansion foam* — Special-purpose foam concentrates that are similar to Class A foams. Because they have a low-water content, they minimize water damage. Their low-water content is also useful when runoff is undesirable. Using high-expansion finished foam outside is generally not recommended because the slightest breeze may remove the foam blanket bubbles and reexpose the hazard to potential ignition sources. High-expansion foam concentrates have the following three basic applications:

    — Concealed spaces such as shipboard compartments, basements, coal mines, and other subterranean spaces

    — Fixed-extinguishing systems for specific industrial uses such as aircraft hangers and rolled or bulk paper storage

    — Class A fire applications (slow draining)

    High-expansion foam concentrates have the following characteristics:

    — Poor heat resistance because air-to-water ratio is very high

    — Expansion ratios of 200:1 to 1,000:1 for high-expansion uses and 20:1 to 200:1 for medium-expansion uses (Whether the finished foam is used in either a medium- or high-expansion capacity is determined by the type of application device.)

Emulsifiers

Emulsifiers are foam concentrates that are used with either Class A or Class B fires. Unlike finished foam that blankets the fuel, an emulsifier is designed to mix with the fuel, breaking it into small droplets and encapsulating them. The resulting emulsion is rendered nonflammable. However, the following drawbacks exist when using emulsifiers:

- Emulsifiers should only be used with fuels that are 1 inch (25 mm) or less in depth. When the fuel is deeper, it is almost impossible to mix the emulsifier thoroughly with the fuel.

- Once the emulsifier is thoroughly mixed with the fuel, it renders the fuel unsalvageable.

- Emulsifiers do not work effectively with water-soluble or water-miscible fuels because an emulsion cannot be formed between the concentrate and the fuel.

- Emulsifiers can have a negative effect on fish, aquatic life, and bodies of water. When emulsifiers are used for Class A fires and Class B spills, the effects of run-off must be taken into consideration.

Foam concentrates vary in their finished-foam quality and, therefore, their effectiveness for suppressing vapors. Manufacturers and suppliers will be able to provide information about freeze protected versions of foams. Foam quality is measured in terms of its 25-percent-drainage time and its expansion ratio. **Drainage time** is the time required for one-fourth (25 percent or quarter) of the total liquid solution to drain from the foam. **Expansion ratio** is the volume of finished foam that results from a unit volume of foam solution. In general, the required application rate for applying foam to control an unignited liquid spill is substantially less than that required to extinguish a spill fire.

Drainage Time — The amount of time it takes foam to break down or dissolve; also called drainage rate, drainage dropout rate, or drainage.

Expansion Ratio — Ratio of the finished foam volume to the volume of the original foam solution.

Roll-On Application Method (Bounce) — Method of foam application in which the foam stream is directed at the ground at the front edge of the unignited or ignited liquid fuel spill; foam then spreads across the surface of the liquid.

Bank-Down Application Method (Deflect) — Method of foam application that may be employed on an un-ignited or ignited Class B fuel spill. The foam stream is directed at a vertical surface or object that is next to or within the spill area. The foam deflects off the surface or object and flows down onto the surface of the spill to form a foam blanket.

Rain-Down Method — This method of foam application directs the stream into the air above the unignited or ignited spill or fire and allows the foam to float gently down onto the surface of the fuel.

Long drainage times result in long-lasting foam blankets. The greater the expansion ratio is, the thicker the foam blanket that can be developed in a given period of time. All Class B foam concentrates, except the special foams made for acid and alkaline spills, may be used for both fire fighting and vapor suppression. Air-aspirating nozzles produce a larger expansion ratio than do water fog nozzles.

First responders must exercise care when applying any of the foams onto a spill. All foams (except fluoroprotein types, which are suitable for subsurface injection) should not be plunged directly into the spill but applied onto the ground at the edge of the spill and rolled gently onto the material, known as the **roll-on application method**. If the spill surrounds some type of obstacle, the foam can be banked off the obstacle, known as the **bank-down application method (Figure 10.12)**. Another foam application method is the **rain-down method**. Foam is sprayed into the air over the target area in a fog pattern. The tiny foam droplets *rain* down over the spill. As the foam bubbles burst, the foam bleeds together to form a film over the fuel. This method is best used with AFFF; it is considerably less effective with other types of foam.

Figure 10.12 The bank-down method of applying foam.

When possible, first responders should use air-aspirating nozzles for vapor suppression rather than water fog nozzles because aerated foam maintains the vapor suppressive blanket for a longer period of time. However, non-aerated AFFF can be effective in suppressing flammable liquid fires.

Selection of the proper foam concentrate for vapor suppression is important. Because finished foam is composed principally of water, it should not be used to cover water-reactive materials. Some fuels destroy foam bubbles, so a foam concentrate must be selected that is compatible with the liquid as mentioned earlier. Other points to consider when using foam for vapor suppression are as follows:

- Water destroys and washes away foam blankets; do not use water streams in conjunction with the application of foam.
- A material must be below its boiling point; foam cannot seal vapors of boiling liquids.
- The film that precedes the foam blanket (such as with AFFF blankets), is not a reliable vapor suppressant. Reapply aerated foam periodically until the spill is completely covered.

First responders must be trained in the techniques of vapor suppression. Training for extinguishment of flammable liquid fires does not necessarily qualify a first responder to mitigate vapors produced by haz mat spills. More detailed information about using Class B foam at incidents involving flammable and combustible liquid spills is provided in the IFSTA manual, **Principles of Foam Fire Fighting**, 2nd edition. See **Skill Sheet 10-6** for more information about performing vapor suppression.

Vapor Dispersion

Vapor dispersion is the action taken to direct or influence the course of airborne hazardous materials. Pressurized streams of water from hoselines or unattended master streams may be used to help disperse vapors **(Figure 10.13)**. These streams create turbulence, which increases the rate of mixing with air and reduces the concentration of the hazardous material. After using water streams for vapor dispersion, it is necessary for first responders to confine and analyze runoff water for possible contamination. **Skill Sheet 10-7** provides a set of steps for performing basic vapor dispersion.

Figure 10.13 Vapor dispersion uses pressurized streams of water from hoselines or unattended master streams to disperse vapors.

CAUTION

ABC (monoammonium phosphate) dry chemical and some sodium bicarbonate-based BC dry chemical agents will destroy a foam blanket and should not be used in conjunction with foam. Other agents such as potassium based dry chemical are compatible with foam.

Before use, check with manufacturers for information on issues of compatibility.

Ventilation

Ventilation involves controlling the movement of air by natural or mechanical means. Ventilation is used to remove and/or disperse harmful airborne particles, vapors, or gases when spills occur inside structures. The same ventilation techniques used for smoke removal can be used for haz mat incidents (see IFSTA's **Essentials of Fire Fighting** manual). When conducting negative-pressure ventilation, fans and other ventilators must be compatible with the

Explosion-Proof Equipment — Encased in a rigidly built container so it withstands an internal explosion and also prevents ignition of a surrounding flammable atmosphere; designed to not provide an ignition source in an explosive atmosphere.

Dispersion — Act or process of being spread widely.

atmosphere where they are being operated. Equipment must be **explosion proof** in a flammable atmosphere. When choosing the type of ventilation to use, remember that positive-pressure ventilation is usually more effective than negative-pressure ventilation when it comes to removing atmospheric contaminants. Some experts consider ventilation to be a type of vapor dispersion.

Dispersion

Dispersion (as opposed to *vapor dispersion*) involves breaking up or dispersing a hazardous material that has spilled on a solid or liquid surface. Both chemical and biological agents have been used for this purpose, usually on hydrocarbon spills such as oceanic crude oil. Dispersion often has the unfortunate effect of spreading the material over a wide area, however, and the process itself may cause additional problems. Because of these problems, the use of dispersants may require the approval of government authorities.

Dilution

Dilution is the application of water to a water-soluble material to reduce the hazard. Dilution of liquid materials rarely has practical applications at haz mat incidents in terms of spill control; dilution is often used during decontamination operations. The amount of water needed to reach an effective dilution increases overall volume and creates a runoff problem that may be difficult to contain. This situation is especially true of water-soluble liquids. Dilution may be useful when very small amounts of corrosive material are involved such as in cases of irregular dispersion or a minor accident in a laboratory, but even then, it is generally considered for use only after other methods have been rejected. A simple set of steps for performing dilution are provided in **Skill Sheet 10-8**.

Dissolution

Dissolution — Act or process of dissolving one thing into another; process of dissolving a gas in water.

The process of dissolving a gas in water is called **dissolution**. This tactic can only be used on water-soluble gases such as anhydrous ammonia or chlorine and is generally conducted by applying a fog stream to a breach in a container or directly onto the spill. Ideally, the escaping gas then passes through the water and dissolves. When considering this option, first responders must remember that it may create additional problems with contaminated runoff water and other issues. For example, using water spray at a chlorine incident to bring the vapors to the ground may have the beneficial effect of reducing or eliminating a toxic plume, but it may also create hydrochloric acid on the ground with all the complications associated with that chemical.

Neutralization

Some hazardous materials may be neutralized to minimize the amount of harm they do upon contact. Usually, *neutralization* involves raising or lowering the pH of corrosive materials to render them *neutral* (pH 7) **(Figure 10.14)**. However, the term can be applied to any chemical reaction that reduces the hazard of the material. Neutralization is a difficult process; for example, adding too much of a neutralizer can cause a pH shift in the opposite direction. Neutralization should only be conducted (with few exceptions) under the direction of a hazardous materials technician, allied professional, or standard operating procedures.

Figure 10.14 Neutralization can be difficult because it is tricky to correctly adjust the pH of corrosive materials to 7.

Leak Control

A leak involves the physical breach in a container through which product is escaping. The goal of leak control is to stop or limit *the escape* or to *contain* the release either in its original container or by transferring it to a new one. It is often referred to as **containment**. The type of container involved, the type of breach, and properties of the material determine tactics and tasks relating to leak control. Leak control and containment are generally considered offensive actions. Normally, personnel trained below the Technician Level do not attempt offensive actions. There are a few, notable exceptions such as turning off a remote valve or situations dealing with gasoline, diesel, liquefied petroleum gas (LPG), and natural gas fuels **(Figure 10.15)**. With these fuels, Operational-Level responders can take offensive actions provided they have appropriate training, procedures, equipment, and PPE.

Containment — Act of stopping the further release of a material from its container.

Figure 10.15 Ops-Level responders may shut off valves to natural gas lines.

Figure 10.16 Ops-Level responders may also operate emergency remote shutoff valves on cargo tank trucks.

Leak control dictates that personnel enter the hot zone, which puts them at great risk. The IC must remember that the level of training and equipment provided to personnel are limiting factors in performing leak control.

In some situations it may be safe and acceptable for Operations-Level responders to operate emergency remote shutoff valves on cargo tank trucks (specifically, low-pressure liquid tanks and high-pressure tanks) **(Figure 10.16)**. The locations and types of remote shutoff valves vary depending on the truck **(Figure 10.17a-d)**. Nonpressure liquid tanks (MC/DOT-306/406), low-pressure chemical tanks (MC/DOT-307/407), and high pressure tanks (MC-331) will have an emergency shutoff switch on the left front corner of the tank (behind the driver). Some will also have one on either the right or the left rear corner. For example, MC-331s of 3,500 gallon (13 249 L) capacity or larger should have two emergency shutoff valves located remotely from each other, typically in this configuration (one on the tank behind the driver, the other on the rear of

Figure 10.17a Nonpressure liquid tanks (MC/DOT-306/406) will have emergency shutoff valves on the left front corner (behind the driver). They may also have them located on the rear of the tank, in the center of the tank near the valves, or the valve box. *Courtesy of Rich Mahaney.*

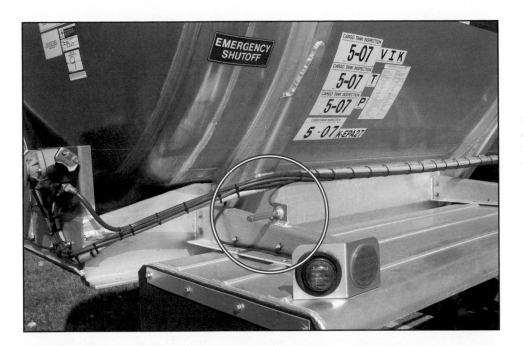

Figure 10.17b Low-pressure chemical tanks (MC/DOT-307/407) will have emergency shutoff valves on the left front corner (behind the driver). *Courtesy of Rich Mahaney.*

Figure 10.17c MC-331s will have two emergency shutoff valves located remotely from each other, one on the left front corner of the tank, the other typically on the right rear.

Figure 10.17d Shutoff valves can be switches, handles, pull levers, or fusible devices. Often, they will be labeled or painted red.

the tank, often on the passenger side). MC-331s of this capacity may also have an electronically operated shut-down device that can be activated 150 feet (46 m) from the vehicle. This device may also stop the engine and perform other functions. Some cargo tanks may have emergency shutoffs in the center of the tank near valves and piping, or built into the valve box. Activation of these shutoff valves varies by device but is usually as simple as flipping a switch (or handle) or breaking off a fusible device.

NOTE: Corrosive liquid tanks (MC/DOT 312) do not normally have emergency shutoff valves.

Piping systems and pipelines carrying hazardous materials may have remote shutoff or control valves that can be operated to stop the flow of product to the incident area without entering the hot zone. For example, by closing remote valves feeding a broken natural gas line, the flow of gas to the break can be stopped. Depending on the diameter and length of piping, a significant amount of product may release for some time before the flow stops. In most cases, onsite maintenance personnel or local utility workers know where these valves are located and can be given the authority and responsibility for closing them under the direction of the IC. They generally will understand the proper procedures and consequences of closing the valve. However, Operations-Level first responders who are trained and authorized to operate shutoff valves at their facilities in the event of emergency may do so in accordance with their SOPs. **Skill Sheet 10-9** covers steps for shutting off a remote valve.

Fire Control

Fire control is the strategy of minimizing the damage, harm, and effect of fire at a haz mat incident. If a fire is present in addition to a release (spill or leak), the incident is considerably more complicated. Based on risk and hazard assessment, a decision must be made whether to extinguish the fire, and if so, how. If the products of combustion are less of a hazard than the leaking chemical or extinguishment efforts will place firefighters in undue risk, the best course of action may be to protect exposures and let the fire burn until the fuel is consumed. If there is a threat of catastrophic container failure, boiling liquid expanding vapor explosion (BLEVE) or other explosion or if the resources needed to control the incident are unavailable, *withdrawal* may be the safest (and best) tactical option.

When containers or tanks of flammable liquids or gases are exposed to flame impingement, water streams should be deployed for maximum effective reach in order to prevent a BLEVE **(Figure 10.18)**. This cooling is commonly best achieved by directing a stream (or streams) at areas where there is direct flame impingement on the tank as well as along the tank's top so that water runs down both sides. This water stream cools the tank's vapor space. The piping and steel supports under tanks should also be cooled to prevent their collapse.

Gas fed fires burning around relief valves or piping should not be extinguished unless turning off the supply can stop the leaking product. An increase in the intensity of sounds or fire issuing from a relief valve indicates pressure within the container is increasing and container failure may be imminent.

WARNING!
Do not assume that relief valves are sufficient to safely relieve excess pressures. The rupture of flammable liquid and gas storage vessels has killed responders.

Figure 10.18 When flammable liquid and pressurized gas containers are impinged by flames, they should be cooled with water such as unmanned master streams. *Courtesy of Rich Mahaney.*

Improperly used water streams used inappropriately on flammable liquid fires can dramatically increase the size and intensity of a fire. When extinguishing agents are applied to burning tanks, they could overflow and threaten adjacent containers. Runoff from water streams applied to hazardous materials needs to be confined until it can be analyzed. Confinement can be accomplished through damming, diking, and retention.

Proper use of hose lines and extinguishing agents is critical to safely controlling flammable liquid and gas fires. If extinguishing agents are applied incorrectly, fuel may be pushed to unwanted locations, endangering people or exposures. Also, if the wrong agent is applied, firefighters may be placed in hazardous locations without the possibility of controlling the fire.

When firefighters are confronted with pressurized storage vessels, attention should be given to cooling all areas of the tank that are in the vapor space, if this is an appropriate tactic for the incident. Water streams should be deployed to keep the entire tank's surface wet **(Figure 10.19)**. Fire control operations may need to assign spotters that can visually confirm the entire surface of the tank is being cooled. If initial water streams are inadequate to cool both the pressurized tank and exposures, priority must be given to the involved container.

WARNING!
If initial water streams are inadequate to cool both the pressurized tank and exposures, priority must be given to the involved container. Failure to maintain the integrity of the tank will place the lives of everyone present in jeopardy!

Figure 10.19 It is best to keep the entire tank's surface wet.

When water streams are used to disperse gas being released under pressure, the mass and velocity of the water streams must exceed the mass and velocity of the escaping gas. Water streams must be delivered in a manner that disperses or disrupts escaping gas **(Figure 10.20)**. Any break in the water stream pattern will allow burning fuel to break through to hose line crews.

Figure 10.20 If water streams are used to disperse gas being released under pressure, the mass and velocity of the water streams must exceed the mass and velocity of the escaping gas in order to achieve sufficient dispersion.

Foam, dry chemical, and water are commonly used when flammable liquids are involved in fire. Foam should be selected when the fuel can be blanketed, separating the fuel from its air supply. As when using foam for vapor suppression, foam must be compatible with the fuel that is burning, and it must applied at a sufficient rate to extinguish the fire **(Figure 10.21)**. Protein, fluroprotein, and aqueous film forming foam (AFFF) have been the mainstay of flammable liquid fire fighting for years. High expansion foam is used for

Figure 10.21 Rain-down method of applying foam to extinguish a flammable liquid fire.

specific firefighting hazards. However, with increased production and use of alternative fuels such as ethanol, alcohol resistant foams are commonly required. The U.S. Pipeline and Hazardous Materials Safety Administration (PHMSA) recommends using *ERG* guide 127 for these materials (see **Appendix I**). More detailed information about required flow rate for foam application, necessary duration, and the logistics necessary to support foam firefighting operations is provided in the IFSTA manual, **Principles of Foam Fire Fighting**, 2nd edition.

Summary

Spill, leak, and fire control tactics play an important role in haz mat incident mitigation. Operations-level personnel may use defensive spill control tactics to confine released materials, preventing them from spreading and coming into contact with (and harming) people, property, or the environment. They may also attempt to contain a release by operating remote emergency shutoff devices. Fire control tactics may involve fighting fires, protecting exposures, and suppression of flammable vapors. Responders must know the correct application of various foam concentrates for operations involving Class B flammable and combustible materials.

Review Questions

1. Define spill control.
2. What is the difference between absorption and adsorption?
3. What is vapor suppression?
4. With which types of Class B foam should firefighters be familiar?
5. Describe the characteristics of each type of Class B foam.
6. What are the disadvantages to using emulsifiers?
7. What sorts of tactics are aimed at reducing the amount of harm caused by the material?
8. What is the goal of leak control?
9. What is fire control?
10. How can a BLEVE be avoided?

WARNING! Hazardous materials incidents can be extremely dangerous. Hazardous materials can cause serious injury or fatality. Appropriate personal protective equipment (PPE) must be worn and safety precautions must be followed. The following skill sheet demonstrates general steps; specific haz mat incidents may differ in procedure. Always follow departmental procedures for specific incidents.

Note: Prior to performing absorption/adsorption, the incident commander or other qualified responder must identify the material and determine the appropriate level of PPE required at the incident based on the hazardous material, training of responders, terrain, weather, and other size-up factors.

Step 1: Verify that all responders involved in the control function are wearing appropriate PPE for performing absorption/adsorption operations and that appropriate hand tools have been selected.

Step 2: Select a location to efficiently and safely perform the absorption/adsorption operation.

Note: The deployment of sorbent/adsorbent materials is usually conducted in locations where the hazardous liquid has pooled or for small spills where the hazardous liquid threatens to enter areas where it will be difficult to control (sewers or a water course).

Step 4: Deploy the sorbent/adsorbent in a manner that most efficiently controls the spill.
Note: The hazardous chemical is absorbed/adsorbed within the immediate location of the spill without additional environmental damage.

Step 5: Upon mitigation of the incident, place any contaminated material, such as clothing, in an approved container for transportation to a disposal location. Seal and label the container and document appropriate information for department records.

Step 6: Decontaminate tools.

Step 7: Advance to decontamination line for decontamination.

Step 3: Select the most appropriate sorbent/adsorbent.

Note: When conducting adsorption, consider compatibility of adsorption material and released product.

WARNING! Hazardous materials incidents can be extremely dangerous. Hazardous materials can cause serious injury or fatality. Appropriate personal protective equipment (PPE) must be worn and safety precautions must be followed. The following skill sheet demonstrates general steps; specific haz mat incidents may differ in procedure. Always follow departmental procedures for specific incidents.

Note: Prior to performing the defensive control function, the incident commander or other qualified responder must identify the material and determine the appropriate level of PPE required at the incident based on the hazardous material, training of responders, terrain, weather, and other size-up factors.

Step 1: Verify that all responders involved in the control function are wearing appropriate PPE for performing damming operations and that appropriate hand tools have been selected.

Step 2: Select a location to efficiently and safely perform the damming operation.

Step 3: Construct the dam in a location and manner that most efficiently controls the spill.

Step 4: Upon mitigation of the incident, place any contaminated material, such as clothing, in an approved container for transportation to a disposal location. Seal and label the container and document appropriate information for department records.

Step 5: Decontaminate tools.

Step 6: Advance to decontamination line for decontamination.

WARNING! Hazardous materials incidents can be extremely dangerous. Hazardous materials can cause serious injury or fatality. Appropriate personal protective equipment (PPE) must be worn and safety precautions must be followed. The following skill sheet demonstrates general steps; specific haz mat incidents may differ in procedure. Always follow departmental procedures for specific incidents.

Note: Prior to performing the defensive control function, the incident commander or other qualified responder must identify the material and determine the appropriate level of PPE required at the incident based on the hazardous material, training of responders, terrain, weather, and other size-up factors.

Step 1: Verify that all responders involved in the control function are wearing appropriate PPE for performing diking operations and that appropriate hand tools have been selected.

Step 2: Select a location to efficiently and safely perform the diking operation.

Note: The construction of a dike is usually done when the hazardous liquid spill threatens to contaminate a large area if it is not contained, or if the spill can be more easily managed at another location further downstream. It is used to direct the flow to another location. Most often the location of the dike is determined by the natural terrain where the spill has occurred relative to potential exposure and damage if the spill threatens a high value location.

Step 3: Construct the dike in a location and manner that most efficiently controls and directs the spill to a desired location.

Step 4: Upon mitigation of the incident, place any contaminated material, such as clothing, in an approved container for transportation to a disposal location. Seal and label the container and document appropriate information for department records.

Step 5: Decontaminate tools.

Step 6: Advance to decontamination line for decontamination.

WARNING! Hazardous materials incidents can be extremely dangerous. Hazardous materials can cause serious injury or fatality. Appropriate personal protective equipment (PPE) must be worn and safety precautions must be followed. The following skill sheet demonstrates general steps; specific haz mat incidents may differ in procedure. Always follow departmental procedures for specific incidents.

Note: Prior to performing the defensive control function, the incident commander or other qualified responder must identify the material and determine the appropriate level of PPE required at the incident based on the hazardous material, training of responders, terrain, weather, and other size-up factors.

Step 1: Verify that all responders involved in the control function are wearing appropriate PPE for performing diversion operations and that appropriate hand tools have been selected.

Step 2: Select a location to efficiently and safely perform the diversion operation.

Note: Hazardous liquid spills are usually diverted when they threaten to contaminate a stream or other high value area. Diversion is used to divert the flow to another location, to buy time. Most often the diversion is used to divert the flow to a containment area.

Step 3: Construct the diversion in a location and manner that most efficiently controls and directs the spill to a desired location. Working as a team, use hand tools to break the soil, remove the soil, pile the soil, and pack the soil tightly.

Step 4: Upon mitigation of the incident, place any contaminated material, such as clothing, in an approved container for transportation to a disposal location. Seal and label the container and document appropriate information for department records.

Step 5: Decontaminate tools.

Step 6: Advance to decontamination line for decontamination.

WARNING! Hazardous materials incidents can be extremely dangerous. Hazardous materials can cause serious injury or fatality. Appropriate personal protective equipment (PPE) must be worn and safety precautions must be followed. The following skill sheet demonstrates general steps; specific haz mat incidents may differ in procedure. Always follow departmental procedures for specific incidents.

Note: Prior to performing the defensive control function, the incident commander or other qualified responder must identify the material and determine the appropriate level of PPE required at the incident based on the hazardous material, training of responders, terrain, weather, and other size-up factors.

Step 1: Verify that all responders involved in the control function are wearing appropriate PPE for performing retention operations and that appropriate hand tools have been selected.

Step 2: Select a location to efficiently and safely perform the retention operation.

Step 3: Evaluate the rate of flow of the leak to determine the required capacity of the retention vessel.

Step 5: Upon mitigation of the incident, place any contaminated material, such as clothing, in an approved container for transport to a disposal location. Seal and label the container and document appropriate information for department records.

Step 6: Decontaminate tools.

Step 7: Advance to decontamination line for decontamination.

Step 4: Working as a team, retain the hazardous liquid so that it can no longer flow.

WARNING! Hazardous materials incidents can be extremely dangerous. Hazardous materials can cause serious injury or fatality. Appropriate personal protective equipment (PPE) must be worn and safety precautions must be followed. The following skill sheet demonstrates general steps; specific haz mat incidents may differ in procedure. Always follow departmental procedures for specific incidents.

Note: Prior to performing the defensive control function, the incident commander or other qualified responder must identify the material and determine the appropriate level of PPE required at the incident based on the hazardous material, training of responders, terrain, weather, and other size-up factors.

Step 1: Ensure that all responders involved in control functions are wearing appropriate PPE for performing vapor suppression operations.

Step 2: Select a location to efficiently and safely perform the vapor suppression operation.

Step 3: Evaluate the quantity and surface area of the hazardous material that has leaked.

Step 4: Determine the appropriate type of foam for the type of hazardous material present.

Step 7: Apply finished foam in an even layer covering the entire hazardous material spill area.

Step 8: Upon mitigation of the incident, place any contaminated material in an approved container for transportation to a disposal location.

Step 9: Seal, label, and manifest the container for removal.

Step 10: Decontaminate tools.

Step 11: Advance to decontamination line for decontamination.

Step 5: Working as a team, deploy the foam eductor and foam, and advance the hoseline and foam nozzle to a position from which to apply the foam.

Step 6: Flow hoseline until finished foam is produced at the nozzle.

WARNING! Hazardous materials incidents can be extremely dangerous. Hazardous materials can cause serious injury or fatality. Appropriate personal protective equipment (PPE) must be worn and safety precautions must be followed. The following skill sheet demonstrates general steps; specific haz mat incidents may differ in procedure. Always follow departmental procedures for specific incidents.

Note: Prior to performing the defensive control function, the incident commander or other qualified responder must identify the material and determine the appropriate level of PPE required at the incident based on the hazardous material, training of responders, terrain, weather, and other size-up factors.

Step 1: Ensure that all responders involved in the control function are wearing appropriate PPE for performing vapor dispersion operations.

Step 2: Select a location to efficiently and safely perform the vapor dispersion operation.

Step 4: Constantly monitor the leak concentration, wind direction, exposed personnel, environmental impact, and water stream effectiveness.

Step 5: Upon mitigation of the incident, place any contaminated material, such as clothing, in an approved container for transportation to a disposal location. Seal and label the container and document appropriate information for department records.

Step 6: Decontaminate tools.

Step 7: Advance to decontamination line for decontamination.

Step 3: Working as a team, advance the hoseline to a position to apply agent through vapor cloud to disperse vapors.

WARNING! Hazardous materials incidents can be extremely dangerous. Hazardous materials can cause serious injury or fatality. Appropriate personal protective equipment (PPE) must be worn and safety precautions must be followed. The following skill sheet demonstrates general steps; specific haz mat incidents may differ in procedure. Always follow departmental procedures for specific incidents.

Note: Prior to performing the defensive control function, the incident commander or other qualified responder must identify the material and determine the appropriate level of PPE required at the incident based on the hazardous material, training of responders, terrain, weather, and other size-up factors.

Step 1: Verify that all responders involved in the control function are wearing appropriate PPE for performing dilution operations.

Step 2: Select a location to efficiently and safely perform dilution operations.

Step 3: Evaluate the rate of flow of the leak to determine the required capacity of the retention area and the quantity of water required to dilute the material.

Step 4: Working as a team, monitor and assess the leak, and advance hose lines and tools to retention area.

Step 6: Monitor any diking or dams to ensure integrity of retention area.

Step 5: Flow water to dilute spilled material.

Step 7: Upon mitigation of the incident, place any contaminated material, such as clothing, in an approved container for transportation to a disposal location. Seal and label the container and document appropriate information for department records.

Step 8: Decontaminate tools.

Step 9: Advance to decontamination line for decontamination.

WARNING! Hazardous materials incidents can be extremely dangerous. Hazardous materials can cause serious injury or fatality. Appropriate personal protective equipment (PPE) must be worn and safety precautions must be followed. The following skill sheet demonstrates general steps; specific haz mat incidents may differ in procedure. Always follow departmental procedures for specific incidents.

Note: Prior to performing the defensive control function, the incident commander or other qualified responder must identify the material and determine the appropriate level of PPE required at the incident based on the hazardous material, training of responders, terrain, weather, and other size-up factors.

Step 1: Ensure that all responders involved in control functions are wearing appropriate PPE for performing remote valve shutoff operations.

Step 2: Identify and locate the emergency remote shutoff device.

Step 3: Operate the emergency remote shutoff device properly.

Step 4: Notify the Incident Commander of the completed objective.

Step 5: Document the activity log.

Air Monitoring and Sampling

Chapter Contents

Divider page photo courtesy of the U.S. Air Force, Photo by Senior Airman Taylor Marr.

chapter 11

Key Terms

Competencies

NFPA® 472:	6.7.3.2	6.7.3.4	6.7.4.2
6.7.3.1	6.7.3.3	6.7.4.1	

Air Monitoring and Sampling

Learning Objectives

1. Discuss air monitoring and sampling.

2. Discuss concentrations and exposure limits.

3. Explain the basics of air monitoring. [NFPA® 472, 6.7.3.3, 6.7.3.4]

4. Describe the selection and maintenance of detection and monitoring devices. [NFPA® 472, 6.7.3.1]

5. Explain how to detect specific hazards. [NFPA® 472, 6.7.3.2]

6. Perform a pH test on an unknown liquid. [Skill Sheet 11-1; NFPA® 472, 6.7.4.1, 6.7.4.2]

7. Perform air monitoring with a multi-gas meter. [Skill Sheet 11-2; NFPA® 472, 6.7.4.1, 6.7.4.2]

8. Perform air monitoring with a photoionization detector. [Skill Sheet 11-3; NFPA® 472, 6.7.4.1, 6.7.4.2]

9. Perform air monitoring with colorimetric indicator tubes. [Skill Sheet 11-4; NFPA® 472, 6.7.4.1, 6.7.4.2]

10. Detect radiation using a gas-filled detector. [Skill Sheet 11-5; NFPA® 472, 6.7.4.1, 6.7.4.2]

11. Describe other technologies used to detect hazardous materials.

Chapter 11
Air Monitoring and Sampling

Case History

In 1984, the owner of a bulk petroleum storage facility in Phoenix, Arizona, discovered that a toluene storage tank (10 feet [3 m] in diameter and 20 feet [6 m] in height) was contaminated and would have to be drained and cleaned. On November 15th, a supervisor and an unskilled laborer drained the tank to its lowest level - leaving approximately 3 inches (76 mm) of sludge and toluene in the bottom - and prepared for an entry into the tank via the top access portal.

The supervisor rented an SCBA from a local rental store and instructed the laborer in use of the SCBA and in the procedure they intended to follow. Since a ladder would not fit into the 16-inch (406 mm) diameter access hole, the supervisor secured a knotted, 1/4-inch (6 mm) rope to the vent pipe on top of the tank and lowered the rope into the hole. The 16-inch (406 mm) diameter opening on the top of the tank was not large enough to permit the laborer to enter wearing the SCBA. Therefore, it was decided the SCBA would be loosely strapped to the laborer so it could be held over his head until he cleared the opening. Once entry had been made, the supervisor was to lower the SCBA onto the laborer's back so it could be properly secured.

Immediately before the incident, both employees were on top of the tank. The laborer was sitting at the edge of the opening. The supervisor turned to pick up the SCBA. While he was picking up the unit, he heard the laborer in the tank. He turned and looked into the opening and saw the laborer standing at the bottom of the tank. He told the laborer to come out of the tank, but there was no response. The supervisor bumped the rope against the laborer's chest attempting to get his attention. The laborer was mumbling, but was still not responding to his supervisor's commands. At this point, the supervisor pulled the rope out of the tank, tied the SCBA to it and lowered the unit into the tank. Again, he yelled to the laborer in the tank, bumped him with the unit and told him to put the mask on. There was still no response. The laborer fell to his knees, then fell onto his back, and continued to mumble. At this point, the supervisor told the facility manager (who was on the ground) to call the fire department.

The first call went to the police department who relayed it to the fire department. Included in the fire department response was a hazardous materials team because of the information received about the material in the tank. The fire department (including the rescue and the hazardous materials teams) arrived on the scene approximately 10 minutes after the initial notification. After assessing the situation, fire officials decided to implement a rescue procedure rather than a hazardous materials procedure. Therefore, removal of the disabled person inside the tank was given top priority.

The 16-inch (406 mm) diameter opening at the top of the tank was not large enough to lower a firefighter donned in full rescue gear. Therefore, it was decided to cut through the side of the tank to remove the victim. The firefighters were aware of the contents of the tank (toluene) and the possibility of an explosion.

The fire department devised a cutting procedure that involved making two 19-inch (483 mm) vertical cuts and a 19-inch (483 mm) horizontal cut with a gasoline-powered disc saw. After the cuts were completed, the steel flap would be pulled down and the victim removed.

While the hazardous materials team was cutting, other firefighters were spraying water on the saw from the exterior to quench sparks. Two other firefighters were spraying water on the interior cut from the top opening. Three firefighters with the hazardous materials team were doing the actual cutting; they were alternately operating the saw because of the effort required to cut through the 1/4-inch (6 mm) thick steel. Sometime during the horizontal cut a decision was made to bring the two firefighters off of the top, which meant no water spray on the interior. Simultaneously, the exterior water spray was removed to extinguish flammable liquid burning on the ground as a result of the shower of sparks from the saw. Thus, at the precise time of the explosion, no water was being sprayed on the saw/cut from exterior or interior. Both vertical cuts were completed and the horizontal cut was 95 percent complete when the explosion occurred.

One firefighter was killed instantly from the explosion and several were injured. The man inside the tank was presumed to be already dead at the time of the explosion.

The following factors may have contributed to the rescue effort fatality and injuries:

- The condition of the person inside the tank was not known.
- The location and size of the only access portal on the tank precluded entry by a rescuer wearing full protective clothing and equipment.
- The fire department's confined space entry procedures precluded entry into a confined space containing hazardous materials without full protective clothing and equipment.
- The choice of methods to open the tank for rescue entry introduced an ignition source to an atmosphere which was known to be potentially explosive.
- The use of water sprays to prevent ignition of a flammable/explosive atmosphere in a confined space may not be effective under certain conditions.
- There were combustible materials on the ground surrounding the tank which ignited before the explosion and necessitated removal of exterior water spray away from the saw/cut.
- The fire department chain of command possibly created confusion when orders were given without full knowledge of the situation.
- The number of fire department personnel in the immediate area may have been excessive.
- The victim (firefighter) was directly in front of the cut during the cutting procedure and when the explosion occurred.

The NIOSH report of the incident came to the following recommendations based on their investigation of the incident:

- While cutting the tank and assisting fellow firefighters who were cutting, one firefighter stood directly in front of the opening, rather than to the side. This maximized the impact the victim received from the explosion. It is recommended that procedures be outlined that minimize such risk by firefighters.
- When hazardous tasks are performed only essential personnel should be in the immediate area, regardless of perceived risk by firefighters. Nonessential personnel should be permitted only after the hazardous task(s) has been completed.

- More extensive departmental procedures for efforts involving responses to explosive environments and hazardous materials are needed. Procedures should include command responsibilities, determinations of and distinctions between rescue and recovery efforts, uses of potential sources of ignition, methods to minimize risks of ignition, etc.

- City fire departments should establish a registry of confined spaces and toxic/explosive substances for specific companies within the area in which they serve. Such a registry should provide not only the name of the substance, but should also provide sufficient information so that emergency response personnel will have one comprehensive source that provides information sufficient to safely effect a rescue effort. *

This case illustrates the importance of detection and sampling the atmosphere before entering a confined space or potentially hazardous environment. It also demonstrates the importance of continuous monitoring while operations are being conducted because actions taken at the scene may alter atmospheric conditions.

Courtesy of NIOSH

Detection and monitoring devices allow responders to detect, identify, and measure hazardous materials **(Figure 11.1)**. Different devices incorporate different technologies to do different things; one device may be designed to measure radiation while another may determine the percentage of flammable vapors in the air. ***No one device or instrument will detect, identify, and measure all hazardous materials, and all devices and instruments have their strengths and limitations.*** With proper training, Ops-Level personnel may perform air monitoring and sampling missions at haz mat and WMD incidents under the direction of qualified personnel (such as a haz mat technician or allied professional) or written guidance such as standard operation procedures.

Figure 11.1 First responders can use detection and monitoring devices to detect, identify, and measure hazardous materials. *Courtesy of MSA.*

Air monitoring and sampling can be an important aspect of mitigation, assisting in the following tasks:

- Identifying hazards (what materials are involved in the incident and the concentrations present) **(Figure 11.2)**

- Determining appropriate PPE, tools, and equipment

- Determining perimeters and the scope of the incident (how far the materials have traveled, areas that have been contaminated, and/or areas that are free of contamination and may be safe)

- Checking the effectiveness of defensive operations

- Ensuring that decon operations are effective

- Detecting leaks from containers or piping systems

- Monitoring decon runoff for contamination levels

Figure 11.2 Sampling can be conducted to identify unknown substances. *Courtesy of Sherry Arasim.*

Most agencies that use air detection and monitoring devices will have a variety of equipment used to detect a number of different materials and hazards. The responders assigned to use them must have a good understanding of the devices' capabilities, and they must understand what is being measured and how the instrument relays that information to the user. If responders do not understand how to use or understand the devices correctly, they could easily jeopardize their safety and the safety of others. For example, it may take several seconds for a device to draw in a sample and analyze it (known as the **instrument reaction time**). If responders move too quickly, they may find themselves in a situation where the concentration of the hazardous material is much higher than what the meter is telling them because they have moved beyond the area where their meter took the sample. No instrument is any better than the knowledge, skills, and ability of the individual using it. They become a team and must practice and work together.

Instrument Reaction Time — Elapsed time between the movement of an air sample into a monitoring/detection device and the reading provided to the user. Also called Response Time.

Responders must use their understanding of the behavior of hazardous materials coupled with an understanding of the detection device in order to be successful in their mission. For example, it is important to recognize that most gases are heavier than air, while only a few are lighter than air. Detection and monitoring then, must be conducted at different heights and levels **(Figure 11.3)**.

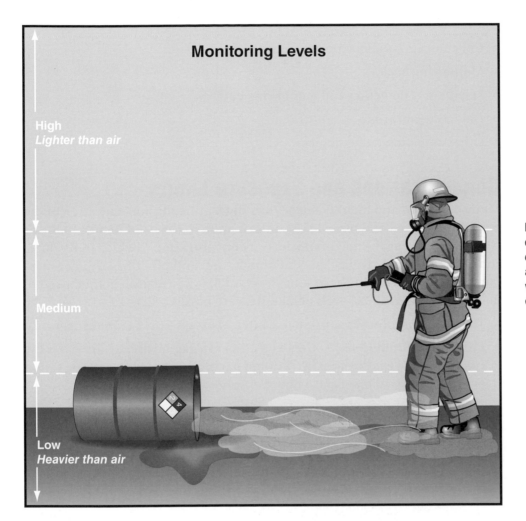

Monitoring Levels

High
Lighter than air

Medium

Low
Heavier than air

Figure 11.3 Because vapor densities vary and air currents can move hazardous gases and vapors in unexpected ways, samples must be taken at different heights.

Responders assigned detecting, monitoring, and sampling duties must be trained to correctly use the instruments available to them. They must understand the capabilities and limitations of each device. They must be able to interpret the data provided to them accurately. Additionally, they must be able to properly maintain, field test, and calibrate the devices per manufacturers' instructions. Finally, they must be able to use the devices in accordance with predetermined procedures based on the availability, capabilities, and limitations of personnel, appropriate personal protective equipment, and other resources available at the incident in accordance with the incident action plan. If possible, monitoring and sampling results should be verified using more than one sampling method and more than one technology.

At WMD or criminal incidents, personnel who conduct sampling activities must ensure that evidence preservation and chain of custody is followed because it is a very important aspect of crime scene investigation. They must

WARNING!
All personnel assigned to conduct detection, monitoring, and sampling must have proper training to do so, and they must wear appropriate PPE when operating in potentially hazardous areas.

follow appropriate protocols in regard to chain of custody, packaging, labeling, and transportation of evidence to the testing authority. The results of detection and monitoring activities must be recorded. Law enforcement assistance will be required to assist in this process.

It is beyond the scope of this manual to thoroughly cover all the detection and monitoring devices available to responders. However, many of the devices that agencies commonly use will be addressed. This chapter will also discuss the following:

- Concentrations and exposures limits
- Monitoring basics
- Selection of detection and monitoring devices
- Detecting hazards
- Other detection technologies

Concentrations and Exposure Limits

Many detection and monitoring devices determine the concentration of the material detected and provide this information to the user. Exposure limits (values expressing the maximum **dose** or **concentration** to which individuals should be exposed given a specific time frame) and other hazard measurements are expressed in these same terms. This information is then used to determine if hazardous levels of materials are present.

The concentration of a substance is often expressed in the following terms:

- *Parts per million (ppm)* — Concentration of a gas or vapor in air; parts (by volume) of the gas or vapor in a million parts of air; also the concentration of a particular substance in a liquid or solid
- *Parts per billion (ppb)* — Concentration of a gas or vapor in air; parts (by volume) of the gas or vapor in a billion parts of air; usually used to express extremely low concentrations of unusually toxic gases or vapors; also the concentration of a particular substance in a liquid or solid
- *Milligrams per cubic meter (mg/m³)* — Unit for expressing concentrations of dusts, gases, or mists in air
- *Grams per kilogram (g/kg)* — Expression of dose used in oral and dermal toxicology testing to denote grams of a substance dosed per kilogram of animal body weight
- *Milligrams per kilogram (mg/kg)* — Expression of toxicological dose to denote milligrams of a substance dosed per kilogram of animal body weight
- *Micrograms of material per liter of air (µg/L)* — Unit for expressing concentrations of chemicals in air
- *Milligrams per liter (mg/L)* — Unit for expressing concentrations of chemicals in water

Exposure limits may be expressed in terms such as *threshold limit value (TLV®), short-term exposure limit (STEL), threshold limit value-ceiling (TLV®-C),* and *permissible exposure limit (PEL)*. Concentrations that are high enough to kill or cause serious injury or illness are expressed in terms of *immediately dangerous to life or health (IDLH)*. **Table 11.1** provides definitions and exposure

Dose — Quantity of a chemical material ingested or absorbed through skin contact for purposes of measuring toxicity.

Concentration — (1) Quantity of a chemical material inhaled for purposes of measuring toxicity. (2) Percentage (mass or volume) of a material dissolved in water (or other solvent).

Table 11.1
Exposure Limits

Term	Definition	Exposure Period	Organizaion
IDLH Immediately Dangerous to Life or Health	An atmospheric concentration of any toxic, corrosive, or asphyxiating substance that poses an immediate threat to life. It can cause irreversible or delayed adverse health effects and interfere with the individual's ability to escape from a dangerous atmosphere.*	Immediate (This limit represents the maximum concentration from which an unprotected person can expect to escape in a 30-minute period of time without suffering irreversible health effects.)	**NIOSH** National Institute for Occupational Safety and Health
IDLH Immediately Dangerous to Life or Health	An atmosphere that poses an immediate threat to life, would cause irreversible adverse health effects, or would impair an individual's ability to escape from a dangerous atmosphere.	Immediate	**OSHA** Occupational Safety and Health Administration
LOC Levels of Concern	10% of the IDLH		
PEL Permissible Exposure Limit**	A regulatory limit on the amount or concentration of a substance in the air. PELs may also contain a skin designation. The PEL is the maximum concentration to which the majority of healthy adults can be exposed over a 40-hour workweek without suffering adverse effects.	8-hours Time-Weighted Average (TWA)*** (unless otherwise noted)	**OSHA** Occupational Safety and Health Administration
PEL (C) PEL Ceiling Limit	The maximum concentration to which an employee may be exposed at any time, even instantaneously.	Instantaneous	**OSHA** Occupational Safety and Health Administration
STEL Short-Term Exposure Limit	The maximum concentration allowed for a 15-minute exposure period.	15 minutes (TWA)	**OSHA** Occupational Safety and Health Administration
TLV® Threshold Limit Value†	An occupational exposure value recommended by ACGIH® to which it is believed nearly all workers can be exposed day after day for a working lifetime without ill effect.	Lifetime	**ACGIH®** American Conference of Governmental Industrial Hygienists
TLV®-TWA Threshold Limit Value-Time-Weighted Average	The allowable time-weighted average concentration.	8-hour day or 40-hour workweek (TWA)	**ACGIH®** American Conference of Governmental Industrial Hygienists
TLV®-STEL Threshold Limit Value-Short-Term Exposure Limit	The maximum concentration for a continuous 15-minute exposure period (maximum of four such periods per day, with at least 60 minutes between exposure periods, provided the daily TLV®-TWA is not exceeded).	15 minutes (TWA)	**ACGIH®** American Conference of Governmental Industrial Hygienists

Continued

Table 11.1 (continued)

Term	Definition	Exposure Period	Organizaion
TLV®-C Threshold Limit Value-Ceiling	The concentration that should not be exceeded even instantaneously.	Instantaneous	**ACGIH®** American Conference of Governmental Industrial Hygienists
BEIs® Biological Exposure Indices	A guidance value recommended for assessing biological monitoring results.		**ACGIH®** American Conference of Governmental Industrial Hygienists
REL Recommended Exposure Limit	A recommended exposure limit made by NIOSH.	10-hours (TWA) ††	**NIOSH** National Institute for Occupational Safety and Health
AEGL-1 Acute Exposure Guideline Level-1	The airborne concentration of a substance at or above which it is predicted that the general population, including "susceptible" but excluding "hypersusceptible" individuals, could experience notable discomfort. †††	Multiple exposure periods: 10 minutes 30 minutes 1 hour 4 hours 8 hours	**EPA** Environmental Protection Agency
AEGL-2 Acute Exposure Guideline Level-2	The airborne concentration of a substance at or above which it is predicted that the general population, including "susceptible" but excluding "hypersusceptible" individuals, could experience irreversible or other serious, long-lasting effects or impaired ability to escape. Airborne concentrations below AEGL-2 but at or above AEGL-1 represent exposure levels that may cause notable discomfort.	Multiple exposure periods: 10 minutes 30 minutes 1 hour 4 hours 8 hours	**EPA** Environmental Protection Agency
AEGL-3 Acute Exposure Guideline Level-3	The airborne concentration of a substance at or above which it is predicted that the general population, including "susceptible" but excluding "hypersusceptible" individuals, could experience life-threatening effects or death. Airborne concentrations below AEGL-3 but at or above AEGL-2 represent exposure levels that may cause irreversible or other serious, long-lasting effects or impaired ability to escape.	Multiple exposure periods: 10 minutes 30 minutes 1 hour 4 hours 8 hours	**EPA** Environmental Protection Agency
ERPG-1 Emergency Response Planning Guideline Level 1	The maximum airborne concentration below which it is believed nearly all individuals could be exposed for up to one hour without experiencing other than mild transient adverse health effects or perceiving a clearly defined objectionable odor.	Up to 1 hour	**AIHA** American Industrial Hygiene Association

Continued

Table 11.1 (concluded)

Term	Definition	Exposure Period	Organizaion
ERPG-2 Emergency Response Planning Guideline Level 2	The maximum airborne concentration below which it is believed nearly all individuals could be exposed for up to one hour without experiencing or developing irreversible or other serious health effects or symptoms that could impair an individual's ability to take protective action.	Up to 1 hour	**AIHA** American Industrial Hygiene Association
ERPG-3 Emergency Response Planning Guideline Level 3	The maximum airborne concentration below which it is believed nearly all individuals could be exposed without experiencing or developing life-threatening health effects.	Up to 1 hour	**AIHA** American Industrial Hygiene Association
TEEL-0 Temporary Emergency Exposure Limits Level 0	The threshold concentration below which most people will experience no appreciable risk of health effects.		**DOE** Department of Energy
TEEL-1 Temporary Emergency Exposure Limits Level 1	The maximum concentration in air below which it is believed nearly all individuals could be exposed without experiencing other than mild transient adverse health effects or perceiving a clearly defined objectionable odor.		**DOE** Department of Energy
TEEL-2 Temporary Emergency Exposure Limits Level 2	The maximum concentration in air below which it is believed nearly all individuals could be exposed without experiencing or developing irreversible or other serious health effects or symptoms that could impair their abilities to take protective action.		**DOE** Department of Energy
Teel-3 Temporary Emergency Exposure Limits Level 3	The maximum concentration in air below which it is believed nearly all individuals could be exposed without experiencing or developing life-threatening health effects.		**DOE** Department of Energy

* It should be noted that the NIOSH definition only addresses airborne concentrations. It does not include direct contact with liquids or other materials.

** PELs are issued in Title 29 *CFR* 1910.1000, particularly Tables Z-1, Z-2, and Z-3, and are enforceable as law.

*** Time-weighted average means that changing concentration levels can be averaged over a given period of time to reach an average level of exposure.

† TLVs® and BEIs® are guidelines for use by industrial hygienists in making decisions regarding safe levels of exposure. They are not considered to be consensus standards by the ACGIH®, and they do not carry the force of law unless they are officially adopted as such by a particular jurisdiction.

†† NIOSH may also list STELs (15-minute TWA) and ceiling limits.

††† Airborne concentrations below AEGL-1 represent exposure levels that could produce mild odor, taste, or other sensory irritation.

periods for the various limits as well as the organizations responsible for establishing them. The NIOSH Pocket Guide and other resources provide exposure limits for many materials.

Rather than providing concentrations of specific materials, some meters will provide the percentage of a material in the atmosphere, for example, the percentage of oxygen in the air, or percentage of the LEL for flammable gases and vapors. As with toxic materials, responders must know precisely what the meter is telling them, and what this information means in terms of their safety.

Safety Points to Remember
Keep these following points in mind:

- Wear appropriate PPE for potential and suspected hazards until detection and monitoring determines the actual hazards present.
- The lower the exposure limit(s), the more potentially harmful the substance is.
- As long as your exposure level never exceeds the lowest number indicated by any of the listed exposure levels for that material, you should be safe from any toxic effects.
- Exit IDLH atmospheres immediately, and *NEVER* enter them without appropriate PPE, which must include self-contained breathing apparatus (SCBA).

Monitoring Basics

The hazards present at the incident and the mission of responders will dictate the detection and monitoring strategies and tactics used at the incident as well as the PPE required to perform them. If the mission is rescue, these may be significantly different than if the mission is defensive mitigation of a release. Both require size-up and risk assessment. Rescue tactics will be discussed in Chapter 12, Rescue.

OSHA Monitoring Requirements
In its standard, 29 *CFR* 1910.120, U.S. OSHA requires detection and monitoring for flammables, radiation, and oxygen levels before entry into potentially hazardous atmospheres. Detection and monitoring shall be conducted to identify IDLH conditions, exposure over permissible exposure limits or published exposure levels, exposure over a radioactive material's dose limits or other dangerous condition such as the presence of flammable atmospheres, and/or oxygen-deficient environments. OSHA also requires periodic monitoring as conditions warrant (for example, changing environmental conditions or moving to a different location in the work site). See 29 *CFR* 1910.120(h)(2) for more information about OSHA's initial entry monitoring requirements.

It is helpful if responders have some idea of what hazardous materials are involved in the incident. Selecting PPE and monitoring and sampling equipment become easier if this information is known or suspected. Resources may be consulted to understand the hazards and properties of the materials involved, with a risk-based response developed appropriately.

When materials are unknown, it is necessary to take an analytic approach in attempting to identify and characterize the hazards present. It is important to monitor for oxygen levels (even if SCBA is worn, many detection devices require sufficient oxygen in order to function correctly), radiation, corrosives, flammables, oxidizers (and explosives), toxics, and exothermic reactions.

Before beginning detection and monitoring activities, have a monitoring plan which considers the following questions:

- Why is monitoring being conducted (ie. to determine hazard control zones or identify the hazards present)?

- What readings are anticipated? (If unanticipated readings are detected, consider the possibility of instrument failure, but do not ignore results.)

- Which instruments will be able to detect the materials anticipated under the current incident conditions?

- Is more than one hazard present?

- How will current conditions affect monitoring devices (rain, humidity, temperature, etc.)?

To stay safe while monitoring, do the following:

- Operate under the direction of a haz mat technician, specialist, or allied professional or follow written procedures.

- Understand the limitations of the instruments and detection devices being used, and follow manufacturers' instructions for **calibration** and usage.

- Take any damaged device out of service immediately because it may no longer be intrinsically safe, and it may provide false readings.

- Always wear appropriate PPE.

- Always work with a buddy and have a back-up team waiting in appropriate PPE.

- Approach the hazard area from upwind.

- Follow local protocols and SOPs established for what to do should an alarm sound or a hazardous material be detected.

- Pay particular attention to low lying areas, confined spaces, and containers where vapors and gases are likely to concentrate **(Figure 11.4, p. 520)**.

- Move slowly, making allowances for instruments with significant response times.

- Monitor for vapors and gases at ground level, waist level, and above the head.

It is also important to accurately document detection and sampling results such as the time of the reading, location and level of the reading, the reading obtained, and the instrument used. Record this information in a notebook immediately. Follow local protocols for reporting readings to the Incident Commander.

WARNING!
Never rely on one type of instrument exclusively.

WARNING!
Always assume more than one hazard is present!

Calibration — Set of operations used to standardize or adjust the values of quantities indicated by a measuring instrument.

Figure 11.4 Since most vapors and gases are heavier than air, they are likely to concentrate in low-lying areas and confined spaces. *Courtesy of MSA.*

Determining PPE

Even if the hazardous materials are unidentified, understanding the influence of such factors as state of matter, reactivity, vapor pressure, vapor density, and chemical properties can be very helpful in determining what PPE you need to wear during detection and monitoring operations. For example, in general, hazardous solids are much easier to protect against than hazardous materials in other forms. While solids may be reactive, toxic, biological, explosive, or radioactive, once released from their container, they tend not to travel far unless blown by wind or dispersed by an action such as an explosion. Solids have no vapor pressure, and, except for dusts with very small particle size, they tend not to stay suspended in air for long periods of time compared to gases and vapors. The majority of releases are easily controlled, and firefighter protective equipment with SCBA or Level C PPE typically provide sufficient protection for monitoring and sampling activities **(Figure 11.5)**.

Hazardous liquids may present a variety of hazards and will release vapors depending on temperature and vapor pressure. Vapors will travel according to vapor density, prevailing air movements, and topography. Vapors can travel from their point of origin making them much more difficult to control and potentially more difficult to protect against than solids. Firefighter protective equipment with SCBA will typically provide sufficient protection for short term exposures and Ops-Level monitoring and sampling activities involving most liquid releases. Mitigation efforts in the hot zone may require respiratory protection and chemical protective clothing (Level B PPE). The key is to:

- Provide respiratory protection
- Avoid direct contact with spills and product **(Figure 11.6)**
- Protect against fire hazards

Hazardous gases present the greatest threat when released. Gases expand rapidly, filling enclosed spaces and potentially traveling great distances from their source. Corrosive gases such as ammonia, chlorine, hydrogen chloride, methylamine, fluorine, and phosgene present an extreme threat to health and safety. Corrosive gases can damage and penetrate firefighter protective clothing and SCBA, representing both a skin contact and respiratory hazard, and therefore require the highest levels of PPE (Level A PPE) **(Figure 11.7, p. 522)**. Toxic gases can kill quickly if inhaled. Flammable gases present an extreme fire hazard while oxidizing gases can cause explosions. Inert gases can displace oxygen levels in enclosed spaces causing an asphyxiation hazard. If released gases are known or suspected to be involved in an incident you should operate with extreme caution during monitoring and sampling activities, for example, if a high pressure container is involved.

Figure 11.5 Firefighter protective clothing with SCBA will be sufficient to conduct detection and monitoring activities at many haz mat incidents. *Courtesy of the U.S. Coast Guard, photo by Petty Officer First Class Paul Roszkowski.*

Figure 11.6 It is important to avoid direct contact with the hazardous material as much as possible. *Courtesy of the U.S. Navy, photo by Photographer's Mate Airman Jason Frost.*

Figure 11.7 Incidents involving ton cylinders, compressed gas cylinders, and high pressure vessels need to be evaluated carefully in order to select appropriate PPE for detection and monitoring activities since they are likely to contain materials that will release as a gas. For example, corrosive gases present a high degree of threat to responders, requiring Level A PPE. *Courtesy of the U.S. Coast Guard, photo by Telfair H. Brown.*

Selection and Maintenance of Detection and Monitoring Devices

Many issues must be considered when determining what tools to use for detection and monitoring operations including the following:

- *Mission of the operation* — For example, is the priority of the operation rescue or product control?

- *Suspected hazards involved* — For example, monitoring for radiation at explosive incidents or using pH paper when corrosives are involved.

- *Portability and user friendliness* — Some instruments weigh more than others; some are bulky or difficult to use. It is important to determine if responders will be able to use the devices correctly given the mission and PPE **(Figure 11.8)**.

- *Instrument reaction time* — There is a delay ranging from seconds to minutes between the time the instrument detects the material and the readings are displayed; if the mission is a quick rescue, instruments with long delays may be useless.

Figure 11.8 When purchasing monitoring and detection equipment, it is important to consider portability and user friendliness. A device that can't be operated in the field while wearing PPE may not be of much use at an actual incident.

- *Sensitivity and selectivity* — Some instruments will detect lower concentrations than others, while others will only detect very specific materials. Consideration must be given as to how well the instrument will detect the desired chemical or chemical family and to what degree.

- *Calibration* — Most instruments need to be calibrated prior to use, and many factors can affect this process including temperature, humidity, elevation, and atmospheric pressure. Calibrating an instrument in the field can, in some cases, be difficult.

- *Training* — How difficult is it to learn how to use the instrument? How often does training need to be conducted?

Before purchasing detection and air monitoring equipment, users of prospective equipment should be contacted to get information about the equipment. Durability, dependability, weight, ease of use, and many other factors must be evaluated and only experience using the particular device can accomplish this. The cost of filters, probes, internal parts, and calibration can be expensive. Many factors reduce the reliability of instruments, procedures for use can be lengthy, and the elements (moisture, temperature , atmosphere) must be taken into consideration on many instruments. Being the first to purchase a new type instrument may not be practical. New concepts and methods of detection make some instruments obsolete very quickly. The cost of the instrument does not determine its effectiveness.

Detection and monitoring devices must be calibrated, maintained, and decontaminated in accordance to manufacturers' directions. Improperly calibrated and maintained devices are a safety hazard because readings may be inaccurate and misleading. When performing maintenance and calibration do the following:

- Use calibrant gases recommended by the manufacturer, and calibrate them frequently (**Figure 11.9**).

- Store devices in accordance with manufacturers' recommendations and be mindful of the expiration dates and shelf-life of some sensors, test strips, and colorimetric tubes.

- Test instruments routinely to ensure proper operation.

Figure 11.9 Calibrate instruments frequently, using the calibrant gases recommended by the manufacturer.

Zeroing Instruments in the Field

Calibrating an instrument to a calibrant gas is somewhat different than zeroing an instrument in the field. True calibration is conducted when the instrument is adjusted using a known type and concentration of a calibrant gas in order to standardize the measurements to be taken by the unit. Responders will need to zero many instruments in the field to adjust these units to the existing environment at those locations. Care must be taken to avoid zeroing the instrument in locations with potential contaminants, for example, zeroing 4-gas monitors near running vehicles where carbon monoxide levels may be high.

Detecting Hazards

The sections that follow will discuss detection and monitoring devices for the following hazards:

- Corrosives
- Oxygen
- Flammables
- Toxics
- Oxidizers
- Radiation
- Reactives

Corrosives

A large percentage of haz mat incidents involve corrosive materials. Corrosive gases and vapors can damage detection and monitoring instruments as well as PPE, so pH should always be one of the first hazards monitored unless hazards are known for certain.

Responders should be familiar with the following properties of corrosives:

- **pH** — The pH of a solution is a measurement of the hydrogen ions in a solution, indicating its strength. The pH scale ranges from 0 to 14. A pH of 7 is neutral. Any substance that is neither acidic nor basic is neutral. A pH less than 7 is acidic, and a pH greater than 7 is basic. In bases, as numbers increase above 7, the compound becomes increasingly more alkaline. Conversely, in acids, as numbers decrease below 7, the compound becomes increasingly acidic.

- **Concentration** — Acids and bases are usually created by dissolving a chemical (usually a liquid or a gas) in water. The degree to which the chemical is diluted determines the solution's *concentration*. Thus, the concentration reflects the amount of acid or base that is mixed with water. A 95-percent solution of formic acid is composed of 95 percent formic acid and 5 percent water. Generally, the higher the concentration, the more damage the acid or base will do (or the more corrosive it will be) relative to itself. For example, a 98-percent solution of sulfuric acid will burn the skin much more quickly and badly than an equal amount of a 1-percent sulfuric acid solution.

- *Strength* — The number of hydrogen ions or hydroxyl ions, respectively, determine the strength of an acid or base. *Details:*
  - The higher the number of hydrogen ions in the solution, the stronger the acid and the more corrosive it will be relative to other acids of equal concentration. For example, a 98-percent solution of hydrochloric acid (a strong acid) is far more corrosive than a 98-percent solution of acetic acid (a much weaker acid).
  - The higher the number of hydroxyl ions produced in making a solution, the stronger the base will be. For example, sodium hydroxide is far more caustic than lime.

Each whole pH value below 7 is 10 times more acidic than the next higher value. For example, a pH of 4 is 10 times more acidic than a pH of 5 and 100 times (10 × 10) more acidic than a pH of 6. The same holds true for pH values above 7, each of which is 10 times more alkaline than the value below it. For example, a pH of 10 is 10 times more alkaline than a pH of 9. At the extremes of the scale, a pH of 1 is 1,000,000 times more acidic than a substance with a pH of 7, and a pH of 14 is 10,000,000 times more alkaline. Pure water is neutral, with a pH of 7.0 (**Figure 11.10**).

pH Concentration

Concentration of Hydrogen Ions Compared to Distilled Water	pH
10,000,000	0
1,000,000	1
100,000	2
10,000	3
1,000	4
100	5
10	6
1	7
$1/10$	8
$1/100$	9
$1/1,000$	10
$1/10,000$	11
$1/100,000$	12
$1/1,000,000$	13
$1/10,000,000$	14

Figure 11.10 A pH of 1 is 1,000,000 times more acidic than a substance with a pH of 7, and a pH of 14 is 10,000,000 times more alkaline.

The primary tools used to detect and measure corrosivity are pH meters and pH paper (**Figure 11.11a and b**). Flouride test paper designed to monitor fluorine levels are also discussed in the sections that follow. Fluorides are another concern and should be monitored separately using fluoride test paper when dealing with unknowns.

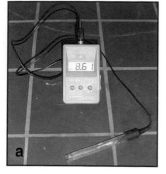

Figure 11.11a and b pH meters **(a)** and pH papers **(b)** are used to detect and measure corrosivity.

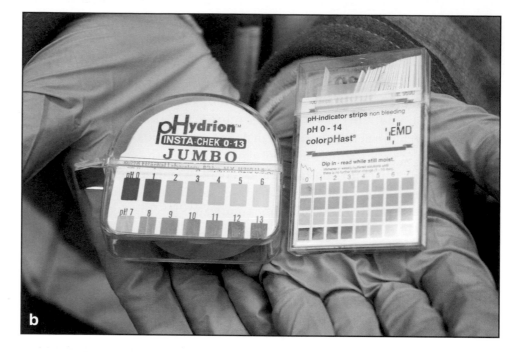

pH Paper and Meters

pH paper is designed to change color when it comes into contact with corrosive materials; the color indicates the pH. There is not a standard color system established for pH paper, so different brands may use different colors and configurations (see Figure 11.11b). Often, pH paper provides a scale from 0 to 13, but some brands may be more or less specific. pH paper may also have expiration dates after which it should not be used. Responders should be familiar with the following pH associations:

- pH 0-3 = strong acids (**Figure 11.12**)
- pH 7 = neutral (water)
- pH 10-14 = strong base (**Figure 11.13**)
- pH paper stripped or bleached = oxidizers and organic peroxides

Hydrocarbons may appear to give a reading between pH 4-6, however, this is not the true pH of the material (**Figure 11.14**). One way to test this reading is to wave the pH paper in the vapor space above the material. However, this will only work for materials that have a fairly high vapor pressure.

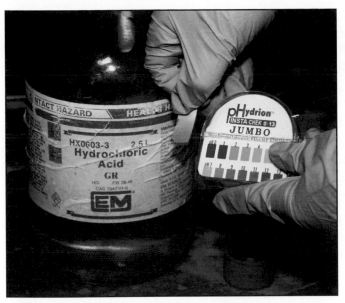

Figure 11.12 A pH of 0-3 indicates a strong acid.

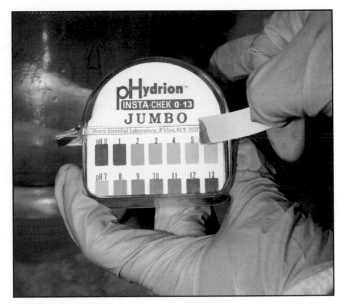

Figure 11.13 A pH 10-14 indicates a strong base.

Figure 11.14 Many hydrocarbons may appear to give a pH reading between 4-6, but this is not a true measurement of corrosivity.

Limitations of pH Paper

Limitations of pH paper include the following:

- Need for close proximity and/or contact with the hazardous material in order to conduct testing
- Inability to detect the concentration of the material
- Difficulty reading the paper if the material sampled is contaminated with oil, mud, or other opaque materials
- Difficulty reading the paper if the material sampled chemically strips the paper or alters it in unexpected ways (such as with highly concentrated acids and bases, certain oxidizers, and hydrocarbons)

Responders can attach a strip of pH paper to their SCBA mask, PPE, a pole, or other instrument (for example, a probe) to ensure that corrosive atmospheres are detected quickly. However, before doing so, most pH papers require that they are wetted with distilled water. Dry pH paper is very slow to react to atmospheric conditions (if it ever does) and therefore if this step is overlooked, there is a potential for first responders to unknowingly wander into extremely corrosive environments. If pH paper starts to change color, evacuation and reevaluation of the situation may be necessary, particularly if firefighter protective clothing is being worn and/or other detection and monitoring devices are being used or carried. **Skill Sheet 11-1** shows how to perform a pH test on an unknown liquid.

pH meters provide more exact readings than pH paper, however, they must be calibrated before each use, with the probe being rinsed with distilled water before and after calibration. To use a pH meter, insert the probe into the material being tested. pH meter readings can be affected by oils, temperature, and other contaminants.

Fluoride Test Paper

Fluorine is the most reactive of all the chemical elements, and compounds containing fluorine are called fluorides. One widely used fluoride is hydrogen fluoride (HF) which is extremely corrosive, toxic, and mildly reactive. Because of extreme health effects, the highest level of PPE (Level A) is required to work with HF, and in situations where the hazardous materials are unidentified or unknown, testing for fluorides (HF in particular) is recommended to protect responders.

WARNING!
If responders are not wearing Level A PPE and fluoride test strips change color, evacuate the area immediately!

Fluoride test papers can be used to determine the presence of fluoride ions and gaseous hydrogen fluoride. The pinkish-red paper turns yellowish-white in the presence of fluorides. Additionally, chlorates, bromates and sulphates in significant amounts will cause white discolorations of the paper. Like pH strips, fluoride test paper strips can be attached to PPE or other detection equipment.

Oxygen

Oxygen meters detect the percentage of oxygen in the air. Below 19.5% oxygen, the atmosphere is considered oxygen deficient and immediately dangerous to life and health (IDLH)(requiring use of supplied air such as SCBA). Above 23.5% the atmosphere is considered oxygen enriched, presenting a potential threat for fire and explosion. While monitoring for oxygen may seem unnecessary if responders are wearing SCBA or other supplied-air respirators, in situations other than rescue, it is required by OSHA 1910.120 to monitor for oxygen prior to entry into potentially hazardous atmospheres **(Figure 11.15)**. Oxygen is required for many other detection devices to function correctly.

CAUTION
Breathing into an oxygen meter to test it will blow carbon dioxide into the sensor which may cause degradation.

Normal air contains 20.9% oxygen, 78.1% nitrogen, and 1% other gases, and any readings below 20.9% indicate that some other contaminant is in the air displacing the oxygen, potentially at toxic or extremely hazardous levels. Keeping in mind that a contaminant will displace air proportionally (not just oxygen will be displaced), a one percent drop in oxygen is equivalent to 50,000ppm of something else in the air **(Figure 11.16)**. Even if oxygen levels are not low enough to trigger an alarm, responders should be aware that reduced levels of oxygen potentially represent a significant hazard in the form of toxic contaminants. SCBA should be worn in these circumstances even if oxygen levels are above 19.5%.

Oxygen sensors degrade quickly, even when they are not in use. Contact with other types of chemicals such as other oxidizers and carbon dioxide can also degrade these sensors. Because of this degradation, these sensors need to be replaced frequently.

Figure 11.15 Many detection devices such as CGIs need oxygen to function correctly, so it important to monitor oxygen levels even when wearing SCBA.

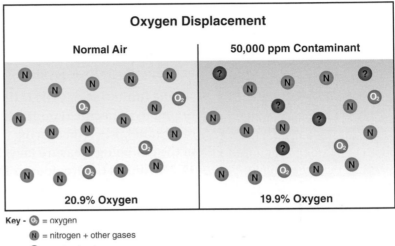

Figure 11.16 For every oxygen molecule in the air, there are approximately 4 molecules of other gases (primarily nitrogen). When a contaminant is introduced into the atmosphere, it doesn't just displace the oxygen; it displaces the nitrogen and other gases as well. Thus, for every oxygen molecule that is displaced, approximately 4 molecules of other gases are also displaced. An oxygen meter will only register the displaced oxygen, but responders need to remember that even a 1% drop in oxygen indicates that there is a significant amount (50,000ppm) of *something else* the air.

Humidity, temperature, and elevation can affect the readings from these sensors. They need to be zeroed in clean air (for example, away from sources of vehicle exhaust) at the elevation they will be used at (since oxygen levels vary at different elevations).

Limitations of Oxygen Meters

Oxygen meters have several limitations including the following:

- Corrosive gases can cause rapid sensor failure.
- Strong oxidizers such as chlorine, bromine, and fluorine can cause abnormally high readings (false positives).
- Sensors deteriorate steadily over time and need frequent replacement.
- Changes in temperature (and temperature extremes), humidity, and atmospheric pressure can affect the monitor.

CAUTION

Responders must understand what type of monitor they have and what it can and cannot do!

WARNING!

Manufacturers provide response curves and conversion factors that are specific to individual meters. Conversion factors for one meter must not be applied to other models or manufacturers' meters because of differences in calibration.

WARNING!

LEL meters will not provide accurate readings in oxygen deficient or enriched atmospheres.

Flammables

Most fire service responders are already familiar with devices such as combustible gas indicators (CGIs) which measure the amount of flammable vapors and gases in the atmosphere in one of these three ways:

1. Percentage of the lower explosive limit (%LEL)

2. Parts per million (ppm)

3. Percentage of gas per volume of air

Most meters measure the LEL, and for this reason they are sometimes called LEL meters rather than CGIs. Typically, LEL meters will sound an alarm at 10% of the LEL of the calibration gas, however this is not always the case. Even low percentages of LEL indicate that something is in the air, potentially at dangerously toxic levels.

One limitation of CGIs is the calibration issue. Because the meter is calibrated to a specific flammable gas (commonly methane, pentane, propane, or hexane), when measuring other flammable gases/vapors, the actual LEL of the gas being measure may differ from the reading being provided **(Table 11.2)**. **Table 11.3** provides examples of conversion factors (also called multipliers or response curves) for various gases. Responders using LEL meters must be aware of these potential discrepancies and make allowances to LEL readings accordingly.

Another potential limitation of CGIs is sensitivity to oxygen levels. Because many of these meters use a combustion chamber to burn the flammable gas, the atmosphere must have enough oxygen to support combustion in order for the instrument to function correctly. Too much oxygen can exaggerate readings or even damage sensors. Therefore, concurrent monitoring for oxygen levels is recommended when using CGIs **(Figure 11.17)**.

Other factors which can influence CGI readings include:

- Catalyst poisons
- Concentrations exceeding 100% of the LEL
- Concentrations exceeding the UEL
- Chlorinated hydrocarbons
- Oxygen-acetylene mixtures

Limitations of CGI Meters

Limitations of CGI meters include the following:

- Sensitivity to battery life; as batter power decreases, the meter may lose responsiveness.
- Corrosive gases may damage sensors.
- Meter response may be sluggish in extremely cold weather.
- Cell phones, magnetic fields, high voltage lines, radios, and static electricity may interfere with readings.
- Sensitivity to oxygen levels; too little or too much oxygen will interfere with accurate readings.

Table 11.2
Comparison of Actual LEL and Gas Concentrations with Typical Instrument Readings

Gas Type	Actual % LEL	Actual Gas Concentration	Typical Display Reading (% LEL)
Pentane	50%	0.07%	50%
Methane	50%	2.50%	100%
Propane	50%	1.05%	63%
Styrene	50%	0.55%	26%

Source: *Courtesy of MSA*

Table 11.3
Sample Conversion Factors

Gas or Vapor	Factor
Hexane	0.68
Hydrogen	0.39
Isopropyl Alcohol	0.73
Methyl Ethyl Ketone	0.90
Methane	0.38
Methanol	0.58
Mineral Spirits	1.58
Nitro Propane	0.95
Octane	1.36
Pentene	0.86
Iso-Pentene	0.86
Isoprene	0.58
Propane	0.56
Styrene	1.27
Vinyl Acetate	0.70
Vinyl Chloride	1.06
O-Xylene	1.36

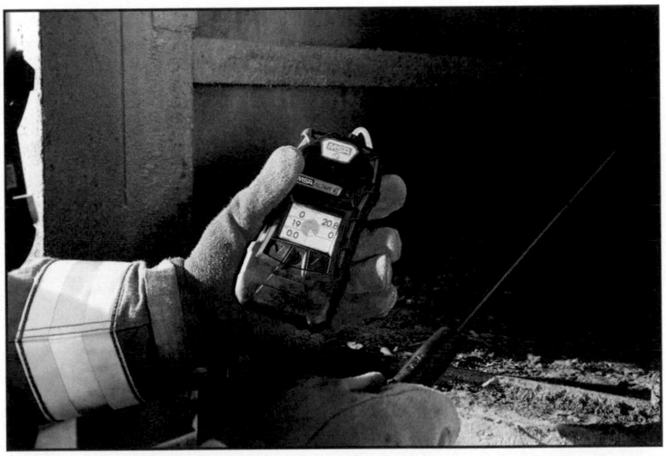

Figure 11.17 CGIs are sensitive to oxygen levels, therefore concurrent monitoring should be conducted. Most multi-gas meters monitor for flammables and oxygen at the same time. *Courtesy of MSA.*

Toxics

Many different technologies are used to detect toxic materials. Some are chemical specific (such as a carbon monoxide detector) while others are designed to identify the presence of large groups of chemicals such as organic gases and vapors. Some are very simple, while others are quite complex. This section will discuss the following detection devices:

- Chemical specific electrochemical cells
- Test strips and papers
- Photoionization detectors (PIDs)
- Colorimetric indicator tubes
- Flame ionization detectors (FIDs)
- Raman spectrometers

Determining Toxicity

The effect produced by a toxic compound is primarily a function of the dose (amount of a substance ingested or administered through skin contact) and the concentration (amount of the substance inhaled in this context) of the compound **(Figure 11.18)**. This principle, termed the *dose-response relationship*, is a key concept in toxicology. Many factors affect the normal dose-response relationship, but typically, as the dose increases, the severity of the toxic response increases **(Figure 11.19, p. 534)**. For example, people exposed to 100 parts per million (ppm) of tetrachloroethylene (a solvent commonly used for dry-cleaning fabrics) may experience relatively mild symptoms such as headache and drowsiness. However, exposure to 200 ppm of tetrachloroethylene can result in a loss of motor coordination in some individuals, and exposure to 1,500 ppm for 30 minutes may result in a loss of consciousness. The severity of the toxic effect also depends on the duration of exposure, a factor that influences the dose of the compound in the body. **Table 11.4, p. 535** shows factors that influence toxicity.

Toxicity is also a factor of exposure over time. In other words, an exposure to a compound that begins to have toxic effects at 200 ppm will be exaggerated if the exposure takes place in 10 minutes versus one that takes place over 48 hours where as there may not be any noticeable effect.

Poisons and the measurements of their toxicity are often expressed in terms of *lethal dose* (LD) for amounts ingested and *lethal concentration* (LC) for amounts inhaled. These terms are often found on safety data sheets. As a general rule, the smaller the value presented, the more toxic the substance is. Obviously, the lower the dose or concentration needed to kill, the more dangerous it is. These values are normally established by testing the effects of exposure on animals (rats or rabbits) under laboratory conditions over a set period of time. Dose terms are defined as follows:

- *Lethal dose (LD)* — Minimum amount of solid or liquid that when ingested, absorbed, or injected through the skin will cause death. Sometimes the lethal dose is expressed in conjunction with a percentage such as LD_{50} (most common) or LD_{100}. The number refers to the percentage of an animal test group that the listed dose killed (usually administered orally).

- *Median lethal dose (LD_{50})* — Statistically derived single dose of a substance that can be expected to cause death in 50 percent of animals when administered orally. The LD_{50} value is expressed in terms of weight of test substance per unit weight of test animal (mg/kg). The term LD_{50} does not mean that the other half of the subjects are necessarily completely well. They may be very sick or almost dead, but only half will actually die.

- *Lethal dose low (LD_{LO} or LDL)* — Lowest administered dose of a material capable of killing a specified test species.

Concentration terms are defined as follows:

- *Lethal concentration (LC)* — Minimum concentration of an inhaled substance in the gaseous state that will be fatal to the test group (usually within 1 to 4 hours). Like lethal dose, the lethal concentration is often expressed as LC_{50}, indicating that concentrations at the listed value killed half of the test group. The 50 percent of the population not killed may suffer effects ranging from no response to severe injury. LC may be expressed in the following ways:

— Parts per million (ppm)

— Milligrams per cubic meter (mg/m^3)

— Micrograms of material per liter of air (μg/L)

— Milligrams per liter (mg/L) (see information box)

- **Lethal concentration low (LC$_{LO}$ or LCL)** — Lowest concentration of a gas or vapor capable of killing a specified species over a specified time.

While lethal dose and lethal concentration values are obtained under laboratory conditions using test animals, emergency responders should be aware that exertion, stress, and individual metabolism or chemical sensitivities (allergies) may make persons more vulnerable to the harmful effects of hazardous materials.

The incapacitating dose (ID) for an organism (such as a human being) is expressed similarly to lethal dose and lethal concentration. Incapacitation can vary from moderate (unable to see, breathless) to severe (convulsions). The term is often used in the context of chemical warfare agents. Incapacitating doses are expressed as follows:

- **ID$_{50}$** — Dose that incapacitates 50 percent of the population of interest
- **ID$_{10}$** — Dose that incapacitates 10 percent of the population of interest

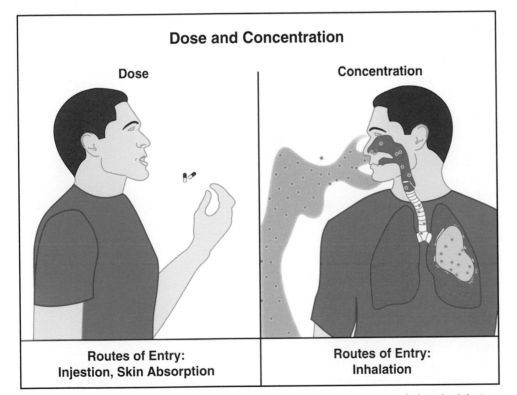

Dose and Concentration

Dose

Concentration

Routes of Entry:
Injestion, Skin Absorption

Routes of Entry:
Inhalation

Figure 11.18 For purposes of measuring toxicity, a *dose* is the amount of chemical that gets into the body through ingestion or skin contact (such as being absorbed through the skin). *Concentration* refers to the amount of a chemical inhaled.

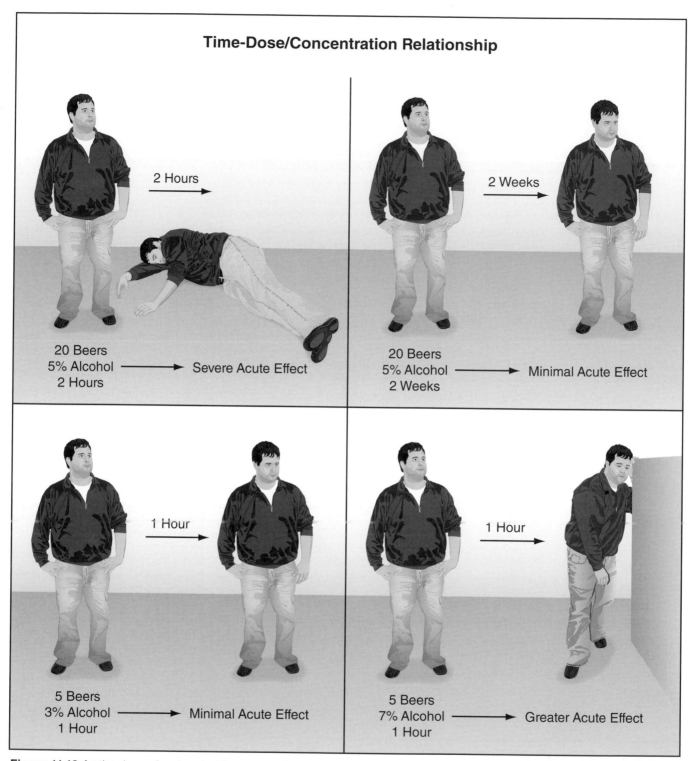

Figure 11.19 As the dose of a chemical increases, the severity of the toxic response also increases. Likewise, the duration of exposure has an effect on the toxic response. A dose spread over a long period of time may have less effect than the same dose administered over a shorter period.

Table 11.4
Factors Influencing Toxicity

Type of Factor	Examples
Factors related to the chemical	Composition (salt, freebase, etc.), physical characteristics (size, liquid, solid, etc.), physical properties (volatility, solubility, etc.), presence of impurities, breakdown products, carriers
Factors related to exposure	Dose, concentration, route of exposure (inhalation, ingestion, etc.), duration
Factors related to person exposed	Heredity, immunology, nutrition, hormones, age, sex, health status, preceding diseases
Factors related to environment	Media (air, water, soil, etc.), additional chemicals present, temperature, air pressure

Source: U.S. Centers for Disease Control and Prevention (CDC).

Chemical Specific Detectors

Some chemical monitors use sensors designed to detect a single chemical such as carbon monoxide, hydrogen sulfide, ammonia, chlorine, hydrazine, ethylene oxide, hydrogen cyanide, and phosgene. Sometimes these sensors are combined with a CGI and an oxygen sensor to form 2-, 3-, or 4- gas monitors **(Figure 11.20)**. A typical 4-gas monitor will detect LEL, oxygen, carbon monoxide, and hydrogen sulfide. Usually, an alarm is sounded when hazardous or potentially hazardous levels are detected. Four-gas monitors may sometimes be combined with photoionization detectors (PIDs, see next section) and are sometimes referred to as 5 gas monitors. The sensors in these devices are prone to degrade over time, and they may be affected by temperature and humidity. **Skill Sheet 11-2** demonstrates the steps for performing air monitoring using a multi-gas meter.

Figure 11.20 Multi-gas monitors combine sensors for different gases into one device.

Photoionization Detectors (PIDs)

PIDs use an ultraviolet lamp to ionize samples of gaseous materials **(Figure 11.21)**. They are used to detect low to very low concentrations of many organic and some inorganic gases and vapors, and they make good general survey instruments, capable of detecting and measuring concentrations in real-time. Although they cannot identify the material(s) present, they are very effective at determining that contaminants of some sort are present. **Skill Sheet 11-3** demonstrates the steps necessary to use a PID to detect contaminants. PIDs should be used in the following situations:

- At the edge of a release, where concentrations may be too low to be detected by a CGI

- When atmospheric contamination is suspected involving either flammable and/or non-flammable atmospheres

- Investigating complaints about odors or strange smells

- Locating low-volume chemical releases

- Evaluating the extent of contamination from a release and assessing risk to the public and environment

Figure 11.21 PIDs can detect many organic and some inorganic gases at the same time, making them good general survey instruments.

What This Means To You

At a gasoline spill, a CGI will detect the presence of flammable atmospheres, helping you to avoid explosions. A PID will detect the presence of toxic materials, helping you prevent future illness from exposures, such as cancer from exposure to benzene. For example, the LEL of gasoline is 1.4% or 10,400 ppm. The IDLH of Benzene is 500 ppm and cannot be read by a CGI. A PID, however, will let you know that contaminants are in the air at potentially dangerous levels.

Limitations of PIDs

The following are limitations of PIDs:

Ionization Potential (IP) – The energy required to free an electron from its atom or molecule.

- Certain models of PID instruments are not intrinsically safe, so they must be used in conjunction with a CGI.

- They cannot be used to identify unidentified/unclassified substances.

- Most PID's use several different lamps (or bulbs) to measure the **ionization potential (IP)** of a material. Operators need to have some suspicion of what material they are looking for and ensure that the proper lamp is in the instrument they are using **(Table 11.5)**. The NIOSH Pocket Guide provides the ionization potential numbers for many materials.

- They must be calibrated to isobutylene.

- They do not respond to any products with IP's greater than the ultraviolet lamp that is being used.

- They should not be used in rain or high humidity environments.

- As with all electronic meters, they must not be used in situation where liquids can accidentally be drawn into the probe.

- They are expensive to purchase and maintain.

- Electrical power lines, voltage sources, radio transmissions, and transformers may cause measurement errors.

- The lamp window must be periodically cleaned to ensure ionization of the new compounds by the probe (i.e., new air contaminants).

- As with all meters, high winds and humidity may affect readings because they both cause dilution of the product, skewing the readings.

- Tiny particulates of dust may affect readings.

Table 11.5
Ionization Potential and Lamp Strength

Chemical Name	Ionization Potential (IP)	Necessary Lamp Strength*
Isopropyl Ether	9.20 eV	9.5eV - 11.8eV
Methyl Ethyl Ketone	9.54 eV	10.0 eV - 11.8eV
Cyclohexane	9.88 eV	10.0 eV- 11.8eV
Ethyl Alcohol	10.47 eV	10.6 eV- 11.8eV
Acetylene	11.40 eV	11.7eV or 11.8eV
Hydrogen Cyanide	13.60 eV	Undetectable
Hydrogen Fluoride	15.98 eV	Undetectable

*Typical Lamp Strengths: 9.5eV, 10.0 eV, 10.2 eV, 10.6 eV, 11.7eV, 11.8eV; lamp strength must be higher than IP for the material to be detected

Colorimetric Indicator Tubes

Reagent — A chemical that is known to react to another chemical or compound in a specific way, often used to detect or synthesize another chemical.

Colorimetric tubes are sealed glass tubes filled with a **reagent** that will change color when exposed to a specific chemical family. Colorimetric tubes are chemical specific, but react to and rely on chemical cross reactivity **(Figure 11.22)**. They are useful for confirming the presence of a suspected material, but not very useful for determining or detecting unidentified products. These devices are known to have a significant error rate of 25%-35%.

Figure 11.22 Colorimetric tubes are chemical specific, so they are most effective when the hazardous material involved is known or suspected. *Courtesy of MSA.*

Reagents

Reagents are substances that are known to react to other chemicals in a specific way. Reagents used for the purpose of detection typically undergo a known color change when coming into contact with a specific substance. Reagents can be impregnated on strips, badges, or patches used to detect liquid and/or gaseous materials **(Figure 11.23)**. They can also come in liquid or solid form, and they have to be mixed with, or come in contact with, the material being analyzed.

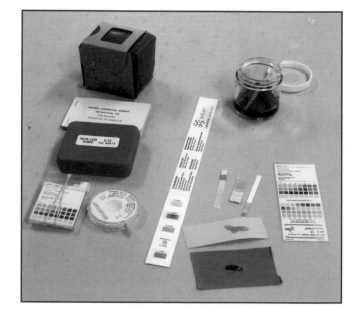

Figure 11.23 Reagents that undergo a known color change when exposed to certain materials are used in a variety of detection strips, badges, and patches.

Colorimetric tubes are simple to use. First, the tips are broken off. Next, one end of the tube is attached to a motorized pump, bellows, piston, or hand pump to draw air into the tube **(Figure 11.24)**. The number of pump strokes needed will vary depending upon the type of tube and material, but each stroke will draw a specific, measurable quantity of air through the tube and the reagent. If the material is present in the air, the tube will change color. The greater the length of stain in the tube or the more intense the color, the greater the amount of material in the air. Some manufacturers have instruments which contain a chip measurement system (CMS) that will interpret the readings and provide a digital readout. **Skill Sheet 11-4** demonstrates how to use colorimetric tubes to perform air monitoring. **It is very important to follow the manufacturer's instructions on how to use these devices.**

Figure 11.24 The tube is attached to a bellows or pump in order to draw air through the reagent inside. *Courtesy of MSA.*

Limitations of Colorimetric Tubes

The following are limitations of colorimetric tubes:

- They are cross sensitive (they may react to other chemicals in the same chemical "family").

- They have a limited shelf life.

- They have a significant error rate (25%-35%), and some will give false positives for products other than their labeled use.

- To get accurate readings, the number of pump strokes must be exact and complete to ensure full air movement through the tube.

- Temperature may affect the tubes (most are calibrated to work at 68°F to 77°F [20°C to 25°C], if temperatures are above or below this range, readings must be corrected by using a correction table, conversion factors, or a different number of pump strokes).

- Humidity and atmospheric pressure may affect the tubes, and require the use of a correction table, conversion factors, or a different number of pump strokes to obtain accurate readings.

- Exposure to light, even for a few hours, can cause the tubes to degrade and become non-functioning.

- Reagents may react with other vapors and gases present, causing unanticipated and unusual color changes.

- Responders with impaired color vision may not be able to accurately read colorimetric tubes.

- It may be difficult to read the color change in certain kinds of lighting.

Flame Ionization Detectors

Like PIDs, flame ionization detectors (FIDs) are used when relatively small concentrations of contaminants are known or suspected **(Figure 11.25)**. Calibrated to methane gas, they have a detection range from 1.0 to 10,000ppm for methane, and .2 to 1,000ppm for other materials. FIDs utilize a hydrogen flame to ionize gaseous materials. These devices will detect organic gases and vapors at lower concentration than PIDs, however, they will NOT detect inorganic materials at all. FIDs can be combined with gas chromatographs (CG) to determine amounts of individual components in gaseous mixtures.

Figure 11.25 FIDs detect organic gases and vapors at lower concentration than PIDs, but they don't detect inorganic materials at all.

Limitations of FIDs

The following are limitations of FIDs:

● While designed to be intrinsically safe, because they use a hydrogen flame, FIDs may not be safe to operate in hazardous (explosive) atmospheres if they are damaged (it is important to read the manufacturer's instructions on this issue).

● They do not identify what the material is.

● They do not detect inorganic materials.

● Because of calibration issues, the concentrations present may be much higher than the meter reading.

● As with CGIs, oxygen deficient and oxygen enriched atmospheres will cause FIDs to malfunction (most will alert the user if the hydrogen flame goes out).

● Large concentrations of flammable gases may increase the size of the hydrogen flame. This in turn may cause an increase in ion production followed by an automatic shutdown of the hydrogen supply. This causes a meter reading of 0.

● FIDs should be operated in moderate temperatures (40°F to 50°F [5°C to 10°C]). Temperatures above and below this range may cause the meter to provide erroneous readings.

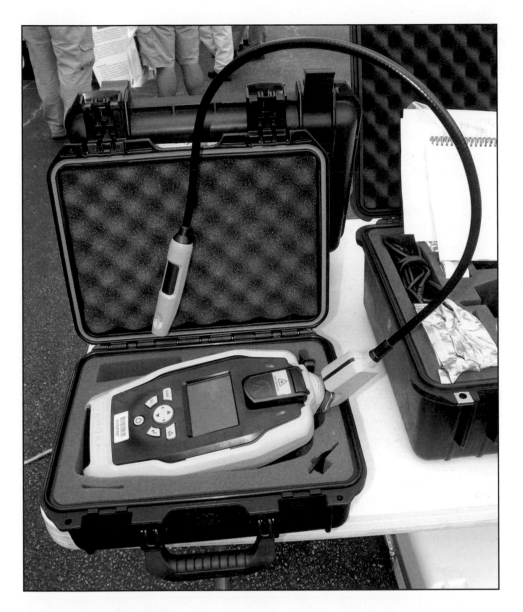

Figure 11.26 This device uses Raman spectroscopy to identify hazardous materials.

Raman Spectrometers

Similar to infrared spectroscopy, raman spectroscopy uses light (typically a laser) to illuminate a sample thereby creating a spectral signature that is unique to each material **(Figure 11.26)**. By comparing the signatures to a library of known signatures, identification of the material can be made. Raman spectrometers are used to identify liquids and solids (including many mixtures), and some have the benefit of being able to identify substances through clear containers such as glass bottles or plastic bags. They also don't destroy the material during testing. However they are not very effective at identifying residue or very small (non-visible) amounts of material. Raman spectrometers do not detect water, so they can identify materials dissolved in water, and they are effective at identifying many solutions.

While some raman spectrometers can be used to test materials through containers, responders may need to get very close for the laser to work, in some cases almost touching. Additionally, there is some concern that the laser may cause a reaction in some reactive materials, so it is best to test a small amount of material if possible.

Limitations of Raman Spectrometers

The following are limitations of raman spectrometers:

- They are limited to identifying visible quantities of liquids and solids.

- Some instruments can be used to take readings through clear glass containers, however, they cannot penetrate opaque containers (for example, brown bottles). Other instruments need a sample placed directly into the device in order to be analyzed.

- They cannot be used to identify dark or black materials because they absorb the laser light.

- The laser can ignite flammable or explosive materials (particularly dark materials such as black powder).

- Unshielded lasers may present an eye hazard.

Oxidizers

Organic peroxides can initiate explosive polymerization in certain materials, and they are components of improvised explosives such as triacetone tiperoxide (TATP) and hexamethylene triperoxide diamine (HMTD). Peroxide test strips use a reagent to detect the presence of these materials. A color change to blue after contact for 15 seconds indicates the presence of organic peroxides.

Potassium iodide (KI) starch paper can be used to test the oxidizing potential of unknown chemicals **(Figure 11.27)**. This paper changes color from white to blue/violet, purple, or black when contacting oxidizing materials (nitrites and free chlorine). The faster the color change, the greater the oxidizing potential. A rapid change may indicate an explosive.

One limitation of these test strips is that the responder must be in close proximity to the material in order to use them. If a peroxide or potential explosive is detected, responders need to withdraw immediately and contact bomb disposal technicians.

Figure 11.27 Potassium iodide (KI) starch paper can be used to test the oxidizing potential of unknown chemicals.

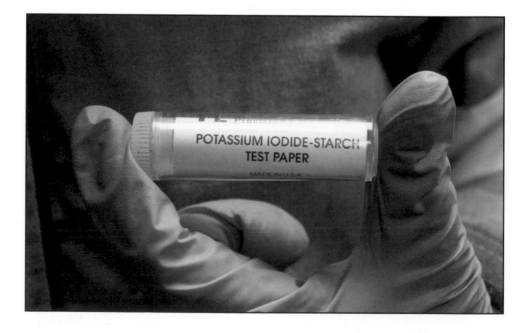

Radiation

Units called **curies (Ci)** and **becquerels (Bq)** are used to measure radioactivity, indicating the number of nuclear decays/disintegrations a radioactive material undergoes in a certain period of time. A Ci represents a large amount of activity, whereas a Bq represents a small amount of activity. These units come from two different measurement systems: English System and SI (**International System of Units**). Curies, part of the English System, are still commonly used in the U.S., whereas becquerels are more commonly used in other parts of the world.

The amount of radiation exposure is usually expressed in a unit called *rem* (for a large amount of radiation) or *millirem (mrem)*. However, several terms are used to express radiation dose and exposure with which first responders should be familiar. These units may be used on radiation dose instruments (dosimeters) and radiation survey meters.

As with curies and becquerels, two systems of units are used to measure and express radiation exposure and radiation dose (energy absorbed from the radiation). The U.S. still commonly uses the English System, and these units are as follows:

- *Roentgen (R)* — Used for measuring exposure, applied only to gamma and X-ray radiation; the unit used on most U.S. dosimeters; R per hour (R/hr) is used on radiation survey meters.

- *Radiation absorbed dose (rad)* — Used to measure the amount of radiation energy absorbed by a material. This unit applies to any material and all types of radiation, but it does *not* take into account the potential effect that different types of radiation have on the human body. For example, 1 rad of alpha radiation is more damaging to the human body than 1 rad of gamma radiation.

- *Roentgen equivalent in man (rem)* — Used for the absorbed dose equivalence as pertaining to a human body; applied to all types of radiation. This unit takes into account the energy absorbed (as measured in rad) and the biological effect on the body due to different types of radiation. Rem is used to set dose limits for emergency responders.

For gamma and X-ray radiation, a common *conversion factor* between exposure, absorbed dose, and dose equivalent is as follows:

1 R = 1 rad = 1 rem

The SI unit used to measure absorbed dose is called *gray (Gy)* whereas the unit for dose equivalence is *seivert (Sv)*. Seivert is used on some newer radioactive survey meters and meters outside the U.S. The U.S. Department of Defense also uses survey meters that read in units of gray per hour (Gy/hr).

Depending on the type of incident, it may not be obvious to emergency responders that they are arriving on a scene that may expose them to radiation or contamination. They will not smell, taste, feel, or see radiation. Therefore it is imperative that some form of detection instrumentation is available to test for the presence of radiation and contamination, particularly at explosive incidents that may be terrorist attacks (**Figure 11.28, p. 544**). Emergency responders must always check for radiation at explosive incidents.

Curie (Ci) — English System unit of measurement for radioactivity, indicating the number of nuclear decays/disintegrations a radioactive material undergoes in a certain period of time.

Becquerel (Bq) — International System unit of measurement for radioactivity, indicating the number of nuclear decays/disintegrations a radioactive material undergoes in a certain period of time.

International System of Units (Système International d'unités) — Modern metric system, based on units of ten.

Roentgen (R) — English System unit used to measure radiation exposure, applied only to gamma and X-ray radiation; the unit used on most U.S. dosimeters.

Roentgen Equivalent in Man (rem) — English System unit used to express the radiation absorbed dose (rad) equivalence as pertaining to a human body; applied to all types of radiation; used to set radiation dose limits for emergency responders.

Radiation Absorbed Dose (rad) — English System unit used to measure the amount of radiation energy absorbed by a material. Its International System equivalent is gray (Gy).

Figure 11.28 Radiation should be monitored for at all incidents involving explosions. *Courtesy of the U.S. Department of Energy.*

The simplest and most affordable tool available for first responders to detect radiation and contamination are hand-held portable survey instruments. Understanding the capabilities, limitations, and operational techniques of radiological survey instruments is critical to this task. Local ongoing training should be administered routinely on this topic. If a department has no instrumentation, it should consult a governmental radiation authority for guidance on life-saving operations.

There are many different models and types of radiological survey instruments with a variety of features and controls. Like other instruments used in hazard identification, each is designed for a specific use and each has its limitations. Radiological instruments can be divided into three groups:

- Instruments used for measuring radiation exposure
- Instruments used to detect contamination
- Instruments used for dose monitoring and personal dosimetry

Contamination can emit alpha, beta, gamma, or a combination of these types of radiation. Many commonly available survey instruments commonly allow the user the option of changing the detector or probe depending on the intended use of the instrument. Selecting the proper instrument for a specific task depends upon understanding the different types of probes or detectors and how its use affects the operating characteristics of the instrument. Attaching different types of probes to the survey meter can change many radiological survey instruments from radiation detection instruments into contamination detection instruments.

There are two general categories of detectors: gas-filled detectors and scintillation detectors. Both of these detector types are used for radiation survey instruments as well as contamination survey instruments. The sections that follow will discuss these two types in addition to personal dosimetry devices.

Gas-Filled Detectors

With a gas-filled detector, radiation ionizes the gas inside the detection chamber and the instrument's electronics measure the quantity of ions created. Ion chambers and Geiger-Mueller (GM) detectors are common examples of gas-filled detectors **(Figure 11.29)**.

An ion chamber is a very simple type of gas-filled detector that often uses regular ambient air as the detection gas, which can cause them to be affected by temperature and humidity. Ion chambers are often calibrated so that the response is directly related to the intensity of the radiation, making them reliable instruments when encountering radiations with varying energies.

The GM detector, or GM tube, was originally developed in 1928. GM detectors are sealed from outside air and are not typically affected by temperature or humidity. GM tubes with a very thin window may be capable of detecting alpha, beta, and gamma radiation, making them useful for detecting radiological contamination. GM tubes that use a sealed metal body are better suited for measuring penetrating gamma radiation that can be an external exposure hazard. The metal case makes this type of probe less suitable for use in detecting radiological contamination. **Skill Sheet 11-5** explains the steps for using a gas-filled detector to detect radiation.

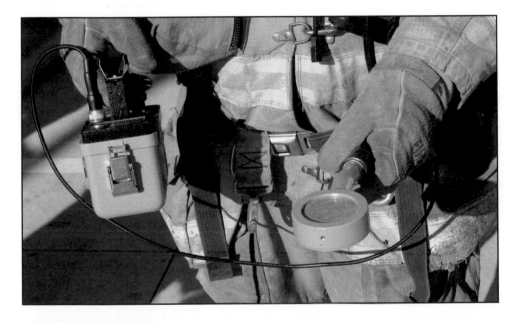

Figure 11.29 A Geiger-Mueller (GM) detector is a common example of a gas-filled detectors.

Scintillation Detectors

In scintillation detectors, radiation interacts with a crystal such as sodium iodide, cesium iodide, or zinc sulfide to produce a very small flash of light **(Figure 11.30, p. 546)**. The electronics of the instrument amplify this light pulse thousands of times in order to produce a useful signal that can be processed. A device called a *photomultiplier tube* that is attached to the crystal provides the light amplification. Some scintillation detectors have a thin Mylar® covering over the crystal making them useful for detecting radiological contamination. Generally speaking, scintillation detectors are most useful when detection of very small amounts of radiation is required.

Scintillation crystals that are sealed in a metal body are better suited for measuring penetrating gamma radiation. Scintillation detector probes are usually larger than gas-filled detector probes because of the photomultiplier tube. They are also susceptible to breakage if not handled properly. Dropping the instrument can shatter the crystal, photomultiplier tube, or both.

Figure 11.30 A scintillation detector.

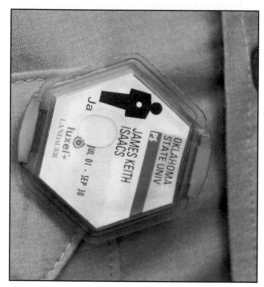

Figure 11.31 A dosimeter like this radiation badge can keep track of an individual's total radiation dose accumulated over a long period of time.

Monitoring and Personal Dosimetry Devices

Dosimetry devices are useful for keeping track of the wearer's total accumulated radiation dose. A dosimeter is like the odometer on a vehicle. Whereas the odometer measures total miles (kilometers) traveled, the dosimeter measures the total amount of dose received. There are several different types of **dosimeters** available **(Figure 11.31)**. Some commonly used personal dosimeters are self-reading, meaning they do not require processing at a lab to retrieve dose information.

A self-reading dosimeter (SRD) measures the radiation dose in roentgens (R), milliroentgens (mR), sieverts (Sv), or gray (Gy). Generally SRDs only measure gamma and X-ray radiation. SRDs are called by many names: direct reading dosimeter (DRD), pocket ion chamber (PIC), and pencil dosimeters. To read the dosimeter, hold it up to a light source and look through the eyepiece. Responders should always record the SRD reading before they enter a radiation field (hot zone). Periodically (at 15- to 30-minute intervals), read the SRD while working in the hot zone, and read it again upon exit from the hot zone. If a higher-than-expected reading is indicated, or if the SRD reading is off-scale, responders should:

- Notify others in the hot zone
- Have them check their SRDs
- Exit the hot zone immediately
- Follow local procedures

Dosimeter — Device used to measure an individual's exposure to an environmental hazard such as radiation or sound.

⚠️ **Protective Action Recommendations**

In August, 2008, U.S. DHS/FEMA published the *Planning Guidance for Protection and Recovery Following Radiological Dispersal Device (RDD) and Improvised Nuclear Device (IND) Incidents*, in the *Federal Register, Vol. 73, No. 149*. **Table 11.6, p. 548** summarizes the recommended public protection guides for radiological and improvised nuclear device incidents. In the early phase of an incident, sheltering in place or evacuation are recommended if radiation dose levels are predicted to be between 1 and 5 *rem* (0.01–0.05 *Sv*).

Table 11.7, p. 549 provides emergency worker guidelines for the early phase of an incident. While the guidelines suggest a 5 *rem* (0.05 *Sv*) occupational limit for most situations (with 10 and 25 *rem* [0.1 and 0.25 *Sv*] exceptions provided for extraordinary circumstances), these limits are not inflexible. As always, the IC, local SOPs, and the circumstances of the incident determine the appropriate response. Using the ALARA (as low as reasonably achievable) principle, it should be possible to conduct many operations, including rescues, below the 5 *rem* (0.05 *Sv*) limit.

Reactives

While there is not a meter or device designed to detect reactive materials as such, chemical reactions that are of concern to emergency responders will emit heat. For example, if a hazardous material in a container begins to polymerize, it will produce heat. Responders, then, can aim an **infrared thermometer** (or temperature gun) directly at the container to look for rising temperatures indicating a reaction in progress or a temperature increase that may lead to pressure change.

Infrared thermometers must be read within the context of the environment and the incident. A metal container sitting on an asphalt pad on a hot day may be quite warm. Responders need to look for rising temperatures or heat readings that cannot be accounted for given the surrounding environment. Thermal imaging cameras may also be used to detect heat and liquid levels in containers at haz mat incidents.

Infrared Thermometer — A non-contact measuring device that detects the infrared energy emitted by materials and converts the energy factor into a temperature reading. Also called a Temperature Gun.

Other Technologies Used to Detect Hazardous Materials

The following is a list of other detection and monitoring devices based on the substance being detected and/or the technology being used in the device:

- *Biological immunoassay indicators* – These detectors indicate the presence of biological agents and toxins by detecting the presence of specific antibodies. Most of these detectors require an active sample to be placed in the device in order to be analyzed.

- *Chemical agent monitors (CAMs)* – Chemical agent monitors utilize various technologies to specifically detect chemical warfare agents. Most of these devices require the responder to come in close contact with the material that is suspected to be a chemical agent.

Table 11.6
Protective Action Guides for RDD and IND Incidents

Phase	Protective Action Recommendation	Protective Action Guide
Early	Sheltering-in-place or evacuation of the Public[a]	1 to 5 *rem* (0.01–0.05 *Sv*) projected dose[b]
	Administration of prophylactic drugs — potassium iodide[c, e] Administration of other prophylactic or decorporation agents[d]	5 *rem* (0.05 *Sv*) projected dose to child thyroid[c, e]
Intermediate	Relocation of the Public	2 *rem* (0.02 *Sv*) projected dose first year. Subsequent years, 0.5 rem/y (0.005 *Sv/y*) projected dose[b]
	Food interdiction	0.5 *rem* (0.005 *Sv*) projected dose, or 5 *rem* (0.05 *Sv*) to any individual organ or tissue in the first year, whichever is limiting
	Drinking water interdiction	0.5 *rem* (0.005 *Sv*) projected dose in the first year

[a] Should normally begin at 1 *rem* (0.01 *Sv*); take whichever action (or combination of actions) that results in the lowest exposure for the majority of the population. Sheltering may begin at lower levels if advantageous.

[b] Total Effective Dose Equivalent (TEDE)—the sum of the effective dose equivalent from external radiation exposure and the committed effective dose equivalent from inhaled radioactive material.

[c] Provides thyroid protection from radioactive iodine only.

[d] For other information on other radiological prophylactics and medical countermeasures, refer to *http://www.fda.gov/cder/drugprepare/default.htm, http:/www.bt.cdc.gov/radiation, or http://www. orau.gov/reacts*.

[e] Committed Dose Equivalent (CDE). FDA understands that a KI administration program that sets different projected thyroid radioactive dose thresholds for treatment of different population groups may be logistically impractical to implement during a radiological emergency. If emergency planners reach this conclusion, FDA recommends that KI be administered to both children and adults at the lowest intervention threshold (*i.e.,* >5 *rem* (0.05 *Sv*) projected internal thyroid dose in children) (FDA 2001).

Source: U.S. DHS/Federal Emergency Management Agency's *Planning Guidance for Protection and Recovery Following Radiological Dispersal Device (RDD) and Improvised Nuclear Device (IND) Incidents*, published in the *Federal Register, Vol. 73, No. 149*, Friday, August 1, 2008

- *Fourier Transform Infrared (FTIR) spectrometer* – Devices using infrared spectroscopy technology compare the infrared spectra of chemical samples against a library of known spectral signatures. When the signature of the sample matches the signature of a known chemical in the library, identification is made. These devices cannot identify biological materials, but they can determine if a material is chemical or biological in nature.

- *Photoacoustic IR spectroscopy (PIRS)* – This detection technology uses a highly sensitive microphone to essentially listen to the sound made when a material absorbs infrared radiation. By recognizing the sound unique to a particular material, the material is then identified.

Table 11.7
Emergency Worker Guidelines in the Early Phase

Total Effective Dose Equivalent (TEDE)[a] Guideline	Activity	Condition
5 *rem* (0.05 *Sv*)	All occupational exposures	All reasonably achievable actions have been taken to minimize dose.
10 *rem* (0.1 *Sv*)	Protecting valuable property necessary for public welfare (e.g., a power plant)	• All appropriate actions and controls have been implemented; however, exceeding 5 *rem* (0.05 *Sv*) is unavoidable. • Responders have been fully informed of the risks of exposures they may experience. • Dose >5 *rem* (0.05 *Sv*) is on a voluntary basis. • Appropriate respiratory protection and other personal protection is provided and used. • Monitoring available to project or measure dose.
25 *rem* (0.25 *Sv*)[b]....	Lifesaving or protection of large populations. It is highly unlikely that doses would reach this level in an RDD incident; however, worker doses higher than 25 rem (0.25 Sv) are conceivable in a catastrophic incident such as an IND incident.	• All appropriate actions and controls have been implemented; however, exceeding 5 *rem* (0.05 *Sv*) is unavoidable. • Responders have been fully informed of the risks of exposures they may experience. • Dose >5 *rem* (0.05 *Sv*) is on a voluntary basis. • Appropriate respiratory protection and other personal protection is provided and used. • Monitoring available to project or measure dose.

a The projected sum of the effective dose equivalent from external radiation exposure and committed effective dose equivalent from internal radiation exposure.

b EPA's 1992 PAG Manual states that "Situations may also rarely occur in which a dose in excess of 25 rem for emergency exposure would be unavoidable in order to carry out a lifesaving operation or avoid extensive exposure of large populations." Similarly, the NCRP and ICRP raise the possibility that emergency responders might receive an equivalent dose that approaches or exceeds 50 rem (0.5 Sv) to a large portion of the body in a short time (Limitation of Exposure to Ionizing Radiation, National Council on Radiation Protection and Measures, NCRP Report 116 (1993a). If lifesaving emergency responder doses approach or exceed 50 rem (0.5 Sv) emergency responders must be made fully aware of both the acute and the chronic (cancer) risks of such exposure.

- *Ion-mobility spectrometers* – These devices use a radioactive source to ionize samples in order to determine their spectra. Currently, these devices are designed to detect chemical warfare agents and explosives (**Figure 11.32, p. 550**).

- *Mass spectrometers* – Mass spectrometers ionize samples in order to determine their composition. Like other spectroscopy devices, they compare test results to a library of known measurements in order to make a positive identification. Mass spectrometry may be combined with Gas Chromatography (referred to as GCMS) whose role is mainly to separate the various components of the mixture of chemicals.

- *Gas chromatography (GC) detectors* – GC detectors are a chemical analysis instrument used to separate and identify chemicals in a complex sample. A gas chromatograph uses a flow-through narrow tube known as the *column*, through which different chemical constituents of a sample pass in a gas stream at different rates depending on their various chemical and physical properties and their interaction with a specific column filling. As the chemicals exit the end of the column, they are detected and identified electronically.

Figure 11.32 Ion-mobility spectrometry technology uses a radioactive source to ionize samples in order to determine its spectra.

Figure 11.33 SAW technology is primarily used to detect chemical warfare agents, but may be combined with other sensor technology to increase detection capabilities.

- *Wet chemistry testing kits* – These kits contain portable chemistry sets designed to enable logical and progressive testing of a sample in order to identify it. (Many response organizations use field test kits for white powder incidents.)

- *DNA fluoroscopy devices* – Detectors using DNA fluoroscopy technology have the ability to identify specific DNA sequences, thereby detecting and identifying types of biological agents.

- *Polymerase chain reaction devices (PCRs)* – PCR technology is used to identify DNA. Devices utilizing this technology are used to detect and identify biological agents and toxins).

- *Surface acoustical wave devices (SAWs)* – Devices utilizing surface acoustical wave technology are used to detect nerve agents and blister agents **(Figure 11.33)**.

- *Thermoelectric Conductivity* – The absorption of various chemicals can change the electrical conductivity of materials such as heated metals oxide semiconductors and room temperature conductive polymers. By measuring the change in sensor conductivity using an electronic circuit, the chemicals can be detected.

- *Other personal detection devices* — Include organic vapor badges (or film strips, wrist bands, stick pins, etc.), mercury badges, and formaldehyde badges or strips used to measure individual exposure to certain chemicals.

Responders should understand that technologies are continually evolving, and new instruments are constantly being developed. New analytical methods frequently appear on the market, and responders should evaluate these thoroughly before investing in new technologies and devices.

Summary

First responders given the mission of conducting monitoring and detection at haz mat/WMD incidents must be thoroughly trained in the operation, limitations, and maintenance of the devices they use. It is important to always follow manufacturers' instructions regarding the use, operation, and maintenance of the equipment used. At a minimum, responders should monitor for corrosives, oxygen levels, flammables, toxics, and radiological materials. They may also monitor for oxidizers and detect exothermic chemical reactions by checking for rising temperatures. Responders conducting monitoring and detection tasks must don appropriate PPE for the mission and follow all applicable safety procedures such as operating in a buddy system and acting in accordance with SOPs should a device sound an alarm or hazardous materials are detected.

Review Questions

1. How does air monitoring and sampling assist responders?

2. What terms are often used to express the concentration of a substance?

3. What factors should be considered when creating a monitoring plan?

4. What are some guidelines to stay safe while monitoring?

5. What factors must be considered when selecting detection and monitoring devices?

6. What is pH? What does a pH of 7 identify?

7. What are the limitations of oxygen meters?

8. What is the difference between lethal dose and lethal concentration?

9. In what situations would a PID be used?

10. Describe different types of radiation detectors.

WARNING! Hazardous materials incidents can be extremely dangerous. Hazardous materials can cause serious injury or fatality. Appropriate personal protective equipment (PPE) must be worn and safety precautions must be followed. The following skill sheet demonstrates general steps; specific haz mat incidents may differ in procedure. Always follow departmental procedures for specific incidents.

Note: Prior to performing monitoring and detection activities, the incident commander or other qualified responder must attempt to identify the hazardous material present and determine the appropriate level of PPE required at the incident based on the training of responders, terrain, weather, and other size-up factors. Operations-Level responders performing monitoring and detection activities must do so under the guidance of a haz mat technician, allied professional, or SOPs.

Step 6: If vapors do not exist, take sample of liquid product using a pipette without personally coming into contact with the material.

Note: If necessary, attach strip to long rod or pole to ensure that user does not come into contact with material.

Step 1: Ensure that all responders are wearing appropriate PPE for performing testing operations.

Step 2: Select appropriate monitor for specific situation.

Step 3: Consider expiration date of pH papers. Prepare tape or strips for testing.

Step 4: Approach the product from uphill and upwind. Make certain that no responders come in direct contact with the spilled product.

Step 5: When approaching the product, determine the presence of corrosive vapors by waving wetted test paper in atmosphere.

Step 7: Compare results to pH paper chart to determine if the product is an acid, a base, or neutral.

Note: Confirmation of a corrosive atmosphere will eliminate the use of electronic meters for further testing.

Step 8: Dispose of contaminated test paper accordingly.

Step 9: Advance to decontamination line for decontamination.

WARNING! Hazardous materials incidents can be extremely dangerous. Hazardous materials can cause serious injury or fatality. Appropriate personal protective equipment (PPE) must be worn and safety precautions must be followed. The following skill sheet demonstrates general steps; specific haz mat incidents may differ in procedure. Always follow departmental procedures for specific incidents.

Note: Prior to performing monitoring and detection activities, the incident commander or other qualified responder must attempt to identify the hazardous material present and determine the appropriate level of PPE required at the incident based on the training of responders, terrain, weather, and other size-up factors. Operations-Level responders performing monitoring and detection activities must do so under the guidance of a haz mat technician, allied professional, or SOPs.

Step 1: Ensure that all responders are wearing appropriate PPE for performing testing operations.

Step 2: Select appropriate monitor for specific situation. Note that electronic meters cannot be used in corrosive atmospheres.

Step 3: Ensure that the multi-gas meter has been maintained and appropriately calibrated according to manufacturer's instructions.

Step 4: Start up multi-gas meter and acquire a fresh-air calibration.

Step 5: Approach the product from uphill and upwind. Before testing, consider the response time of multi-gas meter. Consider also the rising or sinking of particular materials in atmosphere. Responders should test atmosphere at various levels for appropriate readings.

Note: Ensure that the meter remains a safe distance from the material at all times. If the meter is used to analyze a liquid, it must stay at least 1 inch (25 mm) away from product.

Step 6: Approach the release and determine presence of:

a. Flammable atmosphere (10% of LEL)

Note: Responders using LEL meters must be aware that there are potential discrepancies in meter readings, and must make allowances to LEL readings accordingly (multipliers, response curves relative response, or correction factors).

b. Oxygen-rich atmosphere (above 23.5%)

c. Oxygen-deficient atmosphere (below 19.5%)

d. Other toxic materials, depending on sensor (CO, H_2S, NH_3, PH_3, etc.) being used

Note: Be aware that some values may be read as percentage, while others may be read as ppm. Understand how your specific meter should be read.

Step 7: If meter indicates an IDLH atmosphere, take appropriate actions according to AHJ.

Step 8: Decontaminate meter according to manufacturer's instructions.

Step 9: Advance to decontamination line for decontamination.

WARNING! Hazardous materials incidents can be extremely dangerous. Hazardous materials can cause serious injury or fatality. Appropriate personal protective equipment (PPE) must be worn and safety precautions must be followed. The following skill sheet demonstrates general steps; specific haz mat incidents may differ in procedure. Always follow departmental procedures for specific incidents.

Note: Prior to performing monitoring and detection activities, the incident commander or other qualified responder must attempt to identify the hazardous material present and determine the appropriate level of PPE required at the incident based on the training of responders, terrain, weather, and other size-up factors. Operations-Level responders performing monitoring and detection activities must do so under the guidance of a haz mat technician, allied professional, or SOPs.

Step 1: Ensure that all responders are wearing appropriate PPE for performing testing operations.

Step 2: Select appropriate monitor for specific situation. Note that electronic meters cannot be used in corrosive atmospheres.

Note: If material is known, determine ionization potential.

Note: Determine electron volt strength of specific meter being used.

Note: Determine if monitor reads in PPM or PPB.

Step 3: Ensure that the photoionization detector has been maintained and appropriately calibrated according to manufacturer's instructions.

Step 4: Start up PID meter and acquire a fresh-air calibration.

Step 5: Approach the product from uphill and upwind. Before testing, consider the response time of the photoionization detector. Consider also the rising or sinking of particular materials in atmosphere. Responders should test atmosphere at various levels for appropriate readings.

Note: Ensure that the meter remains a safe distance from the material at all times. If the meter is used to analyze a liquid, it must stay at least 1 inch (25 mm) away from product.

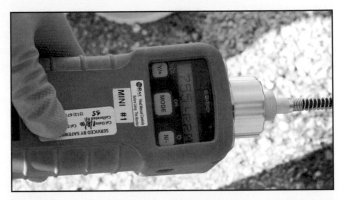

Step 6: Approach release and determine level of toxic gas or vapor in the atmosphere. Compare specific atmospheric levels according to established exposure values.

Note: Responders using PID meters must be aware that there are potential discrepancies in meter readings, and must make allowances to PID readings accordingly (multipliers, response curves relative response, or correction factors).

Step 7: Decontaminate meter according to manufacturer's instructions.

Step 8: Advance to decontamination line for decontamination.

WARNING! Hazardous materials incidents can be extremely dangerous. Hazardous materials can cause serious injury or fatality. Appropriate personal protective equipment (PPE) must be worn and safety precautions must be followed. The following skill sheet demonstrates general steps; specific haz mat incidents may differ in procedure. Always follow departmental procedures for specific incidents.

Note: Prior to performing monitoring and detection activities, the incident commander or other qualified responder must attempt to identify the hazardous material present and determine the appropriate level of PPE required at the incident based on the training of responders, terrain, weather, and other size-up factors. Operations-Level responders performing monitoring and detection activities must do so under the guidance of a haz mat technician, allied professional, or SOPs.

Step 1: Ensure that all responders are wearing appropriate PPE for performing testing operations.

Step 2: Ensure that tubes are within expiration dates.

Step 3: If material is known, select appropriate tube. If material is unknown, use various clues at the scene to determine appropriate tube.

Step 4: Consult manufacturer's instructions for specific applications and use of each tube.

Step 5: Approach the product from uphill and upwind. Consider the rising or sinking of particular materials in atmosphere. Responders should test atmosphere at various levels for appropriate readings.

Step 6: Break tips of both ends of tube and insert into colorimetric pump unit or insert colorimetric chip cartridge according to manufacturer's instructions. Perform testing of atmosphere.

Note: Ensure that testing with individual tubes is performed in only ONE location. Movement during the testing process can result in skewed results.

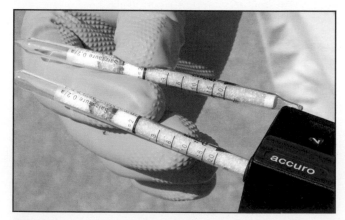

Step 7: Compare results to instructions provided by manufacturer.

Note: It may be necessary to compare used tube to unused tube to verify that color change occurred.

Step 8: Decontaminate according to manufacturer's recommendations. Dispose of used colorimetric tubes according to AHJ.

Step 9: Advance to decontamination line for decontamination.

WARNING! Hazardous materials incidents can be extremely dangerous. Hazardous materials can cause serious injury or fatality. Appropriate personal protective equipment (PPE) must be worn and safety precautions must be followed. The following skill sheet demonstrates general steps; specific haz mat incidents may differ in procedure. Always follow departmental procedures for specific incidents.

Note: Prior to performing monitoring and detection activities, the incident commander or other qualified responder must attempt to identify the hazardous material present and determine the appropriate level of PPE required at the incident based on the training of responders, terrain, weather, and other size-up factors. Operations-Level responders performing monitoring and detection activities must do so under the guidance of a haz mat technician, allied professional, or SOPs.

Step 1: Ensure that all responders are wearing appropriate PPE for performing testing operations.

Step 2: Select the appropriate monitor for the specific situation.

Step 3: Ensure that the detector has been maintained and appropriately calibrated according to manufacturer's instructions.

Step 4: Start up the gas-filled meter and acquire background radiation levels (CPM for alpha and beta, or MREM for gamma).

Step 5: Approach the product from uphill and upwind. Consider the rising or sinking of particular materials in atmosphere. Responders should test atmosphere at various levels for appropriate readings.

Step 6: Read the meter accordingly.

Note: Meters may read radiation using various terms. Understand how your specific meter reads radiation rates.

Step 7: Determine the presence of gamma radiation.

Note: Reading should be considered within the hot zone if it indicates two times above background.

Step 8: Determine presence of alpha or beta radiation.

Note: Reading should be considered within the hot zone if it indicates 200-300 CPM above background.

Step 9: Compare radiation values to AHJ's SOP values to determine appropriate actions.

Step 10: Decontaminate meter according to manufacturer's recommendations.

Step 11: Advance to decontamination line for decontamination.

Victim Rescue and Recovery

Chapter Contents

Divider page photo courtesy of David Lewis.

chapter 12

Key Terms

Competencies

NFPA® 472:	6.8.3.1(2)(b)	6.8.3.1(2)(e)	6.8.3.2	6.8.4.1(3)
6.8.3.1(1)	6.8.3.1(2)(c)	6.8.3.1(3)	6.8.4.1(1)	6.8.4.1(4)
6.8.3.1(2)(a)	6.8.3.1(2)(d)	6.8.3.1(4)	6.8.4.1(2)	6.8.4.1(5)

Victim Rescue and Recovery

Learning Objectives

1. Discuss rescue operations. [NFPA® 472, 6.8.3.1(1), 6.8.3.1(2)(a-d), 6.8.3.1(3-4), 6.8.4.1(1), 6.8.4.1(5)]

2. Conduct a triage. [Skill Sheet 12-1; NFPA® 472, 6.8.4.1(4)]

3. Identify rescue tools and equipment. [NFPA® 472, 6.8.3.2, 6.8.4.1(2)]

4. Describe various rescue methods.

5. Demonstrate the incline drag. [Skill Sheet 12-2; NFPA® 472, 6.8.4.1(3)]

6. Demonstrate the blanket drag. [Skill Sheet 12-3; NFPA® 472, 6.8.4.1(3)]

7. Demonstrate the webbing drag. [Skill Sheet 12-4; NFPA® 472, 6.8.4.1(3)]

8. Demonstrate the cradle-in-arms lift/carry – One-rescuer method. [Skill Sheet 12-5; NFPA® 472, 6.8.4.1(3)]

9. Demonstrate the seat lift/carry – Two-rescuer method. [Skill Sheet 12-6; NFPA® 472, 6.8.4.1(3)]

10. Demonstrate the extremities lift/carry – Two-rescuer method. [Skill Sheet 12-7; NFPA® 472, 6.8.4.1(3)]

11. Demonstrate the chair lift/carry method 1 – Two rescuers. [Skill Sheet 12-8; NFPA® 472, 6.8.4.1(3)]

12. Demonstrate the chair lift/carry method 2 – Two rescuers. [Skill Sheet 12-9; NFPA® 472, 6.8.4.1(3)]

13. Discuss recovery operations. [NFPA® 472, 6.8.3.1(1), 6.8.3.1(2)(e)]

Chapter 12
Victim Rescue and Recovery

Case History

About 5:03 a.m., central daylight time, on June 28, 2004, a westbound Union Pacific Railroad (UP) freight train traveling on the same main line track as an eastbound BNSF Railway Company (BNSF) freight train struck the midpoint of the 123-car BNSF train as the eastbound train was leaving the main line to enter a parallel siding. The accident occurred at the west end of the rail siding at Macdona, Texas, on the UP's San Antonio Service Unit. The collision derailed the 4 locomotive units and the first 19 cars of the UP train as well as 17 cars of the BNSF train.

As a result of the derailment and pileup of railcars, the 16th car of the UP train, a pressure tank car loaded with liquefied chlorine, was punctured. Chlorine escaping from the punctured car immediately vaporized into a cloud of chlorine gas that engulfed the accident area to a radius of at least 700 feet (213 m) before drifting away from the site. Three persons, including the conductor of the UP train and two local residents, died as a result of chlorine gas inhalation. The UP train engineer, 23 civilians, and 6 emergency responders were treated for respiratory distress or other injuries related to the collision and derailment. Damages to rolling stock, track, and signal equipment were estimated at $5.7 million, with environmental cleanup costs estimated at $150,000.

The initial notification to local emergency response authorities came via a 9-1-1 call placed at 5:06 a.m. from a residence on Nelson Road to the Bexar County 9-1-1 Emergency Call Center. The caller reported difficulty breathing and the presence of white smoke outside the residence. The caller also, in what could be described as a weak voice, referred to a train derailment. The 9-1-1 operator heard the word "smoke" and understood that the caller was experiencing breathing difficulty but apparently did not recognize the words "train derailment," and the caller was transferred to a fire department dispatcher. The caller again reported "train derailment" and "smoke," but the fire dispatcher also did not recognize that the incident involved a train derailment. The response was mistakenly processed as a "difficulty breathing and smoke in the residence" response action.

Southwest Volunteer Fire Department emergency responders were dispatched to the Nelson Road residence at 5:08 a.m., followed shortly thereafter by Bexar County Sheriff's Office patrol units that had been dispatched for support. None of the responders were yet aware that they were responding to a train accident and that their path to the Nelson Road residence was blocked by derailed equipment.

When fire department responders approached the accident site in darkness at about 5:15 a.m., they began to have difficulty breathing as they became exposed to the vapor cloud of chlorine from the punctured railcar. They immediately withdrew from the scene and requested mutual aid from other agencies. Some of the firefighters obtained protective clothing and self-contained breathing apparatus, and commencing about 5:40 a.m., reentered the scene to conduct a search for survivors.

The initially requested mutual aid resources began to respond to the scene shortly thereafter, which included the Bexar County Office of Emergency Management. About 6:10 a.m., the Bexar County Office of Emergency Management established the unified incident command system, activated the Bexar County Emergency Operations Center, and initiated the Bexar County emergency management plan. Additional mutual aid resources were also being dispatched to the scene, including the San Antonio Fire Department.

About 6:15 a.m., Southwest Volunteer Fire Department officers, who were advancing west along Nelson Road, came upon an individual who would later be identified as the UP train engineer stumbling along the roadway about 240 feet (73 m) east of the grade crossing. He was in respiratory distress and was transported from the scene for medical attention. A short time later, responders determined that the derailment wreckage at the grade crossing prevented access to residences at the west end of Nelson Road, one of which was their dispatch destination. The obstructed grade crossing also prevented the immediate rescue of three individuals who were reported to be trapped in their residence by the vapor cloud several hundred feet to the south of the emergency dispatch destination. Responders told investigators that early in the response effort they considered, then rejected, a plan to use a helicopter to access Nelson Road south of the grade crossing and evacuate the two occupied residences there. They decided that such a plan was ill-advised until the gas plume had reduced or stabilized because of the vulnerability of the helicopter equipment and crew and the possibility that rotor wash could spread the gas.

At about 6:33 a.m., hazardous materials response contractors retained by the UP began to arrive and conduct a technical assessment of the chlorine release. Further access to the accident site through the wreckage pileup at the grade crossing area was restricted until the assessment could be made and firefighters and hazardous materials personnel could don the appropriate personal protective equipment (principally Level-A hazardous materials suits). Sunrise occurred at 6:37 a.m., and the wind was moderate but steady toward the northwest.

An evacuation zone was established around the accident site with a radius of about 2 miles (3 k). Other than assisting with the evacuation of residents within the evacuation zone and assisting hazardous materials release mitigation responders with the technical assessment of the chlorine gas release, no further direct rescue attempt activity of the responding firefighters and mutual aid responders for the three individuals who were reported to be trapped in their residence by the vapor cloud was documented for about the next 3 hours.

During this time, pursuant to 9-1-1 call center instructions, the three residents were attempting, without success, to flee and find a safe shelter from the chlorine fumes that had engulfed their residence. Also during this time, according to postaccident interviews, the principals of the primary responding emergency services agencies (the San Antonio Fire Department, the Bexar County Office of Emergency Management, and the Southwest Volunteer Fire Department) were involved in what was described as a certain amount of discordant debate regarding jurisdictional boundaries and incident command authority.

About 9:45 a.m., with a preliminary technical assessment of the chlorine gas release having been completed, the first of three firefighter entry teams entered the accident area to attempt a rescue of three trapped persons on Nelson Road who were unable to escape the chlorine vapor cloud that had enveloped their residence. This first entry team, however, became disoriented while attempting to advance through the wreckage pileup and inadvertently diverted down the wrong roadway (actually a long driveway leading to another residence) and away from their objective. Along that roadway, the team encountered the body of a person who was later identified as the UP train conductor. Shortly thereafter, one of the entry team firefighters showed signs of dehydration, prompting the dispatch of a second entry team to come to the aid of the first. About 10:12 a.m., a third entry team, consisting of two firefighters and one UP employee, was dispatched to carry on with the rescue mission that had been aborted by the first entry team. This team successfully advanced through the wreckage pileup and, about 10:55 a.m., reached the three persons who had been trapped at their residence. All three were found to be in considerable respiratory distress. About 11:46 a.m., after the responders had revived and stabilized them, the three individuals were transported by helicopter to a local hospital for medical attention. About 11:55 a.m., the entry team entered another Nelson Road residence and found two persons who had sustained fatal injuries.

Three persons died from chlorine gas inhalation as a result of the accident. One was the UP train conductor; the other two were occupants of a residence about 220 feet (67 m) south of the grade crossing on Nelson Road.

The UP train engineer, though critically injured, survived the collision and the exposure to the chlorine vapor. Investigators reviewed medical records to identify 23 civilians who were treated at local medical facilities for respiratory distress or other ailments likely associated with the gas release or its aftermath. Local media reported other individuals as receiving medical treatment after the accident, but investigators were unable to locate medical records that would confirm the count.

Investigators identified six emergency responders who were either treated at the scene or transported to local medical facilities. Two of these were firefighters who were treated at a local hospital for respiratory difficulties and released. Two other firefighters were treated on scene for dehydration. One Bexar County Sheriff's Office deputy officer was treated for respiratory difficulty at a local hospital and released. The sixth injured responder, an ambulance attendant, was treated at a local hospital and released. Two UP technical support employees received treatment at local medical facilities for respiratory distress most likely associated with exposure to the chlorine gas.

Source: National Transportation Safety Board

Operations-Level responders may be called upon to rescue victims at haz mat incidents. Developing and implementing a successful rescue and recovery operation requires training, a rescue plan, a comprehensive understanding of the rescue process, information about local capabilities and facilities, and the skills necessary to perform rescues safely and efficiently. Knowledge and flexibility are critical elements in the successful rescue and recovery of victims from contaminated environments. Responders performing rescue and recovery operations must be prepared for the following:

- Direct exposure to the hazards in the hot zone

Figure 12.1 Responders conducting rescues must be prepared for the dangerous and unstable conditions presented by the incident. *Courtesy of the U.S. Marine Corps, photo by Sgt. Christopher D. Reed.*

CAUTION

Operations-Level responders should only enter potentially contaminated areas to perform rescue of known living victims or to perform an immediate reconnaissance to determine if victims are still alive.

Line of Sight — The unobstructed, imaginary line between an observer and the object being viewed.

- Dangers posed by the unstable physical environment of the incident **(Figure 12.1)**

- Stress from working in protective clothing

- Emotional trauma of a situation involving injured victim(s)

- Decontamination of themselves and the person or persons rescued

Rescue and recovery of victims in a hazardous materials/WMD release require various tactics and safety procedures depending on the type of incident, number of living victims, location of victims, and whether the victims are ambulatory or non-ambulatory. Responders must be able to determine if rescue of victims is feasible and ensure they are not endangering themselves in the process.

This chapter will discuss safety considerations for conducting rescues and briefly touch on rescue methods. Additionally, it will explore the four incident response conditions based upon the status of the victim that affect rescue and recovery at a hazardous materials/WMD release:

- **Line-of-Sight** with Ambulatory Victims

- Line-of-Sight with Non-Ambulatory Victims

- Non-Line-of-Sight with Ambulatory Victims

- Non-Line-of-Sight with Non-Ambulatory Victims

Rescue Operations

As with all operations at haz mat incidents, the risks identified during the initial size-up will dictate many elements of the response. Responders must operate within the structure of the incident command system, following established procedures and local emergency response plan guidelines for conducting rescues **(Figure 12.2)**. Operations-Level responders must work under the control of a Technician-Level responder, an Allied Professional who can continuously assess and observe actions and provide immediate feedback, or under the guidance of SOPs that clearly spell out what actions Operations-Level responders can and cannot perform.

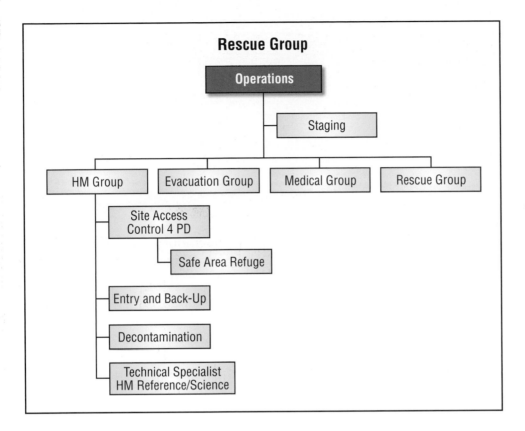

Figure 12.2 IMS configuration of the logistics section at a haz mat incident with a designated rescue group.

Responders operating under SOPs must have a thorough understanding of their responsibilities and the correct protocols for implementing them. SOPs should outline detailed task requirements, clearly indicating which tasks are beyond the Operations-Level responder as well as the PPE needed to perform the tasks they are assigned (see the Information Box for a sample Rapid Extraction policy).

The Incident Commander will formulate the incident action plan (IAP) at all haz mat incidents, including the rescue plan. After the situation has been assessed and a need for rescue and recovery has been established, emergency responders will immediately implement the rescue and recovery plan in accordance with local procedures. When assigned rescue and recovery responsibilities, responders must know the plan and follow it. This will ensure they are fully prepared, protected, and have the necessary resources to save lives.

The response scenario will directly influence initial response operations. If the number of victims is small (one or two victims), emergency responders trained in victim rescue and recovery should be able to handle the incident requirements with a single entry operation. However, if the number of victims is well beyond the number of available responders, emergency responders will be challenged with simultaneously rescuing and removing victims, establishing

Figure 12.3 Incidents with large numbers of victims will present a challenge for first-arriving units. Training for mass casualty incidents will help responders maintain control in such situations until additional help arrives.

a safe haven or area of safe refuge until EMS operations can commence, and making multiple entries in a short period of time **(Figure 12.3)**. The effects of physical exertion and heat stress will add to the challenges.

Responders conducting rescue missions may come from a variety of emergency responder disciplines and may include firefighters, hazardous material/WMD responders, EMS personnel, law enforcement officers, or a combination of these personnel assembled into a team. In addition to SOPs, responders should be familiar with the SOPs for mutual-aid response in their area.

Sample Rapid Extraction Policy

One of the primary missions that first responders may need to accomplish is rescue of exposure victims. The objective of rescue at hazardous material incidents is to rapidly extract people from unsafe areas and deliver them for decontamination and treatment, while controlling the associated risks to firefighters.

To differentiate victim rescue from rescue of first responders, the term used to describe actions directed at retrieving civilian exposure victims is "rapid extraction." Incident Commanders may assemble units to perform extraction missions in Extraction Crews or Groups.

Under ideal conditions, everyone in the area of a hazardous materials release would self-rescue, gather in a location upwind and uphill, and initiate self-decontamination. In the case of industrial accidents, responders may find that the initial reaction by trained people on the scene of the release may closely match the ideal. This will not be the case, however, where the release is committed intentionally or accidentally on an unsuspecting and unprepared population. In the event of such a criminal act, the need to perform extraction of victims is more likely.

Types of Extraction

Responders have a choice among different extraction methods, and should focus on using a method that minimizes risk but still allows the mission to be completed.

Extractions can be categorized by who performs the rescue: self-rescue or bystander rescue, which represents the least risk to the responder; responder rescue, where the level of risk and potential benefit may be initially unknown; and technical rescue, such as that performed by USAR or the Hazardous Materials Task Force (HMTF).

Assessment of Risk Vs. Benefit

Type	Risk	Benefit
Self-Rescue	Low	High
Bystander Rescue	Low	High
Responder Rescue	Low-High	Unknown
Technical Rescue	Med-High	Low

Using remote methods to contact the people in the unsafe area, such as the apparatus PA, a bullhorn, or in-place public address systems is the least-risk method to get people to leave an unsafe area. Where this method is employed, those people less impacted by the release should be encouraged to assist those viable victims who have been impacted to a greater degree by the release (bystander rescue).

In some cases, responders may find that victims, though ambulatory, may need assistance to get to a safe environment due to the symptoms associated with the exposure, such as dimness of vision or confusion. This type of assisted rescue may require responders to be in proximity or even direct contact with the victims.

Extraction of unresponsive victims may result in exposure to levels of contamination that exceed the capability of the protective clothing available to first responders. As a result, this type of physical rescue or extrication may require the specialized equipment and PPE available to the Hazardous Materials Task Force.

Risk Assessment

First responders must quickly assess whether there are victims needing extraction, the number of victims, and what resources will be needed. In order to accomplish this assessment, Incident Commanders should assign a company to this task as soon as possible, in spite of the demands of the incident.

Since the primary protective equipment assigned to firefighters consists of structural fire fighting PPE and SCBA, risk assessment should determine whether this level of protection is appropriate for the environment.

The primary factor in determining whether the environment is appropriate for structural PPE and SCBA is the viability of the victims. Where the hazardous material is unknown, exposure victims are considered viable when they are able to control their movement (standing, sitting, or kneeling), or are able to respond verbally or physically.

Those victims who are unable to respond, or are unable to voluntarily control movement, may still be viable under certain conditions, e.g., the number of victims is low, they are not located in an enclosed space or the enclosed space can be adequately ventilated, and access to decontamination and treatment is immediately available.

The assessment to determine the presence of viable victims may include direct observation, interview of witnesses or employees exiting the venue, and reports from other first responders. The existence of unprotected viable victims is a strong indication that the toxicity of the environment is below that which would impact a firefighter protected by full structural PPE and an SCBA for short exposures.

Sample Rapid Extraction Policy (concluded)

Even though an Incident Commander may determine that there are viable victims who need extraction, the scale of the incident will directly impact the decision process. Similar to multi-casualty incidents, the priority must be on providing the most good for the most people. Directing initial arriving units to provide emergency decontamination to self and bystander-rescued exposure victims may produce a better outcome than devoting all units to extraction efforts.

Rapid Extraction Guidelines

A successful extraction will minimize the exposure for both the victims and the rescuers, by limiting exposure time and providing immediate decontamination. To accomplish this, decontamination must be available prior to entry, and extraction teams shall use no more than one 45-minute SCBA bottle prior to evaluation for exposure by Haz Mat personnel.

Prior to initiating rapid extraction, first responders shall take the following actions:

1. Establish that there are viable victims and a need for extraction operations.

2. Establish decontamination. At minimum, a charged hoseline in the designated location.

3. Establish rapid intervention. A single engine company, following the 2-in 2-out rule, can fulfill the rapid intervention requirement. At larger incidents, Incident Commanders may designate entire companies as Rapid Intervention (may be assigned to Hazardous Materials Task Force).

4. Announce the intent to initiate rapid extraction to IC or incoming companies:

   a. Identify the access point

   b. Identify the number of personnel on team

5. Maintain continuous communications during extraction efforts.

Upon encountering any of the following exit conditions, first responders must immediately leave the rescue site, make appropriate notifications, and proceed to decontamination:

1. Any signs or symptoms of exposure for any members of an extraction team;

2. Encountering an area with ONLY non-viable victims;

3. The first SCBA audible alarm (approximately 25% reserve for exiting and decontamination).

Until the exit conditions are encountered, and while viable victims remain, first responders may continue extraction operations.

Upon completion of the extraction missions, first responders shall:

1. Proceed through decontamination. Be guided by directions from the Hazardous Materials Task Force;

2. Consider PPE used during extraction as contaminated, and keep it segregated from other equipment until it is determined to be safe;

3. Be examined by ALS personnel for post-entry symptoms and vitals;

4. Proceed to Rehab, as needed.

Courtesy of the Los Angeles Fire Department

Determining Feasibility of Rescues

Before implementing any IAP involving victims trapped in a contaminated environment, responders must be able to answer these basic questions to determine the feasibility of conducting rescue operations:

- What products are involved? The type of hazardous material will affect decisions about PPE, viability of victims, and a host of other considerations that will dictate tactics for the operation. If products are unidentified, this too, will affect strategies and tactics. In some cases, it may be determined that the product involved and circumstances of the incident are too hazardous to risk a rescue with the personnel, PPE, and equipment available.

- What are the other known factors about the incident? Have witnesses provided additional information that might be useful to the decision-making process? Are there other hazards present unrelated to the hazardous material?

- Are victims within line of sight or will a search be needed? Conducting a search for potential victims who are not in sight may increase the risk to responders. Responders conducting searches will have to extend their time in the hot zone to conduct the search, and they may be exposed to hazards that are not detectable from outside the search area.

- Is it a rescue operation or a recovery operation? Rescue operations are a high priority and may be conducted without complete mitigation of risk. Recovery operations should be conducted only after the risk to responders has been minimized or eliminated.

- Do on-scene emergency responders have the necessary PPE and training to perform the mission? Responders must have the PPE and training necessary to enter the hot zone to make the rescue safely **(Figure 12.4)**.

Figure 12.4 Rescuers must have appropriate PPE to enter the hot zone safely. *Courtesy of the U.S. Marine Corps.*

- Do on-scene emergency responders have the necessary equipment to perform the mission? If needed equipment is not available, responders must wait until it is obtained or arrives on scene.

- Are there enough personnel available to conduct a rescue safely? First arriving units may need to wait for additional personnel to arrive in order to conduct rescue operations. At a minimum, five people are needed: two entry team members, two backup team members, and at least one person to staff a decon station.

- Is there monitoring available that can be used to assist in classifying the environment and aid in providing additional safety for rescue teams? Many fire departments and industry employers now provide basic monitoring and detection devices for their Operations-Level employees **(Figure 12.5)**. If responders are trained in their use (and SOPs allow), these devices should be used during rescue operations.

Figure 12.5 Detection and monitoring equipment can help to classify potentially hazardous atmospheres and should be used during rescue operations if available. *Courtesy of the U.S. Marine Corps, photo by Cpl. Scott M. Biscuiti.*

Rescue Safety

Rescuers must always consider their own safety first. ICs also must consider the hazards to which rescuers may be exposed while conducting search and rescue operations. Safety is the primary concern of rescuers because hurried, unsafe search and rescue operations may have serious consequences for rescuers as well as victims. Safety of responders depends upon rescuers and their officers making a good initial size-up, continuing the size-up throughout the operation, and performing a risk/benefit analysis before each major step in the operation.

Planning Rescues

When planning a rescue operation, first responders should consider all hazards that are or may potentially be present during the incident. Every effort must be made to identify if hostile human threats, improvised explosive devices, or other devices utilized in an intentional release incident are still present. If any of these conditions apply, rescue and recovery must delayed until the situation can be stabilized.

The number of available responders, their level of training, the circumstances of the incident, and the type of PPE available all directly affect initial rescue and recovery operations **(Figure 12.6)**. For example, a single fire unit assigned to rescue activities and provided with chemical protective clothing will require at least five trained members (two for entry, two for back up, and one for decontamination) in order to enter the affected area. The necessary number of trained individuals must be assembled and a decontamination station must be established and staffed before entry is made. In the U.S., other required OSHA positions must be staffed in addition to the rescue team, for example, IC and Safety Officer.

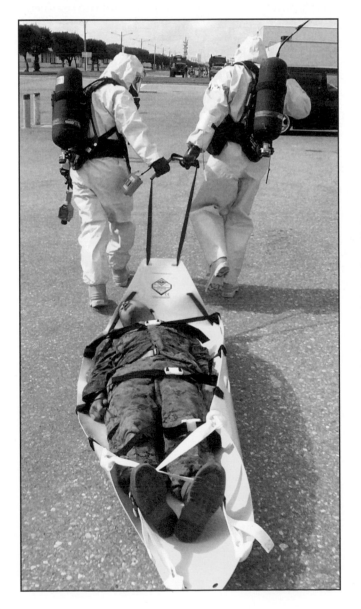

Figure 12.6 Responders must have appropriate PPE, necessary equipment, and enough personnel to conduct a rescue. At least five trained members are needed to enter the hot zone (two for entry, two for back up, and one for decontamination). *Courtesy of the U.S. Marine Corps.*

All unit leaders and supervisors within the ICS organization are responsible for providing proper supervision and overseeing the safety of all entry teams. Risk assessment and the selection of control options are typically the responsibility of a Technician-Level responder or Allied Professional.

Since the majority of Operations-Level responders are assigned to individual companies or units, the following information is specifically directed towards Operations-Level responders assigned in this manner. After the initial briefing and situation size-up, the IC or the unit leader should establish a hot zone entry point. Once this has been established, an assembly area is then established adjacent to the entry point. The unit leader's responsibilities include the following:

- Designating team assignments and team radio contact members
- Briefing unit members of the objective and required tasks
- Continually maintaining both immediate and functional supervision over teams to ensure their safety during the operation
- Ensuring accountability and tracking of personnel
- Relaying any critical or pertinent information received from the teams up the chain of command
- Ensuring a decontamination station is established
- Ensuring a back-up team is in position prior to entry

The unit leader does not don a chemical protective suit unless the situation indicates that a suit is required in order to properly supervise the unit's operation. For safety reasons, if possible, add additional responders positioned in areas so that line of sight can be maintained. Situations that would require the unit leader to don chemical protective equipment include the following:

- Entry team would not be in line of sight.
- Incident requires a complex entry team operation.
- Operation requires several entry teams.

A decontamination station must be established prior to any entry of CPC-equipped personnel into hazard zones **(Figure 12.7)**. The primary function of this station is to provide immediate and adequate decontamination to entry team members so they can be safely removed from CPC with minimum exposure to any material they may have had contact with. It can also be used to decontaminate first responders and/or civilians until shower units arrive on the scene.

Figure 12.7 A decontamination station of some sort must be established before rescuers enter the hot zone to assist victims. *Courtesy of the U.S. Air Force, photo by Master Sgt. Jim Varhegyi.*

The entry team must consist of at least two trained members in the appropriate level of PPE. Entry team members will perform the actual search, rescue, and removal of victims from the hazardous area. While operating in the hot zone, they must stay alert for any clues, signs, or causes that could have contributed to the spill or release. This information must be reported to the unit leader immediately.

Team members must work in close proximity of one another at all times (buddy system) and maintain radio contact. If one member must leave the operating zone the other team member must also leave. If one of the entry team members becomes incapacitated for an unknown reason, the other team member must notify the back-up team for the need of an immediate rescue.

The back-up team consists of two members trained in the appropriate level of PPE, on standby to perform a single task: the removal of a down member **(Figure 12.8)**. Their operating time should be less than that of the entry team, and they must be under the functional supervision of the unit leader at all times.

Figure 12.8 Back-up team members must wear the same level of PPE as the entry team, and their sole task at the incident is to remain on standby in case an entry team member needs to be rescued.

When entering the hot zone, responders must be alert to their surroundings at all times. These basic guidelines should always be followed:

- Immediately exit any area where evidence of chemical contamination is encountered and no living victims are identified.

- Avoid contact with any hazardous or unidentified materials.

- Undergo emergency decontamination immediately upon exiting the hazardous area.

- Immediately obtain medical assistance when needed.

- If conditions in excess of IDLH are found, attempt to change the environment (ventilation, vapor dispersion/suppression) to enable other responders to assist.

Conducting Rescues

Rescue priorities at haz mat incidents may differ from other emergency incidents in which efforts are made to reach the more severely injured victims first in order to save them if possible. At haz mat incidents, rescuers (with appropriate training and PPE for the hazards present) typically start at the outer edges of the incident and work their way in, following these priorities:

1. Assist ambulatory casualties to save themselves (direct them to an area of safe refuge to await decontamination).

2. Evacuate non-ambulatory casualties showing signs of life.

3. Evacuate non-ambulatory casualties showing signs of life from the hot zone.

4. Recover the dead

NOTE: Always follow procedures established by the authority having jurisdiction when conducting the activities listed above!

First efforts should be directed at visible victims, followed by victims out of sight. The sections that follow will discuss the four incident response situation based upon the status of the victim that effect rescue and recovery.

Line-of-Sight with Ambulatory Victims

Ambulatory victims within the line of sight should be addressed first. These victims are generally the farthest away from the release, have experienced the lowest exposure and related dose, and require the least amount of time to remove. If possible, these individuals should be directed to a safe haven or area of safe refuge within the warm zone until a determination is made regarding the need for decontamination **(Figure 12.9)**. These victims may be directed with verbal instructions, signs, hand signals, whistles, or light sources (at night). Responders must also have plans for managing non-cooperative victims.

Failure to effectively control contaminated individuals can severely disrupt the best scene management plan. An example of this would be to discover that individuals with radioactive dust were allowed to bypass decon and enter the medical treatment area. Such a discovery would not only extend the hot and warm zone areas unnecessarily but would require completely reestablishing the medical treatment area and possible decontamination or abandonment of medical supplies. First responders should also avoid touching these victims. At criminal events they may need to be interviewed by law enforcement as witnesses.

Figure 12.9 Ambulatory victims should be directed to an area of safe refuge or decontamination line. It is important to keep potentially contaminated victims from leaving the scene prior to undergoing decon if at all possible.

Line-of-Sight with Non-Ambulatory Victims

After entry teams have removed the ambulatory victims within the line of sight, they can turn their attention to the non-ambulatory victims within line of sight **(Figure 12.10)**. Part of the planning phase is to anticipate what the tools and equipment that will be needed upon entry. Follow local SOPs to determine when, the extent, and the type of triage to be conducted in the hot zone in order to determine which victims should be moved to the decontamination area **(Figure 12.11)**. Typically, unconscious victims who respond to touch stimulation will be moved to the decon area by responders trained to these mission-specific competencies so that medical personnel can treat critical injuries and medically monitor victims as decontamination takes place (or after, as SOPs and training dictate). More triage and treatment will be conducted after victims have been decontaminated.

Figure 12.10 Once ambulatory victims have been removed from the hot zone, rescuers can begin removing visible non-ambulatory victims. *Courtesy of the U.S. Navy, photo by Photographer's Mate 2nd Class Phillip A. Nickerson Jr.*

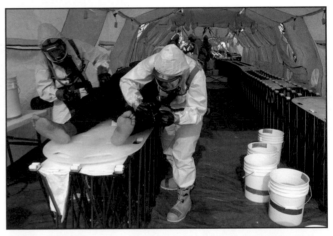

Figure 12.11 Follow local SOPs to determine which non-ambulatory victims should be moved to the decontamination area. *Courtesy of the U.S. Air Force, photo by Spc. Kurt Lamel.*

Non-Line-of-Sight with Ambulatory Victims

Ambulatory victims who are not in the line of sight are generally closer to the incident or source of the release, and have experienced a greater exposure and related dose. If possible, these individuals should be directed to a safe haven or area of safe refuge where they can be assessed for decontamination need or medical treatment. Responders may need to enter the hazard area to find these victims, so there is an increased level of risk which must be reflected in planning and safety measures.

Non-Line-of-Sight with Non-Ambulatory Victims

Non-ambulatory victims not within line of sight are the last to be rescued from the hot zone. These victims are generally the closest to the hazardous materials/WMD event and have experienced the greatest exposure and related dose. Rescue and removal of these victims typically poses the greatest danger to emergency response personnel and require increased planning and resources to carry out.

Conducting Triage

A cursory triage of injured victims may be conducted after victims have been moved from the hot zone into an area of refuge in order to determine priority for decon. A more thorough triage will be conducted after victims have gone through the decon process **(Figure 12.12)**. One common system requires responders to quickly assess a victim's status and assign them into four basic categories:

- *Priority 1* — Life-threatening injuries and illnesses (highest priority)
- *Priority 2* — Serious, but not life-threatening injuries
- *Priority 3* — Minor injuries
- *Priority 4* — Dead or fatally injured

Responders must be familiar with the triage system used by their agency. **Skill Sheet 12-1** demonstrates the use of one common triage system.

Figure 12.13 Preprinted triage tags can assist in the triage process.

Figure 12.12 Thorough triage and medical treatment is provided to non-ambulatory victims after they have been decontaminated.

Stokes Basket — Basket-type litter — wire or plastic — suitable for transporting patients from locations (such as a pile of rubble, structural collapse, or the upper floor of a building) where a standard litter would not be easily secured; may be used with a harness for lifting.

Rescue Equipment

Performing rescue and recovery operations requires specialized tools and equipment. These include:

- Personal protective clothing and equipment suitable for the hazards present
- Triage tags **(Figure 12.13)**
- Equipment such as backboards and **stokes baskets** to quickly package non-ambulatory victims for removal

- **SKEDs®**, carts, buggies, and similar devices to move non-ambulatory victims **(Figure 12.14a and b)**

- Extrication equipment

- Technology used to search for victims such as heat-sensing devices (thermal imagers) and fiber optic cameras

SKED® — Compact device for patient removal from a confined space with spinal immobilization; may be used as a lifting device with a harness.

Responders must be trained in the correct use of such equipment, and the equipment must be maintained in good condition. For more information about equipment used for rescue, see IFSTA's **Essentials of Firefighting** and **Urban Search and Rescue in Collapsed Structures** manuals.

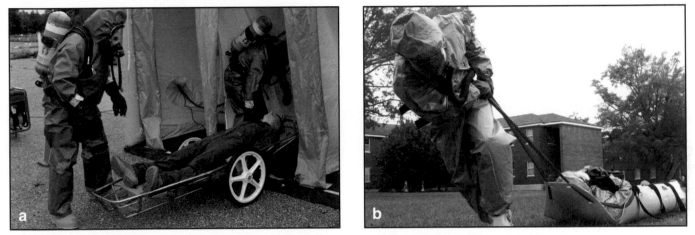

Figure 12.14a and b Rescue equipment may include **(a)** carts and **(b)** SKEDs®. Both are used to move non-ambulatory victims. (b)*Courtesy of the U.S. Marine Corps.*

Rescue Methods

An uninjured victim (or one with minor injuries) may be directed using hand signals or verbal instructions to walk to safety. If physical assistance is needed, one or two rescuers may be needed, depending on how much help is available and the size and condition of the victim.

The chief danger in moving an injured victim quickly from the hot zone is the possibility of aggravating a spinal injury. However, in an extreme emergency, such as an IDLH atmosphere or threat of explosion, the possible spinal injury becomes secondary to the goal of preserving life. In these cases, the victim should be pulled in the direction of the long axis of the body — not sideways.

It is always better to have two or more rescuers when attempting to move an adult. One rescuer can safely carry a small child, but two, three, or even four rescuers may be needed to move a large adult. There are a number of carries and drags that may be used to move a victim from an area quickly; some of the more common ones are described in the sections that follow. Realistically, in very hostile environments, victims are removed by whatever means are available at the moment. Sometimes that means grabbing an arm, leg, clothes, belt, hair, or whatever works.

Incline Drag

This drag can be used by one rescuer to move a victim up or down a stairway or incline and is very useful for moving an unconscious victim. **Skill Sheet 12-2** describes the method of performing the incline drag.

Blanket Drag

This drag can be implemented by one rescuer using a blanket, rug, or sheet. **Skill Sheet 12-3** describes the procedure for the blanket drag.

Webbing Drag

This drag can be implemented by one rescuer using a loop of 1-inch (25 mm) webbing. **Skill Sheet 12-4** describes the procedure for the webbing drag.

Cradle-in-Arms Lift/Carry

This lift and carry is effective for carrying children or very small adults if they are conscious. It is usually not practical for carrying an unconscious adult because of the weight, relaxed condition of the body, and the difficulty in supporting the head and neck. **Skill Sheet 12-5** describes the procedure for the cradle-in-arms lift/carry.

Seat Lift/Carry

This lift/carry can be used with a conscious or an unconscious victim and is performed by two rescuers. **Skill Sheet 12-6** describes the procedure for the seat lift/carry.

Three-Person Lift/Carry

Many victims are more comfortable when left in a supine position, and this lift/carry is an effective way to lift a victim who is lying down. The three-person lift/carry is often used for moving a victim from a bed to a gurney, especially when the victim is in cramped quarters.

Extremities Lift/Carry

The extremities lift/carry can be used with either a conscious or an unconscious victim. This technique requires two rescuers. **Skill Sheet 12-7** describes the procedure for the extremities lift/carry.

Chair Lift/Carry

The chair lift/carry can be used with either a conscious or an unconscious victim. Be sure that the chair used is sturdy; do not attempt this carry using a folding chair. **Skill Sheets 12-8** and **12-9** describe two methods of performing the Chair Lift/Carry.

Moving a Victim onto a Long Backboard or Litter

At haz mat incidents, rescuers will often have the advantage of being able to use some type of litter or SKED® (or other sliding device or material) to remove a victim. There are many different types of litters such as the standard ambulance cot, army litter, scoop stretcher, basket litter, and long backboard **(Figure 12.15)**. For more information on using these devices, see IFSTA's **Essentials of Firefighting** manual, 5th edition.

Figure 12.15 Victims can be moved on basket litters, backboards, or stretchers. *Courtesy of the U.S. Navy, photo by Mass Communication Specialist 2nd Class Kirk Worley.*

Recovery Operations

Rescue operations should be performed on all victims before **recovery** begins. Rescue is the removal of both ambulatory and non-ambulatory victims who are still living and who have a likelihood of surviving their injuries or exposure. Recovery is removal of the dead.

Recovery operations are a lower priority and should be coordinated by the IC with law enforcement or coroner personnel. In general, recovery operations are not implemented until all living, presumed to be alive (both ambulatory and non-ambulatory) have been rescued **(Figure 12.16)**. Reducing the hazards to create a safer environment to operate is essential when performing victim recovery. It may be necessary for bodies and human remains to remain in place until law enforcement or investigation efforts are completed. The remains of deceased victims are recovered for body identification.

Recovery — Situation in which the victim is dead, and the goal of the operation is to recover the body.

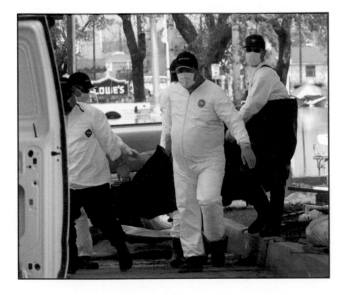

Figure 12.16 Recovery operations do not begin until all living victims have been rescued. At crime scenes, bodies should not be disturbed until released by investigators. *Courtesy of Chris Mickal.*

Summary

All rescue operations must be conducted by responders who have appropriate training and equipment to do so safely. They must operate within the IMS structure, in accordance with the rescue plan established at the scene. Whenever entering the hot zone, rescuers must operate with a buddy, and backup personnel must be assigned. A decon station must be established so both rescuers and victims can be decontaminated as quickly as possible.

Rescuers should help visible ambulatory victims first. These individuals can be directed to an area of safe refuge within the warm zone to await decontamination and further assistance as necessary. Visible non-ambulatory victims should be assisted next. A variety of rescue methods can be used to move these victims, including common lifts, drags, and carries. As operations progress and conditions safely permit, victims who are not in the line of sight may be searched for and rescued if necessary. Ambulatory victims can be directed to an area of safe refuge to await decontamination. Rescue of non-ambulatory victims who are not visible typically poses the greatest danger to responders and should only be conducted after careful planning and a thorough assessment of the situation.

Review Questions

1. What questions should be asked to determine the feasibility of conducting rescue operations?

2. What are the unit leader's responsibilities when planning a rescue?

3. What is the purpose of a decontamination station?

4. What guidelines should be followed when entering the hot zone?

5. What are the priorities when conducting a rescue?

6. What are the categories when conducting triage?

7. What types of tools and equipment may be required for rescue and recovery?

8. Describe various rescue methods.

9. What is the difference between rescue and recovery?

10. When does recovery begin?

Step 4a: Attempt to open the airway. If breathing cannot be started by opening the airway, mark Priority 0.

Step 4b: If patient starts breathing, or their respiratory rate is more than 30/minute, mark Priority 1.

Step 4c: If respiratory rate is less than 30/minute, go to Step 5.

Step 5: Assess patient's radial pulse.

Step 1: Ensure scene safety and proper PPE.

Step 2: Identify patients to be triaged.

Step 3: Assess patient's mobility.

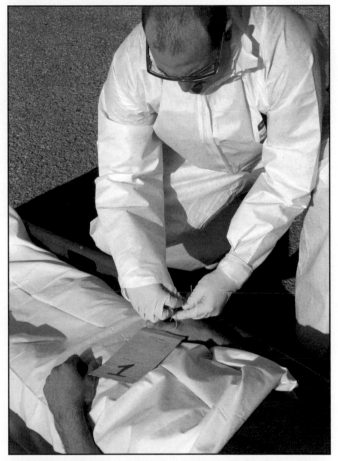

Step 5a: If patient is breathing, but has no radial pulse, mark Priority 1.

Step 5b: If patient is breathing and has a pulse, go to Step 6.

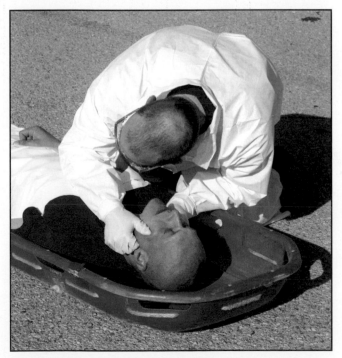

Step 4: Assess patient's respiration.

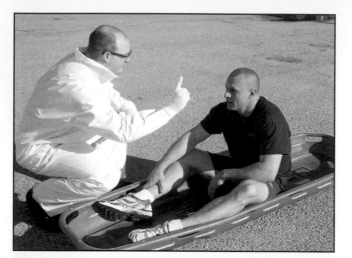

Step 6: Assess patient's level of consciousness.

Step 6a: If patient is alert (able to follow simple commands), mark Priority 2.

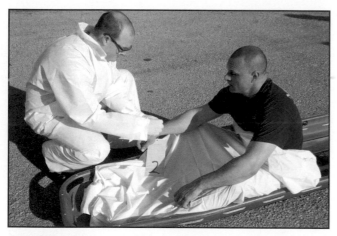

Step 6b: If any altered mental status, mark Priority 1.

Step 7: Re-triage the Priority 3 "Walking Wounded" patients. Check for any change in medical condition, i.e. shock, mental status, etc.

Step 1: Turn the victim (if necessary) so that the victim is supine.

Step 2: Kneel at victim's head.

Step 3: Support the victim's head and neck.

NOTE: If head or neck injury is suspected, provide appropriate support for head during movement.

Step 4: Lift the victim's upper body into a sitting position.

Step 5: Reach under the victim's arms.

Step 6: Grasp the victim's wrists.

Step 7: Stand. The victim can now be eased down a stairway or ramp to safety.

12-3

Demonstrate the blanket drag.

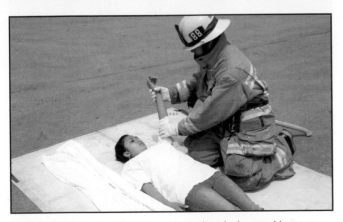

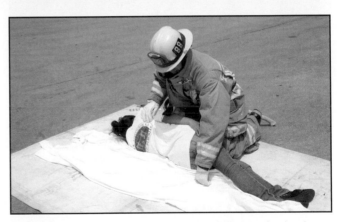

Step 1: Spread a blanket next to the victim, making sure that it extends above the victim's head.

Step 2: Kneel on both knees at the victim's side opposite the blanket.

Step 3: Extend the victim's nearside arm above his or her head.

Step 4: Roll victim against your knees.

Step 5: Pull the blanket against the victim, gathering it slightly against the victim's back.

Step 6: Roll victim gently onto the blanket.

Step 7: Straighten the blanket on both sides.

Step 8: Wrap the blanket around the victim.

Step 9: Tuck the lower ends around the victim's feet if enough blanket is available.

Step 10: Pull the end of the blanket at the victim's head.

Step 11: Drag the victim to safety.

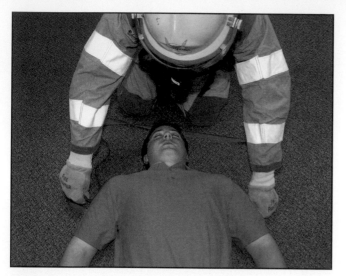

Step 1: Place the victim on his or her back.

Step 2: Slide the large webbing loop under victim's head and chest so the loop is even with their arm pits. Position the victim's arms so that they are outside the webbing.

Step 3: Pull the top of the large loop over the victim's head so that it is just past their head.

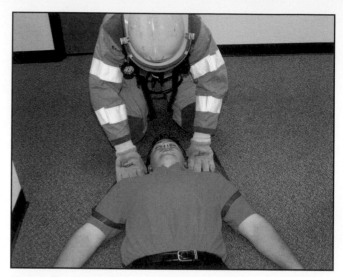

Step 4: Reach down through the large loop and under the victim's back and grab the webbing.

Step 5: Pull the webbing up and through the loop so that each webbing loop is drawn snugly around the victim's shoulders.

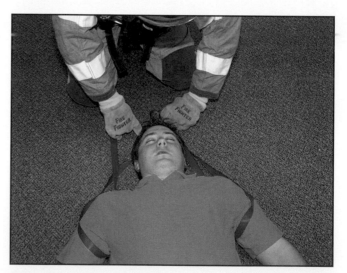

Step 6: Adjust hand placement on the webbing to support the victim's head.

Step 7: Drag the victim to safety by pulling on the webbing loop.

12-5

Demonstrate the cradle-in-arms lift/carry — One-rescuer method.

Skill SHEETS

Step 1: Place one arm under the victim's arms and across the back.

Step 2: Place the other arm under the victim's knees.

Step 3: Lift the victim to about waist height while keeping your back straight.

Step 4: Carry the victim to safety.

12-6

Demonstrate the seat lift/carry — Two-rescuer method.

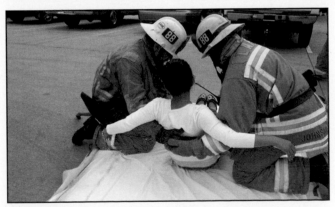

Step 1: Raise the victim to a sitting position.

Step 2: Link arms across the victim's back.

Step 3: Reach under the victim's knees to form a seat.

Step 4: Lift the victim using your legs. Keep your back straight while lifting.

Step 5: Move the victim to safety.

Step 1: *Both Rescuers:* Turn the victim (if necessary) so that the victim is supine.

NOTE: Keep head and neck stabilized during rolling to prevent spinal injury.

Step 2: *Rescuer #1:* Kneel at the head of the victim.

Step 3: *Rescuer #2:* Stand between the victim's knees.

Step 4: *Rescuer #1:* Support the victim's head and neck with one hand and place the other hand under the victim's shoulders.

Step 5: *Rescuer #2:* Grasp the victim's wrists.

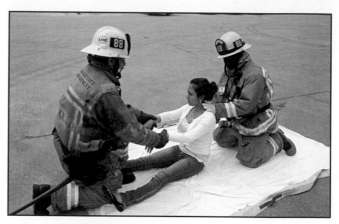

Step 6: *Rescuer #2:* Pull the victim to a sitting position.

Step 7: *Rescuer #1:* Push gently on the victim's back.

Step 8: *Rescuer #1:* Reach under the victim's arms and grasp the victim's wrists as Rescuer #2 releases them. Grasp the victim's left wrist with the right hand and right wrist with the left hand.

Step 9: *Rescuer #2:* Turn around, kneel down, and slip hands under the victim's knees.

Step 10: *Both Rescuers:* Stand and move the victim on command from Rescuer #1.

Step 1: *Both Rescuers:* Turn the victim (if necessary) so that the victim is supine.

NOTE: Keep head and neck stabilized during rolling to prevent spinal injury.

Step 2: *Rescuer #1:* Lift the victim's knees until the knees, buttocks, and lower back are high enough to slide a chair under the victim.

Step 3: *Rescuer #2:* Slip a chair under the victim.

Step 4: *Both Rescuers:* Raise the victim and chair to a 45-degree angle.

Step 5: *Both Rescuers:* Lift the seated victim with one rescuer carrying the legs of the chair and the other carrying the back of the chair.

12-9

Demonstrate the chair lift/carry method 2 — Two rescuers.

Step 1: *Rescuer #1:* Place the victim in a sitting position.

Step 2: *Rescuer #1:* Reach under the victim's arms and grasp the victim's wrists.

Step 3: *Rescuer #2:* Position the chair next to the victim.

Step 4: *Rescuer #2:* Grasp the victim's legs under the knees.

Step 5: *Both Rescuers:* Lift gently and place the victim onto the chair.

Step 6: *Both Rescuers:* Raise the victim and chair to a 45-degree angle.

Step 7: *Both Rescuers:* Lift the seated victim with one rescuer carrying the legs of the chair and the other carrying the back of the chair.

Evidence Preservation and Sampling

Chapter Contents

chapter 13

Key Terms

Competencies

NFPA® 472:	6.5.3.1(1)(b)	6.5.3.1(1)(n)	6.5.3.1(4)(c)	6.5.3.1(7)(c)	6.5.4.1(7)
6.5.2.1(1)(a)	6.5.3.1(1)(c)	6.5.3.1(1)(o)	6.5.3.1(4)(d)	6.5.3.1(7)(d)	6.5.4.1(8)
6.5.2.1(1)(b)	6.5.3.1(1)(d)	6.5.3.1(2)(a)	6.5.3.1(5)(a)	6.5.3.1(7)(e)	6.5.4.1(9)
6.5.2.1(1)(c)	6.5.3.1(1)(e)	6.5.3.1(2)(b)	6.5.3.1(5)(b)	6.5.3.1(8)	6.5.4.1(10)
6.5.2.1(1)(d)	6.5.3.1(1)(f)	6.5.3.1(2)(c)	6.5.3.1(5)(c)	6.5.3.1(9)	6.5.4.1(11)
6.5.2.1(1)(e)	6.5.3.1(1)(g)	6.5.3.1(2)(d)	6.5.3.1(5)(d)	6.5.3.2	6.5.4.1(12)
6.5.2.1(2)(a)	6.5.3.1(1)(h)	6.5.3.1(3)(a)	6.5.3.1(6)(a)	6.5.4.1(1)	6.5.4.1(13)
6.5.2.1(2)(b)	6.5.3.1(1)(i)	6.5.3.1(3)(b)	6.5.3.1(6)(b)	6.5.4.1(2)	6.5.4.1(14)
6.5.2.1(2)(c)	6.5.3.1(1)(j)	6.5.3.1(3)(c)	6.5.3.1(6)(c)	6.5.4.1(3)	6.5.4.2
6.5.2.1(2)(d)	6.5.3.1(1)(k)	6.5.3.1(3)(d)	6.5.3.1(6)(d)	6.5.4.1(4)	
6.5.2.1(2)(e)	6.5.3.1(1)(l)	6.5.3.1(4)(a)	6.5.3.1(7)(a)	6.5.4.1(5)	
6.5.3.1(1)(a)	6.5.3.1(1)(m)	6.5.3.1(4)(b)	6.5.3.1(7)(b)	6.5.4.1(6)	

Evidence Preservation and Sampling

Learning Objectives

1. Discuss various hazards at crimes involving hazardous materials or weapons of mass destruction (WMD). [NFPA® 472, 6.5.2.1(1), 6.5.2.1(2)]

2. Discuss the first responder's role in investigation.

3. Describe the different response phases at criminal hazardous materials/WMD incidents.

4. Explain the FBI's twelve-step process for collecting evidence. [NFPA® 472, 6.5.3.1(1-9)]

5. Demonstrate evidence preservation and sampling. [Skill Sheet 13-1; NFPA® 472, 6.5.4.1(1-14)]

Chapter 13
Evidence Preservation and Sampling

Case History

A deputy sheriff noticed a vehicle at a vacant house in an unincorporated area of his county and stopped to investigate. He noted a strange odor coming from the house and knocked on the door. As he knocked, a man ran from the rear of the house. The deputy sheriff apprehended the man and questioned him. During questioning, the man admitted that he had a meth lab in the house.

The deputy notified the dispatcher, and the county volunteer fire department arrived at the scene. The house was in an isolated area, so the fire department did not enter the house. Instead, they waited for the county haz mat team to arrive.

Once the county haz mat team arrived, they donned appropriate PPE and entered the house and observed the lab equipment and chemicals in the lab. They made original notes, photographed, and sketched the scene. The county laboratory technician arrived and assisted the hazmat team in collecting, packaging, and labeling the evidence.

After all the evidence was collected, the proper case report was completed, and it was then submitted to the District Attorney. Later, it was presented to the Grand Jury. The facts were found to be true, and the suspect was tried in a criminal court. The presiding judge questioned all the evidence, collection methods, and sample analyses and ultimately accepted them.

To protect the public, Operations-Level responders may be required to collect samples of contaminants or suspected contaminants to support medical treatment, determine how to best mitigate the situation, and to determine the type of decontamination required. However, when responding to incidents occurring from criminal intent, sampling concerns extend beyond public health, and the samples become evidence. This evidence is crucial to law enforcement personnel for criminal investigation and prosecution **(Figure 13.1, p. 592)**. In fact, at criminal haz mat/WMD incidents, evidence preservation and sampling is the top priority after life preservation, hazard mitigation, and the environment. Following proper procedures enables responders to collect evidence that can help apprehend suspects and prevent incident recurrences. Anything at the scene may be considered evidence, and may not be recognized until trained evidence technicians arrive and evaluate the situation.

When responders observe indicators of criminal activity, they must work with various law enforcement agencies to preserve and protect any potential evidence. They may also be called to assist in gathering evidence for the case against the perpetrator of the crime.

Figure 13.1 First responders conducting operations at crime scenes need to be mindful of preserving evidence such as the aircraft engine parts which are located in the debris on the left side of this photo, taken at the Pentagon after the 9/11 terrorist attack. *Courtesy of FEMA News Photos, photo by Jocelyn Augustino.*

Figure 13.2 Responders who perform evidence sampling tasks must be trained to do so in accordance with the requirements of their jurisdiction, and they must operate under the guidance of a hazardous materials technician, allied professional, or standard operating procedures.

Operations-Level responders assigned to perform evidence preservation and sampling activities must be trained to do so in accordance with the requirements of their jurisdiction **(Figure 13.2)**. They must also operate under the guidance of a hazardous materials technician, allied professional, or standard operating procedures.

This chapter will discuss the following topics:

- Hazards at crime scenes involving haz mat/WMDs
- Investigative authority
- Response phases at criminal haz mat/WMD incidents
- FBI Twelve-Step Process of crime scene investigation

Hazards at Crime Scenes Involving Haz Mat/WMD

Criminal haz mat/WMD incidents, environmental crimes, and illicit labs differ greatly in their characterization, location, and associated hazards. Given these differences, responders may need to alter their response actions at each one. However, all operations must be performed within the framework of the IMS, in accordance with the principles of risk-based response discussed throughout this manual.

Information about types of criminal incidents and tactics to use when responding to these incidents was provided in Chapter 7, Terrorist Attacks, Criminal Activities, and Disasters. But some crime scenes include hazards such as armed individuals, booby traps, or explosives, which bomb squad personnel or other specialists must render safe before proceeding **(Figure 13.3)**. Once these hazards are eliminated, actions such as securing the scene and preserving evidence remain the same.

Figure 13.3 Before the crime scene can be processed, hazards such as bombs and booby traps must be removed. *Courtesy of the FBI.*

The same factors discussed in previous chapters (Chapter 8, Personal Protective Equipment and Chapter 12, Air Monitoring and Sampling) will determine the necessary personal protective equipment (PPE) based on the risks the incident presents (as determined by assessment of intelligence, warning signs, and detection clues) and the mission being performed **(Figure 13.4)**. Decontamination operations must be performed in accordance with procedures discussed in Chapter 9, Decontamination, with care being given to preserving evidence as much as possible.

Before arriving at a potential criminal haz mat/WMD incident, responders should already have planned possible responses that follow jurisdictional laws and account for available personnel and equipment. By carrying out duties as planned, responders can achieve the goals of preserving life, stabilizing the incident, and obtaining **forensic** evidence.

Forensic Science — Application of scientific procedures to the interpretation of physical events such as those that occur at a fire scene; the art of reconstructing past events and then explaining that process and one's findings to investigators and triers of fact; criminalistics.

Figure 13.4 Like all haz mat incidents, the PPE needed to operate at a crime scene will be determined by the hazards present and the mission being performed. *Courtesy of MSA.*

Figure 13.5 Illicit labs may contain a variety of extremely hazardous materials. *Courtesy of MSA.*

Illicit Laboratories

The people who run illicit drug labs have a vested interest in maintaining their livelihoods as creators and distributors of illegal drugs. As a result, armed guards, attack dogs, and booby traps are used to protect the labs. All of these must be neutralized before evidence preservation and sampling activities begin.

Labs that create substances used in criminal events or terrorism may also have dangerous outer security measures, and both types of labs potentially contain chemical, biological, radiological, and explosive agents that could be accidentally released during a raid **(Figure 13.5)**. It is important that tactical teams and bomb squad personnel secure human threats and render explosive hazards safe before beginning tasks. Additional information about operations at illicit labs is provided in Chapter 14, Illicit Laboratories.

Release or Attack with a WMD Agent

Explosives, biological toxins, toxic industrial chemicals, biological pathogens, radioactive sources, chemical warfare agents, and nuclear devices are just some of the items used in haz mat/WMD attacks. See Chapter 7, Terrorist Attacks, Criminal Activities, and Disasters, for more information about operations at these incidents. The release/attack site should be cleared or rendered safe of WMD hazards before beginning evidence collection tasks.

Environmental Crimes

Environmental crimes involve the illegal use and disposal of hazardous substances and waste which pollute the air, water, or soil, and can cause serious injury, chronic illness, or even death. Environmental crime scenes may include additional hazards in the form of armed owners/operators or explosive devices. Tactical teams and bomb squads must remove these hazards before beginning tasks.

Suspicious Letters and Packages

Incidents involving suspicious letters and packages may involve explosives, biological materials, hazardous chemicals, or even radiological materials. Most frequently, suspicious letters and packages involve explosives (mail bombs) and white powders **(Figure 13.6)**. White powder incidents are usually hoaxes, but both anthrax and ricin have been sent through the mail as white powder. The International Association of Fire Chiefs' (IAFC) *Model Procedures for Responding to a Suspicious Package* provides excellent guidance for responding to these types of incidents (see information box). If a package is thought to be suspicious, responders should contact bomb squad personnel to investigate further.

Figure 13.6 Most letter or package incidents involve white powders. However, letter and package bombs are also a potential threat.

Model Procedures for Responding to Suspicious Letters and Packages

The first step in IAFC's *Model Procedures for Responding to a Suspicious Package* is to assess the hazard posed by the suspicious letter, package, or container. You can gain a lot of up front information by interviewing the individual who reported the event, witnesses, bystanders, or other first responders and identify any individuals who had physical contact with the package.

Next, visually evaluate the suspicious package. You may want to do this from a safe distance (i.e., using binoculars) if unsure of the contents.

During your assessment, you will want to answer the following questions.

- Was the package accompanied by a verbal or written threat?
- Is the package open, leaking, giving off an odor or have any suspicious markings?
- If the package is open, was any substance released from the package?
- Is anyone who touched the package feeling ill?
- Is the package making noise?
- Are there any wires protruding?

Depending on the answers to the questions, responders will follow specific response procedures. If the answer to all of the questions is "No," responders should discuss the outcome of the assessment with law enforcement and bomb squad personnel and determine whether the suspicious package should be disposed of or removed. Responders should place and seal the package in a plastic bag, and double-bag it for security.

If the answer to any of the questions is "Yes," responders should take initial actions and request assistance. Initial actions should include:

- Establishing hot, warm, and cold zones based upon the assessed threat
- Shutting down the building ventilation system if there has been a substance release
- Turning off any high-speed mail processing equipment that may have handled the suspicious package
- Isolating anyone exposed or potentially exposed and consider shelter-in-place as an initial tactical consideration (evacuation should only be considered if there is an immediate threat)
- Assessing anyone who physically contacted the package to determine the need for decontamination and medical assistance
- Gathering non-victims together for information from public health, other officials, or interviews

Following SOPs, all victims are normally decontaminated before transport to a medical facility. Only transport a contaminated victim if the victim is in immediate need of medical treatment which cannot be accomplished in the field. When possible, contaminated victims should be placed in a protective envelope (such as a Tyvek® suit) to minimize secondary contamination of the general population and/or the environment. An alternative is to protect the ambulance or transport vehicle from potential cross-contamination from the victim.

Investigative Authority

Responders assigned to perform evidence preservation and sampling tasks must be able to identify the investigative authority at haz mat/WMD crimes. Typically, haz mat crime scenes will initially fall under the jurisdiction of the local law enforcement agency. Depending on the type of crime, location, materials involved, and other factors, that authority may shift to other agencies. For example, as discussed previously, if a crime in the U.S. is determined to be a terrorist attack, investigative authority will shift to the FBI. Depending on the drug and quantities involved, the Drug Enforcement Administration (DEA) will investigate crimes involving illegal drugs **(Figure 13.7)**. Environmental crimes may fall under the jurisdiction of the EPA, again depending on the quantity and types of materials involved. The Postal Inspection Service will investigate incidents involving suspicious letters or packages. **Table 13.1** provides a list of U.S. Federal Agencies that may have investigative authority at crimes involving haz mat/WMDs.

Figure 13.7 In the U.S., the DEA may have investigative authority at incidents involving illegal drugs, depending on the quantity and type involved. *Courtesy of the DEA.*

CAUTION

All evidence collection and sampling must be conducted in coordination with law enforcement and other members of the criminal justice system!

In some cases, more than one agency may be involved in an investigation. For example, at an incident involving a suspicious letter or package, the following agencies may be involved in the investigation:

- Postal Inspection Service
- FBI
- ATF
- Civil support teams
- Public health departments
- Local law enforcement agencies
- Bomb squads/explosive ordnance disposal (EOD) personnel
- Others

Local emergency response plans should detail appropriate procedures for requesting assistance from other agencies such as the bomb squad. Local investigative guidelines are complex and dynamic, and local responders must

Table 13.1
U.S. Federal Agencies with Investigative Authority

FBI	The FBI has investigative jurisdiction over violations of more than 200 categories of federal crimes. Top priority has been assigned to five areas: (1) Counterterrorism (2) Drugs/organized crime (3) Foreign counterintelligence (4) Violent crime (5) White-collar crime In addition, the FBI is authorized to investigate matters where prosecution is not contemplated. For example, under the authority of several executive orders, the FBI is responsible for conducting background security checks concerning nominees to sensitive government positions. The FBI also has been directed or authorized by presidential statements or directives to obtain information about activities suspected of jeopardizing the security of the nation.
EPA	EPA agents investigate the most significant and egregious violators of environmental laws which pose significant threats to human health and the environment.
DEA	DEA investigations occur within the context of DEA's responsibilities to enforce the provisions of federal laws and regulations concerning controlled substances, chemical diversion, and drug-trafficking.
ATF	ATF investigations occur within the context of its responsibilities to enforce laws regarding, for example, distilled spirits, beer, wine, tobacco, firearms, explosives, and arson. More specifically, ATF responsibilities include the enforcement of the Gun Control Act of 1968, Title XI of the Organized Crime Control Act of 1970, the National Firearms Act, the Arms Export Control Act, Chapters 51 and 52 of the Internal Revenue Code of 1986, and the Federal Alcohol Administration Act.
Postal Inspection Service	The Postal Inspection Service is responsible for investigating violations of about 200 federal statutes that deal with the integrity and security of mail; the safeguarding of postal employees, property, and the work environment; and the protection of Postal Service revenue and assets.
NPS	The National Park Service (NPS) is responsible for investigating offenses against the United States committed within the national park system in the absence of an investigation by any other federal law enforcement agency. NPS also has authority on and within roads, parks, parkways, and other federal reservations within the District of Columbia. The types of investigations in which NPS is involved include Assimilated Crimes Act investigations, drug enforcement, environmental crimes, crimes against persons, and resource-related crimes, such as plant and wildlife poaching, archaeological site looting, vandalism of historical sites, and simple theft of resources.

Source: U.S. Government Accounting Office

be familiar with their jurisdictional procedures. Law enforcement authorities (such as the District, County, or City Attorney) should be consutled to determine if Ops-Level personnel (for example, the local hazmat team) are within their legal authority to perform evidence preservation and sampling at the scene.

NOTE: The qualifications, experience, and knowledge of individuals conducting sampling and preservation will normally be scrutinized closely in criminal cases.

Response Phases at Criminal Haz Mat/WMD Incidents

The U.S. FBI identifies four response phases at criminal haz mat/WMD incidents. Understanding these phases can provide a context for first responders to understand, from a law enforcement perspective, how a crime scene involving hazardous materials/WMD is managed and where evidence is collected and sampled in relation to other operations. Establishing unified command early at suspected crimes is critical for ensuring that the incident is managed in a fashion that will maximize the goals of all agencies involved.

The four response phases are as follows:

- *Tactical Phase* — During this phase, law enforcement removes hostile threats **(Figure 13.8)**. For example, subjects are arrested, booby traps are neutralized, and explosives are removed.

Figure 13.8 In the tactical phase, hostile threats are removed. *Courtesy of August Vernon.*

- *Operational Phase* — During this phase, life safety objectives are met and the scene is stabilized and secured. While the first priority is safety, responders should take measures to not disturb the scene. Limited sampling may be conducted during this phase to identify hazards, to help make evacuation decisions, and to assist in the medical treatment of exposed victims. It is important to follow evidentiary protocol during collection, since the sample may be treated as evidence when the Crime Scene Phase begins.

- *Crime Scene Phase* — During this phase, evidence is recovered and packaged for transport, and the crime scene is processed **(Figure 13.9)**. These actions must be conducted in accordance with appropriate law enforcement protocols since materials gathered may be used to prosecute a subject. Law enforcement is responsible for obtaining search warrants and/or consent when necessary. During this phase the critical tasks are:

  — Maintaining personal and public safety

  — Protecting samples

  — Preserving samples

  — Documenting the evidence accurately

  — Maintaining the chain of custody

Figure 13.9 Evidence is collected and processed by investigators during the crime scene phase.

- *Remediation Phase* — In this phase, operations to mitigate any remaining hazards are conducted in order to bring the scene back to a safe condition. Contractors or appropriate federal authorities typically perform remediation.

If the incident involves WMDs, there are specific things responders should and should not do when operating during the Crime Scene phase at a WMD incident. Responders **should**:

- Protect lives and property
- Secure the area
- Contact the FBI (in the U.S., usually done through local law enforcement)
- Limit the number of personnel into the scene and record their names
- Wear appropriate PPE
- Recognize potential evidence
- Identify and mark potential evidence
- Preserve the evidence for future collection
- Collect samples for public safety/health
- Field screen samples prior to lab admittance

NOTE: Some of these tasks need to be done under the guidance of law enforcement and/or evidence technicians; procedures and policies may vary in different jurisdictions.

When operating during the Crime Scene phase at a WMD incident, responders **should NOT** do the following:

- Attempt to process a WMD crime scene without contacting (or receiving approval from) the appropriate law enforcement authority
- Take samples with the intent of giving them to the appropriate law enforcement authority as evidence
- Linger in the site once public safety issues are addressed unless law enforcement requests a site presence

The true value of evidence can be realized only if proper care has been used in observing the simple scientific and legal rules (SOPs) that should govern the journey of physical evidence from its discovery to its final appearance as a court exhibit. The steps in this procedure may be described as follows:

1. Security of the scene and the evidence
2. Discovery of the evidence
3. Documentation of the evidence (via photography or sketches)
4. Collection of the object(s) or sample(s)
5. Packaging of the evidence (to include properly sealing the packaging)
6. Submission to the laboratory
7. Laboratory examination
8. Custody of the evidence pending trial
9. Transportation to court
10. Exhibition in court

The FBI Twelve-Step Process

While not all haz mat crime scenes will fall under the jurisdiction of the FBI, to give responders a general context for evidence preservation and sampling, it is useful to examine the FBI's Twelve-Step Process of crime scene management for collecting and preserving evidence **(Figure 13.10)**. This process is used for collecting evidence in both contaminated and non-contaminated environments during the Crime Scene Phase of the incident. The twelve steps are:

Step 1:   Make Preparations

Step 2:   Approach the Scene

Step 3:   Secure and Protect the Scene

Step 4:   Perform a Preliminary Survey

Step 5:   Evaluate Evidence Possibilities

Step 6:   Prepare a Narrative Description

Step 7:   Photograph the Scene

Step 8:   Prepare Diagrams and Sketches

Step 9:   Conduct Detailed Search

Step 10:   Collect Evidence

Step 11:   Perform the Final Survey

Step 12:   Release the Crime Scene

The sections that follow will discuss these steps in greater detail. These steps represent a law enforcement agency's approach to crime scene evidence and collection during the Crime Scene Phase. Most operations and tasks relating to protecting public safety should already have been completed during the Tactical and Operations Phases of the incident, including sampling activities conducted to assist in size-up, hazard identification and assessment, treatment of victims, and decontamination operations. However, sampling and collection activities during the Operations Phase should be conducted (as much as practical given life safety concerns) in accordance with the details discussed

Figure 13.10 The U.S. FBI follows a twelve-step process to manage crime scenes. *Courtesy of the FBI.*

in the section, Collecting Evidence. For more information about crime scene investigation, see the U.S. Department of Justice's document, *Crime Scene Investigation, A Guide for Law Enforcement.*

Making Preparations

Like all emergency responders, investigators must maintain training, supplies, and equipment to ensure a proper response. Good preparation ensures that all the equipment needed to conduct the investigation (from lighting equipment to evidence packaging and documentation) is brought to the scene in good condition.

Also during the preparation stage, investigators will begin to determine which laws have potentially been violated and the specific **charges** to be **prosecuted**. Evidence will be collected in order to prove these charges are true and to document the facts at the scene. This will aid in criminal prosecution. Applicable **warrants** (such as **search warrants**) and **affidavits** must be obtained, and information and intelligence discussed. A site safety plan must be formulated, and an evidence recovery plan created. Investigators should determine which laboratories will be sent evidence and samples (see information box).

Evidence Laboratories

Most law enforcement and fire service providers in larger urban areas have already identified which labs have which capability and simply refer to a resource list. When investigators have a choice of labs available to them, they need to determine in advance the capabilities of labs based upon the following questions:

- How does the lab prefer samples be collected, packaged and documented?
- What are the facility's storage capabilities?
- What are the lab's testing procedures?
- What are the labs processing time frames?
- Does the lab receive samples 24-hours a day or not?
- Does the lab only accept samples from certain agencies (for example, law enforcement) but not from others?
- Does the lab provide chain of custody sealing labels for sample containers?

Charge — The law that the law enforcement agency believes the defendant has broken.

Prosecute — To charge someone with a crime; a prosecutor tries a criminal case on behalf of the government.

Warrant — A writ (written order) issued by a competent magistrate (or judge) authorizing an officer to make an arrest, a seizure, or a search or to do other acts incident to the administration of justice. Some states identify the type of magistrate who can issues certain types of warrants and what is needed for a warrant.

Search Warrant — Written order, in the name of the People, State, Province, Territory, or Commonwealth, signed by a magistrate (or judge), commanding a peace officer to search for personal property or other evidence and return it to the magistrate.

Affidavit — A written statement of facts confirmed by the oath of the party making it, before a notary or officer having authority to administer oaths.

Approaching the Scene

Upon initial arrival, the safety of personnel is of primary concern. The hazards at the scene and the tactical requirements they present will determined the necessary PPE for the incident **(Figure 13.11)**. If applicable warrants are required, they will be obtained and executed. (Warrants may not be needed if a *consent to search form* is signed by the appropriate person.) Notes should be taken at this time to document initial observations.

Figure 13.11 In addition to wearing PPE to protect themselves, responders may need to wear gloves or protective booties to protect the crime scene. Wearing protective booties prevents responders from bringing contaminants from other incidents into the crime scene on their boots.

Documentation

Documentation can be done in many different ways. For example video, photographs, stills, sample logs, the incident action plan, and the site safety plan are all potential documentation tools. However, local procedures for documentation vary from jurisdiction to jurisdiction, so make sure the response plan is in line with these procedures or requirements.

All documentation of actions and observations at the haz mat/WMD scene will be passed to crime scene investigators. Individual responders may even be called to testify about these observations. Emergency responders must coordinate with the law enforcement agency having jurisdiction in order to avoid problems with inadmissible evidence. It is important to work with the local district attorney in order to determine what procedures must be followed. This guidance must be obtained before the process is started and should be incorporated into SOPs.

The investigator(s) in charge of the crime scene shall compile reports and other documentation pertaining to the crime scene investigation into a **case file** by **(Figure 13.12)**. This file shall be a record of the actions taken and evidence collected at the scene. This documentation shall allow for independent review of the work conducted.

Documentation — Written notes, audio/videotapes, printed forms, sketches and/or photographs that form a detailed record of the scene, evidence recovered, and actions taken during the search of the crime scene.

Case File — The collection of documents comprising information concerning a particular investigation.

The investigator(s) in charge should obtain the following for the crime scene case file:

- Initial responding officer(s') documentation
- Fire and hazmat personnel reports
- Emergency medical personnel (including hospital staff) documents
- Entry/exit documentation
- Photographs/videos
- Crime scene sketches/diagrams
- Evidence documentation
- Record of consent form or search warrant
- Reports such as forensic/technical reports (when they become available)

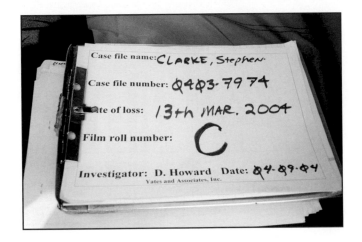

Figure 13.12 A case file is kept to document each investigation. *Courtesy of Donny Howard, Agent, Oklahoma State Fire Marshal's Office.*

Securing and Protecting the Scene

If it has not already been done, responders need to establish a security perimeter with an access control point. Controlling, identifying, and removing persons at the crime scene is a very important aspect of protecting the crime scene. Once boundaries have been established, the entry and exit of all people entering and leaving the scene should be documented. This documentation also aids the use of a personnel accountability system because it requires full identification of all personnel in the crime scene.

To aid in securing and protecting the scene, responders should do the following:

- Prevent individuals from altering/destroying physical evidence by restricting their movement, location, and activity while ensuring and maintaining safety at the scene.

- Identify all individuals at the scene who may be suspects or **witnesses** and secure and separate them. (First responders on the scene are potential witnesses and will be noted by law enforcement.)

- Determine if bystanders are witnesses, and, if not, remove them from the scene.

Witness — A person called upon to provide factual testimony before a judge or jury.

Transient Evidence — Evidence which by its very nature or the conditions at the scene will lose its evidentiary value if not preserved and protected (for example, blood in the rain).

Trace Evidence — Physical evidence that results from the transfer of small quantities of materials (for example, hair, textile fibers, paint chips, glass fragments, gunshot residue particles).

- Exclude unauthorized and nonessential personnel (for example, the media) from the scene **(Figure 13.13)**.
- Take note of any type of evidence, taking special care to document and protect **transient** and **trace evidence**.
- Consult with law enforcement and/or evidence technicians before disturbing transient or trace evidence.
- Document any items (or victims) within the perimeter that have been disturbed or moved by emergency responders.

Figure 13.13 Scene security is extremely important. Unauthorized personnel such as the media and bystanders must be kept behind the outer perimeter.

Investigation Scene Procedures

It is important to maintain a crime scene as found and keep it uncontaminated. **DO NOT** do any of the following:

- Smoke or chew tobacco
- Use the telephone or bathroom
- Eat or drink
- Move any items including weapons (unless necessary for the safety and well-being of persons at the scene)
- Adjust the thermostat
- Turn any equipment, lights, appliances, etc. on or off
- Open doors or windows
- Touch anything unnecessarily
- Reposition moved items
- Litter or spit within the established boundaries of the scene

Walk-Through — An initial assessment conducted by carefully walking through the scene to evaluate the situation, recognize potential evidence, and determine resources required; also, a final survey conducted to ensure the scene has been effectively and completely processed.

Performing a Preliminary Survey

After securing the scene, investigators will perform a preliminary survey by conducting a **walk-through** or reconnaissance survey. This survey is an initial assessment conducted to evaluate the situation, recognize potential evidence, and determine resources required to process the scene. Photos and notes will be taken, and the environment assessed to determine hazards. The evidence perimeter will be adjusted as necessary, and personnel and equipment needs will be determined. A general theory about the crime may be developed at this time based on the information obtained.

During the scene walk-through, investigators should act as follows:

- Avoid contaminating the scene by using the established path of entry.
- Prepare preliminary documentation of the scene as observed.
- Identify and protect fragile and/or perishable evidence, ensuring that all evidence that may be compromised is immediately documented, photographed, and collected.

Evaluating Evidence Possibilities

Anything and everything at a crime scene is considered to be evidence until evidence technicians eliminate it. Anything and everything must therefore be treated accordingly. After the initial walk-through, investigators will determine what is important, plan for evidence collection, and determine how to file the case. These determinations are performed in the cold zone and take the most time. All items of evidence will have a collection plan developed for them.

When evaluating evidence possibilities, investigators should act as follows:

- Conduct a careful and methodical evaluation considering all physical evidence possibilities (such as biological fluids, fingerprints, trace evidence).
- Focus first on the easily accessible areas in open view and proceed to out-of-view locations.
- Select a systematic search pattern for evidence collection based on the size and location of the scene(s).
  - Select a progression of processing/collection methods so that initial techniques do not compromise subsequent processing/collection methods.
  - Concentrate on the most transient evidence and work to the least transient forms of physical evidence.
- Move from least intrusive to most intrusive processing/collection methods.
- Continually assess environmental and other factors that may affect the evidence.
- Be aware of multiple scenes such as victims, suspects, vehicles, and locations.
- Recognize other, available methods to locate, technically document, and collect evidence such as alternate light sources, enhancement, blood pattern documentation, and projectile trajectory analysis.

Preparing a Narrative Description

In this step, investigators prepare a narrative description describing the scene in detail, moving from general to specific. This description includes information such as the condition and position of evidence, weather, and lighting. The narrative is very systematic, and photographs and sketches will be used to supplement it.

Photographing the Scene

It is important to photograph the scene as soon as possible (**Figure 13.14, p. 606**). These photographs must be recorded in a photographic log. Every piece of evidence will be photographed at different distances: overall, medium range, and close up, from all angles. A scale for perspective will be provided as necessary **(Figure 13.15)**. Fragile areas of the crime scene should be photographed first. The overall crime scene will be photographed, as well, both interior and exterior as applicable. Photographs of the address of the structure and the neighborhood can be very beneficial.

NOTE: It is important to determine in advance what the court/legal system will allow in terms of photographic images and evidence and how these photos are submitted in terms of chain-of-custody. There are many variables involved (resolution, digital vs. non-digital, color vs. black and white, cell phone pictures, video, GPS verification, etc.). This information should be discussed with local prosecutors (for example, the local district attorney) in advance.

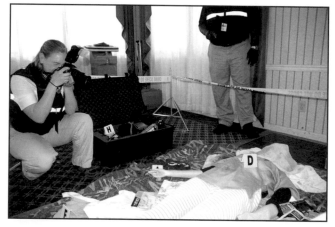

Figure 13.14 Photographs are an important element of documenting the scene and should be taken as soon as possible. *Courtesy of the U.S. Army, photo by Kevin Stabinsky.*

Figure 13.15 In addition to photographing items from different angles and different distances, in some cases, a scale must be added to provide perspective. *Courtesy of the Iowa Fire Service Training Bureau.*

Preparing Diagrams and Sketches

In this step, investigators prepare detailed diagrams and sketches of evidence and the crime scene to supplement photographs. The diagrams and sketches should assist in detailing distance and size relationships, and contain measurements and scale **(Figure 13.16)**. Sketches will document the following:

- Immediate area of the scene, noting case identifiers and indicating north on the sketch

- Relative location of items of evidence (correlating evidence items with evidence records)

- Evidence prior to movement

- Rooms, furniture, and or other objects

- Distance to adjacent buildings or other landmarks

Conducting a Detailed Search

Before beginning evidence collection, investigators conduct a detailed search for evidence using a search grid or pattern. This search is a very thorough investigation of the crime scene during which a complete evaluation is made. Responders should wear appropriate clothing and protection (for example, latex gloves to avoid leaving fingerprints) in order to avoid contamination

Figure 13.16 The exact dimensions of the crime scene and distances between objects will be recorded in crime scene sketches. *Courtesy of the U.S. Army.*

Figure 13.17 Wearing latex gloves prevents investigators from contaminating the crime scene with their own fingerprints. *Courtesy of the U.S. Department of Defense.*

of evidence **(Figure 13.17)**. If it is a large area or there is limited law enforcement personnel, other emergency response personnel may be asked to assist in this search.

Collecting Evidence

Processing physical evidence at the crime scene is one of the most important parts of the investigation. Evidence must be documented, collected, preserved, and packaged with attention to scene integrity and protection from contamination or harmful change. During the processing of the scene and following documentation, evidence should be appropriately packaged, labeled, and maintained in a secure, temporary manner until final packaging and submission to a secured evidence storage facility or the crime laboratory.

Investigators should act as follows:

- Maintain scene security throughout processing and until the scene is released.
- Layout the investigative evidence collection grid system.
- Document the collection of evidence by recording its location at the scene, date of collection, and who collected it.
- Collect each item identified as evidence.
- Establish chain of custody.
- Obtain standard/reference samples from the scene (sometimes called *control samples* or *blank samples*).
- Immediately secure electronically recorded evidence (for example, answering machine tapes, surveillance camera videotapes, computers) and move them from the scene to a secure area.

- Identify and secure evidence in containers (for example, label, date, initial container) at the crime scene; different types of evidence require different containers (for example, porous, nonporous, and crushproof); follow the requirements of the lab which will be analyzing the samples in regard to the types of containers and the amount of sample to collect **(Figure 13.18)**.

Figure 13.18 Follow lab requirements as to the type of evidence containers to use and how much sample to collect. All samples must be appropriately sealed, labeled, packaged, and documented.

- Package items to avoid contamination and cross-contamination.
- Document the condition of firearms/weapons prior to rendering them safe for transportation and submission.
- Avoid excessive handling of evidence after it is collected.
- Maintain evidence at the scene in a manner designed to diminish degradation or loss.
- Transport and submit evidence items for secure storage.

The sections that follow will discuss many elements of evidence collection and processing including the following:

- Chain of custody
- Site characterization
- Sampling plans
- Protecting samples and evidence
- Sampling methods and equipment
- Field screening
- Decontamination
- Labeling and packaging

NOTE: Skill Sheet 13-1 demonstrates the steps for collecting and processing evidence.

Chain of Custody

All evidence must be handled and moved to an evidence custodian for documentation into the evidence chain in accordance with the law enforcement AHJ's, **chain-of-custody** procedures Chain of custody is the practice of tracking an item of evidence from the time it is found until it is ultimately disposed of or returned. The chain-of-custody is a written history that must include each person who maintains visual or physical control over the item throughout the process **(Figure 13.19)**. Each person in the chain-of-custody is a candidate for subpoena to court.

Chain of Custody — A process used to maintain and document the chronological history of the evidence; documents should include name or initials of the individual collecting the evidence, each person or entity subsequently having custody of it, dates the items were collected or transferred, agency and case number, victim's or suspect's name, and a brief description of the item.

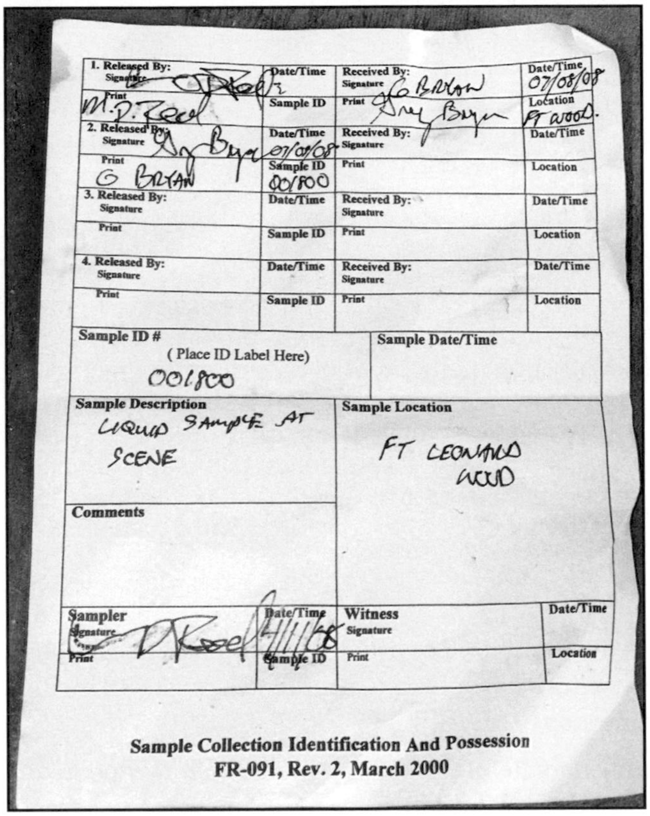

1. Released By: Signature	Date/Time	Received By: Signature	Date/Time 07/08/08
Print M. D. Reed	Sample ID	Print	Location FT WOOD
2. Released By: Signature	Date/Time 07/08/08	Received By: Signature	Date/Time
Print G BRYAN	Sample ID 001800	Print	Location
3. Released By: Signature	Date/Time	Received By: Signature	Date/Time
Print	Sample ID	Print	Location
4. Released By: Signature	Date/Time	Received By: Signature	Date/Time
Print	Sample ID	Print	Location

Sample ID # (Place ID Label Here) 001800

Sample Date/Time

Sample Description: LIQUID SAMPLE AT SCENE

Sample Location: FT LEONARD WOOD

Comments

Sampler Signature	Date/Time	Witness Signature	Date/Time
Print	Sample ID	Print	Location

Sample Collection Identification And Possession
FR-091, Rev. 2, March 2000

Figure 13.19 A chain-of-custody form may be used to document who maintained control of a sample throughout its processing.

Although the ultimate responsibility for the chain of custody is placed upon the law enforcement AHJ, as a responder taking public health samples, you must be prepared to provide the information required to establish this chain.

To ensure that the chain of custody is successfully maintained, responders should follow SOPs and be prepared to provide the following information:

- Name, ID, and rank
- Agency
- Date
- Time that evidence changed possession
- Specific location
- Incident number or **case identifiers** (if appropriate)
- When and to whom the sample was released (some chain-of-custody forms ask who evidence was received from as well)

Case Identifiers — The alphabetic and/or numeric characters assigned to identify a particular case.

NOTE: If evidence is locked in a vehicle, this should be noted on the evidence form including information as to how long it was there and how secure the vehicle was during that time.

Site Characterization

Site Characterization —
The size up and evaluation of hazards, problems, and potential solutions of a site.

Before determining a sampling plan, initial monitoring should be performed to **characterize the site** and the possible threats. Haz mat monitoring teams should be prepared to monitor for any of the following:

- Radiation
- Corrosives (corrosive atmosphere and corrosive liquids)
- Flammables
- Percentage of oxygen
- Toxic industrial or military chemicals

OSHA Requirements

Per OSHA, a preliminary evaluation of a site's characteristics shall be performed prior to site entry by a qualified person in order to aid in the selection of appropriate protection methods prior to site entry. Immediately after initial site entry, a more detailed evaluation of the site's specific characteristics shall be performed by a qualified person in order to further identify existing site hazards and to aid in the selection of the appropriate engineering controls and personal protective equipment for the tasks to be performed.

Sampling Plans

Evidence should be collected in accordance with the sampling plan. Although sample plans will vary (following local or federal protocols), most will include all of the following sampling steps:

1. Preparing evidence containers before entering the exclusion zone, using the system agreed upon by the AHJ and the receiving laboratory
2. Recording the sample location, conditions, and other pertinent information in a field notebook
3. Confirming the sample container number agrees with the overpack container number
4. Wrapping the sample container in absorbent material
5. Placing the sample in an overpack container and sealing it with tamper-proof tape
6. Completing the chain-of-custody form
7. Placing the sample and chain-of-custody form in an approved transport container

The sample plan should also include sampling protocols for:

- Protecting samples and evidence
- Field screening samples
- Labeling and packaging
- Decontaminating samples and evidence

CAUTION

Follow the AHJ's written sampling protocols that outline sampling techniques, types of containers to be used, the sealing process to be used, and other specific procedures!

A minimum of two individuals are recommended for a sampling team:

- A primary sampler (or *dirty* sampler) who takes the samples and handles all sample equipment (sampling tools, container, and sample)

- An assistant to the sampler (or *clean* person) who handles only clean equipment and provides it to the sampler when needed

When possible, a third individual can be added to the sampling team. This individual is also clean and can provide a variety of assistance by documenting, photographing, and even acting as an Assistant Safety Officer in the hot zone. Safe external packaging and transportation of evidence is the responsibility of the law enforcement AHJ, in cooperation with the receiving laboratory and the operator of the transport vehicle. Responders must be trained in sampling and evidence collection methods and the equipment used during sampling and collection activities. Many states have documents that provide guidance on evidence collection and sampling, and this information is available on the Internet.

Protecting Samples and Evidence

Cross-contamination can happen easily if precautions are not taken to prevent it. If sampling equipment touches a non-sterile surface other than the sample itself, it can no longer be used. Each sampling process should be started with new, sterile equipment. Additionally, samplers must change gloves between each separate sample. Protocols may require control samples to be taken from the contaminated area.

Sampling Methods and Equipment

Samples and evidence can come in many forms. The particular type of material present, and the amount of the material will determine sampling method and equipment required **(Figure 13.20)**. In the case of suspicious letters and packages, the entire letter, envelope, or package (as well as the hazardous materials contained within) should be treated as physical evidence.

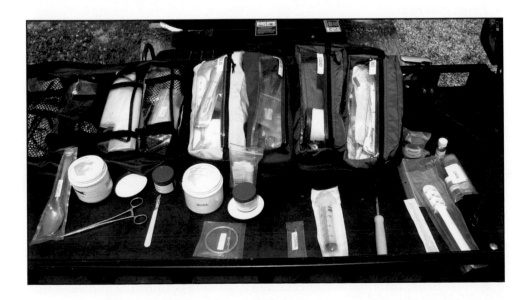

Figure 13.20 Evidence collection kits should contain the equipment necessary to collect a variety of evidence, including both wet and dry samples.

Initial monitoring results should indicate the general characteristics of the potential threat and type of contaminant that is present (radiological, biological, chemical, or combination). With this information, investigators can better determine the correct sampling method.

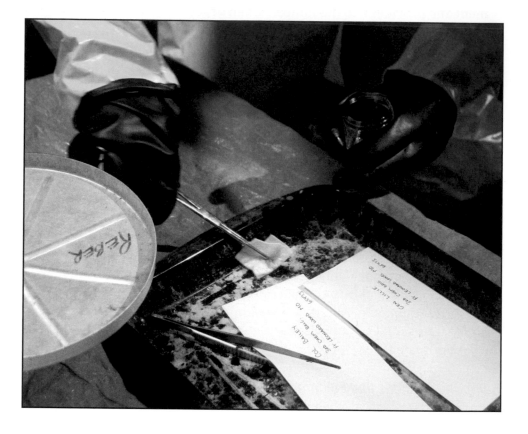

Figure 13.21 Wipe samples are used to collect visible or suspected contaminants on surfaces.

Wipe Sample — Sample collected by wiping a representative surface of a known area.

Wipe Samples. **Wipe samples** are used when contaminants are visible or suspected on surfaces **(Figure 13.21)**. Typical equipment found in a wipe sample kit include the following:

- Latex gloves
- Paper and glass fiber filters
- Cotton swabs
- Forceps
- Sterile sample bags
- Dropper bottles
- Tape
- Marking pen
- Solvents such as deionized water (type of purified water with mineral ions [salts] removed) and isopropyl alcohol

When taking a wipe sample, use the following techniques and precautions:

- Attempt to wipe a one-square foot area, pressing hard on the sample surface.
- Be sure to use information determined from field monitoring and detection equipment since chemical, radiological, and biological sample require different wipe sample materials.

- Use gauze pads for large areas and swabs for smaller areas. Hard to reach areas can often be reached by using a clip and a wire.
- Store samples in a sterile container or bag **(Figure 13.22a and b)**.

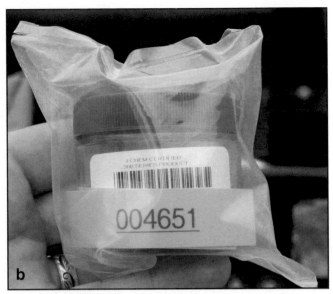

Figure 13.22a and b Swabs or pads used to collect samples should be placed directly into sterile containers.

Liquid Samples. Many sampling kits come with specific instructions for testing various suspected contaminants. Liquid sample kits may include the following items:

- Transfer pipettes (glass or plastic)
- Syringes and tubing
- Certified clean jars
- Sterile vials

NOTE: Local protocols should dictate how much liquid should be collected when taking liquid samples **(Figure 13.23)**. Be careful not to contaminate the outside of the clean sample container when placing the sample inside.

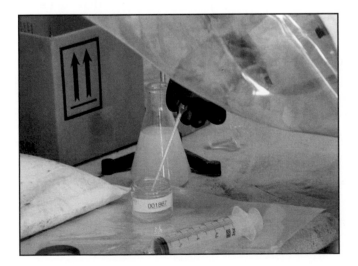

Figure 13.23
Laboratories and local protocols should dictate how much fluid to collect when taking liquid samples.

Solid Samples. Solid samples of evidence may include granular solids, powders, and even scraping from solid materials such as paint or wood. Equipment found in a wipe sample kit could include the following items:

- Latex gloves
- Paper and glass fiber filters
- Cotton swabs
- Scoops
- Spatulas
- Forceps
- Dropper bottles
- Tape
- Sterile sample bags
- Marking pen
- Certified clean jars
- Sterile vials
- Solvents such as deionized water and isopropyl alcohol

The procedures for obtaining solid samples on surfaces are similar to those for wipe samples. Smaller amount of powder may be collected using a swab, while a scoop or spatula is better for collecting granular solids. Scalpels are used for scraping wood or paint that may have absorbed a contaminant and scissors are useful for cutting material or fabric that may have absorbed a vapor or liquid contaminant.

Field Screening Samples

Once samples have been collected, responders must field screen them to eliminate specific hazards before they can be sent to a laboratory. These tests are necessary to ensure safety for individuals involved with packing, transporting, and performing lab tests on the samples. A haz mat team or other qualified personnel should perform all field screenings. Once field screening has ruled out these hazards, the materials may be moved and transported to the appropriate laboratory.

Before field screening, responders should ensure that bomb squad personnel have cleared any potential explosives. Responders should also check for crystallized materials around caps and containers as these are indicators of potentially shock sensitive explosives and reactive chemicals.

To field screen, responders will need to establish a suitable work area that is well-ventilated, potentially outdoors. When using destructive field screening techniques, be mindful of the need to leave enough material to be tested by the laboratory for evidentiary purposes.

Field screen samples to test for, at a minimum:

- ***Explosivity*** — Have bomb squad personnel check for explosive materials/ devices
- ***Radioactivity*** — Check for alpha, beta, and gamma radiation
- ***Flammability*** — Use a combustible gas meter to check flammability
- ***Oxygen*** — Use an oxygen monitor to detect oxygen levels

- *Corrosivity* — On liquids, check for corrosives using pH paper
- *Volatility* — Use a PID to detect **volatile organic compounds (VOCs)**

Volatile Organic Compounds (VOCs) — Organic chemicals that have high vapor pressures under normal conditions.

Decontaminating Samples and Evidence

Decontamination of evidence will remove contamination from the exterior evidence packaging only; exterior evidence packaging should not be opened for the purpose of decontaminating interior evidence packaging. Responders should take care during decontamination to preserve the integrity of evidence (such as fingerprints). Many evidence containers will be double-bagged or placed inside multiple containers for protection of the samples and the safety and health of the people handling them. Follow laboratory instructions and procedures for decontamination of evidence packages.

Labeling and Packaging

Law enforcement/or laboratories (for example, laboratories in the Laboratory Response Network, see information box) usually have established labeling and packaging protocols. At a minimum, all samples must be labeled with date, time, sample number, sample, and locations of sample site. A seal number is placed on a tamper-proof sample container seal **(Figure 13.24)**. Sample numbers are assigned in the sample log; the log also describes each sample taken.

Depending on requirements, samples may be packaged with commercial hazardous sample packaging systems or non-commercial systems **(Figure 13.25)**. In the U.S. the commercial systems require Department of Transportation (DOT) certification. DOT and International Air Transportation Association (IATA) provide standardized commercial packaging equipment including certified safe containers, labels, absorbent, and documentation forms.

NOTE: Special containment vessels must be used for highly toxic liquid.

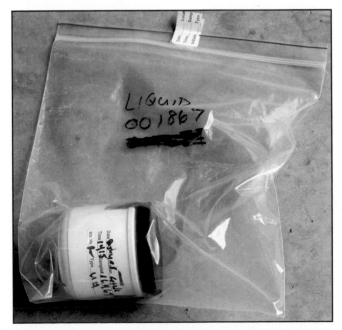

Figure 13.24 All samples must be labeled with date, time, sample number, sample, and location, then sealed with a tamper-proof sample container seal.

Figure 13.25 Extremely toxic or radiological materials may need to be packaged in special containers, such as this one used for radiological materials.

The DOT does regulate non-commercial packaging systems; however, the liability is placed on the employer. Therefore, the employers should take precautions to ensure safe packaging and prepare an emergency action plan.

To prevent cross-contamination between samples, act as follows:

- Have two clean sampling team members.
- Label each sample carefully with a sample and seal number.
- Use a commercial hazardous sample packaging system.
- Discard any sampling equipment that touches a non-sterile surface other than the sample itself.

The U.S. Laboratory Response Network (LRN)

The Laboratory Response Network (LRN) and its partners maintain an integrated national and international network of laboratories that are fully equipped to respond quickly to acts of chemical or biological terrorism, emerging infectious diseases, and other public health threats and emergencies. The LRN was established by the Department of Health and Human Services, Centers for Disease Control and Prevention (CDC) in accordance with Presidential Decision Directive 39, which outlined national antiterrorism policies and assigned specific missions to federal departments and agencies.

Through a collaborative effort involving LRN founding partners, the FBI and the Association of Public Health Laboratories, the LRN became operational in August 1999. Its objective was to ensure an effective laboratory response to bioterrorism by helping to improve the nation's public health laboratory infrastructure, which had limited ability to respond to bioterrorism.

Today, the LRN is charged with the task of maintaining an integrated network of state and local public health, federal, military, and international laboratories that can respond to bioterrorism, chemical terrorism and other public health emergencies. The LRN is a unique asset in the nation's growing preparedness for biological and chemical terrorism. The linking of state and local public health laboratories, veterinary, agriculture, military, and water- and food-testing laboratories is unprecedented.

In the years since its creation, the LRN has played an instrumental role in improving the public health infrastructure by helping to boost laboratory capacity. Laboratories are better equipped, their staff levels are increasing, and laboratories are employing advanced technologies.

Public health infrastructure refers to essential public health services, including the people who work in the field of public health, information and communication systems used to collect and disseminate accurate data, and public health organizations at the state and local levels.

Source: CDC

Performing the Final Survey

Final survey of the crime scene ensures that evidence has been collected and the scene has been processed prior to release. In addition, a systematic review of the scene ensures that evidence, equipment, or materials that the investigation generates are not inadvertently left behind and any dangerous materials or conditions have been reported and addressed.

The investigator(s) in charge should ensure the following:

- Each area identified as part of the crime scene is visually inspected
- All evidence collected at the scene is accounted for
- All equipment and materials generated by the investigation are removed
- Any dangerous materials or conditions are reported and addressed

Releasing the Crime Scene

After the final survey, investigators will release the scene to the owner or responsible party. Once released, reentry may require new warrants, so be certain that all evidence and information has been collected. Documentation should include date and time and to whom the scene was released.

Summary

Ops-Level first responders with appropriate training may be asked to assist with evidence preservation and sampling activities at criminal incidents involving hazardous materials. Such incidents may include illicit laboratories, attacks with WMD agents, environmental crimes, or suspicious letters or packages. Each incident must be evaluated individually to determine investigative authority, and a variety of agencies may be involved.

Haz mat crimes typically go through four response phases: the tactical phase, the operational phase, the crime scene phase, and the remediaton phase. During the tactical phase, hostile threats are removed by law enforcement. During the operational phase, the scene is secured and stabilized, and life safety objectives are met. Evidence is collected and the crime scene is processed during the crime scene phase. The scene is brought back to safe condition for release during the remediation phase.

The FBI Twelve-Step Process is one model for crime scene investigation, but responders should always follow local SOPs for conducting investigations. When collecting samples and evidence, it is extremely important to follow the sampling plan, using techniques that will withstand the scrutiny of the legal process. Evidence must be carefully documented, collected, preserved, packaged, and protected from contamination or harmful change.

Review Questions

1. What are some specific hazards found at illicit laboratories, WMD incidents, environmental crimes, and incidents involving suspicious letters and packages?

2. What agencies may be involved in investigating hazardous materials or WMD incidents?

3. What are the four response phases at criminal hazardous materials or WMD incidents?

4. What are the twelve steps of the FBI's process for collecting evidence?

5. What are the steps for collecting samples?

Step 1: Prepare an evidence collection plan and evidence collection kit for use.

Step 2: Follow all safety procedures to ensure safe entry into hot zone.

Step 3: Ensure that all responders involved in evidence collection are wearing appropriate PPE for performing evidence collection operations in the hot zone.

Step 4: Document all personnel entering hot zone to ensure proper documentation for chain of custody purposes.

Step 5: Enter the scene. Document evidence using photography, sketches, and/or video as determined by the AHJ's SOPs.

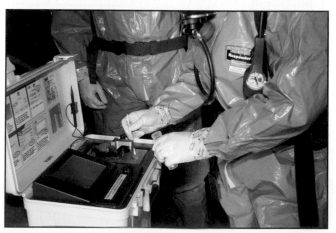

Step 6: Collect sample and prepare for field screening of corrosivity, flammability, oxidation, radioactivity, and volatile organic compounds, following AHJ's SOPs for evidence sampling and field screening.

Note: Field screening for explosives or materials that can cause violent or toxic reactions should be conducted by bomb squad personnel. If explosives are found, withdraw and follow bomb squad instructions on how to proceed.

Step 7: Seal sample container with tamper-proof seal. Label the seal with date, time, and initials/name of person collecting sample. Document sample location through photographic and/or written documentation.

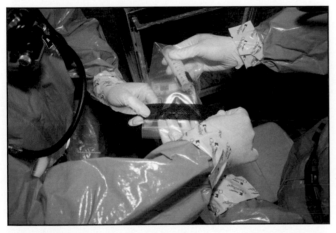

Step 8: Put sample into secondary container, such as zip-top bag. Label secondary container.

Step 9: Proceed to decontamination line for decontamination.

Step 10: Decontaminate exterior of secondary container while proceeding through decontamination.

Note: If any evidence changes custody during decontamination, this must be documented on chain of custody form.

Step 11: Follow laboratory instructions for packaging evidence for transportation, ensuring that documentation of chain of custody is performed.

Illicit Laboratories

Chapter Contents

Divider page photo courtesy of MSA.

Key Terms

Competencies

NFPA® 472:	6.9.3.1	6.9.3. 2.2(5)	6.9.3.4.1	6.9.3.5	6.9.4.1.2(1)
6.9. 2.1(1)	6.9.3. 2.1	6.9.3. 2.2(6)	6.9.3.4.2(1)	6.9.4.1	6.9.4.1.2(2)
6.9. 2.1(2)	6.9.3. 2.2(1)	6.9.3.3(1)	6.9.3.4.2(2)	6.9.4.1.1(1)	6.9.4.1.2(3)
6.9. 2.1(3)	6.9.3. 2.2(2)	6.9.3.3(1)(a)	6.9.3.4.2(3)	6.9.4.1.1(2)	6.9.4.1.3
6.9. 2.1(4)	6.9.3. 2.2(3)	6.9.3.3(1)(b)	6.9.3.4.2(4)	6.9.4.1.1(3)	6.9.4.1.4
6.9. 2.1(5)	6.9.3. 2.2(4)	6.9.3.3(1)(c)	6.9.3.4.2(5)	6.9.4.1.1(4)	6.9.4.1.5

Illicit Laboratories

Learning Objectives

1. Discuss general hazards at illicit laboratories. [NFPA® 472, 6.9.2.1(4), 6.9.4.1.1(3)]

2. Identify and avoid booby traps at illicit laboratories. [Skill Sheet 14-1; NFPA® 472 6.9.4.1.1(3)]

3. Discuss drug labs. [NFPA® 472, 6.9.2.1(1), 6.9.4.1.2(1), 6.9.4.1.2(3)]

4. Describe chemical agent labs. [NFPA® 472, 6.9.2.1(2)]

5. Describe explosives labs. [NFPA® 472, 6.9.4.1.2(2-3)]

6. Discuss biological laboratories. [NFPA® 472, 6.9.2.1(3)]

7. Discuss operations at illicit labs. [NFPA® 472, 6.9.2.1(5), 6.9.3.1, 6.9.3.2.1, 6.9.3.2.2(1-6), 6.9.3.3(1) (a-c), 6.9.3.4.1, 6.9.3.4.2(1-5), 6.9.3.5, 6.9.4.1, 6.9.4.1.1(1), 6.9.4.1.1(4), 6.9.4.1.3]

8. Explain remediation of illicit labs. [NFPA® 472, 6.9.4.1.4, 6.9.4.1.5]

Chapter 14
Illicit Laboratories

Case History

In July 2009, police in Cedar Rapids, Iowa, arrested a 41-year old driver in an apartment complex parking lot for possession of precursors (lithium, anhydrous ammonia, and pseudophedrine), aiding and abetting the manufacture of methamphetamine, possession of cocaine, and driving with a suspended license. A mobile meth lab was found in his vehicle.

While handling the meth lab, a cap popped open, spraying a police officer in the face with anhydrous ammonia. The officer was taken to a local hospital where his face was flushed to remove contamination, and his condition was monitored to ensure that inhalation of the chemical did not cause harm to his lungs and respiratory tract. He was released from the hospital later that day.

Illicit laboratories which produce illegal drugs such as methamphetamine (meth) are scattered around the country. Less likely (but more hazardous) are illicit labs that manufacture weapons of mass destruction. These labs can be anywhere, in hotel rooms, travel trailers and campers, barns, warehouses, storage units and sheds, cars, and homes of all kinds **(Figure 14.1a-c)**. These labs are dangerous places, because of both the substances held within and the operators who run the labs.

Figure 14.1a – c Illicit labs can be located virtually anywhere. Responders should be wary of booby traps and secondary devices protecting lab locations.

A response to an illicit lab can be very dangerous. In many cases, responders may be called for another type of an event (such as a fire), only to discover that an illicit lab is involved. Therefore it is very important to recognize the indicators for these types of labs and understand the hazards associated with them. Developing and implementing a successful response to an incident at an illicit lab requires fast and accurate analysis and a quickly developed, workable plan that accounts for potential hazards and jurisdictional responsibilities.

It is very important to identify the kinds of activities taking place in the lab as quickly as possible. Specialized teams exist to assist in this assessment and intelligence process. Examples of these teams in the U.S. include:

- Drug Enforcement Agency (DEA) Clandestine Lab Teams
- Local or State Law Enforcement Lab Teams for the illegal manufacture of drugs
- Federal Bureau of Investigation (FBI) Hazardous Materials Response Unit for the manufacture of WMD materials

General Hazards at Illicit Laboratories

Because of the nature of the items stored in illicit labs, the items being produced, and the lab operator's vested interest in the protection of their product and eluding law enforcement, responses at these labs are very dangerous. **Table 14.1** provides a summary of hazards and the agencies responsible for the tactics to mitigate the hazards.

Table 14.1 Illicit Lab Response - Tactical Guidelines	
Hazard	**Responsibility**
Operator present within the laboratory, with access to weapons	Law enforcement tactical teams specifically trained to operate within a hazardous environment
Anti-personnel devices (booby-traps) around and within the laboratory	Bomb Squad personnel trained for these procedures
Hazardous materials/WMD within the illicit laboratory	Technician- and Operation-Level responders

Lab Operators

Lab operators are likely to be hostile and potentially armed. Law enforcement tactical teams specifically trained to operate within a hazardous environment must conduct interactions with the lab operator. This team should have a level of protection based on an assessment of the intelligence information on materials being produced, and information including any protective clothing the operator used, activity of animals in the laboratory, interviews with neighbors, and other intelligence-gathering methods.

Booby Traps

Booby-traps are common at illicit labs. The operators of illicit labs have a lot invested in their labs, including procedures and devices to protect the lab and its contents. Specific information on potential booby-traps should be obtained from bomb squads, but the information should be used for awareness only. Bomb technicians specifically trained in searching for and dismantling booby-traps should complete this task.

Booby-traps can be inside or outside of the lab, and they may include any of the following **(Figure 14.2, p. 626)**:

- Explosives (including grenades and dynamite)
- Wires attached to explosives or alerting devices
- Weapons tied to doors
- Bottles that will break thereby mixing chemicals to produce toxic fumes
- On/off switches that have been reversed
- Holes in floors (trap doors to snake pits)
- Electrified door handles
- Exposed wiring
- Animals (such as dogs and poisonous snakes)
- Spikes
- Hooks
- Acid

Responders should always maintain good situational awareness and stay alert for booby traps. Properly trained bomb squad personnel should clear potential anti-personnel devices. Other actions to take to avoid booby traps include the following:

- Maintaining situational awareness
- Limiting personnel entering a suspected hazardous area
- Using explosion proof equipment
- Taking aerial photographs for reconnaissance prior to entry
- Avoiding complacency
- Refraining from handling, touching or moving items in the lab
- Checking doors and openings for wires and/or traps before opening

Turning equipment found at an illicit lab on or off may also be a trigger for booby traps. In addition, electrical pumps such as those used in cooling baths in red phosphorous meth labs should be left on to continue the circulation of cooling water. Interrupting this flow may result in an overheated reaction and ignition of nearby combustibles. See **Skill Sheet 14-1** for methods to identify and avoid booby traps at illicit labs.

What This Means To You

If you're the first to discover or detect the presence of an illicit lab, don't disturb the lab in any way. Don't flip switches or turn on or off lights. Don't shut off the electricity to the facility. Use extreme caution in your movements, being mindful of the potential for booby traps. Just back out the way you came in, evacuate the surrounding area, and request appropriate resources. If it is on, leave it on. If it is off, leave it off. Wait until appropriate instructions are given.

Booby Traps

Figure 14.2 Examples of booby traps found at illicit labs.

Hazardous Materials

In illicit labs, both the final product and production materials are harmful. These will vary depending on the type of lab, but one can expect a variety of flammable, toxic, and biological hazards (the latter primarily in biological labs) **(Figure 14.3)**. Additionally, environmental crimes are frequently associated with illicit labs. Operators may do any of the following:

- Illegally dispose of hazardous waste

- Utilize improper, unapproved processes and locations

- Release hazardous vapors into residential areas

- Pour hazardous waste down sanitary sewers (where flammable liquids could ignite)

- Move unmarked hazardous materials (no placards) in unapproved containers

Responders must stay alert for these kinds of crimes, as well as other hazards associated with labs. Following sections will discuss individual types of labs with their associated hazards in greater detail.

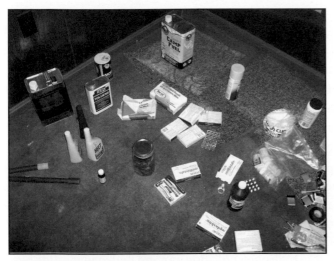

Figure 14.3 Labs may contain a variety of hazards, including highly toxic and flammable materials. *Courtesy of August Vernon.*

Figure 14.4 Meth labs are the most common type of drug lab. *Courtesy of MSA.*

Drug Labs

It is estimated that a significant majority (80 to 90%) of all illegal clandestine drug labs are set up to produce **methamphetamine (meth) (Figure 14.4)**. These labs may also make ecstasy, cocaine, phenyl-2-propanone (P2P), phencyclidine (PCP), heroin, LSD, amphetamines, and other illegal drugs. However, because meth labs represent the most common type of clandestine drug lab, this section primarily focuses on them.

Meth is easy to make and uses a variety of ingredients commercially available in local stores. Because of the increasing hazard of meth labs, some U.S. states have placed restrictions on the purchase of items used in making meth. The process of making meth is called *cooking*, and many different *recipes* or methods exist. Three of the most common are known as the *One/Single Bottle* method, *Red P* method, and *Nazi/Birch* method **(Figure 14.5, p. 628)**.

Methamphetamine (Meth) — A central nervous system stimulant drug that is similar in structure to amphetamine.

Figure 14.5 An example of a *Red P* meth lab. *Courtesy of MSA.*

The various recipes differ slightly in the process and the chemicals used, but all of them are potentially very dangerous because the chemicals are often highly flammable, corrosive, and toxic **(Figures 14.6 a and b)**. A summary of the products commonly used in making meth and the hazards associated with them is provided in **Table 14.2, p. 630**. Meth labs present a danger to the meth cook, community surrounding the lab, and emergency response personnel who discover the lab.

Flammability is perhaps the most serious hazard associated with meth labs, and many labs are discovered only after a fire or explosion has occurred. Other products used in making meth are highly corrosive acids or bases, while others are extremely toxic. One of the byproducts, **phosphine** gas, is sometimes classified as a chemical warfare choking agent. Some products are oxidizers. Meth production processes generate hydrogen chloride gas and hydrogen iodide gas. Meth lab locations may remain a serious health and environmental hazard for years after the lab is removed unless they are properly decontaminated (often an extremely expensive process).

People not wearing PPE who enter a meth lab before it has been properly decontaminated and ventilated may experience headaches, nausea, dizziness, fatigue, shortness of breath, coughing, chest pain, lack of coordination, burns, and even death **(Figure 14.7)**. Risk of injury or toxicity from chemical exposure varies depending on the toxic properties of the chemicals or byproducts, their quantity and form, their concentrations, the duration of exposure, and the route of exposure.

Chemicals and products typically found in meth labs include the following:

- Pseudoephedrine (from antihistamines such as Sudafed®)
- Red phosphorus
- Iodine crystals
- Sodium metal or lithium (from batteries, pellets, or wire solids)

Phosphine — A colorless, flammable, and toxic gas with an odor of garlic or decaying fish which can ignite spontaneously on contact with air; phosphine is a respiratory tract irritant that attacks primarily the cardiovascular and respiratory systems causing peripheral vascular collapse, cardiac arrest and failure, and pulmonary edema.

Figure 14.6a and b Products and equipment commonly used in meth labs.

Figure 14.7 Responders must wear appropriate PPE in illicit labs. *Courtesy of MSA.*

- Anhydrous ammonia
- Starting fluid or Coleman® fuel
- Ethyl ether
- Sulfuric acid or rock salt
- Hydrochloric acid
- Sodium hydroxide

CAUTION

The presence of sodium, lithium, and other reactive substances can complicate fire suppression activities at illicit laboratories because they react with water from hose streams and/ or sprinkler systems.

Table 14.2
Methamphetamine Sources and Production Hazards

Chemical Name	Common Sources/Uses	Hazards	Production Role
Acetone	• Paint solvent • Nail polish remover	• Highly flammable • Vapor is irritating to eyes and mucous membranes • Inhalation may cause dizziness, narcosis, and coma • Liquid may do damage upon contact • Ingestion may cause gastric irritation, narcosis, and coma	• Pill extraction • Cleaning glassware • Cleaning finished methamphetamine (meth)
Anhydrous Ammonia	• Sold as fertilizer; also used as a refrigerant gas • Stolen from farms and other locations for illegal meth production • Often stored in propane tanks or fire extinguishers at illegal meth labs, which causes the fittings of the tank or extinguisher to turn blue	• Toxic • Corrosive • Flammable • Severe irritant; may cause severe eye damage, skin burns and blisters, chest pain, cessation of breathing, and death	Meth production process
Ethyl Alcohol/ Denatured Alcohol/ Ethanol/ Grain Alcohol	• Sold as solvents • Is the alcohol found in beverages at greatly reduced concentrations	• Highly flammable • Toxic; may cause blindness or death if swallowed • Inhalation may affect central nervous system causing impaired thinking and coordination • Skin and respiratory tract irritant (may be absorbed through the skin) • May affect the liver, blood, kidneys, gastrointestinal tract, and reproductive system	• Used with sulfuric acid to produce ethyl ether (see Ethyl Ether/Ether entry) • Cleaning glassware
Ephedrine	Over-the-counter cold and allergy medications	Harmful if swallowed in large quantities	Primary precursor for meth

Continued

Table 14.2 (continued)

Chemical Name	Common Sources/Uses	Hazards	Production Role
Ethyl Ether/Ether	Starting fluids	• Highly flammable • Oxidizes readily in air to form unstable peroxides that may explode spontaneously • Vapors may cause drowsiness, dizziness, mental confusion, fainting, and unconscious at high concentrations	Separation of the meth base before the *salting-out* process begins, primarily in the *Nazi/Birch* method
Hydrochloric Acid/Muriatic Acid (Other acids can be used as well, including sulfuric acid and phosphoric acid)	Commerical or industrial strength cleaners for driveways, pools, sinks, toilets, etc.	• Toxic; ingestion may cause death • Corrosive; contact with liquid or vapors may cause severe burns • Inhalation may cause coughing, choking, lung damage, pulmonary edema, and possible death • Reacts with metal to form explosive hydrogen gas	Production of water-soluble salts
Hydrogen Peroxide	• Common first aid supply • Used for chemical manufacturing, textile bleaching, food processing, and water purification	• Strong oxidizer • Eye irritant	Extrication of iodine crystals from Tincture of Iodine
Hypophos-phorous Acid	Laboratory Chemical	• Corrosive • Toxic • Generates deadly phosgene during initial reaction	Source of phosphorous in *Red P* method
Iodine	Tincture of iodine	• Toxic • Vapors irritating to respiratory tract and eyes • May irritate eyes and burn skin	• Meth production process • Can be mixed with hydrogen sulfide to make hydriodic acid (strong reducing agent) • Can be mixed with red phosphorus and water to form hydriodic acid
Isopropyl Alcohol	Rubbing Alcohol	• Flammable • Vapors in high concentrations may cause headache and dizziness • Liquid may cause severe eye damage	• Pill extraction • Cleaning finished meth

Continued

Table 14.2 (continued)

Chemical Name	Common Sources/Uses	Hazards	Production Role
Lithium Metal	Lithium batteries	• Flammable solid • Water-reactive (reacts with water to form lithium hydroxide, which can burn the skin and eyes)	Reacts with anhydrous ammonia and ephedrine pseudoephedrine in the *Nazi/Birch* method
Methyl Alcohol	HEET® Gas-Line Antifreeze and Water Remover	• Highly flammable • Vapors may cause headache, nausea, vomiting, and eye irritation • Vapors in high concentrations may cause dizziness, stupor, cramps, and digestive disturbances • Highly toxic when ingested	Pill extraction
Mineral Spirits/ Petroleum Distillate	• Lighter fluid • Paint thinner	• Flammable • Toxic when ingested • Vapors may cause dizziness • May affect central nervous system and kidneys	Separation of meth base before *salting-out* process begins
Naphtha	Camping fuel for stoves and lanterns	• Highly flammable • Toxic when ingested • May affect the central nervous system • May cause irritation to the skin, eyes, and respiratory tract	• Separation of meth base before salting-out process begins • Cleaning preparation
Pseudoephedrine	Over-the-counter cold and allergy medications	Harmful if swallowed in large quantities	Production of meth (same as Ephedrine)
Red Phosphorous	Matches	• Flammable solid • Reacts with oxidizing agents, reducing agents, peroxides, and strong alkalis • When ignited, vapors are irritating to eyes and respiratory tract • Heating in a reaction or cooking process generates deadly phosphine gas • Can convert to white phosphorous (air reactive) when overheated	Mixed with iodine in the *Red P* method; serves as a catalyst by combining with elemental iodine to produce hydriodic acid (HI), which is used to reduce ephedrine or pseudoephedrine to meth
Sodium Hydroxide (Other alkaline materials may also be used such as sodium, calcium oxide, calcium carbonate, and potassium carbonate)	Drain openers	• Very corrosive; burns human skin and eyes • Generates heat when mixed with an acid or dissolved in water	After cooking, an alkaline product such as sodium hydroxide turns the very acidic mixture into a base

Table 14.2 (continued)

Chemical Name	Common Sources/Uses	Hazards	Production Role
Sulfuric Acid	Drain openers	• Extremely corrosive • Inhalation of vapors may cause serious lung damage • Contact with eyes may cause blindness • Both ingestion and inhalation may be fatal	Creates the reaction in the salting phase; combines with salt to create hydrogen chloride gas, which is necessary for the *salting-out* phase
Toluene	Solvent often used in automotive fuels	• Flammable • Vapors may cause burns or irritation of the respiratory tract, eyes, and mucous membranes • Inhalation may cause dizziness; severe exposure may cause pulmonary edema • May react with strong oxidizers	Separation of meth base before the *salting-out* process begins
Hydrogen Chloride		• Toxic • Corrosive • Eye irritant • Vapor or aerosol may produce inflammation and may cause ulceration of the nose, throat, and larynx	• Created by adding sulfuric acid to rock salt • Used to salt out meth from base solution
Phosphine Gas		• Very toxic by inhalation • Highly flammable; ignites spontaneously on contact with air and moisture, oxidizers, halogens, chlorine, and acids • May be fatal if inhaled, swallowed, or absorbed through skin • Contact causes burns to skin and eyes	• Byproduct • Produced when red phosphorous and iodine are combined during the cooking process
Hydrogen Iodide/ Hydriodic Acid Gas		• Highly toxic • Attacks mucous membranes and eyes	• Byproduct • Produced when red phosphorous and iodine are combined during the cooking process • Causes the reddish/orange staining commonly found on the walls, ceilings, and other surfaces of meth labs
Hydriodic Acid		• Corrosive • Causes burns if swallowed or comes in contact with skin	• Byproduct • Produced when red phosphorous and iodine are combined during the cooking process

In addition to recognizing the types of chemicals typically found in meth labs, first responders should also be familiar with the types of equipment used in the process. They may expect to find the following items:

- **Condenser tubes** — Used to cool vapors produced during cooking
- **Filters** — Coffee filters, cloth, and cheesecloth
- **Funnels/turkey basters** — Used to separate layers of liquids
- **Gas containers** — Propane cylinders, fire extinguishers, self-contained underwater breathing apparatus (SCUBA) tanks, plastic drink bottles (often attached to some sort of tubing) **(Figure 14.8)**

Figure 14.8 Anhydrous ammonia will turn the brass fittings on propane cylinders and other containers blue.

- **Glassware** — Particularly Pyrex® or Visions® cookware, mason jars, and other laboratory glassware that can tolerate heating and violent chemical reactions
- **Heat sources** — Burners, hot plates, microwave ovens, and camp stoves **(Figure 14.9)**

Figure 14.9 Extra heat sources such as Bunsen burners and hot plates may be evidence of a meth lab. *Courtesy of Joan Hepler.*

- *Grinders* — Used to grind up ephedrine or pseudoephedrine tablets
- *pH papers* — Used to test the pH levels of the reactions
- *Tubing* — Glass, plastic, copper, or rubber

Other clues to the presence of meth labs in structures include the following:

- Windows covered with plastic or tinfoil
- Knowledge of renters who pay landlords in cash
- Unusual security systems or other devices
- Excessive trash **(Figure 14.10)**

Figure 14.10 Excessive amounts of trash may be a lab indicator, particularly if meth product containers are included. *Courtesy of MSA.*

- Increased activity, especially at night
- Unusual structures
- Discoloration of structures, pavement, and soil
- Strong odor of solvents
- Smell of ammonia, starting fluid, or ether
- Iodine- or chemical-stained bathroom or kitchen fixtures

It is estimated that for every pound (0.5 kg) of meth produced, 6 pounds (3 kg) of hazardous waste is generated. This waste may then be dumped in the regular residential trash, down the drain to the septic system, beside roadways, on vacant properties, and in streams or ponds/lakes. It may also be buried. Disposal of this waste is very expensive, and the cleanup process is potentially very dangerous. Many law enforcement departments have contracts with private hazardous materials waste disposal contractors to handle the cleanup and decon of seized illegal meth labs and dumps.

Chemical Agent Labs

Some chemical warfare agents can be made in illicit laboratories. While the recipes may be easy to find, the actual materials necessary to make them may not be. Some ingredients may be common, but access to others is restricted. The following may be indicators of a chemical lab:

- Military manuals

Organophosphate Pesticides — Group of chemicals used to kill insects by disrupting their brains and nervous systems; these chemicals inactivate acetylcholinesterase, an enzyme which is essential to nerve function in insects, humans, and many other animals.

- Underground "cookbooks"

- Chemicals such as **organophosphate pesticides** that would not normally be used to make meth or other illegal drugs

- Chemicals such as methyl iodide and phosphorus trichloride (which might indicate attempts to make sarin)

- Lab equipment sophisticated enough to conduct the chemical reactions needed to make chemical agents **(Figures 14.11a and b)**

- Presence of cyanides or acids

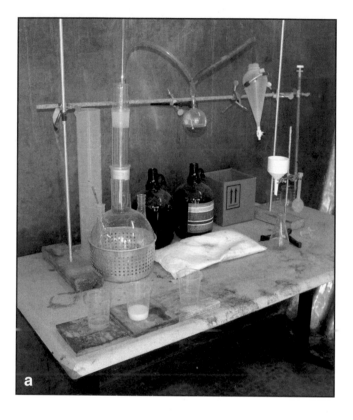

Figure 14.11a and b Clues to a chemical agent lab include sophisticated lab equipment and underground literature.

Explosives Labs

Some explosives labs do not need to heat or cook any of their materials, and therefore may lack the glassware, tubing, Bunsen burners, chemical bottles, and other trappings traditionally associated with laboratories. For example, a work area in a garage used to assemble custom fireworks or pyrotechnics might be considered an explosives lab. However, a lab established to make explosive chemical mixtures might look more like a traditional industrial or university chemistry lab, and labs used to make peroxide-based explosives might look much like a meth or drug lab.

Explosive labs are the second most common type of lab discovered, after drug labs. These can be mistaken for clandestine drug labs due to the presence of household chemicals. Some improvised explosive materials can also be mistaken for narcotics. These labs can be discovered anywhere because they do not require a lot of equipment or resources. Recipes are very easy to find on the Internet, in anarchist literature, and other sources **(Figure 14.12)**. Common explosive materials such as black powder or smokeless powder can be easily incorporated into an IED.

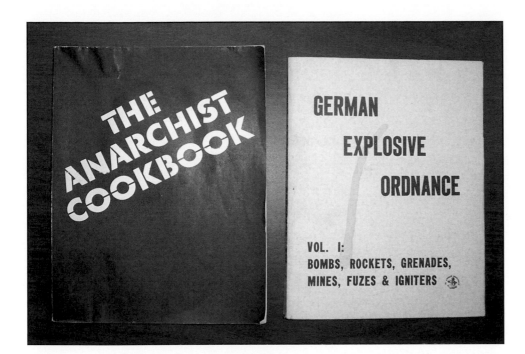

Figure 14.12 Recipes for making explosives are readily available to the public. *Courtesy of August Vernon.*

Some improvised explosive materials such as peroxide-based explosives require a production lab to create. The materials needed to produce these dangerous explosives can be found in hardware and drugstores **(Figure 14.13)**. The basic ingredients are a fuel and oxidizer.

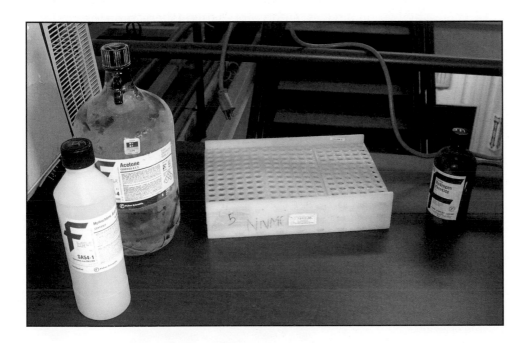

Figure 14.13 Products used to make peroxide explosives (like acetone and hydrogen peroxide) are easily acquired.

Peroxide-based explosives are extremely sensitive to heat, shock, and friction. Some indicators of a possible explosives lab can include:

- Refrigerators/coolers/ice baths
- Glassware and laboratory equipment
- Blenders
- Blasting caps/batteries/fuses/switches

- Pipes/end caps/storage containers
- Shrapnel-type materials
- Strong acidic odors
- Explosives, military ordnance

Some of the common ingredients that may be found in a peroxide based explosives lab include the following:

- Acetone
- Ethanol
- Hexamine (solid fuel for camp stoves)
- Hydrogen peroxide
- Strong or weak acids (such as sulfuric or citric acids)

Once the materials are produced, they can be incorporated into a variety of IEDs. If a material is being transported using a cooling method (such as ice in a cooler), it should be treated with caution, as this is a favorite method of transporting raw materials. If mishandled, the materials in an explosives lab can pose a significant danger.

Other clues to the presence of an explosives lab might include literature on how to make bombs, significant quantities of fireworks, hundreds of matchbooks or flares, ammunition like shotgun shells, black powder, smokeless powder, blasting caps, commercial explosives, or incendiary materials (**Figure 14.14**). Finding these items in conjunction with components that can be used to make IEDs (pipes, activation devices, empty fire extinguishers, propane containers, etc.) would give even more evidence of an explosives lab. Also, electronic components such as wires, circuit boards, cellular phones and other items can point towards the possible design of an IED. **Responders must use caution inside any type of clandestine laboratory.**

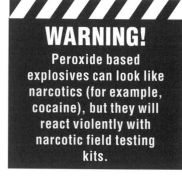

WARNING!
Peroxide based explosives can look like narcotics (for example, cocaine), but they will react violently with narcotic field testing kits.

WARNING!
Do not touch white crystals or powder found in a clandestine laboratory!
They may be extremely sensitive to heat, friction, and shock.

Figure 14.14 The presence of commercial explosives, fireworks, large quantities of black powder, and/or other incendiary materials should raise suspicion of an explosives lab. *Courtesy of the U.S. Bureau of Alcohol, Tobacco, Firearms and Explosives and the Oklahoma Highway Patrol Bomb Squad.*

Biological Labs

First responders should be alert to the indicators of a biological laboratory. Biological labs will have equipment and materials that are quite different from chemical labs including such things as microscopes, growth media, autoclaves, glove boxes, incubators, and refrigerators **(Figure 14.15)**. Most likely they will not have chemicals such as gasoline, propane, anhydrous ammonia, or other flammable and corrosive liquids.

Figure 14.15 An example of a biological lab. While the beans are an obvious clue, also note the absence of condenser tubes, heating sources (such as Bunsen burners), and many of the products commonly used to make illicit drugs and explosives.

Indicators include the following:

- The presence of antibiotics and vaccines

- Personal protective equipment such as respirators (particularly with HEPA filters), rubber gloves, and masks

- Laboratory or test animals and/or related materials, such as cages and food

- Bacterial or viral cultures or related materials, such as growth containers (Petri dishes, agar plates) and culture/growth mediums (agar, meat broth, gelatin) **(Figures 14.16)**

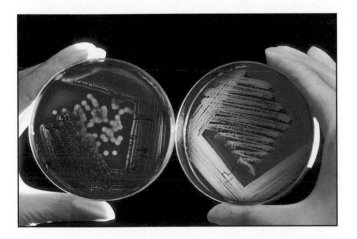

Figure 14.16 Petri dishes/agar plates are used to grow biological cultures. *Courtesy of the National Cancer Institute, photo by Bill Branson.*

Glovebox — A sealed container designed to allow a trained scientist to manipulate microorganisms while being in a different containment level than that of the agent they are manipulating; built into the sides of the glovebox are two glove ports arranged in such a way that one can place their hands into the ports, into gloves and perform tasks inside the box without breaking the seal.

- Presence of biological materials known to be the source of toxins (such as castor beans)
- Biological safety cabinets or **glove boxes** (improvised setups can utilize plastic sheeting, Plexiglas, duct tape, and fans) **(Figure 14.17)**

Figure 14.17 Glove boxes such as this one can be improvised with Plexiglas, plastic sheeting, duct tape, and other common materials. *Courtesy of the CDC, photo by Joel G.Breman, M.D., D.T.P.H.*

- Laboratory equipment such as incubators, refrigerators, and bench top fermentors, (biological labs will typically have fewer pieces of specialized glassware; for example, condenser tubes or distillation and reflux setups would be more indicative of a chemical lab as opposed to a biological lab) **(Figures 14.18)**

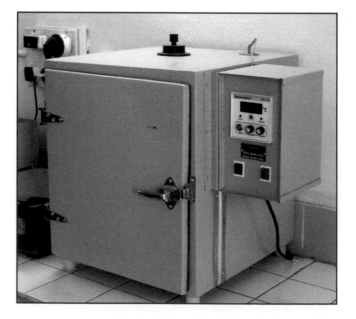

Figure 14.18 The presence of a bacteriological incubator is a good indicator of a biological lab.

- Bleach or other sterilization supplies such as antiseptics and **autoclaves** (or pressure cookers)

- Alterations to building ventilation systems

- Instruction manuals or other books, magazines, and internet resources relating to biological agents

- Sprayers, nebulizers, or other delivery devices

Autoclave — A device that uses high-pressure steam to sterilize objects.

Operations at Illicit Labs

Although an illicit lab may be the scene of an emergency, in the U.S., SOPs for response to these types of labs need to follow rules established by OSHA 1910.120. SOPs should include staffing of positions and activities related to decon, safety, rescue, haz mat, and firefighting operations. These rules apply to all agencies involved in the response (fire service, law enforcement, environmental contractors, etc).

WARNING!
While biological labs may work with biological agents in liquids, slurries, or powders, responders should take special care around powders because they may be more dangerous than other forms.

Responders must also be mindful to preserve evidence because this is a crime scene. A specially trained team will perform any evidence recovery and forensic operations as detailed in Chapter 13, Evidence Preservation and Sampling. The processes involved in illicit drug production change frequently; therefore, haz mat/WMD responders must frequently interact with law enforcement drug response teams.

All on-site activities are directed by the law enforcement agency having jurisdiction. This jurisdiction is normally kept within the parameters established by a search warrant or other protocol in effect that authorizes seizing the illicit laboratory. In order to quickly and accurately analyze a situation at an illicit laboratory, situational awareness should be maintained by receiving regular threat briefings from law enforcement on anticipated threats from haz mat/WMD incidents.

Because so many issues and possible outcomes are at stake when illicit laboratories are discovered, law enforcement jurisdiction, investigative guidelines, and investigative priorities are complicated and ever-changing. Investigative authority at illicit labs may differ based on the type of lab, crime(s) involved, law enforcement jurisdiction, and other factors. Specific jurisdictional situations should be identified before an illicit laboratory is found, and these jurisdictions should be reviewed on a routine basis. In the U.S., jurisdictional confirmation may be needed from:

- Local or state law enforcement authorities

- Federal Bureau of Investigation

- Drug Enforcement Agency

- United States Postal Inspection Service

- Environment Protection Agency

Entering an illicit laboratory should only be done after a careful risk analysis and an effective incident action plan is developed. Creating this plan can be a complicated process that must be completed quickly in order to rapidly execute an effective response.

Planning a response requires coordination among agencies; however, coordination can be difficult because no two illicit labs are alike. Chemicals and biological hazards will vary, as will the hazards that the lab operator has created and any defenses present. Different entities have their own jurisdictions in which they will take lead responsibility.

Coordination challenges that may comprise more than one entity include:

- Securing and preserving the scene with law enforcement
- Site reconnaissance and hazard identification with bomb squad personnel
- Determining atmospheric hazards through air monitoring/detection
- Mitigating immediate hazards while preserving evidence
- Coordinating crime scene operation with the law enforcement agency having investigative authority
- Documenting personnel and scene activities associated with the incident

As discussed in Chapter 13, the Operations-Level responder should become familiar with the local or state and federal procedures for documentation at a crime scene because local procedures vary for crime scene documentation. For example, some jurisdictions do not allow the use of video documentation, while others do not permit digital photographs.

Law enforcement will be responsible for securing and preserving the response scene when illicit laboratories are involved. Law enforcement duties may include the following:

- Neutralizing tactical threats **(Figure 14.19)**

Figure 14.19 Law enforcement is responsible for neutralizing tactical threats. *Courtesy of MSA.*

- Rendering safe any explosive ordnance or booby traps
- Taking full accountability and identifying all personnel in the crime scene
- Documenting any items disturbed within the crime scene
- Protecting evidence from potential damage or destruction

Hazmat and bomb squad teams will be required to work together to resolve situations found within illicit drug or WMD laboratories. In many of these illicit labs, hazards will be present in combinations of devices and materials, such as the following:

- Explosive/chemical
- Explosive/radiological
- Explosive/biological

Teams may also have to work together to clear booby traps and to define the character of explosive ordnance and haz mat hazards. Response agencies should liaise between bomb squad and haz mat assets in order to prepare for situations in which multiple hazards are present.

Figure 14.20 Monitoring and detection must be conducted at illicit labs to assess atmospheric hazards. *Courtesy of MSA.*

Assessing atmospheric hazards is a primary task performed as part of the site risk assessment process **(Figure 14.20)**. Local response plans should ensure the proper equipment is available and used. Equipment selected to determine atmospheric hazards must be capable of assisting in performing the risk assessment. At a minimum, monitoring and detection equipment should include the following:

- CGI
- Oxygen meter
- PID
- pH paper
- Radiological monitoring equipment

Responders should coordinate with appropriate law enforcement agencies before reconnaissance operations are conducted for briefings on intelligence involving the laboratory or site. As in all responses, standard priorities of scene operations apply during responses to illicit drug or WMD laboratories and should be part of the response plan, including the following:

- Life safety

- Incident stabilization

- Protection of property and the environment

It is always critical to mitigate immediate hazards while making best efforts to preserve evidence.

Coordinating crime scene operation with the law enforcement agency having investigative authority is a priority when planning a response to potential illicit drug or WMD laboratories. Responders should look for signs of criminal activity involving chemical, biological, radiological, or explosive materials and devices.

Operations-Level responders will support law enforcement operations within their scope of training. They should also be familiar with crime scene jurisdictions and procedures, such as the following:

- Investigative law enforcement leadership

- Search warrant requirements

- Rules of evidence

- Crime scene documentation

- Policies regarding photography

- Evidence custodial requirements

- Chain of custody

- Specific requirements set forth by the prosecuting attorney

An effective incident action plan at an illicit lab is based on careful analysis of the situation presented to responders. The planned response must take into account the type of laboratory encountered, the hazards presented inside and outside of the lab, and the jurisdictional parameters of the agencies who will aid in preserving the crime scene. There may be unusual circumstances such as innocent people who will need to be protected and/or evacuated. Animal control units may need to be contacted **(Figure 14.21)**. While a response at an illicit lab is inherently a dangerous one, an incident action plan based on solid information will provide the safest response possible. The incident action plan must include considerations for personal protective equipment and decontamination. The protection and proper and legal disposal of lab equipment and chemicals not needed for evidence must be included in the incident action plan.

Figure 14.21 Animal control units may need to be contacted if lab owners have guard dogs or other dangerous animals on the property. *Courtesy of Stillwater Animal Welfare.*

Personal Protective Equipment (PPE)

At illicit lab responses, PPE selection is based upon an assessment of:

- Intelligence about laboratory operations and contents
- Outward warning signs
- Detection clues such as any protective clothing used by the operator, activity of animals in the laboratory and interviews with neighbors

Law enforcement activities may require PPE designed for tactical law enforcement operations. This PPE must be evaluated to ensure it is appropriate for the anticipated hazards as identified during the risk assessment process. Bomb squad operations will require the appropriate level of protective garment. This garment may also need to be augmented by chemical protective clothing appropriate for the anticipated hazard identified during the risk assessment process. Local procedures may also be developed that dictate which PPE are appropriate for each situation.

Decontamination

Decontamination procedures should be based upon the results of the risk assessment process **(Figure 14.22)**. However, tactical entries may require the use of emergency or technical decontamination procedures specifically focused upon the hazards and special needs associated with tactical operations (for example, procedures for decontaminating weapons, ammunitions, and other specialized equipment).

Figure 14.22 Decon will be required at many illicit lab incidents. *Courtesy of Boca Raton Fire Rescue.*

Decontamination for tactical scenarios should be based upon a rapid deployment and anticipating four potential sources requiring decontamination:

- Uninjured tactical operators and their equipment
- Injured tactical operators
- Uninjured suspects
- Injured suspects

Decontamination procedures must be coordinated with law enforcement tactical teams to resolve potential issues such as:

- Handling of weapons
- Pyrotechnic devices from the laboratory
- Suspects in custody

Remediation of Illicit Labs

Remediation — Act of fixing or correcting a fault, error, or deficiency.

Responders must become familiar with local, state, and federal agency policies concerning the **remediation** of illicit drug/WMD scenes. In some cases, private contractors may be hired to perform remediation activities. Assistance and information should come from:

- Local or state health departments
- Emergency management agencies
- DEA
- EPA
- State/local environmental agencies/departments

Regular briefings from these agencies should be part of an on-going education and awareness program. Information gathered from these briefings should be considered when developing response remediation plans and incorporated into agencies' local emergency response plans and SOPs. Training and exercising together increases the success and safety of joint operations. Joint training allows for locating the deficiencies in operation, needed equipment, and/or assistance needed by other agencies before the actual event occurs. An After Action Report should be written after each incident, with input from all involved and a copy furnished to all participating agencies/departments.

Summary

Each type of illicit lab (drug, chemical agent, explosive, or biological lab) has its own unique indicators and associated hazards with which responders must be familiar. However, all illicit labs have the potential for booby traps and hostile operators. For this reason, if an illicit lab (or suspected lab) is discovered, responders should withdraw as appropriate and notify law enforcement. All on-site operations at illicit labs will be directed by the law enforcement agency having jurisdiction.

When operating in or around illicit labs, responders should follow all appropriate safety procedures including wearing appropriate PPE and establishing decon operations prior to entry. When operating in the hot zone, responders should take care not to disturb the scene in any way, except as necessary to perform their assigned tasks (such as detection and sampling). This will help protect evidence, prevent disturbance of potentially volatile chemicals, and protect against activation of secondary devices or booby traps.

Review Questions

1. What are some booby traps that may be found at illicit labs?

2. How can booby traps be avoided?

3. What types of environmental crimes are associated with illicit labs?

4. What percentage of illegal clandestine labs are set up to produce meth?

5. What types of chemicals and products are typically found in meth labs?

6. What types of equipment may be found at meth labs?

7. What are some other clues to the presence of meth labs in structures?

8. What are some indicators of a possible explosives lab?

9. What are some indicators of biological laboratories?

10. What agencies may be involved at illicit lab incidents?

Step 1: Upon suspicion of an illicit lab, notify bomb squad of possible response.

Step 2: Make preparations for safe entry, including appropriate PPE and correct safety procedures.

Step 3: Maintain situational awareness at all times.

Step 4: Approach the scene carefully, looking for anything suspicious or unusual.

Step 5: Before opening doors or windows, examine for any signs of tampering or booby traps. Start low and work upwards, looking for wires, trigger devices, or items that may fall upon opening.

Step 6: If nothing is found, open door slowly and carefully. Proceed cautiously into room.

Step 7: Upon entering, do not touch or change the environment in any way. This includes, but is not limited to, turning lights or HVAC units on or off or turning electricity to building off.

Step 9: If any suspicious items are noticed, back out of the area, retracing your footsteps. Contact the bomb squad immediately.

Step 10: Upon their arrival, brief bomb squad personnel on findings.

Step 11: Follow bomb squad instructions for proceeding.

Step 8: Examine the room in sections; floor to waist, waist to chin, chin to ceiling, and false ceilings if applicable. Look for wires, bottles, pipes, trip wires, or anything out of the ordinary or that arouses your curiosity.

Appendices

Contents

Appendix A
NFPA® 472 Competencies with Chapter and Page References

Awareness-Level Competencies

NFPA® 472 Competencies	Ch. Ref.	Page References	NFPA® 472 Competencies	Ch. Ref.	Page References
4.2.1(1)	1	9	4.2.2(1)	2	64
4.2.1(2)	2	106-123	4.2.2(2)	2	107-118
4.2.1(3)	2	113-123	4.2.2(3)	2	134-136, 148, 150
4.2.1(4)	1	9-10	4.2.3(1)	3	171
4.2.1(5)	2	66-67	4.2.3(2)	3	173
4.2.1(6)	2	71-105	4.4.1(1)	3	169
4.2.1(7)(a)	2	123	4.4.1(2)	1	13-14
4.2.1(7)(b)	2	133-134	4.4.1(3)(a)	3	176
4.2.1(7)(c)	2	140, 142-143	4.4.1(3)(b)	3	176
4.2.1(7)(d)	2	134-137	4.4.1(3)(c)	1	15-32
4.2.1(7)(e)	2	140-141	4.4.1(3)(d)	1	32-33
4.2.1(7)(f)	2	129-132	4.4.1(4)(a)	3	175-176
4.2.1(8)	2	133-134	4.4.1(4)(b)	3	174
4.2.1(9)	2	109-124	4.4.1(4)(c)	3	173-174, 177-179
4.2.1(10)(a)	2	148	4.4.1(5)(a)	3	174
4.2.1(10)(b)	2	148-149	4.4.1(5)(b)	3	174
4.2.1(10)(c)	2	144-147	4.4.1(5)(c)	3	174
4.2.1(10)(d)	2	145	4.4.1(5)(d)	3	174
4.2.1(10)(e)	2	145	4.4.1(6)(a)	3	181-182
4.2.1(10)(f)	2	145	4.4.1(6)(b)	3	175
4.2.1(10)(g)	2	145	4.4.1(6)(c)	3	175
4.2.1(11)	2	150-152	4.4.1(7)	3	173-174, 177-179
4.2.1(12)	2	150	4.4.1(8)	3	177
4.2.1(13)	2	153-161	4.4.1(9)(a)	3	175, 177
4.2.1(14)	2	154-157	4.4.1(9)(b)	3	173-174
4.2.1(15)	2	154-156	4.4.1(10)	3	173-174, 177-179
4.2.1(16)	2	156-157	4.4.1(11)	3	181-182
4.2.1(17)	2	157-158	4.4.1(12)	3	182-183
4.2.1(18)	2	159-161	4.4.2	3	170
4.2.1(19)	2	158-159			
4.2.1(20)	2	161-162			

Operations-Core Competencies

NFPA® 472 Competencies	Ch. Ref.	Page References	NFPA® 472 Competencies	Ch. Ref.	Page References
5.1.2.2(1)(a)	2	65-162	5.2.1.3.1(1)	2	140-141
5.1.2.2(1)(b)	2	144-150	5.2.1.3.1(2)	2	140-141
5.2.1	2, 4	65-162, 211-225	5.2.1.3.1(3)	2	140-141
5.2.1.1	2	71-105	5.2.1.3.2(1)	2	141-144
5.2.1.1.1(1)	2	82	5.2.1.3.2(2)	2	141-144
5.2.1.1.1(2)	2	79-80	5.2.1.3.2(3)	2	141-144
5.2.1.1.1(3)	2	80-82	5.2.1.3.2(4)	2	141-144
5.2.1.1.2(1)	2	93-95	5.2.1.3.2(5)	2	141-144
5.2.1.1.2(2)	2	94-95	5.2.1.3.2(6)	2	141-144
5.2.1.1.2(3)(a)	2	94-95	5.2.1.3.3	2	122
5.2.1.1.2(3)(b)	2	94	5.2.1.4	4, 6	218-223, 271-274
5.2.1.1.3(1)	2	90	5.2.1.5	2, 3	144-150, 172
5.2.1.1.3(2)	2	87	5.2.1.6	2	153
5.2.1.1.3(3)	2	89	5.2.2(1)	2	113-118
5.2.1.1.3(4)	2	91	5.2.2(2)	2	148
5.2.1.1.3(5)	2	88	5.2.2(3)(a)	1, 2	15-32, 148-149
5.2.1.1.3(6)	2	86	5.2.2(3)(b)	2	148-149
5.2.1.1.3(7)	2	85	5.2.2(3)(c)	2	148-149
5.2.1.1.4(1)	2	79	5.2.2(3)(d)	2	148-149
5.2.1.1.4(2)	2	73-76	5.2.2(3)(e)	2	148-149
5.2.1.1.4(3)	2	73-74, 78	5.2.2(3)(f)	2	148-149
5.2.1.1.5(1)	2	102-103	5.2.2(3)(g)	2	148-149
5.2.1.1.5(2)	2	102-103	5.2.2(3)(h)	2	148-149
5.2.1.1.5(3)	2	102-103	5.2.2(3)(i)	2	148-149
5.2.1.1.5(4)	2	102, 104	5.2.2(3)(j)	2	148-149
5.2.1.1.5(5)	2	102, 104	5.2.2(4)(a)	3	179-181
5.2.1.1.7(1)	2	104	5.2.2(4)(b)	3	179-181
5.2.1.1.7(2)	2	104	5.2.2(4)(c)	3	179-181
5.2.1.1.7(3)	2	105	5.2.2(5)	2, 3	144-149, 172, 179-181
5.2.1.1.7(4)	2	105	5.2.2(6)	5, 7	241-242, 338, 360-362, 365
5.2.1.1.7(5)	2	105	5.2.2(7)	6	292-293
5.2.1.2	2	132-144	5.2.2(8)	1	18-21
5.2.1.2.1(1)	2	129	5.2.2(8)(a)	1	18-20
5.2.1.2.1(2)	2	132	5.2.2(8)(b)	1	20
5.2.1.2.1(3)	2	130-131	5.2.2(8)(c)	1	20-21
5.2.1.2.2	2	132-144	5.2.2(8)(d)	1	21
5.2.1.2.3	2	140-144, 122	5.2.3(1)	3	171-181

NFPA® 472 Competencies	Ch. Ref.	Page References
5.2.3(1)(a)(i)	4	202
5.2.3(1)(a)(ii)	4	207-211
5.2.3(1)(a)(iii)	1	25-27
5.2.3(1)(a)(iv)	4	198, 200
5.2.3(1)(a)(v)	4	194-196
5.2.3(1)(a)(vi)	4	194
5.2.3(1)(a)(vii)	4	194
5.2.3(1)(a)(viii)	4	207
5.2.3(1)(a)(ix)	4	193-194
5.2.3(1)(a)(x)	1	18-22
5.2.3(1)(a)(xi)	4	205-206
5.2.3(1)(a)(xii)	4	199
5.2.3(1)(a)(xiii)	4	204
5.2.3(1)(a)(xiv)	4	200-202
5.2.3(1)(a)(xv)	4	205
5.2.3(1)(b)(i)	3	176-177
5.2.3(1)(b)(ii)	1	21
5.2.3(1)(b)(iii)	1	15, 21
5.2.3(1)(b)(iv)	1	30
5.2.3(1)(b)(v)	1	15
5.2.3(1)(b)(vi)	1	15
5.2.3(2)	4	215-216
5.2.3(3)	4	216-217
5.2.3(4)	4	217-218
5.2.3(5)	4	218-223
5.2.3(6)	4	224
5.2.3(7)	1	15-32
5.2.3(8)(a)	1	22
5.2.3(8)(b)	1	23
5.2.3(8)(c)	1	27
5.2.3(8)(d)	1	27
5.2.3(8)(e)	1	25-27
5.2.3(8)(f)	1	24-25
5.2.3(8)(g)	1	27
5.2.3(8)(h)	1	28-29
5.2.3(8)(i)	1	24-25
5.2.3(8)(j)	1	24-25
5.2.3(9)(a)	7	341

NFPA® 472 Competencies	Ch. Ref.	Page References
5.2.3(9)(b)	7	354
5.2.3(9)(c)	7	341
5.2.3(9)(d)	7	341
5.2.3(9)(e)	7	341
5.2.3(9)(f)	7	367
5.2.3(9)(g)	7	341
5.2.4(1)	3, 4	171-179, 223
5.2.4(2)	4, 6	218-223, 270-275
5.2.4(3)	2	152
5.2.4(4)	2, 6	152, 270-275
5.2.4(5)	7	367
5.3.1(1)	6	266-286
5.3.1(2)	6	266-286
5.3.1(3)	6	302-303
5.3.1(4)	2	161-162
5.3.2(1)	6	266-306
5.3.2(2)	6	302-303
5.3.3(1)(a)i	8	383-385, 392
5.3.3(1)(a)ii	8	385-386, 392
5.3.3(1)(a)iii	8	383-385, 392
5.3.3(1)(a)iv	8	389-390, 392
5.3.3(1)(a)v	8	386-389, 392
5.3.3(1)(a)vi	8	388, 392
5.3.3(1)(b)	8	392, 419-422
5.3.3(2)(a)	8	398-399
5.3.3(2)(b)i	8	401-404
5.3.3(2)(b)ii	8	400-401
5.3.3(2)(b)iii	8	397-399
5.3.4(1)	9	434
5.3.4(2)	9	434
5.3.4(3)	9	434-457
5.3.4(4)	9	444-445
5.3.4(5)	9	444-445
5.3.4(6)	9	444-445
5.4.1(1)	6	286-292
5.4.1(2)	6	286-292
5.4.1(3)(a)	6	298-301
5.4.1(3)(b)	6	301

Operations-Core Competencies (continued)

NFPA® 472 Competencies	Ch. Ref.	Page References
5.4.1(4)	9	466
5.4.1(5)(a)	5	246-247
5.4.1(5)(b)	5	246-247
5.4.1(6)	6	299
5.4.2	7	368-371
5.4.3(1)	1	14-15
5.4.3(2)	6	275-279
5.4.3(3)	5	235-255
5.4.3(4)(a)	5	245-247
5.4.3(4)(b)	5	253-255
5.4.3(5)	5	247
5.4.3(6)	6	292-293
5.4.3(7)	5	252
5.4.4(1)	6	296
5.4.4(2)	6	296
5.4.4(3)	6	295-298
5.4.4(4)	8	420-421
5.4.4(5)	8	381-421
5.4.4(6)	8	423
5.4.4(7)	8	423
5.5.1(1)	6	286
5.5.1(2)	6	282
5.5.2(1)	5	255-258
5.5.2(2)	6	297-298

Mission-Specific Competencies

NFPA® 472 Competencies	Ch. Ref.	Page References
6.2.3.1(1)	8	409
6.2.3.1(2)(a)	8	416
6.2.3.1(2)(b)	8	416
6.2.3.1(2)(c)	8	416
6.2.3.1(2)(d)	8	416
6.2.3.1(2)(e)	8	416
6.2.3.1(2)(f)	8	416
6.2.3.1(3)(a)i	8	405-406
6.2.3.1(3)(a)ii	8	405-406
6.2.3.1(3)(a)iii	8	405-406
6.2.3.1(3)(b)	8	401-404
6.2.3.1(3)(c)	8	419-422
6.2.3.1(3)(d)i	8	421
6.2.3.1(3)(d)ii	8	421
6.2.3.1(3)(d)iii	8	421
6.2.3.1(3)(d)iv	8	421
6.2.3.1(3)(e)	8	392, 419-421
6.2.3.1(3)(f)	9	445-451, 468
6.2.4.1(1)	8	422-423
6.2.4.1(2)	8	422-423
6.2.4.1(3)	8	425-428
6.2.4.1(4)	9	468
6.2.4.1(5)	8	423
6.2.5.1	8	423
6.3.3.1	9	439
6.3.3.2(1)	9	457
6.3.3.2(2)(a)	9	453
6.3.3.2(2)(b)	9	453
6.3.3.2(2)(c)	9	453
6.3.3.2(3)	9	451
6.3.3.2(4)	9	451-457
6.3.3.2(5)	9	451-453
6.3.4.1	9	451-453
6.3.4.2	9	470-471
6.3.5.1	9	457-458
6.3.6.1(1)	9	464
6.3.6.1(2)	9	464

NFPA® 472 Competencies	Ch. Ref.	Page References
6.3.6.1(3)	9	464
6.3.6.1(4)	9	464
6.4.3.1	9	439
6.4.3.2(1)	9	449-450
6.4.3.2(2)(a)	9	449
6.4.3.2(2)(b)	9	449
6.4.3.2(2)(c)	9	449
6.4.3.2(2)(d)	9	449
6.4.3.2(2)(e)	9	449
6.4.3.2(2)(f)	9	450
6.4.3.2(2)(g)	9	450
6.4.3.2(2)(h)	9	450
6.4.3.2(2)(i)	9	450
6.4.3.2(2)(j)	9	450
6.4.3.2(2)(k)	9	450
6.4.3.2(2)(l)	9	448, 450
6.4.3.2(3)	9	446
6.4.3.2(4)	9	458-461
6.4.3.2(5)	9	464
6.4.3.2(6)	9	461-462
6.4.4.1(1)	9	445-446
6.4.4.1(2)	9	445
6.4.4.2(1)	9	467-469
6.4.4.2(2)	9	467-469
6.4.5.1	9	457-458
6.4.6.1(1)	9	464
6.4.6.1(2)	9	464
6.4.6.1(3)	9	464
6.4.6.1(4)	9	464
6.5.2.1(1)(a)	13	594
6.5.2.1(1)(b)	13	594
6.5.2.1(1)(c)	13	594
6.5.2.1(1)(d)	13	594
6.5.2.1(1)(e)	13	594
6.5.2.1(2)(a)	13	596-597
6.5.2.1(2)(b)	13	596-597
6.5.2.1(2)(c)	13	596-597

NFPA® 472 Competencies	Ch. Ref.	Page References
6.5.2.1(2)(d)	13	596-597
6.5.2.1(2)(e)	13	596-597
6.5.3.1(1)(a)	13	603-604, 610
6.5.3.1(1)(b)	13	602-604
6.5.3.1(1)(c)	13	597
6.5.3.1(1)(d)	13	596-597
6.5.3.1(1)(e)	13	596
6.5.3.1(1)(f)	13	605
6.5.3.1(1)(g)	13	611-614
6.5.3.1(1)(h)	13	611
6.5.3.1(1)(i)	13	602-603
6.5.3.1(1)(j)	13	607-616
6.5.3.1(1)(k)	13	614-615
6.5.3.1(1)(l)	13	615-616
6.5.3.1(1)(m)	13	615
6.5.3.1(1)(n)	13	615-616
6.5.3.1(1)(o)	13	609-610
6.5.3.1(2)(a)	13	593-594
6.5.3.1(2)(b)	13	593, 611-615
6.5.3.1(2)(c)	13	612-614
6.5.3.1(2)(d)	13	614-615
6.5.3.1(3)(a)	13	593, 594
6.5.3.1(3)(b)	13	593, 611-615
6.5.3.1(3)(c)	13	612-614
6.5.3.1(3)(d)	13	614-615
6.5.3.1(4)(a)	13	593-594
6.5.3.1(4)(b)	13	593, 611-615
6.5.3.1(4)(c)	13	612-614
6.5.3.1(4)(d)	13	614-615
6.5.3.1(5)(a)	13	593-594
6.5.3.1(5)(b)	13	593, 611-616
6.5.3.1(5)(c)	13	612-614
6.5.3.1(5)(d)	13	614-615
6.5.3.1(6)(a)	13	593-594
6.5.3.1(6)(b)	13	593, 611-615
6.5.3.1(6)(c)	13	612-614
6.5.3.1(6)(d)	13	614-615

Mission-Specific Competencies (continued)

NFPA® 472 Competencies	Ch. Ref.	Page References	NFPA® 472 Competencies	Ch. Ref.	Page References
6.5.3.1(7)(a)	13	615	6.6.4.1(2)(b)	10	483-484
6.5.3.1(7)(b)	13	614	6.6.4.1(2)(c)	10	484
6.5.3.1(7)(c)	13	614	6.6.4.1(2)(d)	10	485
6.5.3.1(7)(d)	13	614	6.6.4.1(3)(a)	10	496
6.5.3.1(7)(e)	13	615	6.6.4.1(3)(b)	10	496
6.5.3.1(8)	13	614	6.6.4.1(3)(c)	10	497
6.5.3.1(9)	13	610-611	6.6.4.1(3)(d)	10	498
6.5.3.2	13	593	6.6.4.1(3)(e)	10	503
6.5.4.1(1)	13	603-604, 610	6.6.4.1(3)(f)	10	499
6.5.4.1(2)	13	602-604	6.6.4.1(3)(g)	10	500
6.5.4.1(3)	13	597	6.6.4.1(3)(h)	10	504-505
6.5.4.1(4)	13	596-597	6.6.4.1(3)(i)	10	502
6.5.4.1(5)	13	596	6.6.4.1(3)(j)	10	501
6.5.4.1(6)	13	605	6.6.4.1(4)	10	489-492
6.5.4.1(7)	13	618-619	6.6.4.1(5)	10	489-492
6.5.4.1(8)	13	618-619	6.6.4.2	10	477
6.5.4.1(9)	13	618-619	6.7.3.1	11	524-550
6.5.4.1(10)	13	618-619	6.7.3.2	11	518-550
6.5.4.1(11)	13	618-619	6.7.3.3	11	518-521
6.5.4.1(12)	13	618-619	6.7.3.4	11	520-521
6.5.4.1(13)	13	618-619	6.7.4.1	11	552-557
6.5.4.1(14)	13	618-619	6.7.4.2	9, 11	468, 523
6.5.4.2	13	593	6.8.3.1(1)	12	569-570
6.6.3.1(1)	10	476-505	6.8.3.1(2)(a)	12	574
6.6.3.1(2)(a)	10	480	6.8.3.1(2)(b)	12	575
6.6.3.1(2)(b)	10	480	6.8.3.1(2)(c)	12	575
6.6.3.1(2)(c)	10	481-482	6.8.3.1(2)(d)	12	575
6.6.3.1(2)(d)	10	481-482	6.8.3.1(2)(e)	12	579
6.6.3.1(2)(e)	10	488	6.8.3.1(3)	12	569-573
6.6.3.1(2)(f)	10	481-482	6.8.3.1(4)	12	565
6.6.3.1(2)(g)	10	489-492	6.8.3.2	8, 12	407-422, 569
6.6.3.1(2)(h)	10	481-482	6.8.4.1(1)	12	571-573
6.6.3.1(2)(i)	10	487	6.8.4.1(2)	12	576-577
6.6.3.1(2)(j)	10	482-487	6.8.4.1(3)	12	583-587
6.6.3.2	10	477	6.8.4.1(4)	12	581-582
6.6.4.1(1)	10	482-487, 494, 501	6.8.4.1(5)	12	572
6.6.4.1(2)(a)	10	483-484	6.9. 2.1(1)	14	624-635, 641-646

NFPA® 472 Competencies	Ch. Ref.	Page References
6.9. 2.1(2)	14	624-627, 635-636, 641-646
6.9. 2.1(3)	14	624-627, 639-646
6.9. 2.1(4)	14	624-626
6.9. 2.1(5)	14	624, 641-644
6.9.3.1	14	641-644
6.9.3. 2.1	14	641-644
6.9.3. 2.2(1)	14	644
6.9.3. 2.2(2)	14	643
6.9.3. 2.2(3)	14	643
6.9.3. 2.2(4)	14	641-644
6.9.3. 2.2(5)	14	642
6.9.3. 2.2(6)	14	642
6.9.3.3(1)(a)	14	641
6.9.3.3(1)(b)	14	641
6.9.3.3(1)(c)	14	641
6.9.3.4.1	14	641-646
6.9.3.4.2(1)	14	624-640
6.9.3.4.2(2)	14	645
6.9.3.4.2(3)	14	645-646
6.9.3.4.2(4)	14	643
6.9.3.4.2(5)	14	646
6.9.3.5	14	645
6.9.4.1	14	641-646
6.9.4.1.1(1)	3, 14	181-182, 642
6.9.4.1.1(2)	9, 14	462, 467-468, 645-646
6.9.4.1.1(3)	14	648-649
6.9.4.1.1(4)	14	648-649
6.9.4.1.2(1)	14	627-641
6.9.4.1.2(2)	14	627-641
6.9.4.1.2(3)	14	627-641
6.9.4.1.3	14	641-644
6.9.4.1.4	14	646
6.9.4.1.5	14	645-646

Appendix B
OSHA Plan States

OSHA State-Plan States and Non-state-Plan States

State-Plan States	Non-state-Plan States
Alaska	Alabama
Arizona	Arkansas
California	Colorado
Connecticut (state and local government employees only)	Delaware
Hawaii	District of Columbia
Indiana	Florida
Iowa	Georgia
Kentucky	Guam
Maryland	Idaho
Michigan	Illinois
Minnesota	Kansas
Nevada	Louisiana
New Mexico	Maine
New York (state and local government employees only)	Massachusetts
North Carolina	Mississippi
Oregon	Missouri
Puerto Rico	Montana
South Carolina	Nebraska
Tennessee	New Hampshire
Utah	New Jersey
Vermont	North Dakota
Virginia	Ohio
Virgin Islands	Oklahoma
Washington	Pennsylvania
Wyoming	Rhode Island
	South Dakota
	Texas
	West Virginia
	Wisconsin

Appendix C
Sample Standard Operating Guideline

TUALATIN VALLEY FIRE AND RESCUE
INCIDENT COMMAND MANUAL

SERIES 300X

OPERATIONAL GUIDELINE
HAZARDOUS MATERIALS RESPONSE

PURPOSE

To provide a standard by which companies trained to the "First Responder Operations" level respond to hazardous materials incidents.

DEFINITIONS

First Responder - Operations - A level of training for first responders to hazardous materials incidents, required by federal and state law; as defined in Oregon Administrative Rule (OAR) 437-01-100(q).

Full Protective Clothing - As it relates to hazardous materials response, full protective clothing means turnouts and SCBA.

On-Scene Commander - A level of training for Incident Commanders on hazardous materials incidents, required by federal and state law; as defined by OAR 437-01-100(q).

Responsible Party - Federal and state regulators assign responsibility for incident clean-up (and costs) to the party who is responsible for the hazardous materials incident (i.e., a fixed facility, transportation agent, etc.).

HMRT - Hazardous Materials Response Team.

Hazardous Materials Group Supervisor (HMRT Leader) - HazMat Group Supervisor reports to the Incident Commander (or Operations Section Chief, if staffed) and is responsible for hazardous materials tactical operations. The HazMat Group Supervisor position is staffed by the Hazardous Materials Response Team Leader.

* Emergency Response Guidebook - North American Emergency Response Guidebook; formerly "DOT Emergency Response Guidebook".

PROCEDURES

I. **TRAINING REQUIREMENTS**

    A. All response personnel must meet the training requirements for "First Responder – Operations" level.

    B. Incident Commanders on hazardous materials incidents must meet the training requirements for "On-Scene Commander".

II. INCIDENT COMMANDER

A. All incidents involving hazardous materials in a spill, release or fire, may require an Incident Commander trained to the "On-Scene Commander" level. All Battalion Chiefs and ICs on the Overhead Team are trained and required to maintain qualifications to the "On-Scene Commander" level.

B. The Incident Commander may call for a full or partial HMRT response if incident mitigation is beyond the training and capabilities of a company response. The IC may also call for technical assistance from the HMRT without a response to the incident site, if the situation warrants.

III. COMPANY FUNCTIONS

A. Companies will respond for the purpose of protecting nearby persons, property or the environment from a hazardous materials release.

B. Companies will respond in a *defensive* fashion without coming in contact with the release or taking actions to stop a release that would place them in danger of contact.

C. The primary function of the Operations level responder is to contain the release from a safe distance, keeping it from spreading and protect exposures. The basic functions are:
- isolate the hazard area and control access
- hazard and risk assessment
- basic control, containment and/or confinement procedures appropriate to the level of training and personal protective clothing and equipment.

D. Companies will not take any actions on hazardous materials incidents that cannot be safely performed in full protective clothing.

IV. HAZARDOUS MATERIALS RESPONSE AND OPERATIONS

A. While enroute to the scene:

1. Contact Fire Comm and obtain available information regarding:
   a. The nature of the incident, e.g., fixed facility, transportation related. etc.
   b. The type of product(s) involved, if known.
   c. The best direction for approaching the scene from upwind, upgrade and upstream.
   d. Who is on-scene that may have information on the nature of the incident.
   e. Any information on the incident conditions that may be known and can be provided while enroute to the incident scene.

* 2. The HazMat Team may be contacted via Fire Comm for technical assistance or response, as appropriate.

3. Approach the incident scene with caution.
   a. Approach the incident scene from upwind, upgrade, upstream or at a right angle to the wind direction and/or gradient.
   b. Consider escape routes. Be aware of situations that require entering areas with egress restrictions, such as fenced compounds.
   c. Position vehicle/apparatus headed away from the incident scene at a safe distance.

B. On Arrival

1. Establish Command and give size-up.

2. Establish a Unified Command if multiple agencies/jurisdictions are involved.

3. Ensure a qualified "On-Scene Commander" (i.e., Battalion Chief) is enroute to the scene.

4. Continuously evaluate need for HazMat Team technical assistance or response.

C. Establish Safe Zone and Control Access

1. Determine the hazard area and establish the Hot Zone, Warm Zone and Cold Zone boundaries.
   a. Based on initial observations, identify a safe distance for initial incident isolation to begin. Some recommendations include:
      • Single drum, not leaking - minimum 150' in all directions
      • Single drum, leaking - 500' in all directions
      • Tank car or tank truck with BLEVE potential - half mile in all directions
   b. Isolate and deny entry to:
      • The general public
      • Anyone not in proper protective clothing and equipment
      • Anyone without a specific assignment

2. Establish the Command Post in the Cold Zone.

3. Identify and establish the Staging Area location in the Cold Zone.

4. Communicate the Zone information, Command Post and Staging Area locations to Fire Comm and incoming units.

5. Determine a safe approach for incoming units and direct them to locations at the Safe Zone Perimeter that will facilitate isolation of the incident, i.e., intersections to block and re-direct traffic, etc. All others should be directed to the Staging Area until assigned.

6. Request police assistance as needed to:
   a. Handle Cold Zone Perimeter control to relieve fire units for incident mitigation.
   b. Handle public evacuations.
   c. Handle public notification for sheltering in place

7. While isolating the incident scene:
   a. Treat all vapor clouds as being toxic and handle accordingly.
   b. Do not walk into, through or touch any spilled materials.
   c. Observe local on-site weather and wind conditions and adjust accordingly.
   d. Position at a safe distance and utilize your binoculars!

* D. Attempt to Identify the Product.

If the product is *known*, proceed to Section V and isolate in accordance with appropriate Emergency Response Guidebook recommendations. Record observations on the hazmat incident worksheet. (Provide the diagram to the incoming Battalion Chief or HazMat Response Team.)

If the product is *unknown*, from a safe distance attempt to gather as much information as possible.

Use Emergency Response Guide #111 isolation recommendations until the material is identified. Record observations on the hazmat incident worksheet.

1. Life Safety is the number one priority. Do not rush into the scene to effect a rescue without first identifying the hazards.

2. Attempt to identify outward warning signs that are indicators of the presence of hazardous materials. These include:
   a. Individuals that have collapsed or are vomiting inside the hazardous area (HMRT response).
   b. Any evidence of fire, as indicated by smoke, greatly increases all hazards.
   c. A loud roar of increasing pitch from a container's operating relief valve (HMRT response).
   d. Evidence of a leak, indicated by a hissing sound.
   e. Birds and insects falling out of the sky (HMRT response).

AND/OR

3. Attempt to identify the material(s) involved by using:
   a. Placards/labels
   b. Container markings
   c. Driver/operator provided information including shipping papers.

4. After determining product:
   a. Perform rescue, if needed, using safety guidelines related to that product.
   b. Re-evaluate distances for isolated area.

5. Communicate your observations to Fire Comm.

6. Anticipate shifting winds when establishing perimeters; consult with the weather service to obtain accurate forecasts of changes that might impact your incident scene and perimeters.

7. Eliminate ignition sources if flammable materials are involved. Remember that non-flammable materials, such as anhydrous ammonia are, in fact, flammable, so always identify if the product has a flammable range.

8. Request additional fire, law enforcement and public works resources, as needed, to secure the incident scene and maintain perimeter control.

* 9. If large dikes and dams need to be built to control spill, consider requisition for heavy equipment and/or assistance of public works resources.

E. Conduct a Risk/Benefit Analysis which includes asking the following questions in relation to the incident you are addressing:

1. What would the outcome be if we did absolutely nothing and allowed the incident to go through natural stabilization?

2. Once you have identified the outcomes of natural stabilization, the next question you should ask is "Can we change the outcomes of natural stabilization?"

3. If the answer to this question is "NO", then isolate the hazard area, deny entry, and protect exposures such as people, the environment and adjacent property/ equipment.

4. If the answer to this question is "YES", then the next question to ask is "What is the cost of my intervention?"

IF THE INCIDENT COMMANDER DETERMINES DEFENSIVE OPERATIONS CAN STABILIZE/CONTAIN THE INCIDENT *AND* IT CAN BE DONE IN FULL PROTECTIVE CLOTHING (TURNOUTS AND SCBA), THE IC SHALL CONDUCT OPERATIONS IN ACCORDANCE WITH THE "DEFENSIVE OPERATIONAL GUIDELINES".

V. DEFENSIVE OPERATIONAL GUIDELINES

A. Attempt to stop/slow/control leak using defensive techniques (such as turning off a valve, etc.).

B. If the leak cannot be stopped, utilize an appropriate containment procedure to prevent the material from flowing and increasing the exposed surface area (i.e., using dirt or absorbent).

VI. DECONTAMINATION: Perform field decontamination as directed by the Incident Commander and/or HazMat Response Team.

NOTE: *ALL CONTAMINATED PATIENTS MUST BE DECONTAMINATED OR PACKAGED FOR TRANSPORT IN A WAY TO PREVENT CONTAMINATION OF TRANSPORT UNITS AND HOSPITALS.*

VII. CLEAN-UP

A. If the incident is on a roadway or public access area, the Incident Commander must ensure that a public safety agency (coordinate with law enforcement officials, if available) remains on-scene to continue isolation procedures and standby until the clean-up company arrives.

B. If a responsible party is not on-scene and making arrangements for clean-up and disposal, contact the on-duty HMRT Team Leader for further instructions.

NOTE: *FIRE DEPARTMENT PERSONNEL SHALL NOT ENGAGE IN CLEAN-UP OPERATIONS. THE APPROPRIATE ROLE IS CONTAINMENT/ STABILIZATION. DO NOT TAKE HAZARDOUS MATERIALS FROM AN INCIDENT TO ANY FIRE DISTRICT FACILITY.*

VIII. CONDUCT TERMINATION PROCEDURES

A. Prior to the demobilization and release of any equipment from the scene, conduct a debriefing of all response personnel (including cooperating agencies).

B. An effective debriefing should:
1. Inform *all responders* exactly what hazardous materials were involved and the accompanying signs and symptoms of exposure.
2. Provide information for personal exposure records.
3. Identify equipment damage and unsafe conditions requiring immediate attention or isolation for further evaluation.
4. Conduct a post-incident analysis and critique. This may be done at the station.

HAZARDOUS MATERIALS RESPONSE
CHECKLIST

CHECKLIST USE

The checklist should be considered as a minimum requirement for this position. Users of this manual should feel free to augment this list as necessary. Note that some activities are one-time actions and others are on-going or repetitive for the duration of an incident.

_____ While enroute to the scene, you may utilize the HazMat Team as a technical resource (contact via Fire Comm).

_____ Approach incident cautiously, uphill, upwind, park headed away from incident, consider escape routes.

_____ Establish Command.

_____ Establish and maintain site access control. Establish initial Zones (Hot: min. of 150'; warm; cold). Establish Command Post and Staging locations.

_____ Attempt to identify materials involved by using placards/labels, container markings, shipping papers and driver provided information. Use Guide #111 if spilled product is unknown.

_____ Perform rescue only when the rescue operation can be done safely.

_____ Request additional fire, law enforcement and public works resources as needed. Consider requisition for heavy equipment.

_____ Conduct risk/benefit analysis.

_____ If the Incident Commander determines defensive operations can stabilize/contain the incident, conduct defensive operations.
  • Attempt to stop/slow/control leak using defensive techniques (such as turning off a valve, etc.).
  • If the leak cannot be stopped, utilize an appropriate containment procedure to prevent the material from flowing and increasing the exposed surface area.

_____ Perform field decontamination as directed by the Incident Commander.

_____ Clean-Up

   If the incident is on a roadway or public access area, the Incident Commander must ensure that a public safety agency remains on-scene to continue isolation procedures and standby until the clean-up company arrives. If the responsible party is not on-scene and making arrangements for clean-up and disposal, contact the on-duty HMRT Team Leader for further instructions.

   NOTE: Fire Department personnel shall not engage in clean-up operations.

_____ Prior to the demobilization and release of any equipment from the scene, conduct a debriefing of all response personnel (including cooperating agencies).

Appendix D
UN Class Placards and Labels

Table D.1 provides the United Nations (UN) placards and labels required for the transportation of dangerous goods.

Table D.1 UN Class Placards and Labels	
	Class 1: Explosive substances or articles
	Class 2: Gases
	Class 3: Flammable Liquids
	Class 4: Flammable solids; substances liable to spontaneous combustion; substances, which, in contact with water, emit flammable gases
	Class 5: Oxidizing substances and organic peroxides
	Class 6: Toxic and infectious substances

Continued

| | Class 7:
Radioactive material |
|---|---|
| RADIOACTIVE I / RADIOACTIVE II / RADIOACTIVE III / FISSILE / RADIOACTIVE | |
| | **Class 8:**
Corrosive substances |
| | **Class 9:**
Miscellaneous dangerous substances and articles |

Appendix E
Hazardous Materials Incident Commander Checklist

❑ Establish Command
- *Consider Unified Command if appropriate*
- *Determine the need to staff Public Information Officer (PIO) position*

❑ Ensure appropriate notifications have been made
- *Haz mat response team to notify appropriate state and federal agencies*
- *Ensure notification to local affected agencies (city, public works, police, Cleanwater, etc.)*
- *Contact with "responsible party" if available*

❑ Ensure a site access control plan has been established
- *Limit and control site access*
- *Establish hot zone, warm zone, cold zone*

❑ Ensure use of appropriate personal protective equipment (PPE)
- *Approve use of appropriate PPE for ALL responders (including haz mat response team, fire, law enforcement, public works personnel)*

❑ Develop Action Plan (written is optimal)
- *Offensive or defensive operation*
- *Appropriate Operational/Technician Level actions*
- *Appropriate protective actions (evacuation or shelter in place)*
- *Appropriate decontamination procedures*
- *Coordination of all on-scene response agencies*
- *If fixed facility, coordinate with facility representative and/or emergency response team*

❑ Safety
- *Provide safety briefing for responders*
- *Ensure Safety Officer is trained to Technician Level*
- *Ensure medical surveillance of personnel in the hot zone*

❑ Ensure appropriate incident termination procedures are carried out
- *Provide ALL responders with information on signs/symptoms of exposure*
- *Provide information for personal exposure records*
- *Determine if any equipment/apparatus exposure occurred; identify follow-up actions if needed*
- *Ensure postincident analysis takes place*
- *Determine need for critical incident stress debriefing*
- *Identify any transition issues before transfer of command*
- *Transfer command to appropriate agency or company (get contact name/number)*

Problem	Strategies	Tactics
Access: Access problems may be related to gaining access or denying access (to civilians or unprotected responders). Generally the first problem presented is limiting access to civilians and unprotected responders.	Isolate and deny entry	• Establish control zones (Hot and Cold) • Control traffic
Container Under Stress: The two types of container stress that responders can readily affect are generally thermal stress (heating) and mechanical stress (due to overpressure).	Ignore	Protect exposures (protective actions only)
	Cool	• Use master stream • Use hoseline
	Extinguish fire	• Remove fuel • Use master stream • Use hoseline • Use foam master stream • Use foam hoseline
	Release pressure	• Transfer product • Release product to atmosphere • Vent and burn
Container Breach/Release: Active strategies to manage a breach/release generally require operations inside the hazard area (Hot Zone).	Ignore	Protect exposures (protective actions only)
	Contain	• Close valve(s) • Tighten attachments • Plug • Patch • Transfer product • Decontaminate (required for entry)

continued

Problem	Strategies	Tactics
Dispersion: Active strategies to control dispersion may be either offensive or defensive (depending on where they are performed). Dispersion control strategies are driven by the form of the material that has been (or is being) released.	Ignore	Protect exposures (protective actions only)
	Confine: Solid	Cover
	Confine: Liquid	• Adsorb or absorb • Dike (Circle or *V*-shape) • Divert • Retain • Dam (underflow or overflow) • Suppress vapor (foam)
	Confine: Energy	Shield
	Disperse: Gas	Disperse vapor (water fog or blower)
Fire: The fire problem includes a direct threat to life safety and exposures, potential to affect container integrity, and release of toxic products of combustion. However, in some cases (pesticides), fire may present less threat than fire-control operations.	Ignore	Protect exposures (protective actions only)
	Extinguish	• Use master stream • Use hoseline • Use foam master stream • Use foam hoseline • Use dry chemical • Use specialized extinguishing agent
Possible Victims: Possible victims may be reported (definitely a known imminent life threat) or inferred based on incident conditions. Victims removed from the hazard area (Hot Zone) may require decontamination.	Determine	Ask
	Notify	• Use public address system • Use telephone
	Locate	• Perform primary search/extraction • Perform decontamination • Perform secondary search
Visible/Known Victims: Victims may be visible or known to be inside the hazard area. These victims may (or may not) be able to rescue themselves. First responders must use care in assessing their capability to affect a rescue (due to limitations in personal protective equipment and training. Victims removed from the hazard area (Hot Zone) may require decontamination.	Rescue	• Rescue themselves • Move to safe refuge • Perform extraction • Perform decontamination

continued

Problem	Strategies	Tactics
Potential Life Exposure: Potential victims may become exposed due to dispersion (downhill or downwind). Responders must consider dispersion, time, and incident conditions in evaluating potential life exposure.	Protect in place	• Notify face to face • Notify by telephone • Notify media
	Evacuate	• Notify face to face • Notify by telephone • Notify media • Shelter • Control traffic • Perform security
Environmental/Property Exposure: Active strategies to minimize environmental/property damage are generally offensive in nature.	Ignore	Self-mitigate
	Control chemical	• Dilute • Neutralize
	Cool	• Use master stream • Use hoseline • Use foam master stream • Use foam hoseline

Appendix G
Emergency Plans

Emergency Plans			
Plan/Predetermined Procedures or Guidelines*	**Developed**	**Authority**	**Who Needs to Know?**
SOPs/SOGs/OIs	By the agency or department before an incident	Agency or Department	First Responders at Awareness and Operational Levels
Pre-Incident Survey/ Plan	By the surveying/planning agency or department before an incident	None	First Responders at Awareness and Operational Levels
Local Emergency Response Plan	By the Local Emergency Planning Committee (LEPC) before an incident	U.S. Environmental Protection Agency (EPA)	First Responders at Awareness and Operational Levels
Incident Action Plan	By the Incident Commander (IC) at the scene of an incident	Incident Management System (IMS)	First Responders at Awareness and Operational Levels
Site Safety and Health Plan	Usually by the Site Safety and Health Supervisor at the scene of hazardous waste sites	Title 29 *CFR* 1910.120, *Hazardous Waste Operations and Emergency Response (HAZWOPER)*	Everyone working at the hazardous waste site
Emergency Response Plan	Part of the Site Safety and Health Plan at the scene of hazardous waste sites	Title 29 *CFR* 1910.120, *Hazardous Waste Operations and Emergency Response (HAZWOPER)*	Everyone working at the hazardous waste site
Site Safety and Control Plan	Part of the Emergency Response Plan (which is part of the Site Safety and Health Plan) at the scene of hazardous waste sites	Title 29 *CFR* 1910.120, *Hazardous Waste Operations and Emergency Response (HAZWOPER)*	Everyone working at the hazardous waste site
Emergency Response Plan	By employers whose employees are engaged in emergency response no matter where it occurs	Title 29 *CFR* 1910.120, *Hazardous Waste Operations and Emergency Response (HAZWOPER)*	First Responders at Awareness and Operational Levels
Emergency Action Plan	By the facility employer before an emergency	U.S. Occupational Safety and Health Administration (OSHA)	All employees at the facility

continued

Emergency Plans (concluded)

Plan/Predetermined Procedures or Guidelines*	Developed	Authority	Who Needs to Know?
Integrated Contingency Plan	By the facility employer before an emergency	Designed to meet the requirements of the required plans of EPA, Department of Transportation (DOT), and OSHA	All employees at the facility

* This table is just a sample of plans that first responders may encounter. There may be other plans on a national, state/province, or local level with which first responders may need to be familiar.

Appendix H

U.S. Bureau of Alcohol, Tobacco, Firearms and Explosives: List of Explosive Materials

A

Acetylides of heavy metals

Aluminum containing polymeric propellant

Aluminum ophorite explosive

Amatex

Amatol

Ammonal

Ammonium nitrate explosive mixtures (cap sensitive)

Ammonium nitrate explosive mixtures (non-cap sensitive)

Ammonium perchlorate having particle size less than 15 microns

Ammonium perchlorate composite propellant

Ammonium perchlorate explosive mixtures

Ammonium picrate [picrate of ammonia, Explosive D]

Ammonium salt lattice with isomorphously substituted inorganic salts

ANFO [ammonium nitrate-fuel oil]

Aromatic nitro-compound explosive mixtures

Azide explosives

B

Baranol

Baratol

BEAF [1, 2-bis (2, 2-difluoro-2-nitroacetoxyethane)]

Black powder

Black powder based explosive mixtures

Blasting agents, nitro-carbo-nitrates, including non-cap sensitive slurry and water gel explosives

Blasting caps

Blasting gelatin

Blasting powder

BTNEC [bis (trinitroethyl) carbonate]

BTNEN [bis (trinitroethyl) nitramine]

BTTN [1,2,4 butanetriol trinitrate]

Bulk salutes

Butyl tetryl

C

Calcium nitrate explosive mixture

Cellulose hexanitrate explosive mixture

Chlorate explosive mixtures

Composition A and variations

Composition B and variations

Composition C and variations

Copper acetylide

Cyanuric triazide

Cyclonite [RDX]

Cyclotetramethylenetetranitramine [HMX]

Cyclotol

Cyclotrimethylenetrinitramine [RDX]

D

DATB [diaminotrinitrobenzene]

DDNP [diazodinitrophenol]

DEGDN [diethyleneglycol dinitrate]

Detonating cord

Detonators

Dimethylol dimethyl methane dinitrate composition

Dinitroethyleneurea

Dinitroglycerine [glycerol dinitrate]

Dinitrophenol

Dinitrophenolates

Dinitrophenyl hydrazine

Dinitroresorcinol

Dinitrotoluene-sodium nitrate explosive mixtures

DIPAM [dipicramide; diaminohexanitrobiphenyl]

Dipicryl sulfone

Dipicrylamine

Display fireworks

DNPA [2,2-dinitropropyl acrylate]

DNPD [dinitropentano nitrile]

Dynamite

E

EDDN [ethylene diamine dinitrate]

EDNA [ethylenedinitramine]

Ednatol

EDNP [ethyl 4,4-dinitropentanoate]

EGDN [ethylene glycol dinitrate]

Erythritol tetranitrate explosives

Esters of nitro-substituted alcohols

Ethyl-tetryl

Explosive conitrates
Explosive gelatins
Explosive liquids
Explosive mixtures containing oxygenreleasing
inorganic salts and hydrocarbons
Explosive mixtures containing oxygenreleasing
inorganic salts and nitro bodies
Explosive mixtures containing oxygenreleasing
inorganic salts and water insoluble fuels
Explosive mixtures containing oxygenreleasing
inorganic salts and water soluble fuels
Explosive mixtures containing sensitized
nitromethane
Explosive mixtures containing tetranitromethane
(nitroform)
Explosive nitro compounds of aromatic
hydrocarbons
Explosive organic nitrate mixtures
Explosive powders

F
Flash powder
Fulminate of mercury
Fulminate of silver
Fulminating gold
Fulminating mercury
Fulminating platinum
Fulminating silver

G
Gelatinized nitrocellulose
Gem-dinitro aliphatic explosive mixtures
Guanyl nitrosamino guanyl tetrazene
Guanyl nitrosamino guanylidene hydrazine
Guncotton

H
Heavy metal azides
Hexanite
Hexanitrodiphenylamine
Hexanitrostilbene
Hexogen [RDX]
Hexogene or octogene and a nitrated
Nmethylaniline
Hexolites
HMTD [hexamethylenetriperoxidediamine]
HMX [cyclo-1,3,5,7-tetramethylene 2,4,6,8-
tetranitramine; Octogen]
Hydrazinium nitrate/hydrazine/ aluminum
explosive system

Hydrazoic acid

I
Igniter cord
Igniters
Initiating tube systems

K
KDNBF [potassium dinitrobenzofuroxane]

L
Lead azide
Lead mannite
Lead mononitroresorcinate
Lead picrate
Lead salts, explosive
Lead styphnate [styphnate of lead, lead
trinitroresorcinate]
Liquid nitrated polyol and trimethylolethane
Liquid oxygen explosives

M
Magnesium ophorite explosives
Mannitol hexanitrate
MDNP [methyl 4,4-dinitropentanoate]
MEAN [monoethanolamine nitrate]
Mercuric fulminate
Mercury oxalate
Mercury tartrate
Metriol trinitrate
Minol-2 [40% TNT, 40% ammonium nitrate, 20%
aluminum]
MMAN [monomethylamine nitrate];
methylamine nitrate
Mononitrotoluene-nitroglycerin mixture
Monopropellants

N
NIBTN [nitroisobutametriol trinitrate]
Nitrate explosive mixtures
Nitrate sensitized with gelled nitroparaffin
Nitrated carbohydrate explosive
Nitrated glucoside explosive
Nitrated polyhydric alcohol explosives
Nitric acid and a nitro aromatic compound
explosive
Nitric acid and carboxylic fuel explosive
Nitric acid explosive mixtures
Nitro aromatic explosive mixtures
Nitro compounds of furane explosive mixtures

Nitrocellulose explosive
Nitroderivative of urea explosive mixture
Nitrogelatin explosive
Nitrogen trichloride
Nitrogen tri-iodide
Nitroglycerine [NG, RNG, nitro, glyceryl trinitrate, trinitroglycerine]
Nitroglycide
Nitroglycol [ethylene glycol dinitrate, EGDN]
Nitroguanidine explosives
Nitronium perchlorate propellant mixtures
Nitroparaffins Explosive Grade and ammonium nitrate mixtures
Nitrostarch
Nitro-substituted carboxylic acids
Nitrourea

O

Octogen [HMX]
Octol [75 percent HMX, 25 percent TNT]
Organic amine nitrates
Organic nitramines

P

PBX [plastic bonded explosives]
Pellet powder
Penthrinite composition
Pentolite
Perchlorate explosive mixtures
Peroxide based explosive mixtures
PETN [nitropentaerythrite, pentaerythrite tetranitrate, pentaerythritol tetranitrate]
Picramic acid and its salts
Picramide
Picrate explosives
Picrate of potassium explosive mixtures
Picratol
Picric acid (manufactured as an explosive)
Picryl chloride
Picryl fluoride
PLX [95% nitromethane, 5% ethylenediamine]
Polynitro aliphatic compounds
Polyolpolynitrate-nitrocellulose explosive gels
Potassium chlorate and lead sulfocyanate explosive
Potassium nitrate explosive mixtures
Potassium nitroaminotetrazole
Pyrotechnic compositions
PYX [2,6-bis(picrylamino)] 3,5- dinitropyridine

R

RDX [cyclonite, hexogen, T4, cyclo-1,3,5,-trimethylene-2,4,6,-trinitramine; hexahydro-1,3,5-trinitro-S-triazine]

S

Safety fuse
Salts of organic amino sulfonic acid explosive mixture
Salutes (bulk)
Silver acetylide
Silver azide
Silver fulminate
Silver oxalate explosive mixtures
Silver styphnate
Silver tartrate explosive mixtures
Silver tetrazene
Slurried explosive mixtures of water, inorganic oxidizing salt, gelling agent, fuel, and sensitizer (cap sensitive)
Smokeless powder
Sodatol
Sodium amatol
Sodium azide explosive mixture
Sodium dinitro-ortho-cresolate
Sodium nitrate explosive mixtures
Sodium nitrate-potassium nitrate explosive mixture
Sodium picramate
Special fireworks
Squibs
Styphnic acid explosives

T

Tacot [tetranitro-2,3,5,6-dibenzo- 1,3a,4,6a tetrazapentalene]
TATB [triaminotrinitrobenzene]
TATP [triacetonetriperoxide]
TEGDN [triethylene glycol dinitrate]
Tetranitrocarbazole
Tetrazene [tetracene, tetrazine, 1(5-tetrazolyl)-4-guanyl tetrazene hydrate]
Tetrazole explosives
Tetryl [2,4,6 tetranitro-N-methylaniline]
Tetrytol
Thickened inorganic oxidizer salt slurried explosive mixture
TMETN [trimethylolethane trinitrate]
TNEF [trinitroethyl formal]
TNEOC [trinitroethylorthocarbonate]
TNEOF [trinitroethylorthoformate]

TNT [trinitrotoluene, trotyl, trilite, triton]
Torpex
Tridite
Trimethylol ethyl methane trinitrate composition
Trimethylolthane trinitratenitrocellulose
Trimonite
Trinitroanisole
Trinitrobenzene
Trinitrobenzoic acid
Trinitrocresol
Trinitro-meta-cresol
Trinitronaphthalene
Trinitrophenetol
Trinitrophloroglucinol
Trinitroresorcinol
Tritonal

U
Urea nitrate

W
Water-bearing explosives having salts of oxidizing
  acids and nitrogen bases, sulfates, or
  sulfamates (cap sensitive)
Water-in-oil emulsion explosive compositions

X
Xanthamonas hydrophilic colloid explosive
  mixture

Approved: March 19, 2004

Appendix I
Emergency Response Involving Ethanol and Gasoline Fuel Mixtures

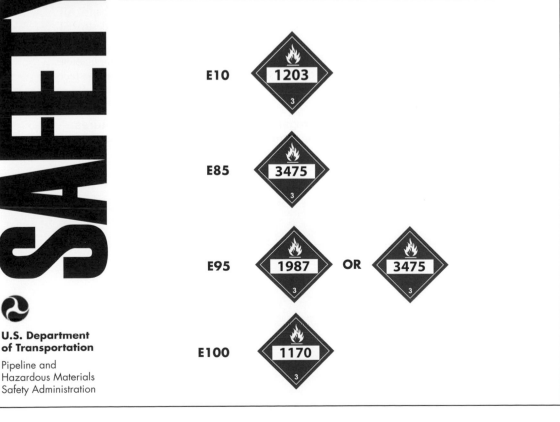

September 19, 2008

SAFETY NEWS

Advisory Guidance:
Emergency Response Involving Ethanol and Gasoline Fuel Mixtures

The Pipeline and Hazardous Materials Safety Administration (PHMSA) is *alerting emergency responders* to new and revised proper shipping names and identification numbers (ID) that may be used on shipping papers for fuel mixtures composed of ethanol (or "ethyl alcohol") and gasoline in various concentrations. The proper shipping names and IDs are added to the ERG2008.

The following chart is provided as guidance in identifying proper shipping names and identification numbers for Ethanol, Gasoline, and gasoline/ethanol fuel blends. Voluntary compliance began January 28, 2008.

Proper Shipping Name and ID	Ethanol Concentrations
Gasohol, NA 1203	E1 thru E10
Gasoline, UN 1203	E1 thru E10
Ethanol and gasoline mixture, UN 3475	E11 thru E99
Denatured alcohol, NA 1987	E95 thru E99
Alcohols, n.o.s, UN 1987	E95 thru E99
Ethanol or Ethyl alcohol, UN 1170	E100

E10 — 1203

E85 — 3475

E95 — 1987 OR 3475

E100 — 1170

U.S. Department of Transportation

Pipeline and Hazardous Materials Safety Administration

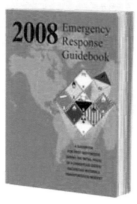

Fires involving ethanol/gasoline mixtures containing more than 10% ethanol, such as E85, should be treated differently than traditional gasoline fires because these mixtures are polar/water-miscible flammable liquids (they mix readily with water) and degrade the effectiveness of non alcohol-resistant fire-fighting foam.

For this reason, PHMSA recommends First Responders refer to Guide 127 of the 2008 Emergency Response Guidebook (ER2008) when responding to incidents involving fuel mixtures known to contain or potentially containing more than 10% alcohol. Guide 127 specifies the use of alcohol-resistant foam.

The International Association of Fire Chiefs (IAFC) recommends the use of alcohol-resistant, aqueous film-forming foam (AR-AFFF) or alcohol-resistant film-forming fluoroprotein foam (AR-FFFP) in application rates* as follows:

Fuel Mixture	Foam Application Rate*
Gas	0.1
Ethanol Spill (not in depth)	0.2
Ethanol Fire (in depth)	0.3
*Application rate expressed in gallons per minute of unexpanded foam solution flow divided by the fire area. (gpm/ft²)	

Gasohol, E10, fires may be extinguished using conventional aqueous film-forming foam (AFFF) or AR-AFFF but increased application rates may be necessary especially for prolonged burn back resistance.

Denatured ethyl alcohol fires, E95, can only be extinguished with AR type foams. All other type of foams or water additives are ineffective as the foam blanket is destroyed when it strikes the fuel surface.

AR type foams must be applied to ethyl alcohol fires using type II gentle application techniques. Direct application to the fuel surface will likely be ineffective unless fuel depth is very shallow.

For additional information regarding ethanol and gasoline mixtures or the HMR contact: Hazmat Information Center:1-800-467-4922 or E-mail: infocntr@dot.gov

Glossary

Glossary

A

Absorbent — Inert material or substance having no active properties that allow another substance to penetrate into the interior of its structure; can be used to pick up a liquid contaminant. An absorbent is commonly used in the abatement of hazardous materials spills. Examples are soil, diatomaceous earth, vermiculite, sand, and other commercially available products. *Also see* Contaminant.

Absorption — (1) Penetration of one substance into the structure of another such as the process of picking up a liquid contaminant with an absorbent. *Also see* Contaminant and Absorbent. (2) Passage of materials (such as toxins) through some body surface into body fluids and tissue. *Also see* Routes of Entry and Toxin.

Acid — Compound containing hydrogen that reacts with water to produce hydrogen ions; a proton donor; a liquid compound with a pH less than 7. Acidic chemicals are corrosive. *Also see* Base and pH.

Activation Energy — Amount of energy that must be added to an atomic or molecular system to begin a reaction.

Acute — (1) Characterized by sharpness or severity; having rapid onset and a relatively short duration. (2) Single exposure (dose) or several repeated exposures to a substance within a short time period. *Also see* Chronic

Acute Exposure Guideline Levels — Airborne concentration of a substance at or above which it is predicted that the general population, including "susceptible" but excluding "hypersusceptible" individuals, could experience notable discomfort; established by the U.S. Environmental Protection Agency (EPA).

Acute Health Effects — Health effects that occur or develop rapidly after exposure to a substance. *Also see* Chronic Health Effects.

Aerosol — Form of mist characterized by highly respirable, minute liquid particles.

Aerosolize — To produce a fine mist or spray characterized by highly respirable, minute particles.

AFFF — Abbreviation for aqueous film forming foam.

Affidavit — A written statement of facts confirmed by the oath of the party making it, before a notary or officer having authority to administer oaths.

Agar — Gelatinous or jelly-like substance used to grow bacterial cultures.

Agency for Toxic Substances and Disease Registry (ATSDR) — Lead U.S. public health agency responsible for implementing the health-related provisions of the Comprehensive Environmental Response, Compensation and Liability Act (CERCLA), and charged with assessing health hazards at specific hazardous waste sites, helping to prevent or reduce exposure and the illnesses that result, and increasing knowledge and understanding of the health effects that may result from exposure to hazardous substances.

Agricultural Terrorism — *See* Agroterrorism.

Agroterrorism — A terrorist attack directed against agriculture, for example, food supplies or livestock. Also called *Agricultural Terrorism.*

Air Bill — Shipping document prepared from a bill of lading that accompanies each piece or each lot of air cargo. *Also see* Bill of Lading and Shipping Papers.

Airline Respirator — *See* Supplied-Air Respirator (SAR).

Air-Purifying Respirator (APR) — Respirator with an air-purifying filter, cartridge, or canister that removes specific air contaminates by passing ambient air through the air-purifying element; may have a full or partial facepiece.

Air-Reactive Material — Substance that ignites when exposed to air at normal temperatures. Also called *pyrophoric. Also see* Reactive Material, Reactivity, and Water-Reactive Material.

Alcohol-Resistant AFFF Concentrate (AR-AFFF) — Aqueous film forming foam that is designed for use with polar solvent fuels. *Also see* Aqueous Film Forming Foam and Foam Concentrate.

Alkali — Strong base. *Also see* Base, Acid, Caustic, and pH.

Allergen — Material that can cause an allergic reaction of the skin or respiratory system.

All-Hazard Concept — Provides a coordinated approach to a wide variety of incidents; all responders use a similar, coordinated approach with a common set of authorities, protections, and resources.

Allied Professional — Individual with the training and expertise to provide competent assistance and direction at haz mat and WMD incidents.

Alpha Particle — Energetic, positively charged particles (helium nuclei) emitted from the nucleus during radioactive decay that rapidly lose energy when passing through matter. *Also see* Alpha Radiation, Beta Particle, and Gamma Ray.

Alpha Radiation — Consists of particles having a large mass and positive electrical charge; least penetrating of the three common forms of radiation. It is normally not considered dangerous to plants, animals, or people unless it gets into the body. *Also see* Beta Radiation, Radiation (2), and Gamma Radiation

American Conference of Governmental Industrial Hygienists® (ACGIH) — Organization that promotes the free exchange of ideas and experiences and the development of standards and techniques in industrial health. *Also see* Biological Exposure Indices (BEI®).

Ammonium Nitrate and Fuel Oil (ANFO) — High explosive blasting agent made of common fertilizer mixed with diesel fuel or oil; requires a booster to initiate detonation. *Also see* High Explosive, Detonation, and Explosive (1).

Analysis — Ability to divide information into its most basic components. *Also see* Cost-Benefit Analysis, and Impact Analysis.

ANFO — Abbreviation for Ammonium Nitrate and Fuel Oil.

Aqueous Film Forming Foam (AFFF) — Synthetic foam concentrate that (when combined with water) can form a complete vapor barrier over fuel spills and fires; highly effective extinguishing and blanketing agent on hydrocarbon fuels. *Also see* Foam Concentrate and Foam System.

Anthrax — A non-contagious potentially fatal disease caused by breathing, eating or absorbing through cuts in the skin bacteria known as *Bacillus anthracis*.

Antibiotic — Type of antimicrobial agent made from a mold or a bacterium that kills, or slows the growth of other microbes, specifically bacteria. Examples include penicillin and streptomycin. Antibiotics are ineffective against viruses.

Antidote — A substance that will counteract the effects of a poison or toxin.

APR — Abbreviation for air-purifying respirator.

ARS — Abbreviation for Acute Radiation Syndrome.

Asphyxia — Suffocation.

Asphyxiant — Any substance that prevents oxygen from combining in sufficient quantities with the blood or from being used by body tissues. *Also see* Chemical Asphyxiant.

Asphyxiation — Condition that causes death because of a deficient amount of oxygen and an excessive amount of carbon monoxide and/or other gases in the blood.

Assessment — Evaluation and interpretation of measurements and other information to provide a basis for decision-making.

ATF — Abbreviation for U.S. Bureau of Alcohol, Tobacco, Firearms and Explosives.

Atmospheric Pressure — Force exerted by the atmosphere at the surface of the earth because of the weight of air. Atmospheric pressure at sea level is about 14.7 psi (101 kPa) {1.01 bar}. Atmospheric pressure increases as elevation decreases below sea level and decreases as elevation increases above sea level.

Atmospheric Storage Tank — Class of fixed facility storage tanks. Pressures range from 0 to 0.5 psi (0 to 3.4 kPa) {0 to 0.03 bar}. Sometimes called *nonpressure storage tank*. *Also see* Lifter Roof Storage Tank, Cone Roof Storage Tank, Floating Roof Storage Tank, Internal Floating Roof Tank, External Floating Roof Tank, Low-Pressure Storage Tank, Horizontal Storage Tank, and Pressure Storage Tank.

ATSDR — Abbreviation for Agency for Toxic Substances and Disease Registry.

Authority Having Jurisdiction (AHJ) — Term used in codes and standards to identify the legal entity, such as a building or fire official, that has the statutory authority to enforce a code and to approve or require equipment. In the insurance industry it may refer to an insurance rating bureau or an insurance company inspection department.

Autoclave — A device that uses high-pressure steam to sterilize objects.

Autoignition — Ignition that occurs when a substance in air, whether solid, liquid, or gaseous, is heated sufficiently to initiate or cause self-sustained combustion without an external ignition source. *Also see* Ignition, Autoignition Temperature, and Ignition Temperature.

Autoignition Temperature — Same as ignition temperature except that no external ignition source is required for ignition because the material itself has been heated to ignition temperature; temperature at which autoignition occurs through the spontaneous ignition of the gases or vapors emitted by a heated material. *Also see* Ignition Temperature, Autoignition, and Ignition.

Autoinjector — A spring-loaded syringe filled with a single dose of a life-saving drug.

Awareness Level — Lowest level of training established by OSHA for personnel at hazardous materials incidents. *Also see* Operations Level.

B

Bacteria — Microscopic, single-celled organisms. *Also see* Virus and Rickettsia.

Baffle — Intermediate partial bulkhead that reduces the surge effect in a partially loaded liquid tank.

Bank-Down Application Method (Deflect) — Method of foam application that may be employed on an un-ignited or ignited Class B fuel spill. The foam stream is directed at a vertical surface or object that is next to or within the spill area. The foam deflects off the surface or object and flows down onto the surface of the spill to form a foam blanket.

Barge — Long, large vessel (usually flat-bottomed, self-propelled, or towed or pushed by another vessel) used for transporting goods on inland waterways. *Also see* Cargo Vessel.

Base — Corrosive water-soluble compound or substance containing group-forming hydroxide ions in water solution that reacts with an acid to form a salt; an alkaline (caustic) substance. *Also see* Acid and pH.

Becquerel (Bq) — International System unit of measurement for radioactivity, indicating the number of nuclear decays/disintegrations a radioactive material undergoes in a certain period of time. *Also see* Radiation (2), Radioactive Material (RAM), and Curie (Ci).

BEI® — Abbreviation for Biological Exposure Indices.

Beta Particle — Particle that is about 1/7,000th the size of an alpha particle but has more penetrating power; has a negative electrical charge. *Also see* Beta Radiation, Alpha Particle and Gamma Ray.

Beta Radiation — Type of radiation that can cause skin burns. *Also see* Alpha Radiation, Gamma Radiation, and Radiation (2).

Bill of Lading — Shipping paper used by the trucking industry (and others) indicating origin, destination, route, and product; placed in the cab of every truck tractor. This document establishes the terms of a contract between shippers and transportation companies; serves as a document of title, contract of carriage, and receipt for goods. *Also see* Shipping Papers and Lading.

Biochemical — Involving chemical reactions in living organisms.

Biological Agent — Viruses, bacteria, or their toxins used for the purpose of harming or killing people, animals, or crops. Also called *Biological Weapon.*

Biological Attack — Intentional release of viruses, bacteria, or their toxins for the purpose of harming or killing citizens. *Also see* Terrorism.

Biological Exposure Indices (BEI®) — Guidance value recommended for assessing biological monitoring results that is established by the American Conference of Governmental Industrial Hygienists (ACGIH).

Biological Toxin — Poison produced by living organisms. Also see Toxin and Poison.

Biological Weapon — *See* Biological Agent.

Blasting Cap — *See* Detonator.

Blast Pressure Wave — Shock wave created by rapidly expanding gases in an explosion.

BLEVE — Acronym for Boiling Liquid Expanding Vapor Explosion.

Blister Agent — Chemical warfare agent that burns and blisters the skin or any other part of the body it contacts. Also called *vesicant. Also see* Chemical Warfare Agent.

Blood Agent/Blood Poison — *See* Chemical Asphyxiant.

Bloodborne Pathogens — Pathogenic microorganisms that are present in the human blood and can cause disease in humans. These pathogens include (but are not limited to) hepatitis B virus (HBV) and human immunodeficiency virus (HIV).

B-NICE — Acronym for Biological, Nuclear, Incendiary, Chemical, and Explosive.

Boiling Liquid Expanding Vapor Explosion (BLEVE) — Rapid vaporization of a liquid stored under pressure upon release to the atmosphere following major failure of its containing vessel; failure is the result of overpressurization caused by an external heat source, which causes the vessel to explode into two or more pieces when the temperature of the liquid is well above its boiling point at normal atmospheric pressure. *Also see* Boiling Point.

Boiling Point — Temperature of a substance when the vapor pressure exceeds atmospheric pressure. At this temperature, the rate of evaporation exceeds the rate of condensation. At this point, more liquid is turning into gas than gas is turning back into a liquid. Also see Vapor Pressure.

Bomb Squad — Crew of emergency responders specially trained and equipped to deal with explosive devices. Also called h*azardous devices unit*s or *explosive ordnance disposal [EOD] personnel.*

Bq — Abbreviation for Becquerel.

Break Bulk Carrier — Ship designed with large holds to accommodate a wide range of products such as vehicles, pallets of metal bars, liquids in drums, or items in bags, boxes, and crates. *Also see* Cargo Vessel.

Bulk Cargo Carrier — Ship carrying either liquid or dry goods stowed loose in a hold and not enclosed in any container. *Also see* Cargo Vessel.

Bulk Container — Cargo tank container attached to a flatbed truck or rail flatcar used to transport materials in bulk. This container may carry liquids or gases. *Also see* Container (1).

Bulk Packaging — Packaging, other than a vessel or barge, including transport vehicle or freight container, in which hazardous materials are loaded with no intermediate form of containment and which has (a) a maximum capacity greater than 119 gallons (450 L) as a receptacle for a liquid, (b) maximum net mass greater than 882 pounds (400 kg) and a maximum capacity greater than 119 gallons (450 L) as a receptacle for a solid, or (c) water capacity greater than 1,000 pounds (454 kg) as a receptacle for a gas. Reference: Title 49 *CFR* 171.8. *Also see* Packaging (1) and Nonbulk Packaging.

Bunsen Burner — A common piece of laboratory equipment that provides a single, gas-fed flame used for heating, combustion, and sterilization.

Bureau of Alcohol, Tobacco, Firearms and Explosives (ATF) — Division of U.S. Department of Treasury that enforces federal laws and regulations relating to alcohol, tobacco, firearms, explosives, and arson.

C

Calibration — Set of operations used to standardize or adjust the values of quantities indicated by a measuring instrument.

Canadian Nuclear Safety Commission — Agency responsible for regulating almost all uses of nuclear energy and nuclear materials in Canada.

Canadian Transport Emergency Centre (CANUTEC) — Canadian center that provides fire and emergency responders with 24-hour information for incidents involving hazardous materials; operated by Transport Canada, a department of the Canadian government. *Also see* Chemical Transportation Emergency Center (CHEMTREC®) and Emergency Transportation System for the Chemical Industry (SETIQ).

CANUTEC — Acronym for Canadian Transport Emergency Centre.

Capacity Indicator — Device installed on a tank to indicate capacity at a specific level.

Capacity Stencil — Number stenciled on the exterior of tank cars to indicated the volume of the tank.

Carbon Dioxide (CO$_2$) — Colorless, odorless, heavier than air gas that neither supports combustion nor burns. CO$_2$ is used in portable fire extinguishers as an extinguishing agent to extinguish Class B or C fires by smothering or displacing the oxygen.

Carbon Monoxide (CO) — Colorless, odorless, dangerous gas (both toxic and flammable) formed by the incomplete combustion of carbon. It combines more than 200 times as quickly with hemoglobin as oxygen, thus decreases the blood's ability to carry oxygen.

Carboy — Cylindrical container of about 5 to 15 gallons (19 L to 57 L) capacity for corrosive or pure liquids; made of glass, plastic, or metal with a neck and sometimes a pouring tip; cushioned in a wooden box, wicker basket, or special drum. *Also see* Container (1).

Carcinogen — Cancer-producing substance.

Cargo Container — *See* Container (1).

Cargo Tank — *See* Cargo Tank Truck.

Cargo Tank Truck — Motor vehicle commonly used to transport hazardous materials via roadway. Also called *tank motor vehicle, tank truck,* and *cargo tank. Also see* Corrosive Liquid Tank, Cryogenic Liquid Tank, Dry Bulk Cargo Tank, High-Pressure Tank, Low-Pressure Chemical Tank, Compressed-Gas Tube Trailer, Nonpressure Liquid Tank, and Elevated Temperature Materials Carrier.

Cargo Vessel — Ship used to transport cargo (dry bulk, break bulk, roll-on/roll off, and container) via waterways. *Also see* Break Bulk Carrier, Bulk Cargo Carrier, Container Vessel, Roll-on/Roll-off Vessel, and Barge.

Case File — The collection of documents comprising information concerning a particular investigation.

Case Identifiers — The alphabetic and/or numeric characters assigned to identify a particular case.

CAS® Number — Number assigned by the American Chemical Society's Chemical Abstract Service that uniquely identifies a specific compound.

Catalyst — A substance that influences the rate of chemical reaction between or among other substances.

Caustic — Substance having the destructive properties of a base. Also see Base, Alkali, and Acid.

CBR — Abbreviation for Chemical, Biological, Radiological.

CBRNE — Abbreviation for Chemical, Biological, Radiological, Nuclear, and Explosive.

CDC — Abbreviation for U.S. Centers for Disease Control and Prevention.

Centers for Disease Control and Prevention (CDC) — U.S. government agency responsible for the collection and analysis of data regarding disease and health trends.

CERCLA — Abbreviation for Comprehensive Environmental Response, Compensation and Liability Act.

CFR — Abbreviation for *Code of Federal Regulations.*

CGA — Abbreviation for Compressed Gas Association.

Chain of Custody — A process used to maintain and document the chronological history of the evidence; documents should include name or initials of the individual collecting the evidence, each person or entity subsequently having custody of it, dates the items were collected or transferred, agency and case number, victim's or suspect's name, and a brief description of the item.

Charge — The law that the law enforcement agency believes the defendant has broken.

Chemical Agent — A chemical substance that is intended for use in warfare or terrorist activities to kill, seriously injure, or incapacitate people through its physiological effects. Also called *chemical warfare agent.*

Chemical Asphyxiant — Substance that reacts to keep the body from being able to use oxygen. Also called *blood poison, blood agent,* or *cyanogen agent. Also see* Asphyxiant.

Chemical Attack — Deliberate release of a toxic gas, liquid, or solid that can poison people and the environment. *Also see* Terrorism and Chemical Warfare Agent.

Chemical Carrier — Tank vessel that transports multiple specialty and chemical commodities.

Chemical Degradation — Process that occurs when the characteristics of a material are altered through contact with chemical substances.

Chemical Properties — Relating to the way a substance is able to change into other substances; reflect the ability to burn, react, explode, or produce toxic substances hazardous to people or the environment.

Chemical Protective Clothing (CPC) — Clothing designed to shield or isolate individuals from the chemical, physical, and biological hazards that may be encountered during operations involving hazardous materials. *Also see* Level A Protection and Personal Protective Equipment (PPE).

Chemical Reaction — Any change in the composition of matter that involves a conversion of one substance into another.

Chemical Transportation Emergency Center (CHEMTREC®) — Center established by the American Chemistry Council that supplies 24-hour information for incidents involving hazardous materials. *Also see* Canadian Transport Emergency Centre (CANUTEC) and Emergency Transportation System for the Chemical Industry (SETIQ).

Chemical Warfare Agent — *See* Chemical Agent.

CHEMTREC® — Acronym for Chemical Transportation Emergency Center.

Choking Agent — Chemical warfare agent that attacks the lungs causing tissue damage. *Also see* Chemical Agent.

Chronic — Of long duration, or recurring over a period of time; (opposite of acute). *Also see* Acute.

Chronic Health Effects — Long-term effects from either a one-time or repeated exposure to a hazardous substance. *Also see* Acute Health Effects.

Ci — Abbreviation for Curie.

Class A Fire — Fire involving ordinary combustibles such as wood, paper, cloth, and the like.

Class A Foam Concentrate — Foam specially designed for use on Class A combustibles. *Also see* Foam Concentrate and Finished Foam.

Class B Fire — Fire involving flammable and combustible liquids and gases such as gasoline, kerosene, and propane.

Class B Foam Concentrate — Foam specially designed for use on ignited or unignited Class B flammable or combustible liquids. *Also see* Foam Concentrate and Finished Foam.

Cleanout Fitting — Fitting installed in the top of a tank to facilitate washing the tank's interior.

Cloud — Ball-shaped pattern of an airborne hazardous material where the material has collectively risen above the ground or water at a hazardous materials incident. *Also see* Cone, Hemispheric Release, and Plume.

Code of Federal Regulations (CFR) — Books or documents containing the specific United States regulations provided for by law; complete body of U.S. federal law.

Cold Zone — Safe zone outside of the warm zone where equipment and personnel are not expected to become contaminated and special protective clothing is not required; the incident command post and other support functions are typically located in this zone; also called the *support zone.*

Colorimetric Tube — Small tube that changes color when air that is contaminated with a particular substance is drawn through it. Also called *detector tube.*

Combination Packaging — Shipping container consisting of one or more inner packagings secured in a nonbulk outer packaging. *Also see* Packaging (1).

Combustible Gas Detector — Indicates the explosive levels of combustible gases.

Combustible Liquid — Liquid having a flash point at or above 100°F (37.8°C) and below 200°F (93.3°C). *Also see* Flammable Liquid.

Command Post — Location at which primary Command functions are executed, usually co-located with the incident base; also called *Incident Command Post (ICP).*

Composite Packaging — Single container made of two different types of material. *Also see* Packaging (1).

Compound — Substance consisting of two or more elements that have been united chemically.

Comprehensive Environmental Response, Compensation and Liability Act (CERCLA) — U.S. law that created a tax on the chemical and petroleum industries and provided broad federal authority to respond directly to releases or threatened releases of hazardous substances that may endanger public health or the environment.

Compressed Gas — Gas that, at normal temperature, exists solely as a gas when pressurized in a container as opposed to a gas that becomes a liquid when stored under pressure. *Also see* Gas, Liquefied Compressed Gas, and Nonflammable Gas.

Compressed Gas Association (CGA) — Trade association that, among other things, writes standards pertaining to the use, storage, and transportation of compressed gases.

Compressed-Gas Tube Trailer — Cargo tank truck that carries gases under pressure; may be a large single container, an intermodal shipping unit, or several horizontal tubes. Also called *tube trailer. Also see* Cargo Tank Truck.

Concentration — (1) Quantity of a chemical material inhaled for purposes of measuring toxicity. (2) Percentage (mass or volume) of a material dissolved in water (or other solvent).

Condensation — Process of going from the gaseous to the liquid state.

Cone — Triangular-shaped pattern of an airborne hazardous material release with a point source at the breach and a wide base downrange. *Also see* Cloud, Hemispheric Release, and Plume.

Cone Roof Storage Tank — Fixed-site vertical atmospheric storage tank that has a cone-shaped pointed roof with weak roof-to-shell seams that are intended to break when excessive overpressure results inside; used to store flammable, combustible, and corrosive liquids. Also called *dome roof tank. Also see* Atmospheric Storage Tank.

Confinement — (1) Process of controlling the flow of a spill and capturing it at some specified location. (2) Operations required to prevent fire from extending from the area of origin to uninvolved areas or structures. *Also see* Containment.

Consist — Rail shipping paper that contains a list of cars in the train by order; indicates the cars that contain hazardous materials. Some railroads include information on emergency operations for the hazardous materials on board with the consist. Also called *train consist. Also see* Shipping Papers and Waybill.

Consumer Product Safety Commission (CPSC) — U.S. government agency charged with protecting the public from unreasonable risks of serious injury or death from more than 15,000 types of consumer products under the agency's jurisdiction, including hazardous materials intended for consumer purchase and use.

Contagious — Capable of being transmitted from one person to another by contact or close proximity.

Container — (1) Article of transport equipment that is (a) a permanent character and strong enough for repeated use; (b) specifically designed to facilitate the carriage of goods by one or more modes of transport without intermediate reloading; and (c) fitted with devices permitting its ready handling, particularly its transfer from one mode to another. The term does not include vehicles. Also called *freight container* and *cargo container.* (2) Boxes of standardized size used to transport cargo by truck or railcar when transported overland or by cargo vessels at sea.

Container Specification Number — Shipping container number preceded by letters *DOT* that indicates the container has been built to U.S. federal specifications.

Container Vessel — Ship specially equipped to transport large freight containers in horizontal or, more commonly, vertical container cells; containers are usually loaded and unloaded by special cranes. *Also see* Cargo Vessel.

Containment — Act of stopping the further release of a material from its container. *Also see* Confinement.

Contaminant — Any foreign substance that compromises the purity of a given substance. *Also see* Contamination

Contamination — Condition of impurity resulting from a mixture or contact with foreign substance. *Also see* Decontamination and Contaminant.

Control Zones — *See* Hazard-Control Zones.

Convulsant — Poison that causes an exposed individual to have convulsions.

Corrosive — *See* Corrosive Material.

Corrosive Liquid Tank — Cargo tank truck that carries corrosive liquids, usually acids. *Also see* Cargo Tank Truck.

Corrosive Material — Gaseous, liquid or solid material that can burn, irritate, or destroy human skin tissue and severely corrode steel. Also called *corrosive. Also see* Hazardous Material.

Covered Floating Roof Tank — *See* Internal Floating Roof Tank.

Covert — Secret, not in the open.

CPC — Abbreviation for Chemical Protective Clothing.

CPSC — Abbreviation for Consumer Product Safety Commission.

Critical Infrastructure — Systems, assets, and networks, whether physical or virtual, so vital that the incapacity or destruction of such systems and assets would have a debilitating impact on security, national economic security, national public health or safety, or any combination of those matters. *Also see* Infrastructure.

Cross Contamination — Contamination of people, equipment, or the environment outside the hot zone without contacting the primary source of contamination; sometimes called *secondary contamination. Also see* Contamination, Decontamination, and Hazard-Control Zones.

Cryogen — Gas that is cooled to a very low temperature, usually below -130°F (-90°C), to change to a liquid. Also called *refrigerated liquid* and *cryogenic liquid. Also see* Cryogenic-Liquid Storage Tank and Cryogenic Liquid Tank.

Cryogenic Liquid — *See* Cryogen.

Cryogenic Liquid Storage Tank — Heavily insulated, vacuum-jacketed tanks used to store cryogenic liquids, equipped with safety-relief valves and rupture disks. *Also see* Cryogen.

Cryogenic Liquid Tank — Cargo tank truck that carries gases that have been liquefied by temperature reduction. *Also see* Cryogen and Cargo Tank Truck.

Curie (Ci) — English System unit of measurement for radioactivity, indicating the number of nuclear decays/disintegrations a radioactive material undergoes in a certain period of time. *Also see* Becquerel, Radioactive Material (RAM), and Radiation (2).

Cyanogen Agent — See Chemical Asphyxiant.

Cyber Terrorism — The premeditated, politically motivated attack against information, computer systems, computer programs, and data which result in violence against noncombatant targets by sub-national groups or clandestine agents.

D

Dangerous Good — (1) Any product, substance, or organism included by its nature or by the regulation in any of the nine United Nations classifications of hazardous materials. (2) Term used to describe hazardous materials in Canada. (3) Term used in the U.S. and Canada for hazardous materials on board aircraft. *Also see* Hazardous Material (1).

Decon — Abbreviation for Decontamination.

Decontamination (Decon) — Process of removing a hazardous, foreign substance from a person, clothing, or area. *Also see* Contamination, Decontamination Corridor, Emergency Decontamination, Technical Decon, Gross Decontamination, and Mass Decontamination.

Decontamination Corridor — Area where decontamination is conducted. Also see Decontamination.

Dedicated Railcar — Car set aside by the product manufacturer to transport a specific product. The name of the product is painted on the car.

Defensive Mode — *See* Defensive Strategy.

Defensive Operations — Operations in which responders seek to confine the emergency to a given area without directly contacting the hazardous materials involved. *Also see* Nonintervention Operations and Offensive Operations.

Defensive Strategy — Overall plan for incident control established by the Incident Commander (IC) that involves protection of exposures as opposed to aggressive, offensive intervention. *Also see* Nonintervention Strategy, Strategy, and Offensive Strategy.

Deflagration — (1) Chemical reaction producing vigorous heat and sparks or flame and moving through the material (as black or smokeless powder) at less than the speed of sound. A major difference among explosives is the speed of the reaction. (2) Can also refer to intense burning; a characteristic of Class B explosives. (3) An explosion involving a chemical reaction in which the reaction (energy front) proceeds at less than the speed of sound.

Degradation — *See* Chemical Degradation.

Dehydration — Process of removing water or other fluids.

Department of Defense (DOD) — Administrative body of the executive branch of the U.S. federal government that encompasses all branches of the U.S. military.

Department of Energy (DOE) — Administrative body of the executive branch of the U.S. federal government that manages national nuclear research and defense programs, including the storage of high-level nuclear waste.

Department of Homeland Security (DHS) — U.S. agency that has the missions of preventing terrorist attacks, reducing vulnerability to terrorism, and minimizing damage from potential attacks and natural disasters; includes the Federal Emergency Management Agency (FEMA), U.S. Coast Guard (USCG), and others.

Department of Justice (DOJ) — Administrative body of the executive branch of the U.S. Federal Government that assigns primary responsibility for operational response to threats or acts of terrorism within U.S. territory to the Federal Bureau of Investigation (FBI). *Also see* Terrorism.

Department of Labor (DOL) — Administrative body of the executive branch of the U.S. Federal Government that is responsible for overseeing U.S. labor laws (labor policy, regulation, and enforcement).

Department of Transportation (DOT) — U.S. Federal agency that regulates the transportation of hazardous materials; responsible for transportation policy, regulation, and enforcement.

Detector Tube — *See* Colorimetric Tube.

Detonation — (1) Supersonic thermal decomposition, which is accompanied by a shock wave in the decomposing material. (2) Explosion with an energy front that travels faster than the speed of sound. *Also see* High Explosive and Explosive (1 and 2).

Detonator — A device used to trigger less sensitive explosives, usually composed of a primary explosive; for example, a blasting cap. Detonators may be initiated mechanically, electrically, or chemically. Also called an *initiator*. Also see Detonation.

DHS — Abbreviation for U.S. Department of Homeland Security.

Diatomaceous Earth — Light siliceous material consisting chiefly of the skeletons (minute unicellular algae); used especially as an absorbent or filter.

Dilution — Application of water to a water-soluble material to reduce the hazard. *Also see* Water Solubility and Dissolution.

Dirty Bomb — *See* Radiological Dispersal Device (RDD).

Dispersion — Act or process of being spread widely. *Also see* Engulf (1) and Vapor Dispersion.

Disseminate — To spread about or scatter widely.

Dissipate — To cause to spread out or spread thin to the point of vanishing.

Dissolution — Act or process of dissolving one thing into another; process of dissolving a gas in water. *Also see* Concentration (2) and Dilution.

Documentation — Written notes, audio/videotapes, printed forms, sketches and/or photographs that form a detailed record of the scene, evidence recovered, and actions taken during the search of the crime scene.

DOD — Abbreviation for the U.S. Department of Defense.

DOE — Abbreviation for the U.S. Department of Energy.

DOJ — Abbreviation for the U.S. Department of Justice.

DOL — Abbreviation for the U.S. Department of Labor.

Dome Roof Tank — *See* Cone Roof Storage Tank.

Dose — Quantity of a chemical material ingested or absorbed through skin contact for purposes of measuring toxicity.

Dosimeter — Device used to measure an individual's exposure to an environmental hazard such as radiation or sound.

DOT — Abbreviation for the U.S. Department of Transportation.

Drainage Time — The amount of time it takes foam to break down or dissolve; also called drainage rate, drainage dropout rate, or drainage.

Dry Bulk Cargo Tank — Cargo tank truck that carries small, granulated solid materials; generally does not carry hazardous materials but in some cases may carry fertilizers or plastic products that can burn and release toxic products of combustion. *Also see* Cargo Tank Truck.

Dust — Solid particle that is formed or generated from solid organic or inorganic materials by reducing its size through mechanical processes such as crushing, grinding, drilling, abrading, or blasting.

Dyspnea — Painful or difficult breathing; rapid, shallow respirations.

E

EBSS — Abbreviation for Emergency Breathing Support System.

Electron — Minute component of an atom that possesses a negative charge.

Element — Most simple substance that cannot be separated into more simple parts by ordinary means.

Elevated Temperature Material — Material that when offered for transportation or transported in bulk packaging is (a) in a liquid phase and at a temperature at or above 212°F (100°C), (b) intentionally heated at or above its liquid phase flash point of 100°F (38°C), or (c) in a solid phase and at a temperature at or above 464°F (240°C). *Also see* Flash Point, Bulk Packaging, and Elevated Temperature Materials Carrier.

Emergency Breathing Support System (EBSS) — Escape-only respirator that provided sufficient self-contained breathing air to permit the wearer to safely exit the hazardous area; usually integrated into an airline supplied-air respirator system.

Emergency Decontamination — The physical process of immediately reducing contamination of individuals in potentially life-threatening situations with or without the formal establishment of a decontamination corridor. Also see *Decontamination and* Decontamination Corridor.

Emergency Medical Services (EMS) — Initial medical evaluation/treatment provided to individuals who become ill or are injured.

Emergency Operations Center (EOC) — Facility that houses communications equipment, plans, contact/notification lists, and staff that are used to coordinate the response to an emergency.

Emergency Response Guidebook (ERG) — Guide developed jointly by Transport Canada (TC), U.S. Department of Transportation (DOT), and the Secretariat of Transport and Communications of Mexico (SCT) for use by firefighters, police, and other emergency services personnel who may be the first to arrive at the scene of a transportation incident involving dangerous goods. It is primarily a guide to aid first responders in quickly identifying the specific or generic hazards of the material involved and protecting themselves and the general public during the initial response phase.

Emergency Response Plan — Document that contains information on the actions that may be taken by a governmental jurisdiction to protect people and property before, during, and after an emergency.

Emergency Transportation System for the Chemical Industry (SETIQ) — Emergency response center for Mexico. *Also see* Canadian Transport Emergency Centre (CANUTEC) and Chemical Transportation Emergency Center (CHEMTREC®).

Emergency Valve — Self-closing tank outlet valve.

EMS — Abbreviation for Emergency Medical Services.

Emulsion — Insoluble liquid suspended in another liquid. *Also see* Insoluble.

Encapsulating — Completely enclosed or surrounded as in a capsule.

Endothermic — Chemical reaction involving the absorption of heat energy.

Engulf — (1) Dispersion of material as defined in the General Emergency Behavior Model (GEBMO). An engulfing event is when matter and/or energy disperses and forms a danger zone. (2) To flow over and enclose; in fire service context, it refers to being enclosed in flames. *Also see* Vapor Dispersion and Dispersion.

Environmental Protection Agency (EPA) — U.S. government agency that creates and enforces laws designed to protect the air, water, and soil from contamination; responsible for researching and setting national standards for a variety of environmental programs.

Environment Canada — Agency responsible for preserving and enhancing the quality of the natural environment (including water, air, and soil quality), conserving Canada's renewable resources, and coordinating environmental policies and programs for the federal government of Canada.

EOC — Abbreviation for Emergency Operations Center.

EPA — Abbreviation for the U.S. Environmental Protection Agency.

Epidemic — The occurrence of more cases of disease than expected in a given area or among a specific group of people over a particular period of time.

ERG — Abbreviation for *Emergency Response Guidebook*.

Etiological Agents — Living microorganisms, like germs, that can cause human disease; a biologically hazardous material.

Evacuation — Controlled process of leaving or being removed from a potentially hazardous location, typically involving relocating people from an area of danger or potential risk to a safer place. *Also see* Shelter in Place.

Evaporation — Process of a solid or a liquid turning into gas.

Evidence — (1) In law, something legally presented in court that bears on the point in question. (2) Information collected and analyzed by an investigator.

Excepted Packaging — Container used for transportation of materials that have very limited radioactivity. *Also see* Packaging (1), Type A Packaging, Type B Packaging, Industrial Packaging, and Strong, Tight Container.

Exothermic — Chemical reaction between two or more materials that changes the materials and produces heat, flames, and toxic smoke.

Expansion Ratio — Ratio of the finished foam volume to the volume of the original foam solution.

Explosion — A physical or chemical process that results in the rapid release of high pressure gas into the environment.

Explosion-Proof Equipment — Encased in a rigidly built container so it withstands an internal explosion and also prevents ignition of a surrounding flammable atmosphere; designed to not provide an ignition source in an explosive atmosphere.

Explosive — (1) Any material or mixture that will undergo an extremely fast, self-propagation reaction when subjected to some form of energy. *Also see* High Explosive and Detonation. (2) Material capable of burning or bursting suddenly and violently. *Also see* Detonation, High Explosive, and Low Explosive.

Explosive Limit — *See* Flammable Limit.

Explosive Range — *See* Flammable Range.

Exposure — (1) Structure or separate part of the fireground to which a fire could spread. (2) People, properties, systems, or natural features that are or may be exposed to the harmful effects of a hazardous materials emergency. (3) Contact with a substance by swallowing, breathing, or touching the skin or eyes. Exposure may be short-term (acute exposure), of intermediate duration, or long-term (chronic exposure).

External Floating Roof Tank — Fixed-site vertical storage tank that has no fixed roof but relies on a floating roof to protect the contents and prevent evaporation. Also called *open-top floating roof tank. Also see* Internal Floating Roof Tank and Floating Roof Storage Tank.

Extremely Hazardous Substance — Chemical determined by the Environmental Protection Agency (EPA) to be extremely hazardous to a community during an emergency spill or release as a result of its toxicity and physical/chemical properties. *Also see* Hazardous Substance.

F

Fallout — The radioactive particles descending from a cloud after a nuclear detonation.

FBI — Abbreviation for Federal Bureau of Investigation.

Federal Bureau of Investigation (FBI) — U.S. agency under the Department of Justice that investigates the theft of hazardous materials, collects evidence for crimes, and prosecutes criminal violation of federal hazardous materials laws and regulations; lead agency on terrorist incident scenes. *Also see* Terrorism.

Federal Emergency Management Agency (FEMA) — U.S. agency that is part of the Department of Homeland Security (DHS) and tasked with responding to, planning for, recovering from, and mitigating against disasters.

FEMA — Acronym for Federal Emergency Management Agency.

FIBC — Abbreviation for Flexible Intermediate Bulk Container.

Finance/Administration Section — Section responsible for all costs and financial actions of the incident; includes the Time Unit, Procurement Unit, Compensation/Claims Unit, and the Cost Unit. *Also see* Incident Command System (ICS).

Finished Foam — Extinguishing agent formed by mixing foam concentrate with water and aerating the solution for expansion. *Also see* Foam Blanket and Foam Concentrate. Also called *foam.*

Fire Point — Temperature at which a liquid fuel produces sufficient vapors to support combustion once the fuel is ignited; usually a few degrees above the flash point. *Also see* Flash Point.

Fissionable — Capable of splitting the atomic nucleus and releasing large amounts of energy.

Flame Impingement — Points at which flames contact the surface of a container or other structure.

Flammability — Fuel's susceptibility to ignition.

Flammable — Capable of burning and producing flames.

Flammable Gas — Any material (except aerosols) that is a gas at 68°F (20°C) or less and that (a) is ignitable at 14.7 psi (101.3 kPa) when in a mixture of 13 percent or less by volume with air or (b) has a flammable range at 14.7 psi (101.3 kPa) by volume with air of at least 12 percent regardless of the lower limit.

Flammable Limit — Percentage of a substance in air that will burn once it is ignited. Most substances have an upper (too *rich*) and lower (too *lean*) flammable limit. Also called *explosive limit. Also see* Flammable Range, Upper Flammable Limit (UFL), and Lower Flammable Limit (LFL).

Flammable Liquid — Any liquid having a flash point below 100°F (37.8°C) and having a vapor pressure not exceeding 40 psi (276 kPa) {2.76 bar}.

Flammable Material — Substance that ignites easily and burns rapidly.

Flammable Range — Range between the upper flammable limit and lower flammable limit in which a substance can be ignited. Also called *explosive range*. *Also see* Flammable Limit.

Flammable Solid — Solid material (other than an explosive) that (a) is liable to cause a fire through friction or retained heat from manufacturing or processing or (b) ignites readily and then burns vigorously and persistently, creating a serious transportation hazard.

Flash Point — Minimum temperature at which a liquid gives off enough vapors to form an ignitable mixture with air near the surface of the liquid. *Also see* Fire Point.

Flexible Intermediate Bulk Container (FIBC) — *See* Intermediate Bulk Container (IBC).

Floating Roof Storage Tank — Atmospheric storage tank that stands vertically; is wider than it is tall. The roof floats on the surface of the liquid to eliminate the vapor space. *Also see* Atmospheric Storage Tank and Lifter Roof Storage Tank.

Foam — *See* Finished Foam.

Foam Blanket — Covering of finished foam applied over a burning surface to produce a smothering effect; can be used on unignited surfaces to prevent ignition. *Also see* Finished Foam and Foam Stability.

Foam Concentrate — (1) Raw chemical compound solution that is mixed with water and air to produce finished foam; may be protein, synthetic, aqueous film forming, high expansion, or alcohol types. (2) Raw foam liquid as it rests in its storage container before the introduction of water and air. *Also see* Class A Foam Concentrate, Class B Foam Concentrate, Aqueous Film Forming Foam, Foam Solution, Foam System, and Finished Foam.

Foam Eductor — Type of foam proportioner used for mixing foam concentrate in proper proportions with a stream of water to produce foam solution. *Also see* Foam Solution and Foam Proportioner.

Foam Proportioner — Device that injects the correct amount of foam concentrate into the water stream to make the foam solution. *Also see* Foam Solution and Foam Eductor.

Foam Solution — Result of mixing the appropriate amount of foam concentrate with water; exists between the proportioner and the nozzle or aerating device that adds air to create finished foam. *Also see* Foam Concentrate, Foam Proportioner, Foam Eductor, and Finished Foam.

Forensic Science — Application of scientific procedures to the interpretation of physical events such as those that occur at a fire scene; the art of reconstructing past events and then explaining that process and one's findings to investigators and triers of fact; criminalistics.

Fragmentation Effect — Process in which bomb parts and components are blown outwards by the explosion in the form of fragments, shards, or shrapnel.

Freezing Point — Temperature at which a liquid becomes a solid at normal atmospheric pressure. *Also see* Melting Point.

Freight Container — *See* Container (1).

Frostbite — Local freezing and tissue damage caused by prolonged exposure to extreme cold. *Also see* Hypothermia.

Fume — Suspension of particles that form when material from a volatilized (vapor state) solid condenses in cool air.

G

Gamma Radiation — Very high-energy ionizing radiation composed of gamma rays. Also see Alpha Radiation, Beta Radiation, and Radiation (2).

Gamma Ray — High-energy photon (packet of electromagnetic energy) emitted from the nucleus of an unstable (radioactive) atom; one of three types of radiation emitted by radioactive materials; most penetrating and potentially lethal of the three types. *Also see* Alpha Particle, Beta Particle, Gamma Radiation, and Radioactive Material (RAM).

Gas — Compressible substance, with no specific volume, that tends to assume the shape of a container. Molecules move about most rapidly in this state. *Also see* Compressed Gas.

Gas Chromatography — Characterizing volatilities and chemical properties of compounds that evaporate enough at low temperatures (about 120°F or 49°C) to provide detectable quantities in the air through the use of instrument analysis in a gas chromatograph.

GEBMO — Acronym for General Emergency Behavior Model.

General Emergency Behavior Model (GEBMO) — Model used to describe how hazardous materials are accidentally released from their containers and how they behave after the release. *Also see* Engulf (1).

Genetic Effect — Mutations or other changes that are produced by irradiation of the germ plasma; changes produced in future generations.

GHS — Abbreviation for Globally Harmonized System of Classification and Labeling of Chemicals (GHS).

Globally Harmonized System of Classification and Labeling of Chemicals (GHS) — International classification and labeling system for chemicals and other hazard communication information such as material safety data sheets.

Global Positioning System (GPS) — System for determining a position on the earth's surface by calculating the difference in time for the signal from a number of satellites to reach a receiver on the ground.

Glovebox — A sealed container designed to allow a trained scientist to manipulate microorganisms while being in a different containment level than that of the agent they are manipulating; built into the sides of the glovebox are two glove ports arranged in such a way that one can place their hands into the ports, into gloves and perform tasks inside the box without breaking the seal.

GPS — Abbreviation for Global Positioning System.

Gross Decontamination — Quickly removing the worst surface contamination, usually by rinsing with water from handheld hoselines, emergency showers, or other water sources. *Also see* Decontamination.

H

Half-life — Time required for half the amount of a substance in or introduced into a living system or ecosystem to be eliminated or disintegrated by natural processes; period of time required for any radioactive substance to lose half of its strength or reduce by one-half its total present energy.

Halogenated Agents — Chemical compounds (halogenated hydrocarbons) that contain carbon plus one or more elements from the halogen series. Also called *halogenated hydrocarbons.*

Halogenated Hydrocarbons — *See* Halogenated Agents.

Hazard — Condition, substance, or device that can directly cause injury or loss; the source of a risk. *Also see* Hazard or Risk Analysis and Hazard Assessment.

Hazard and Risk Assessment — Formal review of the hazards and risk that may be encountered while performing the functions of a firefighter or emergency responder; used to determine the appropriate level and type of personal and respiratory protection that must be worn.

Hazard Area — Established area from which bystanders and unneeded rescue workers are prohibited. *Also see* Hazard-Control Zones.

Hazard Class — Group of materials designated by the U.S. Department of Transportation (DOT) that shares a major hazardous property such as radioactivity or flammability. *Also see* Hazard, Hazardous Material, Hazardous Chemical, and Hazardous Substance.

Hazard-Control Zones — System of barriers surrounding designated areas at emergency scenes intended to limit the number of persons exposed to a hazard and to facilitate its mitigation; major incident has three zones: restricted (hot), limited access (warm), and support (cold). U.S. EPA/OSHA term: *site work zones.* Also called *scene-control zones* and *control zones. Also see* Hazard Area and Initial Isolation Zone.

Hazard Identification — Process of defining and describing a hazard, including its physical characteristics, magnitude and severity, probability and frequency, causative factors, and locations or areas affected.

Hazard or Risk Analysis — Identification of hazards or risks and the determination of the appropriate response to that hazard or risk; combines hazard assessment with risk management concepts. *Also see* Hazard, Hazard Assessment.

Hazardous Atmosphere — Any atmosphere that may or not be immediately dangerous to life or health but that is oxygen deficient, contains a toxic or disease-producing contaminant, or contains a flammable or explosive vapor or gas. *Also see* Immediately Dangerous to Life or Health (IDLH).

Hazardous Chemical — Any chemical that is a physical hazard or health hazard to people; defined by the U.S. Occupational Safety and Health Administration (OSHA). *Also see* Hazardous Material.

Hazardous Material — (1) Any substance or material that possesses an unreasonable risk to health and safety of persons and/or the environment if it is not properly controlled during handling, storage, manufacture, processing, packaging, use, disposal, or transportation. (2) Substance or material in quantities or forms that may pose an unreasonable risk to health, safety, or property when stored, transported, or used in commerce (U.S. Department of Transportation definition). *Also see* Dangerous Good.

Hazardous Materials Incident — Emergency, with or without fire, that involves the release or potential release of a hazardous material. *Also see* Hazardous Material.

Hazardous Materials Regulations (HMR) — Regulations developed and enforced by the U.S. Department of Transportation.

Hazardous Substance — Any substance designated under the U.S. Clean Water Act and the Comprehensive Environmental Response, Compensation and Liability Act (CERCLA) as posing a threat to waterways and the environment when released. *Also see* Hazardous Material and Extremely Hazardous Substance.

Hazardous Waste — Discarded materials with no monetary value that can have the same hazardous properties it had before being used; regulated by the U.S. Environmental Protection Agency (EPA) because of public health and safety concerns. *Also see* Hazardous Material.

Hazardous Waste Operations and Emergency Response (HAZWOPER) — U.S. regulations in Title 29 (Labor) *CFR* 1910.120 for cleanup operations involving hazardous substances and emergency response operations for releases of hazardous substances. *Also see Code of Federal Regulations (CFR).*

HAZWOPER — Acronym for Hazardous Waste Operations and Emergency Response.

Head — (1) Front and rear closure of a tank shell. (2) Alternate term for pressure, especially pressure due to elevation.

Health Canada — Agency responsible for developing health policy, enforcing health regulations, promoting disease prevention, and enhancing healthy living in Canada.

Health Hazard — Material that may directly affect an individual's health once it enters or comes in contact with the body. *Also see* Physical Hazard.

Heat Cramps — Heat illness resulting from prolonged exposure to high temperatures; characterized by excessive sweating, muscle cramps in the abdomen and legs, faintness, dizziness, and exhaustion. *Also see* Heat Exhaustion, Heat Stroke, Heat Rash, and Heat Stress.

Heat Exhaustion — Heat illness caused by exposure to excessive heat; symptoms include weakness, cold and clammy skin, heavy perspiration, rapid and shallow breathing, weak pulse, dizziness, and sometimes unconsciousness. *Also see* Heat Cramps, Heat Stroke, Heat Rash, and Heat Stress.

Heat Rash — Condition that develops from continuous exposure to heat and humid air; aggravated by clothing that rubs the skin; reduces the individual's tolerance to heat.

Heat Stroke — Heat illness caused by heat exposure, resulting in failure of the body's heat regulating mechanism; symptoms include (a) high fever of 105 to 106°F (40.5°C to 41.1°C), (b) dry, red, and hot skin, (c) rapid, strong pulse, and (d) deep breaths or convulsions; may result in coma or possibly death. Also called *sunstroke*. *Also see* Heat Exhaustion, Heat Cramps, Heat Rash, and Heat Stress.

Heavy Metal — Generic term referring to lead, cadmium, mercury, and other elements that are toxic in nature; term may also be applied to compounds containing these elements. Also called *toxic element*.

Hematotoxic Agent (Hemotoxin) — Chemical that damages blood cells.

Hemispheric Release — Semicircular or dome-shaped pattern of airborne hazardous material that is still partially in contact with the ground or water. *Also see* Cloud, Cone, and Plume.

HEPA — Acronym for High Efficiency Particulate Air.

High Efficiency Particulate Air (HEPA) Filter — Respiratory protection filter designed and certified to protect the user from particulates in the air.

High Explosive — Explosive material that detonates at a velocity faster than the speed of sound. *Also see* Explosive (1), Low Explosive, and Detonation.

High-Pressure Tank — Cargo tank truck that carries liquefied gases. *Also see* Cargo Tank Truck.

HMR — Abbreviation for Hazardous Materials Regulations.

Hopper — (1) Any of various receptacles used for temporary storage of a material. (2) Tank holding a liquid and having a device for releasing its contents through a pipe.

(3) Freight car with a floor sloping to one or more hinged doors for discharging bulk contents. (4) Funnel-shaped bin used for the storage of dry solid materials such as corn, which discharges from the bottom.

Horizontal Pressure Vessel — Pressurized storage tank characterized by rounded ends; capacity may range from 500 to 40,000 gallons (1 893 L to 151 416 L). Examples of materials stored: propane, butane, ethane, and hydrogen chloride. *Also see* Pressure Vessel and Spherical Pressure Vessel.

Horizontal Storage Tank — Atmospheric storage tank that is laid horizontally and constructed of steel. *Also see* Atmospheric Storage Tank.

Hot Zone — Potentially hazardous area immediately surrounding the incident site requiring appropriate protective clothing and equipment and other safety precautions for entry; typically limited to technician-level personnel; also called the exclusion zone.

Hydrocarbon Fuel — Petroleum-based organic compound that contains only hydrogen and carbon. *Also see* Polar Solvent Fuel and Liquefied Compressed Gas.

Hydrogen Cyanide (HCN) — Colorless, toxic gas with a faint odor similar to bitter almonds; produced by the combustion of nitrogen-bearing substances.

Hypothermia — Abnormally low body temperature. Also called *systemic hypothermia*. *Also see* Frostbite.

Hypoxia — Condition caused by a deficiency in the amount of oxygen reaching body tissues.

I

IAP — Abbreviation for Incident Action Plan.

IBC — Abbreviation for Intermediate Bulk Container.

IC — Abbreviation for Incident Commander.

ICS — Abbreviation for Incident Command System.

IDLH — Abbreviation for Immediately Dangerous to Life or Health.

IED — Abbreviation for Improvised Explosive Device.

Ignition — Beginning of flame propagation or burning; the start of a fire. *Also see* Autoignition and Ignition Temperature.

Ignition Temperature — Minimum temperature to which a fuel (other than a liquid) in air must be heated in order to start self-sustained combustion independent of the heating source. *Also see* Autoignition Temperature, Autoignition, and Ignition.

Illegal Clandestine Lab — *See* Illicit Laboratory.

Illegal Dump — Site where chemicals are disposed of illegally.

Illicit — Illegal, unlawful.

Illicit Laboratory — Laboratory established to produce or manufacture illegal or controlled substance such as drugs, chemical warfare agents, explosives, or biological agents. Also called *illegal clandestine lab*. *Also see* Meth Lab.

Immediately Dangerous to Life or Health (IDLH) — Any atmosphere that poses an immediate hazard to life or produces immediate irreversible, debilitating effects on health; represents concentrations above which respiratory protection should be required; expressed in parts per million (ppm) or milligrams per cubic meter (mg/m³). Companion measurement to the permissible exposure limit (PEL). *Also see* Hazardous Atmosphere, Permissible Exposure Limit (PEL), Recommended Exposure Limit (REL), Threshold Limit Value (TLV®), and Short-term Exposure Limit (STEL).

Immiscible — Incapable of being mixed or blended with another substance. *Also see* Miscibility, Soluble, and Insoluble.

IMO — Abbreviation for International Maritime Organization.

IMO Type 5 — *See* Pressure Intermodal Tank.

Impingement — Come into sharp contact with.

IM Portable Tank — *See* Nonpressure Intermodal Tank.

Improvised Explosive Device (IED) — Explosive device that is categorized by its container and the way it is initiated; usually homemade, constructed for a specific target, and contained in almost anything. *Also see* Explosive (1 and 2).

Improvised Nuclear Device (IND) — Device that results in the formation of a nuclear-yield reaction (nuclear blast); low-yield device is called a *mininuke*. *Also see* Suitcase Bomb and Radiation (2).

IMS — Abbreviation for Incident Management System.

Incapacitant — Chemical agent that produces a temporary disabling condition that persists for hours to days after exposure has occurred. *Also see* Riot Control Agent.

Incendiary Device — Any mechanical, electrical, or chemical device used intentionally to initiate combustion and start a fire.

Incendiary Thermal Effect — Thermal heat energy resulting from the fireball created by the burning of combustible gases or flammable vapors and ambient air at very high temperatures during an explosion.

Incident Action Plan (IAP) — Written or unwritten plan for the disposition of an incident; contains the overall strategic goals, tactical objectives, and support requirements for a given operational period during an incident; required for all incidents but may not be in writing on relatively small ones.

Incidental Release — Spill or release of a hazardous material where the substance can be absorbed, neutralized, or otherwise controlled at the time of release by employees in the immediate release area, or by maintenance personnel who are not considered to be emergency responders.

Incident Commander (IC) — Person in charge of the incident management system and responsible for the management of all incident operations during an emergency.

Incident Command System (ICS) — Management system of procedures for establishing and maintaining command and control of an incident; developed in California in the early 1970s to address the resource management needs associated with large-scale wildland fires. Also known as the *California FIRESCOPE Incident Command System.*

Incident Management System (IMS) — System described in NFPA® 1561, Standard on Fire Department Incident Management System, that defines the roles, responsibilities, and standard operating procedures used to manage emergency operations. Such systems may also be referred to as Incident Command Systems (ICS). U.S. emergency response agencies comply with Homeland Security Presidential Directive 5, which establishes the National Incident Management System (NIMS) as a standardized approach to incident management.

IND — Abbreviation for Improvised Nuclear Device.

Industrial Packaging — Container used to ship radioactive materials that present limited hazard to the public and the environment (such as smoke detectors). *Also see* Packaging (1), Type A Packaging, Type B Packaging, Strong, Tight Container, and Excepted Packaging.

Infectious — Transmittable, able to infect people.

Infectious Agent — A biological agent that causes disease or illness to its host. *Also see* Biological Attack.

Infrared Thermometer — A non-contact measuring device that detects the infrared energy emitted by materials and converts the energy factor into a temperature reading. Also called a *temperature gun.*

Infrastructure — Public services of a community that have a direct effect on the quality of life (includes communication technologies such as phone lines, vital services such as water supplies, and transportation systems such as airports, highways, and waterways). *Also see* Critical Infrastructure.

Ingestion — Taking in food or other substances through the mouth. *Also see* Routes of Entry.

Inhalation — Taking in materials by breathing through the nose or mouth. *Also see* Routes of Entry.

Inhibitor — Material that is added to products that easily polymerize in order to control or prevent an undesired reaction. Also called *stabilizer. Also see* Polymerization.

Initial Isolation Distance — Distance within which all persons are considered for evacuation in all directions from a hazardous materials incident. *Also see* Protective Action Distance, Protective Action Zone, Isolation Perimeter, and Initial Isolation Zone.

Initial Isolation Zone — Circular zone (with a radius equivalent to the initial isolation distance) within which persons may be exposed to dangerous concentrations upwind of the source and may be exposed to life-threatening concentrations downwind of the source. *Also see* Initial Isolation Distance, Isolation Perimeter, and Hazard-Control Zones.

Injection — Process of taking in materials through a puncture or break in the skin. *Also see* Routes of Entry.

Insoluble — Incapable of being dissolved in a liquid (usually water). Also see Immiscible, Soluble, Emulsion, and Miscibility.

Instrument Reaction Time — Elapsed time between the movement of an air sample into a monitoring/detection device and the reading provided to the user. Also called *response time*.

Intelligence Section — Section may be activated when an incident is heavily influenced by intelligence factors or when there is a need to manage and/or analyze a large volume of classified or highly sensitive intelligence or information; particularly relevant to a terrorism incident for which intelligence plays a crucial role throughout the incident. *Also see* Incident Management System (IMS).

Intermediate Bulk Container (IBC) — Rigid (RIBC) or flexible (FIBC) portable packaging (other than a cylinder or portable tank) designed for mechanical handling with a maximum capacity of not more than three 3 cubic meters (3,000 L, 793 gal, or 106 ft³) and a minimum capacity of not less than 0.45 cubic meters (450 L, 119 gal, or 15.9 ft³) or a maximum net mass of not less than 400 kilograms (882 lbs). *Also see* Packaging (1).

Intermodal Container — Freight container designed and constructed to be used interchangeably in two or more modes of transport. Also called *intermodal tank container*. *Also see* Container (2), Refrigerated Intermodal Container, Intermodal Reporting Marks, and Container Vessel.

Intermodal Reporting Marks — Series of letters and numbers stenciled on the sides of intermodal tanks that may be used to identify and verify the contents of the tank or container. *Also see* Intermodal Container and Railcar Initials and Numbers.

Intermodal Tank Container — *See* Intermodal Container.

Internal Floating Roof Tank — Fixed-site vertical storage tank that combines both the floating roof and closed roof design. Also called *covered floating roof tank*. *Also see* Atmospheric Storage Tank, Floating Roof Storage Tank, and External Floating Roof Tank.

International Maritime Organization (IMO) — Specialized agency of the United Nations devoted to maritime affairs; has developed and promoted the adoption of more than 30 conventions and protocols and 700 codes and recommendations dealing with maritime safety; main purposes are safer shipping and cleaner oceans.

International System of Units (*Système International d'unités*) — Modern metric system, based on units of ten. Also called *SI*.

Interoperability — Ability of two or more systems or components to exchange information and use the information that has been exchanged.

Inverse Square Law — Physical law in which the amount of radiation present is inversely proportional to the square of the distance from the source of radiation.

Inversion — (1) Increase of temperature with height in the atmosphere. Vertical motion in the atmosphere is inhibited allowing for smoke buildup. A *normal* atmosphere has temperature decreasing with height. (2) Atmospheric phenomenon that allows smoke to rise until its temperature equals the air temperature and then spreads laterally in a horizontal layer. Also called *night inversion*. *Also see* Temperature Inversion.

Ion — An atom which has lost or gained an electron and thus has a positive or negative charge.

Ionization — Process in which a charged portion of a molecule (usually an electron) is given enough energy to break away from the atom; results in the formation of two charged particles or ions: (a) molecule with a net positive charge and (b) free electron with a negative charge.

Ionization Potential (IP) — The energy required to free an electron from its atom or molecule.

Ionizing Radiation — Radiation that has sufficient energy to remove electrons from atoms resulting in a chemical change in the atom. *Also see* Radiation (2).

Irritant/Irritating Material — Liquid or solid that upon contact with fire or exposure to air emits dangerous or intensely irritating fumes.

Irritating Agent — *See* Riot Control Agent.

Isolation Perimeter — Outer boundary of an incident that is controlled to prevent entrance by the public or unauthorized persons. *Also see* Initial Isolation Distance and Initial Isolation Zone.

Isotope — Atoms of a chemical element with the usual number of protons in the nucleus, but an unusual number of neutrons; has the same atomic number but a different atomic mass from normal chemical elements. *Also see* Radionuclide.

J

Jacket — Metal cover that protects tank insulation.

Joint Service Lightweight Integrated Suit Technology (JSLIST) — A chemical-protective universal, lightweight, two-piece, front-opening suit that can be worn as an overgarment or as a primary uniform over underwear. The JSLIST liner consists of a non-woven front laminated to activate carbon spheres and bonded to a knitted back that absorbs chemical agents.

JSLIST — Abbreviation for Joint Service Lightweight Integrated Suit Technology.

Jurisdiction — (1) Legal authority to operate or function. (2) Boundaries of a legally constituted entity.

L

Label — Four-inch-square diamond marker required on individual shipping containers containing hazardous materials that are smaller than 640 cubic feet (18 m³). *Also see* Package Marking, Placard, and Marking.

Lading — Freight or cargo that composes a shipment. *Also see* Bill of Lading and Shipping Papers.

LC$_{50}$ — Abbreviation for Lethal Concentration.

LD$_{50}$ — Abbreviation for Lethal Dose.

LEL — Abbreviation for Lower Explosive Limit.

LEPC — Abbreviation for Local Emergency Planning Committee.

LERP — Abbreviation for Local Emergency Response Plan.

Lethal Concentration (LC$_{50}$) — Concentration of an inhaled substance that results in the death of 50 percent of the test population; the lower the value the more toxic the substance; an inhalation exposure expressed in parts per million (ppm), milligrams per liter (mg/liter), or milligrams per cubic meter (mg/m³). *Also see* Concentration (1).

Lethal Dose (LD$_{50}$) — Concentration of an ingested or injected substance that results in the death of 50 percent of the test population; the lower the dose the more toxic the substance; an oral or dermal exposure expressed in milligrams per kilogram (mg/kg). *Also see* Dose.

Level A Protection — Highest level of skin, respiratory, and eye protection that can be given by personal protective equipment (PPE) as specified by the U.S. Environmental Protection Agency (EPA); consists of positive-pressure self-contained breathing apparatus, totally encapsulating chemical-protective suit, inner and outer gloves, and chemical-resistant boots. *Also see* Chemical Protective Clothing (CPC), Special Protective Clothing, and Personal Protective Equipment (PPE).

Level B Protection — Personal protective equipment that affords the highest level of respiratory protection, but a lesser level of skin protection. Consists of positive-pressure self-contained breathing apparatus, totally encapsulating chemical-protective suit, inner and outer gloves, and chemical-resistant boots. *Also see* Chemical Protective Clothing (CPC), Special Protective Clothing, and Personal Protective Equipment (PPE).

Level C Protection — Personal protective equipment that affords a lesser level of respiratory and skin protection than levels A or B. Consists of full-face or half-mask APR, hooded chemical-resistant suit, inner and outer gloves, and chemical-resistant boots. *Also see* Chemical

Protective Clothing (CPC), Special Protective Clothing, and Personal Protective Equipment (PPE).

Level D Protection — Personal protective equipment that affords the lowest level of respiratory and skin protection. Consists of coveralls, gloves, and chemical-resistant boots or shoes.

LFL — Abbreviation for Lower Flammable Limit.

Lifter Roof Storage Tank — Atmospheric storage tank designed so that the roof floats on a slight cushion of vapor pressure; liquid-sealed roof floats up and down with vapor pressure. When the vapor pressure exceeds a designated limit, the roof lifts to relieve the excess pressure. *Also see* Atmospheric Storage Tank and Floating Roof Storage Tank.

Liquefied Compressed Gas — Gas that under the charging pressure is partially liquid at 70°F (21°C). Also called *liquefied gas*. *Also see* Compressed Gas, Liquefied Natural Gas (LNG), Liquefied Petroleum Gas (LPG), and Gas.

Liquefied Flammable Gas Carrier — Tanker used to transport liquefied natural gas (LNG) and liquefied petroleum gas (LPG) (propane and butane for example); generally uses large insulated spherical tanks for product storage. *Also see* Tanker.

Liquefied Gas — *See* Liquefied Compressed Gas.

Liquefied Natural Gas (LNG) — Natural gas stored under pressure as a liquid. *Also see* Liquid Compressed Gas, Gas, and Hydrocarbon Fuel.

Liquefied Petroleum Gas (LPG) — Any of several petroleum products, such as propane or butane, stored under pressure as a liquid. *Also see* Liquefied Compressed Gas, Gas, and Hydrocarbon Fuel.

Liquid Oxygen (LOX) — Oxygen that is stored under pressure as a liquid.

Liquid-Splash Protective Clothing — Chemical-protective clothing designed to protect against liquid splashes per the requirements of NFPA® 1992, *Standard on Liquid Splash-Protective Suits for Hazardous Chemical Emergencies*; part of an EPA Level B ensemble.

Line of Sight — The unobstructed, imaginary line between an observer and the object being viewed.

LNG — Abbreviation for Liquefied Natural Gas.

Local Emergency Planning Committee (LEPC) — Required by SARA Title III, LEPCs are composed of local officials, citizens, and industry representatives with the task of designing, reviewing, and updating a comprehensive emergency plan for an emergency planning district; plans may address hazardous materials inventories, hazardous material response training, and assessment of local response capabilities. *Also see* Local Emergency Response Plan (LERP).

Local Emergency Response Plan (LERP) — Plan required by U.S. Environmental Protection Agency (EPA) that is prepared by the Local Emergency Planning Committee (LEPC) detailing how local emergency response agencies will respond to community emergencies.

Logistics Section — Section responsible for providing facilities, services, and materials for the incident; includes the Communications Unit, Medical Unit, and Food Unit within the Service Branch and the Supply Unit, Facilities Unit, and Ground Support Unit within the Support Branch. *Also see* Incident Command System (ICS).

Lower Explosive Limit (LEL) — Lowest percentage of fuel/oxygen mixture required to support combustion. Any mixture with a lower percentage would be considered "too lean" to burn. Also called *Lower Flammable Limit (LFL)*.

Low Explosive — Explosive material that deflagrates, producing a reaction slower than the speed of sound. *Also see* High Explosive, Explosive (2), and Detonation.

Low-Pressure Chemical Tank — Cargo tank truck designed to carry various chemicals such as flammables, corrosives, or poisons with pressures not to exceed 40 psi (276 kPa) {2.76 bar} at 70°F (21°C). *Also see* Cargo Tank Truck.

Low-Pressure Storage Tank — Class of fixed-facility storage tanks that are designed to have an operating pressure ranging from 0.5 to 15 psi (3.45 kPa to 103 kPa) {0.03 bar to 1.03 bar}. *Also see* Pressure Vessel, Pressure Storage Tank, Noded Spheroid Tank, Spheroid Tank, and Atmospheric Storage Tank.

LOX — Acronym for Liquid Oxygen.

LPG — Abbreviation for Liquefied Petroleum Gas.

M

Manhole — *See* Manway.

Manifold — (1) Top portion of a pump casing. (2) Device used to join a number of discharge pipelines to a common outlet.

Manway — (1) Hole through which a person may go to gain access to an underground or enclosed structure. (2) Openings usually equipped with removable, lockable covers large enough to admit a person into a tank trailer or dry bulk trailer. Also called *manhole*.

Marking — Descriptive name, identification number, weight, or specification along with instructions, cautions, or UN marks required on outer packagings of hazardous materials. *Also see* Package Marking, Label, and Placard.

Mass Decontamination — Conducting gross decontamination of multiple people at one time with or without a formal decon corridor or line. *Also see* Decontamination,

Decontamination Corridor, and Gross Decontamination.

Material — Generic term used by first responders for a substance involved in an incident. *Also see* Product and Hazardous Material.

Material Safety Data Sheet (MSDS) — Form provided by the manufacturer and blender of chemicals that contains information about chemical composition, physical and chemical properties, health and safety hazards, emergency response procedures, and waste disposal procedures of a specified material. *Also see* Safety Data Sheet.

Mechanical Trauma — Injury, such as an abrasion, puncture, or laceration, resulting from direct contact with a fragment or a whole container.

Melting Point — Temperature at which a solid substance changes to a liquid state at normal atmospheric pressure. *Also see* Freezing Point.

Methamphetamine (Meth) — A central nervous system stimulant drug that is similar in structure to amphetamine.

Meth Lab — Illegal clandestine laboratory established to produce illegal methamphetamine (meth). *Also see* Illegal Clandestine Lab.

Miscibility — Two or more liquids' capability to mix together. *Also see* Immiscible, Soluble, and Insoluble.

Mist — Finely divided liquid suspended in the atmosphere; is generated by liquids condensing from a vapor back to a liquid or by breaking up a liquid into a dispersed state by splashing, foaming, or atomizing. *Also see* Aerosol.

Mitigate — (1) To cause to become less harsh or hostile; to make less severe, intense or painful; to alleviate. (2) To take actions to reduce or eliminate long-term risk to human life and property from natural, human-caused, and technological hazards and their effects. (3) One method of sizing up an emergency situation is to locate, isolate, and mitigate.

Mixture — Substance containing two or more materials not chemically united.

Miosis — Abnormal contraction of the pupils resulting in a pinpoint appearance.

Monitor — (1) Measure radioactive emissions from a substance with monitoring device. (2) Closely follow radio communications. (3) Observe and record the activities of an individual performing a function.

MSDS — Abbreviation for Material Safety Data Sheet.

Mutagen — Material that causes changes in the genetic system of a cell in ways that can be transmitted during cell division; effects may be hereditary.

N

National Fire Protection Association (NFPA®) — U.S. nonprofit educational and technical association located in Quincy, MA, devoted to protecting life and property from fire by developing fire protection standards and educating the public.

National Incident Management System - Incident Command System — The U.S. mandated system that defines the roles, responsibilities, and standard operating procedures used to manage emergency operations. Also called NIMS-ICS.

National Institute for Occupational Safety and Health (NIOSH) — U.S. government agency under the Centers for Disease Control and Prevention (CDC), U.S. Department of Health and Human Services, that helps ensure that the workplace and associated equipment are safe; investigates workplaces, recommends safety measures, and reports about on-the-job fire injuries.

National Response Framework (NRF) — Document that provides guidance on how communities, States, the Federal Government, and private-sector and nongovernmental partners conduct all-hazards emergency response.

Nephrotoxic Agent (Nephrotoxin) — Chemical that damages the kidneys.

Nerve Agent — Toxic agent that attacks the nervous system by affecting the transmission of impulses. *Also see* Chemical Agent.

Neurotoxic Agent (Neurotoxin) — Chemical that damages the central nervous system.

Neutron — Part of the nucleus of an atom that has a neutral electrical charge yet produces highly penetrating radiation; ultrahigh energy particle that has a physical mass like alpha or beta radiation but has no electrical charge. *Also see* Radiation (2).

NFPA® — Abbreviation for National Fire Protection Association.

NFPA® 704 Labeling System — System for identifying hazardous materials in fixed facilities. *Also see* NFPA® 704 Placard.

NFPA® 704 Placard — Color-coded, symbol-specific placard affixed to a structure to inform of fire hazards, life hazards, special hazards, and reactivity potential. The placard is divided into sections that identify the degree of hazard according to health, flammability, and reactivity as well as special hazards. *Also see* NFPA® 704 Labeling System.

Night Inversion — *See* Inversion (2).

NIMS — Acronym for National Incident Management System.

NIOSH — Acronym for National Institute for Occupational Safety and Health.

Noded Spheroid Tank — Low-pressure fixed facility storage tank held together by a series of internal ties and supports that reduce stress on the external shell. Also see Pressure Storage Tank, Spheroid Tank, and Low-Pressure Storage Tank.

Nonbulk Packaging — Package that has the following characteristics: (a) maximum capacity of 119 gallons (450 L) or less as a receptacle for a liquid, (b) maximum net mass of 882 pounds (400 kg) or less and a maximum capacity of 119 gallons (450 L) or less as a receptacle for a solid, and (c) water capacity of 1,000 pounds (454 kg) or less as a receptacle for a gas. *Also see* Packaging (1) and Bulk Packaging.

Noncombustible — Incapable of supporting combustion under normal circumstances. *Also see* Combustion and Nonflammable.

Nonflammable — Incapable of combustion under normal circumstances; normally used when referring to liquids or gases. *Also see* Noncombustible and Flammable.

Nonflammable Gas — Compressed gas not classified as flammable. *Also see* Flammable Gas, Compressed Gas, and Gas.

Nonintervention Mode — *See* Nonintervention Strategy.

Nonintervention Operations — Operations in which responders take no direct actions on the actual problem. *Also see* Defensive Operations and Offensive Operations.

Nonintervention Strategy — Overall plan for incident control established by the Incident Commander in which responders take no direct actions on the actual problem. Sometimes referred to as *natural stabilization*. *Also see* Defensive Strategy and Offensive Strategy.

Nonionizing Radiation — Series of energy waves composed of oscillating electric and magnetic fields traveling at the speed of light. Examples: ultraviolet radiation, visible light, infrared radiation, microwaves, radio waves, and extremely low frequency radiation. *Also See* Radiation (2) and Ionizing Radiation.

Nonliquefied Gas — Gas, other than a gas in a solution, that under the charging pressure is entirely gaseous at 70°F (21°C). *Also see* Liquefied Compressed Gas and Gas.

Nonpersistent Agent — Chemical agent that generally vaporizes and disperses quickly (less than 10 minutes). *Also see* Persistent Agent and Vapor Pressure.

Nonpressure Intermodal Tank — Portable tank that transports liquids or solids at a maximum pressure of 100 psi (689 kPa) {6.9 bar}. Also called *IM portable tank*. *Also see* Pressure Intermodal Tank and Intermodal Container.

Nonpressure Liquid Tank — Cargo tank truck used to carry flammable liquids (such as gasoline and alcohol), combustible liquids (such as fuel oil), Division 6.1 poisons, and liquid food products. *Also see* Cargo Tank Truck.

Nonpressure Storage Tank — *See* Atmospheric Storage Tank.

Noxious — Physically harmful or destructive to living beings; unwanted or troublesome.

NRC — Abbreviation for Nuclear Regulatory Commission.

NRF — Abbreviation for National Response Framework.

Nuclear Radiation — *See* Radiation (2).

Nuclear Regulatory Commission (NRC) — U.S. agency that regulates commercial nuclear power plants and the civilian use of nuclear materials as well as the possession, use, storage, and transfer of radioactive materials.

O

Objective — (1) Purpose to be achieved by tactical units at an emergency. (2) Specific, measurable, achievable statement of intended accomplishment.

Occupancy — (1) General fire service term for a building, structure, or residency. (2) Building code classification based on the use to which owners or tenants put buildings or portions of buildings. Regulated by the various building and fire codes. Also called Occupancy Classification.

Occupational Safety and Health Administration (OSHA) — U.S. federal agency that develops and enforces standards and regulations for occupational safety in the workplace.

ODP — Abbreviation for the Office for Domestic Preparedness.

Offensive Mode — *See* Offensive Strategy.

Offensive Operations — Operations in which responders take aggressive, direct action on the material, container, or process equipment involved in an incident. Also see Defensive Operations and Nonintervention Operations.

Offensive Strategy — Overall plan for incident control established by the Incident Commander (IC) in which responders take aggressive, direct action on the material, container, or process equipment involved in an incident. *Also see* Defensive Strategy and Nonintervention Strategy.

Off-Gassing — Emission of toxic gases; the release of chemicals from non-metallic substances under ambient or greater pressure conditions.

Office for Domestic Preparedness (ODP) — Former U.S. agency under the Department of Homeland Security that issues federal emergency responder guidelines for events involving weapons of mass destruction.

OI — Abbreviation for Operating Instruction.

Olfactory Fatigue — Gradual inability of a person to detect odors after initial exposure; may be extremely rapid in the case of some toxins such as hydrogen sulfide.

Operating Instruction (OI) — *See* Predetermined Procedures.

Operations Level — Level of training established by OSHA allowing first responders to take defensive actions at hazardous materials incidents. *Also see* Awareness Level.

Organic Peroxide — Any of several organic derivative of the inorganic compound hydrogen peroxide.

Organophosphate Pesticides — Group of chemicals used to kill insects by disrupting their brains and nervous systems; these chemicals inactivate acetylcholinesterase, an enzyme which is essential to nerve function in insects, humans, and many other animals.

ORM-D — Abbreviation for Other Regulated Materials.

OSHA — Acronym for Occupational Safety and Health Administration.

Other Regulated Material (ORM-D) — Material that does not meet the definition of hazardous material and is not included in any other hazard class but possess enough hazardous characteristics that it requires some regulation; presents limited hazard during transportation because of their form, quantity, and packaging. *Also see* Material and Hazardous Material.

Overturn Protection — Protection for fittings on top of a tank in case of rollover; may be combined with flashing rail or flashing box.

Oxidation — Chemical process that occurs when a substance combines with oxygen; common example is the formation of rust on metal.

Oxidizer — Any substance or material that yields oxygen readily and may stimulate the combustion of organic and inorganic matter. *Also see* Oxidizing Agent.

Oxidizing Agent — A substance that oxidizes another substance; oxdizing agents can cause other materials to combust more readily (or upon contact) or make fires burn more strongly. *Also see* Oxidizer.

P

Package Marking — Descriptive name, instructions, cautions, weights, and specification marks required on the outside of hazardous materials containers. *Also see* Marking, Label, and Placard.

Packaging — (1) Term used by the U.S. Department of Transportation to describe shipping containers and their markings, labels, and/or placards. *Also see* Bulk Packaging, Combination Packaging, Composite Packaging, Excepted Packaging, Individual Container, Industrial Packaging, Intermediate Bulk Container (IBC), Bulk Packaging, Type A Packaging, Type B Packaging, Nonbulk Packaging, and Strong, Tight Container. (2) Readying a victim for transport.

Pandemic — An epidemic occurring over a very wide area (several countries or continents) and usually affecting a large proportion of the population.

PAPR — Abbreviation for powered air-purifying respirator.

Parts Per Million (ppm) — Method of expressing the concentration of very dilute solutions of one substance in another, normally a liquid or gas, based on volume expressed as a ratio of the volume of contaminants (parts) compared to the volume of air (million parts).

PASS — Acronym for Personal Alert Safety System.

Pathogens — Organisms that cause infection such as viruses and bacteria.

Patient Decontamination — Removing contamination from injured patients or victims. *Also see* Decon.

PCB — Abbreviation for Polychlorinated Biphenyl.

PEL — Acronym for Permissible Exposure Limit.

PEL-C — Acronym for Permissible Exposure Limit/Ceiling Limit.

Penetration — Process in which a hazardous material enters an opening or a puncture in a protective material.

Permeation — Process in which a chemical passes through a protective material on a molecular level.

Permissible Exposure Limit (PEL) — Maximum time-weighted concentration at which 95 percent of exposed, healthy adults suffer no adverse effects over a 40-hour workweek; an 8-hour time-weighted average unless otherwise noted; expressed in either parts per million (ppm) or milligrams per cubic meter (mg/m^3). *Also see* Immediately Dangerous to Life or Health (IDLH), Recommended Exposure Limit (REL), Short-Term Exposure Limit (STEL), Threshold Limit Value (TLV®), and Permissible Exposure Limit/Ceiling Limit (PEL-C).

Permissible Exposure Limit/Ceiling Limit (PEL-C) — Maximum concentration to which an employee may be exposed at any time, even instantaneously, as established by Occupational Safety and Health Administration (NIOSH). *Also see* Permissible Exposure Limit (PEL).

Peroxidizable Compound — Material apt to undergo spontaneous reaction with oxygen at room temperature and form peroxides and other products.

Persistence — (1) Length of time a chemical agent remains effective without dispersing. (2) Length of time a chemical remains in the environment. (3) Length of time a chemical agent remains as a liquid; typically a liquid chemical agent is considered persistent if it remains for longer than 24 hours. *Also see* Persistent Agent, Nonpersistent Agent, and Dispersion.

Persistent Agent — Chemical agent that remains effective in the open (at the point of dispersion) for a considerable period of time (more than 10 minutes). *Also see* Persistence, Nonpersistent Agent, and Dispersion.

Personal Alert Safety System (PASS) — Electronic lack-of-motion sensor that sounds a loud tone when an emergency responder becomes motionless; can also be triggered manually. *Also see* Personal Protective Equipment (PPE).

Personal Protective Equipment (PPE) — General term for the equipment worn by fire and emergency services responders; includes helmets, coats, pants, boots, eye protection, gloves, protective hoods, self-contained breathing apparatus (SCBA), and personal alert safety system (PASS) devices. When working with hazardous materials, this may include Chemical Protective Clothing and Special Protective Clothing. Also called *bunker clothes, protective clothing, turnout clothing,* or *turnout gear,* and *full structural protective clothing.*

Person-Borne Improvised Explosives Device (PBIED) — An improvised explosive device carried by a person; includes suicide bombers as well as individuals coerced into carrying the bomb against their will. *Also see* Improvised Explosive Device, High Explosive, and Explosion.

Pesticide — Any substance or mixture of substances intended to be used to control, repel, or kill pests such as insects, mice and other animals, unwanted plants (weeds), fungi, or microorganisms like bacteria and viruses.

Petroleum Carrier — Tank vessel that transports crude or finished petroleum products. *Also see* Tanker.

pH — Measure of acidity of an acid or the level of alkaline in a base. *Also see* Acid and Base.

Phase — Distinguishable part in a course, development, or cycle; aspect or part under consideration.

Phosphine — A colorless, flammable, and toxic gas with an odor of garlic or decaying fish which can ignite spontaneously on contact with air; phosphine is a respiratory tract irritant that attacks primarily the cardiovascular and respiratory systems causing peripheral vascular collapse, cardiac arrest and failure, and pulmonary edema.

Photon — Packet of electromagnetic energy.

Physical Hazard — Material that present a threat to health because of its physical properties. *Also see* Health Hazard.

Physical Properties — Those properties that do not involve a change in the chemical identity of a substance; however, they affect the physical behavior of the material inside and outside the container, which involves the change of the state of the material; examples: boiling point, specific gravity, vapor density, and water solubility.

Pictogram — Drawing or symbol that indicates information.

Placard — Diamond-shaped sign that is affixed to each side of a structure or vehicle transporting hazardous materials to inform people of fire hazards, life hazards, special hazards, and reactivity potential; indicates the primary class of the material and, in some cases, the exact material being transported; required on containers that are 640 cubic feet ($18\,m^3$) or larger.

Plain Language — Communication that can be understood by the intended audience and meets the purpose of the communicator. For the purpose of the *National Incident Management System,* plain language is designed to eliminate or limit the use of codes and acronyms, as appropriate, during incident response involving more than a single agency. Also called *plain text.*

Planning Section — Section responsible for the collection, evaluation, dissemination, and use of information about the development of an incident and status of resources; includes the Situation, Resources, Documentation, and Demobilization Units as well as technical specialists. *Also see* Incident Management System (IMS).

Plume — Irregularly shaped pattern of an airborne hazardous material where wind and/or topography influence the downrange course from the point of release. *Also see* Cloud, Cone, and Hemispheric Release.

Pneumatic — Operated by air or compressed air.

Poison — Any material that when taken into the body is injurious to health. *Also see* Toxin.

Polar Solvents — (1) Flammable liquids that have an attraction for water, much like a positive magnetic pole attracts a negative pole; examples include alcohols, esters, ketones, and amines. (2) A liquid having a molecule in which the positive and negative charges are permanently separated resulting in their ability to ionize in solution and create electrical conductivity. Water, alcohol, and sulfuric acid are examples of polar solvents.

Polychlorinated Biphenyl (PCB) — Toxic compound found in some old oil-filled electric transformers.

Polymerization — Reactions in which two or more molecules chemically combine to form larger molecules; reaction can often be violent. *Also see* Inhibitor.

Pounds Per Square Inch (psi) — Unit for measuring pressure in the English or Customary System. Its International System equivalents are kilopascals (kPa) and bar.

Powered Air-Purifying Respirator (PAPR) — A motorized respirator system that uses a filter to clean surrounding air before delivering to the wearer to breathe; typically includes a blower/battery box worn on the belt, headpiece, and breathing tube.

PPE — Abbreviation for Personal Protective Equipment.

ppm — Abbreviation for Parts Per Million.

Predetermined Procedures — Standard methods or rules in which an organization performs routine functions in addition to operating actions to perform at every possible type of emergency incident. Also known as *standard operating procedure (SOP), standard operating guideline (SOG),* and *operating instruction (OI).*

Pre-Incident Survey — Survey of a facility or location made before an emergency occurs in order to prepare for an appropriate emergency response. Sometimes called Preplan.

Preparedness — Activities to ensure that people are ready for a disaster and respond to it effectively; includes determining what will be done if essential services are lost, developing a plan for contingencies, and practicing the plan.

Preplan — *See* Pre-Incident Survey.

Pressure Intermodal Tank — Liquefied gas container designed for working pressures of 100 to 500 psi (689 kPa to 3 447 kPa) {6.9 bar to 34.5 bar}. Also known as *Spec 51* or *IMO Type 5*. *Also see* Nonpressure Intermodal Tank and Intermodal Container.

Pressure Storage Tank — Class of fixed facility storage tanks divided into two categories: low-pressure storage tanks and pressure vessels. *Also see* Low-Pressure Storage Tank, Pressure Vessel, and Atmospheric Storage Tank.

Pressure Tank Railcar — Tank railcars that carry flammable and nonflammable liquefied gases, poisons, and other hazardous materials. They are recognizable by the valve enclosure at the top of the car and the lack of bottom unloading piping.

Pressure Vessel — Fixed-facility storage tank with operating pressures above 15 psi (103 kPa) {1.03 bar}. *Also see* Low-Pressure Storage Tank, Pressure Storage Tank, and Spherical Pressure Vessel.

Prill — Spherical pellets.

Primary Explosive — High explosive that is easily initiated and highly sensitive to heat; often used as detonators.

Primary Label — Label placed on the container of a hazardous material to indicate the primary hazard. *Also see* Subsidiary Label.

Product — Generic term used in industry to describe a substance that is used or produced in an industrial process. *Also see* Hazardous Material.

Product Identification Number (PIN) — Number assigned by the United Nations and used in the *Emergency Response Guidebook (ERG)* to identify specific product names.

Prosecute — To charge someone with a crime; a prosecutor tries a criminal case on behalf of the government.

Protective Action Distance — Downwind distance from a hazardous materials incident within which protective actions should be implemented. *Also see* Initial Isolation Distance, Protective Action Zone, and Protective Actions.

Protective Action Zone — Area immediately adjacent to and downwind from the initial isolation zone, which is in imminent danger of being contaminated by airborne vapors within 30 minutes of material release. *Also see* Initial Isolation Zone, Protective Action Distance and Protective Actions.

Protective Actions — Steps taken to preserve health and safety of emergency responders and the public. *Also see* Protective Action Distance and Protective Action Zone.

psi — Abbreviation for Pounds Per Square Inch.

Pulmonary Edema — Accumulation of fluid in the lungs.

R

R — Abbreviation for Roentgen.

rad — Acronym for Radiation Absorbed Dose.

Radiation — (1) Transmission or transfer of heat energy from one body to another at a lower temperature through intervening space by electromagnetic waves such as infrared thermal waves, radio waves, or X rays. Also called *radiated heat.* (2) Energy from a radioactive source emitted in the form of waves or particles; emission of radiation as a result of the decay of an atomic nucleus; process know as *radioactivity.* Also called *nuclear radiation. Also see* Ionizing Radiation, Nonionizing Radiation, Radiation Absorbed Dose (rad), Radioactive Material (RAM), Radioactive Particles, Alpha Radiation, Beta Radiation, and Gamma Radiation.

Radiation Absorbed Dose (rad) — English System unit used to measure the amount of radiation energy absorbed by a material. Its International System equivalent is gray. *Also see* Radiation (2) and Radioactive Material (RAM).

Radiation-Emitting Device (RED) — A powerful gamma-emitting radiation source used as a weapon.

Radioactive Material (RAM) — Material whose atomic nucleus spontaneously decays or disintegrates, emitting radiation. *Also see* Radiation (2), Curie (Ci), and Becquerel (Bq).

Radioactive Particles — Particles emitted during the process of radioactive decay; three types: alpha, beta, and gamma. *Also see* Radiation (2)

Radioactivity — *See* Radiation (2).

Radioisotope — An unstable or radioactive isotope (form) of an element that can change into another element by giving off radiation. *Also see* Radionuclide.

Radiological Dispersal Device (RDD) — Device that spreads radioactive contamination over the widest possible area by detonating conventional high explosives wrapped with radioactive material. Also called *dirty bomb.*

Radiological Dispersal Weapons (RDW) — Device that spreads radioactive contamination without using explosives; instead, radioactive contamination is spread using pressurized containers, building ventilation systems, fans, and mechanical devices.

Radionuclide — Any radioactive isotope (form) of any element. *Also see* Radioisotope.

Railcar Initials and Numbers — *See* Reporting Marks.

Rain-Down Method — This method of foam application directs the stream into the air above the unignited or ignited spill or fire and allows the foam to float gently down onto the surface of the fuel.

RAM — Acronym for Radioactive Material.

RDD — Abbreviation for Radiological Dispersal Device.

Reactive Material — Substance capable of or tending to react chemically with other substances; examples: materials that react violently when combined with air or water.

Reactivity/Instability — Ability of two or more chemicals to react and release energy and the ease with which this reaction takes place. *Also see* Chemical Reaction and Reactive Material.

Reagent — A chemical that is known to react to another chemical or compound in a specific way, often used to detect or synthesize another chemical.

Recommended Exposure Limit (REL) — Recommended value expressing the maximum time-weighted dose or concentration to which workers should be exposed over a 10-hour period as established by National Institute for Occupational Safety and Health (NIOSH). *Also see* Permissible Exposure Limit (PEL), Short-Term Exposure Limit (STEL), and Threshold Limit Value (TLV®).

Reconnaissance — Process of examining an area to obtain information about the current specific situation and probable fire behavior and other related fire-suppression information.

Recovery — (1) Process of restoring normal public or utility services following a disaster; activities necessary to rebuild after a disaster such as rebuilding homes and businesses, clearing debris, repairing roads and bridges, and restoring water and other essential services. (2) Situation in which the victim is dead, and the goal of the operation is to recover the body.

RED — Abbreviation for Radiation-Emitting Device.

Redundancy — Secondary or backup systems that allow for uninterrupted use in the event of failure or damage to the primary system.

Reefer — *See* Refrigerated Intermodal Container.

Refrigerated Intermodal Container — Cargo container having its own refrigeration unit; also called *reefer.*

Refrigeration Unit — Cooling equipment used to maintain a constant temperature within a given space.

Regulations — Rules or directives of administrative agencies that have authorization to issue and enforce them.

Rehab — Abbreviation for Rehabilitation.

Rehabilitation (Rehab) — (1) Allowing emergency responders to rest, rehydrate, and recover during an incident; also refers to a station at an incident where personnel can rest, rehydrate, and recover. (2) Activities necessary to repair environmental damage.

REL — Abbreviation for Recommended Exposure Limit.

rem — Acronym for Roentgen Equivalent in Man.

Remediation — Act of fixing or correcting a fault, error, or deficiency.

Reporting Marks — Combination of letters and numbers stenciled on rail tank cars that may be used to get information about the car's contents from the railroad's computer or the shipper. Also called *railcar initials and numbers.*

Resources — All of the immediate or supportive assistance available, or potentially available, for assignment to help control an incident, including personnel, equipment, control agents, agencies, and printed emergency guides.

Response Objectives — Statements based on realistic expectations of what can be accomplished when all allocated resources have been effectively deployed that provide guidance and direction for selecting appropriate strategies and the tactical direction of resources.

RIBC — Abbreviation for Rigid Intermediate Bulk Container.

Rickettsia — Specialized bacteria that live and multiply in the gastrointestinal tract of arthropod carriers (such as ticks and fleas). *Also see* Bacteria and Virus.

Rigid Intermediate Bulk Container (RIBC) — *See* Intermediate Bulk Container (IBC)

Ring Stiffener — Circumferential tank shell stiffener that helps to maintain the tank's cross section.

Riot Control Agent — Chemical compound that temporarily makes people unable to function by causing immediate irritation to the eyes, mouth, throat, lungs, and skin. Also called *tear gas* or *irritating agent*. *Also see* Incapacitant.

Risk — Estimated effect that a hazard would have on people, services, facilities, and structures in a community; likelihood of a hazard event resulting in an adverse condition that causes injury or damage; often expressed as *high, moderate,* or *low* or in terms of potential monetary losses associated with the intensity of the hazard.

Risk Assessment — Process for evaluating risk associated with a specific hazard defined in terms of probability and frequency of occurrence, magnitude and severity, exposure, and consequences.

Risk-Based Response — Method using hazard and risk assessment to determine an appropriate mitigation effort based on the circumstances of the incident.

Risk Management Plan — Written plan that identifies and analyzes the exposure to hazards, selection of appropriate risk management techniques to handle exposures, implementation of chosen techniques, and monitoring of the result of using those techniques. *Also see* Hazard or Risk Analysis and Hazard Assessment.

Roentgen (R) — English System unit used to measure radiation exposure, applied only to gamma and X-ray radiation; the unit used on most U.S. dosimeters. *Also see* Radiation (2), Radioactive Material (RAM), Roentgen Equivalent in Man (rem), Gamma Radiation, and Radiation Absorbed Dose (rad).

Roentgen Equivalent in Man (rem) — English System unit used to express the radiation absorbed dose (rad) equivalence as pertaining to a human body; applied to all types of radiation; used to set radiation dose limits for emergency responders. *Also see* Roentgen (R), Radiation (2), Radioactive Material (RAM), and Radiation Absorbed Dose (rad).

Roll-On Application Method (Bounce) — Method of foam application in which the foam stream is directed at the ground at the front edge of the unignited or ignited liquid fuel spill; foam then spreads across the surface of the liquid.

Roll-On/Roll-Off Vessel — Ship with large stern and side ramp structures that are lowered to allow vehicles to be driven on and off the vessel. *Also see* Cargo Vessel.

Routes of Entry — Pathways by which hazardous materials get into (or affect) the human body; commonly listed routes are inhalation, ingestion, skin contact, injection, and absorption.

RPG — Abbreviation for Rocket-Propelled Grenade.

S

Safety Data Sheet (SDS) — A sixteen-section information sheet provided by a chemical manufacturer or importer that contains information such as the chemical composition, physical and chemical properties, health and safety hazards, emergency response procedures, and transportation and regulatory information of the specified material. The SDS is the globally harmonized version of the *material safety data sheet,* a similar form required by 29 *CFR* 1910.1200, OSHA's Hazard Communications Standard. *Also see* Material Safety Data Sheet.

Safety Officer — (1) Fire officer whose primary function is to administrate safety within the entire scope of fire department operations. Also referred to as the Health and Safety Officer. (2) Member of the IMS Command Staff responsible to the incident commander for monitoring and assessing hazardous and unsafe conditions and developing measures for assessing personnel safety on an incident. Also referred to as the Incident Safety Officer.

Safety Relief Valve — Device on cargo tanks with an operating part held in place by a spring. The valve opens at preset pressures to relieve excess pressure and prevent failure of the vessel.

SAR — Abbreviation for Supplied-Air Respirator.

SARA — Acronym for Superfund Amendments and Reauthorization Act.

Sarin — A fluorinated phosphinate chemical warfare agent classified as a nerve agent, also known by the designation, GB.

SCBA — Abbreviation for Self-Contained Breathing Apparatus.

Scene Control Zones — *See* Hazard-Control Zones.

Search Warrant — Written order, in the name of the People, State, Province, Territory, or Commonwealth, signed by a magistrate (or judge), commanding a peace officer to search for personal property or other evidence and return it to the magistrate.

Secondary Contamination — *See* Cross Contamination.

Secondary Decontamination — Taking a shower after having completed a technical decontamination process. *Also see* Decon and Technical Decontamination.

Secondary Device — Bomb placed at the scene of an ongoing emergency response that is intended to cause casualties among responders; secondary explosive devices are designed to explode after a primary explosion or other major emergency response event has attracted large numbers of responders to the scene.

Secondary Explosive — High explosive that is designed to detonate only under specific circumstances.

Self-Contained Breathing Apparatus (SCBA) — Respirator worn by the user that supplies a breathable atmosphere that is carried in or generated by the apparatus and is independent of the ambient atmosphere. *Also see* Supplied-Air Respirator (SAR).

Self-Presenters — Individuals who seek medical assistance and have not been treated or undergone decontamination at the incident scene.

Sensitizer — *See* Allergen.

SETIQ — Abbreviation for the Emergency Transportation System for the Chemical Industry (Mexico).

Shelter in Place — Having occupants remain in a structure or vehicle in order to provide protection from a rapidly approaching hazard (fire, hazardous gas cloud, etc.). Also called sheltering, protection-in-place, and taking refuge. *Also see* Evacuation.

Shipping Papers — Shipping order, bill of lading, manifest, waybill, or other shipping document issued by the carrier. *Also see* Bill of Lading, Lading, Consist, Waybill, and Air Bill.

Shock Front — The boundary between the pressure disturbance created by an explosion (in air, water, or earth) and the ambient atmosphere, water, or earth.

Short-Term Exposure Limit (STEL) — Fifteen-minute time-weighted average that should not be exceeded at any time during a workday; exposures should not last longer than 15 minutes nor be repeated more than four times per day with at least 60 minutes between exposures. *Also see* Immediately Dangerous to Life or Health (IDLH), Permissible Exposure Limit (PEL), Recommended Exposure Limit (REL), and Threshold Limit Value (TLV®).

Shrapnel Fragmentation — Small pieces of debris thrown from a container or structure that ruptures from containment or restricted blast pressure.

Signal Word — Government-mandated warnings provided on product labels that indicate the level of toxicity, for example CAUTION, WARNING, or DANGER.

Simple Asphyxiant — Any inert gas that displaces or dilutes oxygen below the level needed by the human body. *Also see* Asphyxiant and Inert Gas.

Site Characterization — The size up and evaluation of hazards, problems, and potential solutions of a site.

Site Work Zones — *See* Hazard-Control Zones.

Situational Awareness — An individual's perception and comprehension of the details of their surrounding environment, and the understanding of how events occurring in the moment may affect the future.

Size-Up — Ongoing mental evaluation process performed by the operational officer in charge of an incident that enables him or her to determine and evaluate all existing influencing factors that are used to develop objectives, strategy, and tactics for fire suppression before committing personnel and equipment to a course of action. Size-up results in a plan of action that may be adjusted as the situation changes. It includes such factors as time, location, nature of occupancy, life hazard, exposures, property involved, nature and extent of fire, weather, and fire fighting facilities.

SKED® — Compact device for patient removal from a confined space with spinal immobilization; may be used as a lifting device with a harness.

Skin Contact — Occurrence when a chemical or hazardous material (in any state — solid, liquid, or gas) contacts the skin or exposed surface of the body (such as the mucous membranes of the eyes, nose, or mouth). *Also see* Routes of Entry.

Slurry — (1) Watery mixture of insoluble matter such as mud, lime, or Plaster of Paris. (2) Thick mixture formed when a fire-retardant chemical is mixed with water and a viscosity agent.

Smallpox — A serious, contagious, and sometimes fatal infectious disease. There is no specific treatment for smallpox disease, and the only prevention is vaccination.

SOG — Abbreviation for Standard Operating Guideline.

Soluble — Capable of being dissolved in a liquid (usually water). *Also see* Water Solubility.

SOP — Abbreviation for Standard Operating Procedure.

Sorbent — A material, compound, or system that holds contaminants by adsorption or absorption. With *adsorption,* the contaminant molecule is retained on the surface of the sorbent granule by physical attraction. With *absorption,* a solid or liquid is taken up or absorbed into the sorbent material.

Sorption — Method of removing contaminants; used in vapor- and gas-removing respirators.

Spec 51 — *See* Pressure Intermodal Tank.

Specification Marking — Stencil on the exterior of tank cars indicating the standards to which the tank car was built; specification markings may also be found on intermodal containers and cargo tank trucks.

Specific Gravity — Weight of a substance compared to the weight of an equal volume of water at a given temperature. Specific gravity less than 1 indicates a substances lighter than water; specific gravity greater than 1 indicates a substance heavier than water. *Also see* Physical Properties.

Spherical Pressure Vessel — Round-shaped fixed facility pressure vessel. *Also see* Pressure Vessel.

Spheroid Tank — Round- or oval-shaped fixed facility low-pressure storage tank. *Also see* Low-Pressure Storage Tank, Noded Spheroid Tank, and Pressure Storage Tanks.

Spontaneous Ignition — Combustion of a material initiated by an internal chemical or biological reaction producing enough heat to cause the material to ignite. Also called *spontaneous combustion.*

Stabilization — (1) Stage of an incident when the immediate problem or emergency has been controlled, contained, or extinguished. (2) Process of providing additional support to key places between an object of entrapment and the ground or other solid anchor point to prevent unwanted movement.

Stabilizer — *See* Inhibitor.

Staging Area — Location where incident personnel and equipment are assigned on an immediately available status.

Standard Operating Guideline (SOG) — *See* Standard Operating Procedure (SOP).

Standard Operating Procedure (SOP) — Standard methods or rules in which an organization or a fire department operates to carry out a routine function. Usually these procedures are written in a policies and procedures handbook and all firefighters should be well versed in their content. A SOP may specify the functional limitations of fire brigade members in performing emergency operations. Also called *Standard Operating Guidelines.*

Standard Transportation Commodity Code (STCC Number) — Numerical code used by the rail industry on the waybill to identify the commodity.

STCC Number — *See* Standard Transportation Commodity Code.

STEL — Acronym for Short-Term Exposure Limit.

Stokes Basket — Basket-type litter — wire or plastic — suitable for transporting patients from locations (such as a pile of rubble, structural collapse, or the upper floor of a building) where a standard litter would not be easily secured; may be used with a harness for lifting.

Strategic Goals — Broad statements of desired achievement to control an incident; achieved by the completion of tactical objectives. *Also see* Strategy, Tactical Objectives, and Tactics.

Strategy — Overall plan for incident attack and control established by the Incident Commander (IC). *Also see* Tactics, Nonintervention Strategy, Defensive Strategy, and Offensive Strategy.

Street Clothes — Clothing that is anything other than chemical protective clothing or structural firefighters' protective clothing, including work uniforms and ordinary civilian clothing.

Stress — (1) State of tension put on a shipping container by internal or external chemical, mechanical, or thermal change. (2) Factors that work against the strength of any piece of apparatus or equipment. (3) Any condition causing bodily or mental tension. Also see Critical Incident Stress (CIS) and Posttraumatic Stress Disorder (PTSD).

Strong Oxidizer — Material that encourages a strong reaction (by readily accepting electrons) from a reducing agent (fuel). *Also see* Oxidizer.

Strong, Tight Container — Packaging used to ship materials of low radioactivity. Also see Excepted Packaging, Industrial Packaging, Type A Packaging, Type B Packaging, and Packaging (1).

Structural Firefighters' Protective Clothing — General term for the equipment worn by fire and emergency services responders; includes helmets, coats, pants, boots, eye protection, gloves, protective hoods, self-contained breathing apparatus (SCBA), and personal alert safety system (PASS) devices. Also called *full structural protective clothing* or *bunker gear.*

Subsidiary Label — Label indicating a secondary hazard associated with a material. *Also see* Primary Label.

Suitcase Bomb — Small, suitcase-or backpack-sized nuclear weapon. *Also see* Improvised Nuclear Device.

Superfund Amendments and Reauthorization Act (SARA) — U.S. law that in 1986 reauthorized the Comprehensive Environmental Response, Compensation and Liability Act (CERCLA) to continue cleanup activities around the country and included several site-specific amendments, definitions clarifications, and technical requirements.

Supplied-Air Respirator (SAR) — Atmosphere-supplying respirator for which the source of breathing air is not designed to be carried by the user; not certified for fire-fighting operations. Also known as an *Airline Respirator. Also see* Self-Contained Breathing Apparatus (SCBA).

Syndromic Surveillance — Surveillance using health-related data that precede diagnosis and signal a sufficient probability of a case or an outbreak to warrant further public health response.

Systemic Effect — Something that affects an entire system rather than a single location or entity.

T

Tactical Objectives — Specific operations that must be accomplished to achieve strategic goals. *Also see* Strategic Goals and Tactics.

Tactics — Methods of employing equipment and personnel on an incident to accomplish specific tactical objectives in order to achieve established strategic goals. *Also see* Tactical Objectives, Strategy, and Strategic Goals.

Tanker — Vessel (ship) that exclusively carries liquid products in bulk; also called tank vessel. *Also see* Chemical Carrier, Liquefied Flammable Gas Carrier, and Petroleum Carrier.

Tank Motor Vehicle — *See* Cargo Tank Truck.

Tank Truck — *See* Cargo Tank Truck.

Tank Vessel — *See* Tanker.

TC — Abbreviation for Transport Canada.

Tear Gas — *See* Riot Control Agent.

Technical Assistance — Personnel, agencies, or printed materials that provide technical information on handling hazardous materials or other special problems.

Technical Decontamination — Using chemical or physical methods to thoroughly remove contaminants from responders (primarily entry team personnel) and their equipment; usually conducted within a formal decontamination line or corridor following gross decontamination; also called *formal decontamination*. *Also see* Decontamination, Gross Decontamination, and Decontamination Corridor.

Temperature Inversion — Meteorological condition in which the temperature of the air some distance above the earth's surface is higher than the temperature of the air at the surface. Normally, air temperatures decrease as altitude increases. An air inversion traps air, gases, and vapors near the surface, and impedes their dispersion.

Teratogen — Chemical that interferes with the normal growth of an embryo, causing malformations in the developing fetus.

Terrorism — Unlawful use of force or violence against persons or property for the purpose of intimidating or coercing a government, the civilian population or any segment thereof, in furtherance of political or social objectives; defined by the U.S. Federal Bureau of Investigation (FBI). *Also see* Improvised Explosive Device, Chemical Attack, and Biological Attack.

Threshold Limit Value (TLV®) — Concentration of a given material in parts per million (ppm) that may be tolerated for an 8-hour exposure during a regular workweek without ill effects. *Also see* Threshold Limit Value/Ceiling (TLV®/C), Short-Term Exposure Limit (STEL), Threshold Limit Value/Short-Term Exposure Limit (TLV®/STEL), Permissible Exposure Limit (PEL), Recommended Exposure Limit (REL), and Threshold Limit Value/Time Weighted Average (TLV®/TWA).

Threshold Limit Value/Ceiling (TLV®/C) — Maximum concentration of a given material in parts per million (ppm) that should not be exceeded, even instantaneously. *Also see* Threshold Limit Value (TLV®), Threshold Limit Value/Short-Term Exposure Limit (TLV®/STEL), Short-Term Exposure Limit (STEL), Permissible Exposure Limit (PEL), Recommended Exposure Limit (REL), and Threshold Limit Value/Time Weighted Average (TLV®/TWA).

Threshold Limit Value/Short-Term Exposure Limit (TLV®/STEL) — Fifteen-minute time-weighted average exposure that should not be exceeded at any time nor repeated more than four times daily with a 60-minute rest period required between each STEL exposure. These short-term exposures can be tolerated without suffering from irritation, chronic or irreversible tissue damage, or narcosis of a sufficient degree to increase the likelihood of accidental injury, impair self-rescue, or materially reduce worker efficiency. TLV/STELs are expressed in parts per million (ppm) and milligrams per cubic meter (mg/m³). *Also see* Short-Term Exposure Limit (STEL), Permissible Exposure Limit (PEL), Recommended Exposure Limit (REL), Threshold Limit Value (TLV®), Threshold Limit Value/Ceiling (TLV®/C), and Threshold Limit Value/Time Weighted Average (TLV®/TWA).

Threshold Limit Value/Time-Weighted Average (TLV®/TWA) — Maximum airborne concentration of a material to which an average, healthy person may be exposed repeatedly for 8 hours each day, 40 hours per week without suffering adverse effects; based upon current available data; are adjusted on an annual basis. *Also see* Threshold Limit Value (TLV®), Threshold Limit Value/Ceiling (TLV®/C), Permissible Exposure Limit (PEL), Recommended Exposure Limit (REL), Short-Term Exposure Limit (STEL), and Threshold Limit Value/Short-Term Exposure Limit (TLV®/STEL).

TIC — Abbreviation for Toxic Industrial Chemical.

TIM — Abbreviation for Toxic Industrial Material.

TLV® — Abbreviation for Threshold Limit Value.

TLV®/C — Abbreviation for Threshold Limit Value/Ceiling.

TLV®/STEL — Abbreviation for Threshold Limit Value/Short-Term Exposure Limit.

TLV®/TWA — Abbreviation for Threshold Limit Value/Time-Weighted Average.

Ton Container — Pressurized tank with a capacity of 1 short ton or approximately 2,000 pounds (907 kg or 0.9 tonne).

Topography — Physical configuration of the land or terrain; often depicted using contour lines.

Toxic Atmosphere — Any area, inside or outside a structure, where the air is contaminated by a poisonous substance that may be harmful to human life or health if it is inhaled, swallowed, or absorbed through the skin. *Also see* Toxic Gas.

Toxic Gas — Poisonous gas that contains poisons or toxins that are hazardous to life; many gaseous products of combustion are poisonous; toxic materials generally emit poisonous vapors when exposed to an intensely heated environment. *Also see* Gas, Toxic Material, and Toxin.

Toxic Industrial Chemical (TIC) — *See* Toxic Industrial Material (TIM).

Toxic Industrial Material (TIM) — Industrial chemical that is toxic at a certain concentration, is readily available, and could be used by terrorists to deliberately kill, injury, or incapacitate people. Also called *toxic industrial chemical (TIC)*.

Toxic Inhalation Hazard (TIH) — Volatile liquid or gas known to be a severe hazard to human health during transportation.

Toxicity — Ability of a substance to do harm within the body. *Also see* Toxin.

Toxic Material — Substance classified as a poison, asphyxiant, irritant, or anesthetic that can be poisonous if inhaled, swallowed, absorbed, or introduced into the body through cuts or breaks in the skin. *Also see* Toxic Gas, Asphyxiant, and Toxin.

Toxic Substances Control Act (TSCA) — U.S. law enacted in 1976 to give the Environmental Protection Agency (EPA) the ability to track the 75,000 industrial chemicals currently produced or imported into the United States.

Toxin — Substance that has the property of being poisonous. *Also see* Toxic Material, Biological Toxin, Toxic Gas, Toxicity, Poison, and Toxic Material.

Trace Evidence — Physical evidence that results from the transfer of small quantities of materials (for example, hair, textile fibers, paint chips, glass fragments, gunshot residue particles).

Train Consist — *See* Consist.

Transient Evidence — Evidence which by its very nature or the conditions at the scene will lose its evidentiary value if not preserved and protected (for example, blood in the rain).

Transport Canada (TC) — Agency responsible for developing and administering policies, regulations, and programs for a safe, efficient, and environmentally friendly transportation system in Canada; contributing to Canada's economic growth and social development; and protecting the physical environment.

Trench Foot — Foot condition resulting from prolonged exposure to damp conditions or immersion in water; symptoms include tingling and/or itching sensation, pain, swelling, cold and blotchy skin, numbness, and a prickly or heavy feeling in the foot. In severe cases, blisters can form followed by skin and tissue dying and falling off.

Triage — Screening and classification of sick or injured persons to determine medical priority needs and ensure effective and efficient use of medical personnel, equipment, and facilities.

TSCA — Abbreviation for Toxic Substances Control Act.

Tube Trailer — *See* Compressed-Gas Tube Trailer.

Type A Packaging — Container used to ship radioactive materials with relatively high radiation levels. *Also see* Packaging (1), Excepted Packaging, Industrial Packaging,

Type B Packaging, and Strong, Tight Container.

Type B Packaging — Container used to ship radioactive materials that exceed the limits allowed by Type A packaging such as materials that would present a radiation hazard to the public or the environment if there were a major release. *Also see* Packaging (1), Excepted Packaging, Industrial Packaging, Type A Packaging, and Strong, Tight Container.

U

UC — Abbreviation for Unified Command.

UEL — Abbreviation for Upper Explosive Limit.

UFL — Abbreviation for Upper Flammable Limit.

Unified Command (UC) — Shared command role that allows all agencies with responsibilities for an incident to manage the incident by establishing a common set of incident objectives and strategies.

Universal Precautions — A set of precautions designed to prevent transmission of biological pathogens (specifically, bloodborne pathogens) when providing first aid or health care.

Unstable Material — Material that readily undergoes chemical changes or decomposition.

Upper Explosive Limit (UEL) — Maximum concentration of vapor or gas in air that will allow combustion to occur. Concentrations above this are called "too rich" to burn. Also called *Upper Flammable Limit (UFL)*.

Urban Search and Rescue (US&R) — Operational activities that include locating, extricating, and providing on-site medical treatment to victims trapped in collapsed structures or other environments.

US&R — Acronym for Urban Search and Rescue.

USCG — Abbreviation for U.S. Coast Guard.

U.S. Coast Guard (USCG) — U.S. military, multimission, and maritime service, whose mission is to protect the public, the environment, and U.S. economic interests in U.S. ports and waterways, along the coasts, on international waters, or in any maritime region as required to support national security.

V

Vaccine — Substance given to provide immunization against a specific disease, prepared from the disease-causing biological agent itself or a synthetic substitute.

Vapor — Gaseous form of a substance that is normally in a solid or liquid state at room temperature and pressure; formed by evaporation from a liquid or sublimation from a solid. *Also see* Vapor Density, Vapor Pressure, Vapor Dispersion, Vapor Suppression, and Vaporization.

Vapor Density — Weight of a given volume of pure vapor or gas compared to the weight of an equal volume of dry air at the same temperature and pressure. Vapor density less than 1 indicates a vapor lighter than air; vapor density greater than 1 indicates a vapor heavier than air. *Also see*

Vapor and Physical Properties.

Vapor Dispersion — Action taken to direct or influence the course of airborne hazardous materials. *Also see* Vapor and Dispersion.

Vaporization — Process of evolution that changes a liquid into a gaseous state; rate of vaporization depends on the substance involved, heat, and pressure. *Also see* Vapor and Vapor Density.

Vapor Pressure — (1) Measure of the tendency of a substance to evaporate. (2) The pressure at which a vapor is in equilibrium with its liquid phase for a given temperature. Liquids that have a greater tendency to evaporate have higher vapor pressures for a given temperature. *Also see* Vapor and Boiling Point.

Vapor-Protective Clothing — Gas-tight chemical-protective clothing designed to meet NFPA® 1991, *Standard on Vapor-Protective Suits for Hazardous Chemical Emergencies;* part of an EPA Level A ensemble.

Vapor Suppression — Action taken to reduce the emission of vapors at a hazardous materials spill. *Also see* Vapor and Hazardous Material.

Vector — An animate intermediary in the indirect transmission of an agent that carries the agent from a reservoir to a susceptible host.

Vehicle-Borne Improvised Explosives Device (VBIED) — An improvised explosive device placed in a car, truck, or other vehicle, typically creating a large explosion. Also called a *car bomb* or *vehicle bomb. Also see* Improvised Explosive Device, High Explosive, and Explosion.

Vesicant — *See* Blister Agent.

Violent Rupture — Immediate release of chemical or mechanical energy caused by rapid cracking of the container.

Virus — Simplest type of microorganism that can only replicate itself in the living cells of their hosts; unaffected by antibiotics. *Also see* Bacteria and Rickettsia.

Volatile — Changing into vapor quite readily at a fairly low temperature. *Also see* Vapor and Vaporization.

Volatile Organic Compounds (VOCs) — Organic chemicals that have high vapor pressures under normal conditions.

Volatility — A substance's ability to become a vapor at a relatively low temperature; volatile chemical agents have low boiling points at ordinary pressures and/or high vapor pressures at ordinary temperatures. *Also see* Vapor, Volatile, and Vaporization.

Vomiting agent — Chemical warfare agent that causes violent, uncontrollable sneezing, cough, nausea, vomiting and a general feeling of bodily discomfort. *Also see* Chemical Agent.

W

Walk-Through — An initial assessment conducted by carefully walking through the scene to evaluate the situation, recognize potential evidence, and determine resources required; also, a final survey conducted to ensure the scene has been effectively and completely processed.

Warm Zone — Area between the hot zone and the cold zone, usually containing the decontamination corridor and typically requiring a lesser degree of personnel protection than the Hot Zone; also called *the contamination reduction zone.*

Warrant — A writ (written order) issued by a competent magistrate (or judge) authorizing an officer to make an arrest, a seizure, or a search or to do other acts incident to the administration of justice. Some states identify the type of magistrate who can issues certain types of warrants and what is needed for a warrant.

Water Solubility — Ability of a liquid or solid to mix with or dissolve in water. *Also see* Physical Properties, Dilution, and Soluble.

Water-Reactive Material — Substance, generally a flammable solid, that reacts in varying degrees when mixed with water or exposed to humid air. *Also see* Air-Reactive Material, Reactivity, and Reactive Material.

Waybill — Shipping paper used by a railroad to indicate origin, destination, route, and product. Each car has a waybill that the conductor carries. *Also see* Shipping Papers and Consist.

Weaponize — To alter an agent to improve its ability to be delivered as an effective weapon, for example, reducing the particulate size of the agent to increase the likelihood of inhalation.

Weapons of Mass Destruction (WMD) — Any weapon or device that is intended or has the capability to cause death or serious bodily injury to a significant number of people through the release, dissemination, or impact of one of the following means: toxic or poisonous chemicals (or their precursors), a disease organism, or radiation or radioactivity .

WHMIS — Acronym for Workplace Hazardous Materials Information System.

Wipe Sample — Sample collected by wiping a representative surface of a known area.

Witness — A person called upon to provide factual testimony before a judge or jury.

Workplace Hazardous Materials Information System (WHMIS) — Canadian law requiring that hazardous products be appropriately labeled and marked

WMD — Abbreviation for Weapons of Mass Destruction.

X

X Ray — High-energy photon produced by the interac-

Index

PIRS (photoacoustic IR spectroscopy), 548
Placards
    Canadian, 124
    defined, 110
    ERG information, 187
    hazard classes and divisions, 113–118
    materials requiring placarding, 112
    Mexican, 124, 129
    parts of, 111
    U.S. DOT, 110
Plague, 355
Plain language, 255, 257
Planning Section, 248
Plastic bottle bomb, 333
Plume pattern of material release, 220
Poison. *See also* Toxins
    defined, 16
    fluorine, 16
    placard hazard classes and divisions, 117
    toxic chemicals, 24–25
Polar solvent, 205
Polya, George, 266
Polychlorinated biphenyls (PCBs), 28, 35, 138
Polymerase chain reaction device (PCR), 550
Polymerization, 16, 209, 211
Pool pattern of material release, 222
Position titles, 235
Potassium chlorate, urea nitrate, 328
Potassium-40, 20
Potential Hazards section of ERG, 173
Pound-force per square inch gauge (psig), 73
Pounds per square inch (psi), 73, 200
Powered air-purifying respirator (PAPR), 387, 389–390
PPE. *See* Personal protective equipment (PPE)
Predetermined procedures, 169–170. *See also* Standard
    operating procedures (SOPs)
Pre-incident survey, 66
Preplans, 66. *See also* Pre-incident survey
Pressure containers, 71
Pressure intermodal tank, 94, 95
Pressure measurements and terms, 73
Pressure storage tanks, 74
Pressure tank car, 80–82
Pressure tanks, 73
Primary explosives, 322, 323
Priorities during incidents, 233–234
Private Sector Specialist Employee, 13
Problem-solving process
    analyzing the incident, 270–279
    common elements, 267
    Four-Step (APIE) process, 267, 269
    four-step formula, 266
    Incident Action Plan, 283–286
    models, 267–268
    planning the appropriate response, 279–285
Process Safety Management (PSM) of Highly Hazardous
    Chemicals (HHCs), 42
Procurement Unit, 248
Product control
    fire control, 477, 492–494
    General Hazardous Materials Behavior Model, 476
    leak control

containment, 489
    purpose of, 477, 489
    responder procedures, 489–490
    shutoff valves, 490–492, 504–505
responder procedures, 477
spill control
    absorption, 480, 496
    adsorption, 480, 496
    blanketing/covering, 480–481
    confinement, 477
    dams, 478, 481–482, 497
    dikes, 478, 481–482, 498
    dilution, 488, 503
    dispersion, 488
    dispersion types, 478
    dissolution, 488
    diversion, 481–482, 499
    neutralization, 478, 488
    purpose of, 476, 477
    responder procedures, 478
    retention, 481–482, 500
    tactics, 479
    vapor dispersion, 487, 502
    vapor suppression, 482–487
    ventilation, 487–488
vapor suppression, 485–487
    alcohol-resistant AFFF concentrate, 484
    Aqueous Film Forming Foam, 483, 484
    bank-down application method, 486
    Class B foam concentrate, 482–483
    defined, 482
    drainage time, 485–486
    emulsifiers, 485
    expansion ratio, 485
    fluoroprotein foam, 484, 494
    high-expansion foam, 485, 494–495
    high-quality foam production, 483
    procedures, 501
    rain-down method, 486
    roll-on application method, 486
    selection considerations, 486–487
Progesterone, 28
Project HEROES, 399
Propane bottle bomb, 333
Properties of hazardous materials. See Behavior of hazardous
    materials
Property, protection of, 303–304
Prosecute, defined, 601
Protection
    environmental, 303–304
    property, 303–304
    protecting/defending in place, 301
    of the public, 298–303
        contaminated victims, 301
        evacuation, 298–301
        looting, prevention of, 300
        notification, 299
        protecting/defending in place, 301
        reentry, 300
        relocation facilities and temporary shelters, 299–300
        rescue, 302–303
        self-presenters, 301

Index by Nancy Kopper